Lecture Notes in Physics

Springer-Verlag Berlin Heidelberg GmbH

Editorial Policy

The series *Lecture Notes in Physics* (LNP), founded in 1969, reports new developments in physics research and teaching -- quickly, informally but with a high quality. Manuscripts to be considered for publication are topical volumes consisting of a limited number of contributions, carefully edited and closely related to each other. Each contribution should contain at least partly original and previously unpublished material, be written in a clear, pedagogical style and aimed at a broader readership, especially graduate students and nonspecialist researchers wishing to familiarize themselves with the topic concerned. For this reason, traditional proceedings cannot be considered for this series, though volumes to appear in this series are often based on material presented at conferences, workshops and schools (in exceptional cases the original papers and/or those not included in the printed book may be added on an accompanying CD-ROM, together with the abstracts of posters and other material suitable for publication, e.g. large tables, colour pictures, program codes, etc.).

Acceptance

A project can only be accepted tentatively for publication, by both the editorial board and the publisher, following thorough examination of the material submitted. The book proposal sent to the publisher should consist of at least a preliminary table of contents outlining the structure of the book together with abstracts of all contributions to be included.
Final acceptance is issued by the series editor in charge, in consultation with the publisher, only after receiving the complete manuscript. Final acceptance, possibly requiring minor corrections, usually follows the tentative acceptance unless the final manuscript differs significantly from expectations (project outline). In particular, the series editors are entitled to reject individual contributions if they do not meet the high quality standards of this series. The final manuscript must be camera-ready, and should include both an informative introduction and a sufficiently detailed subject index.

Contractual Aspects

Publication in LNP is free of charge. There is no formal contract, no royalties are paid, and no bulk orders are required, although special discounts are offered in this case. The volume editors receive jointly 30 free copies for their personal use and are entitled, as are the contributing authors, to purchase Springer books at a reduced rate. The publisher secures the copyright for each volume. As a rule, no reprints of individual contributions can be supplied.

Manuscript Submission

The manuscript in its final and approved version must be submitted in camera-ready form. The corresponding electronic source files are also required for the production process, in particular the online version. Technical assistance in compiling the final manuscript can be provided by the publisher's production editor(s), especially with regard to the publisher's own LaTeX macro package which has been specially designed for this series.

Online Version/ LNP Homepage

LNP homepage (list of available titles, aims and scope, editorial contacts etc.):
http://www.springer.de/phys/books/lnpp/

LNP online (abstracts, full-texts, subscriptions etc.):
http://link.springer.de/series/lnpp/

C. Berthier L.P. Lévy G. Martinez (Eds.)

High Magnetic Fields

Applications in Condensed Matter Physics and Spectroscopy

Springer

Editors

Dr. C. Berthier
Prof. L.P. Lévy
Dr. G. Martinez
Grenoble High Magnetic Field Laboratory
BP 166
38042 Grenoble Cedex 9
France

Library of Congress Cataloging-in-Publication Data applied for.

Die Deutsche Bibliothek - CIP-Einheitsaufnahme

High magnetic fields : applications in condensed matter physics and spectroscopy / C. Berthier ... (ed.).
(Lecture notes in physics ; Vol. 595)
(Physics and astronomy online library)
ISBN 978-3-662-14314-8 ISBN 978-3-540-45649-0 (eBook)
DOI 10.1007/978-3-540-45649-0

ISSN 0075-8450

http://www.springer.de

Originally published by Springer-Verlag in 2001
Softcover reprint of the hardcover 1st **edition** 2001

Typesetting: Camera-ready by the authors/editor
Camera-data conversion by Steingraeber Satztechnik GmbH Heidelberg
Cover design: *design & production*, Heidelberg

Printed on acid-free paper
SPIN: 10980057 57/3111 - 5 4 3 2 1

Preface

This book is addressed to all scientists interested in the use of high magnetic fields and in the use of high-field facilities around the world. In particular it will help young scientists and newcomers to the topic to gain a better understanding in areas such as condensed matter physics, in which the magnetic field plays a key role either as a parameter controlling the Hamiltonian, or as an experimental tool to probe the underlying mechanism. This concerns mostly strongly correlated and (or) low dimensional systems. Rather than covering all these subjects in detail, the philosophy here is to give essential physical concepts in some of the most active fields, which have been quickly growing in the last ten to twenty years. Besides its role as a physical parameter in condensed matter physics, a large magnetic field is essential to Electron Paramagentic Resonance (EPR) and Nuclear Magnetic Resonance (NMR) spectroscopies. The state of art of high resolution NMR in liquids and solids and high frequency EPR applied to fields like chemistry and biology are also reviewed in this volume.

The first series of chapters is devoted to the integer and the Fractional Quantum Hall Effects (FQHE) in two-dimensional electron systems. C. Glattli brushes an historical background and a comprehensive review of transport phenomena in these systems, including recent developments on the mesoscopic electronic transport at the edges of quantum Hall samples, chiral Luttinger liquids and fractional excitations. R. Shankar gives a deep introduction to the microscopic theories of FQHE. After a short review of most popular trial wavefunctions and their physical content, he focuses on the Hamitonian theory which is compared to other theoretical approaches and experiments. These two general lectures are followed by the chapter of I. Kukushkin, which reviews magneto-optics experiments on composite fermions, and the chapter of M. Fogler, which surveys the theoretical and experimental state of the art of, stripe, bubble and charge density phases recently discovered in two dimensional electron liquid when many Landau levels are occupied.

The next series of chapters are devoted to quantum magnetism. M. Rice gives a comprehensive review of spin-ladders, while C. Lhuillier emphasizes 2D frustrated magnets, which represent another class of magnets with spin liquid ground states. These theoretical lectures are followed by two experimental reviews: M. Horvatić explores the possibilities of NMR in the investigation of quantum antiferromagnets with carefully selected examples, while C. Broholm gives an extensive review of neutron experiments probing quantum spin chains under high magnetic field.

The physics of quasi-one dimensional conductors is covered in the chapter of C. Bourbonnais, which introduces the properties of the one-dimensional electron gas, tackles the problem of interchain coupling and reviews the various instabilities of the system leading the formation of higher dimensionality states. This theoretical background is compared to experimental results in quasi-one dimensional organic and inorganic conductors. The experimental situation in quasi-one dimensional electronic systems is also developed in the chapter of D. Jérome, which is devoted to the family of the Bechgaard salts and to doped spin-ladders.

Quasi-1D organic conductors are famous for the rich physics they exhibit as a function of the magnetic field: Field Induced Spin Density waves, Lebed angles, dimensionality crossover. Unfortunately, these aspects are not covered in this book, but the reader is encouraged to consult dedicated monographs such as: P. Chaikin, J. Phys. I (France) **6**, 1875 (1996) and "Organic Superconductors "by T. Ishiguro, K. Yamaji and G. Saito, 2nd edition, (Springer Heidelberg, 1998).

The following two chapters are devoted to the problem of superconductivity in high magnetic fields, with the specific problem of reentrance in some low dimensional systems. The problem is treated from the theoretical point of view in the chapter of V. Mineev, and the experimental situation in low dimensional organic salts is reviewed by T. Ishiguro.

The discovery of high temperature superconductors has revitalized the physics of vortices in type II superconductors. This new physics of vortex matter, in which the magnetic field is an external adjustable parameter able to switch from liquid to solid or glassy states, is covered from both theoretical and experimental sides in the chapter written by T. Giamarchi and S. Batthacharya.

The two next chapters are devoted to three-dimensional magnetic systems. T. Kimura and Y. Tokura review the physical properties of manganites and related oxides, famous for their "colossal "magnetoresistance, including the magnetic field induced melting of charge/orbital- ordered state. Some of these manganites belong to the class of materials referred to as "half-metals", the properties of which are described in details in the lecture of J. Coey.

The last two chapters devoted to condensed matter physics bear on the effect of electron–electron interaction near the metal insulator transition, given by Zvi Ovadyahu, and the interference effects in disordered insulators by M. Sanquer. Both involve subtle magneto–transport experiments.

The last four chapters of the book are in a different spirit and describe different applications of resonance spectroscopy. J. Prestegard reviews the recent progress of high resolution NMR in liquids and their application to the determination of biological structures. D. Massiot describes the state of the art of high resolution solid-state NMR in high field, and the possibilities opened for structural studies of various organic and inorganic classes of compounds. D. Gatteschi describes the advantages of high frequency cw EPR, from single ions to integer spin of molecular clusters, whereas M. Fuchs and K. Möbius explore the most recent developments of high field pulsed EPR and their applications to the structure and dynamics of proteins and bio-organic molecules.

The chapters of this book are based on lectures that were given at a two-week International School held in Cargèse, on the Corsica island, in the spring of 2001. It is a great pleasure for us to acknowledge the financial support of the European Community, under the frame of the High Field Infrastucture Cooperative Network, and that of the Centre National de la Recherche Scientifique. We also would like to express our gratitude to all the staff of the Cargèse center and of the GHMFL, who have been of great help in the success of this school, and we are sure that all participants will keep a good memory of the warm welcome of the Cargèse inhabitants. We trust that this book will help many other scientists to benefit from the excellent overviews presented at the International School and that it will contribute in a small way to further advancing this fascinating area of physics.

Grenoble,
April 2002

Claude Berthier
Laurent Lévy
Gérard Martinez

Contents

Quantum Hall Effect: Macroscopic and Mesoscopic Electron Transport

D.C. Glattli

CEA Saclay, Gif-sur-Yvette, France

Abstract. We give here an introduction to the Integer and the Fractional Quantum Hall Effect. Some of the fascinating transport properties of macroscopic samples in this regime are reviewed. Then, the mesoscopic electronic transport at the edge of a quantum Hall sample is described. In particular, the Luttinger liquid physics of fractional edges is discussed.

1 Introduction

The quantum Hall effect is perhaps one of the most beautiful macroscopic manifestation of quantum mechanics in condensed matter after superconductivity and superfluidity. The phenomenon is observed in a two-dimensional electron gas (2DEG) at low temperature in a high perpendicular magnetic field. The physics of the quantum Hall effect is that of a flat macroscopic atom made of up to 10^{11} electrons. Here, the magic quantum numbers are replaced by magic values p/q of filling factor $\nu = n_s/n_\varphi$ which measures the electron density n_s in units of the flux quantum density $n_\varphi = B/\phi_0$ where $\phi_0 = h/e$. Sweeping this parameter, by varying the magnetic field or the density, reveals a series of new quantum fluids signaled by plateaus in the Hall resistance. The plateaus take the remarkable values $\frac{q}{p}\frac{h}{e^2}$ each time ν crosses remarkable values of p/q: an integer [1] ($q = 1$) or a fraction with odd denominator [2,3] ($q = 2s + 1$). For integer values the physical origin of the phenomenon is the Fermi statistics and the cyclotron motion quantization in 2D, while for fractional values an additional ingredient is needed: the Coulomb interaction. These ingredients are the simplest one can imagine. No interaction with the host material is needed as in the case of superconductivity. However, a little amount of disorder is required to reveal the Hall plateaus: it is a rare and unusual case where imperfections help to reveal a fundamental quantum effect! It is remarkable that such a simple system has completely renewed our knowledge of quantum excitations. Topological fractionally charged excitations [3] with fractional anyonic or exclusonic quantum statistics [4], composite fermions [7] or composite bosons [25], skyrmions [7,8], etc., ... are the natural elementary excitations required to understand the quantum Hall effect. The quantum Hall effects have made real some delicate concepts invented for the purpose of particle physics theories or used in mathematical physics for quantum integrable systems [9]. I should emphasize that none of these concepts are complicated or difficult to understand. They all suggest how rich and fascinating is the physics of the quantum Hall effect.

As the quantum Hall effect is a fundamental property of electrons confined in 2D, and not a property of specific materials, any realization of a clean 2D metal should reveal this effect, provided the magnetic field is strong enough to reach a quantum flux density commensurable with the electron density. The highest static magnetic fields presently available require low electrons densities, typically lower than approximately $10^{16}\mathrm{m}^{-2}$ (a field of 20 Tesla corresponds to $n_{\phi} \approx 5.10^{15}\mathrm{m}^{-2}$). Also the experimental discovery of quantum Hall effect would not have been possible without the emergence of high-quality low density electron systems. These samples are a direct product of the semiconductor technology. From the discovery of the integer quantum Hall effect (IQHE) more than twenty years ago [1], followed later by the discovery of the fractional quantum Hall effect (FQHE) [2], to the recent observation of new phases at large filling factors, as those discussed in Fogler's chapter in this book, one can say that each three to five years emerges a new concept associated with the discovery of a new kind of collective quantum state. This regular rate of surprising discoveries is driven by the constant improvement of sample quality. This improvement is illustrated in Fig. 1 where the Hall plateaus observed in two state-of-the-art samples, separated by approximately fifteen years, are displayed.

It is a hard task to give an overview of twenty years of such beautiful physics. There have been already many well documented reviews on this subject written by the most prominent specialists in the field [2,13,8]. As the readership of this book comes from various areas, I will not include the latest developments which would be only appreciated by specialists in the field. Instead, I will try to provide the reader with the knowledge of the basic physics and of the experimental

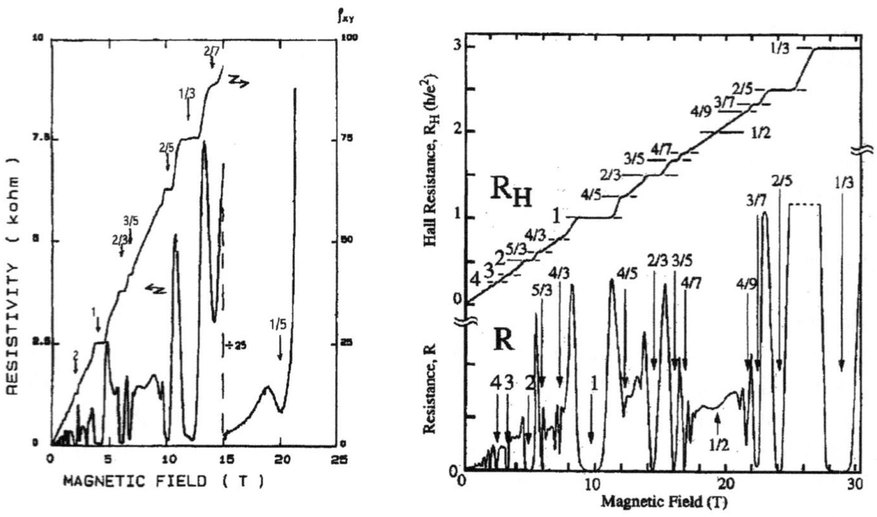

Fig. 1. Hall and longitudinal resistance versus magnetic field measured in typical high mobililty samples. Left: state of the art GaAs/GaAlAs heterojunctions in 1986 (from Ref. [11]); right: today's state of the art sample (from Ref. [2])

manifestations of the quantum Hall effect. I will restrict myself to the transport properties.

After a presentation of the peculiarities of transport in two-dimensions and a presentation of the quantum Hall effect, I will describe the realization of experimental 2D electron systems. Then a chapter will be devoted to the physics of transport in macroscopic samples. The last chapter will address the physics of edge states which carry the current in mesoscopic quantum Hall samples. I will discuss the electronic transport in the light of the Luttinger liquid physics.

1.1 The Hall Effect in 2D

In this section we will introduce some basic definitions using the simple Drude model for the conductivity of a 2D electron gas treated classically. In the presence of a perpendicular magnetic field $B\widehat{z}$ the Lorenz force $e\overrightarrow{v} \times B\widehat{z}$ adds to the longitudinal electric force $e\overrightarrow{E}$. The 2D current density $\overrightarrow{j} = en_s\overrightarrow{v}$ is

$$\overrightarrow{j} = \overrightarrow{j_\mathrm{l}} + \overrightarrow{j_\mathrm{t}} = \sigma_\mathrm{l}\overrightarrow{E} + \sigma_\mathrm{t}\widehat{z} \times \overrightarrow{E} \tag{1}$$

The longitudinal conductivity σ_l (often denoted as σ_{xx}) is associated with dissipation ($\overrightarrow{j_\mathrm{l}}.\overrightarrow{E} \neq 0$), while the transverse or Hall conductivity σ_t (often denoted as σ_{xy}) corresponds to dissipationless transport ($\overrightarrow{j_\mathrm{t}}.\overrightarrow{E} = 0$). Alternatively, the local longitudinal electric field can be related to the transverse and longitudinal current density

$$\overrightarrow{E} = \rho_\mathrm{l}\overrightarrow{j_\mathrm{l}} - \rho_\mathrm{t}\widehat{z} \times \overrightarrow{j_\mathrm{t}} \tag{2}$$

There are useful relations between the ρ's and σ's:

$$\sigma_\mathrm{l} = \frac{\sigma_0}{1+\omega_c^2\tau^2} = \frac{\rho_\mathrm{l}}{\rho_\mathrm{l}^2+\rho_\mathrm{t}^2} \tag{3}$$

$$\sigma_\mathrm{t} = \frac{en_s}{B} - \frac{\sigma_\mathrm{l}}{\omega_c\tau} = \frac{\rho_\mathrm{t}}{\rho_\mathrm{l}^2+\rho_\mathrm{t}^2} \tag{4}$$

In the first relation $\sigma_0 = n_s e^2\tau(B)/m^*$ is the Drude conductivity, $\omega_c = eB/m^*$ is the cyclotron frequency and τ is the B dependent collision time. The quantity ρ_t is the *Hall resistance* R_H

$$\rho_\mathrm{t} = \frac{B}{en_s} \equiv R_\mathrm{H} \tag{5}$$

It has the remarkable property of being independent of material parameters except the electron density. This is not the case for σ_t, but Eq. (4) shows that $\sigma_\mathrm{t} \to \sigma_\mathrm{H} = 1/R_\mathrm{H}$ in the limit of vanishing longitudinal conductance. This limit is reached in the conditions of the Quantum Hall Effect as we will see below.

1.2 What Is Special in 2D?

In two dimensions, the conductivity σ (resistivity) has the dimension of a conductance (resistance). In zero magnetic field, the classical Drude conductivity $\sigma_o = n_s e^2\tau/m^*$ can be rewritten as $\sigma_o = (e^2/h)k_\mathrm{F}l$, where k_F is the Fermi wave vector and l is the mean free path. From this, we see that the classical conductivity can be expressed in terms of quantum units: the universal conductance quantum e^2/h times a dimensionless number which carries information on the microscopic details (here the ratio of the mean free path over the Fermi wavelength).

Similar quantum units can be used for the *classical* Hall conductance in 2D. Indeed, instead of expressing the magnetic field B using conventional units (Tesla, Gauss, ...), we can use *quantum units* for the magnetic field. B can be uniquely defined by the number n_Φ of flux quanta $\Phi_0 = h/e$ per unit area: $B = (h/e)n_\Phi$. The *classical* Hall conductance becomes:

$$\sigma_{\mathrm{H}} = \frac{1}{R_{\mathrm{H}}} = \frac{e^2}{h}\frac{n_s}{n_\Phi} \tag{6}$$

In these units, σ_{H} is just the ratio of the number of electrons over the number of flux quanta in the plane.

Another interesting property in two dimensions is the small sensitivity of the Hall resistance to disorder. Fig. 2 displays schematically a strip of a 2D conductor connected to contacts without (a) and with (b) disorder. We will treat electrons classically. In the limit of $\sigma_{\mathrm{l}} \to 0$, the Hall angle defined as $\tan(\theta_{\mathrm{H}}) = \sigma_{\mathrm{t}}/\sigma_{\mathrm{l}}$ is $90°$ and $\sigma_{\mathrm{t}} = \sigma_{\mathrm{H}}$. The current lines are parallel to the equipotential lines. In a Hall measurement, the Hall voltage V_{H} is defined as the potential difference between the upper and lower edge. If I is the current flowing through the sample, the Hall resistance is given by: $I = (1/R_{\mathrm{H}})V_{\mathrm{H}} = \sigma_{\mathrm{t}} V_{\mathrm{H}}$. More generally, for any point A and B belonging respectively to the lower and upper boundary, one have $I = \int_{A\to B}(\mathbf{j} \times \mathbf{dl}).\mathbf{z} = \sigma_{\mathrm{H}} \int_{A\to B} \mathbf{E}.\mathbf{dl}$. For a uniform sample, the path used to calculate the potential difference is not important. In particular the voltage drop V across the contact is always $V = V_{\mathrm{H}}$ because of the lack of dissipation, assuming no contact resistance. Now lets introduce some defects inside the sample. The defects can be holes in the sample, i.e. insulating regions. They will strongly modify the current lines. However, provided it is possible to find a path $A \to B$ connecting the two contacts following an unperturbed region of *constant*

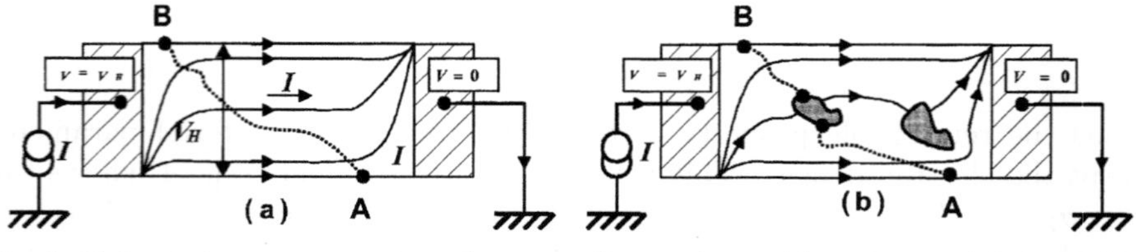

Fig. 2. Schematic representation of current line without (a) and with (b) disorder in a classical dissipationless 2D Hall bar. The path $A \to B$ is arbitrary. In (b) holes,in grey, have been randomly introduced in the sample. Provided one can find a path with constant Hall resistance and $90°$ Hall angle to go from the lower corner of the right contact to the upper corner of the left contact, the voltage drop is not affected by the presence of holes. Note that the current lines correspond here to the net current and not to microscopic electron flow

$\sigma_t = \sigma_H$,with 90 °Hall angle, the voltage drop remains equal to the Hall voltage. The same result would be obtained by filling the holes with a conductor having infinite longitudinal conductance and negligible Hall conductance. This lack of sensitivity to disorder is a property of electromagnetism in 2D. It is topological and survives when quantum mechanics is considered. This is a basic ingredient which makes the Hall resistance of the quantum Hall effect insensitive to disorder on macroscopic scale. Of course, to have a perfectly vanishing longitudinal conductance and to built regions of perfectly constant (quantized) density, i.e. constant σ_H, requires quantum mechanical effects and are unlikely to be observed in classical systems.

1.3 The Surprising Quantizations of the Hall Resistance

Although, there have been already calculations showing that the Shubnikov-de Haas oscillations of the longitudinal resistance and the Hall resistance in 2D was highly non trivial [14], the plateaus of the Hall resistance discovered by K. v. Klitzing, G. Dorda and M. Pepper [1] was not anticipated, nor the accurate (metrological) quantization. The series of striking values of the plateaus which characterizes the Integer quantum Hall effect:

$$R_H = (h/e^2)1/p \tag{7}$$

where p is an integer, tells us that, for some range of magnetic field, the electrons participating to the Hall resistance have a density *pinned* to the flux quantum density:

$$n_s = pn_\Phi \tag{8}$$

as suggested by Eq. (6). The pinning of the density expresses the incompressibility of the 2DEG in the quantum Hall regime. The origin of the incompressibility is the opening of a gap in the excitations. The existence of a gap directly manifests by a vanishing longitudinal resistance on a Hall plateau. Fig. 3 shows typical transport measurements in this regime.

The phenomenon called Integer Quantum Hall Effect can be understood as resulting from the combination of the Landau level formation and of the Fermi-Dirac statistics. We will see that the fact that it is observable on macroscopic scale is due to a zero temperature quantum phase transition: *Localization.* No electron-electron interactions are required at this point, although in real samples (specially very clean) interactions are strongly present.

Interactions have been necessary to explain a second surprising phenomenon observed first at higher magnetic fields: the occurrence of new Hall plateaus for fractional values [2], p being replaced by p/q, with q odd integer, see Fig. 1. This new effect, called the fractional quantum Hall effect comes from the combination of electron-electron interactions and of quantum statistics within a Landau level. New gaps open at fractional values p/q of the ratio n_s/n_Φ. Although the physical origin is different, the experimental manifestations are very similar to those observed in the IQHE regime, as far as transport properties are concerned.

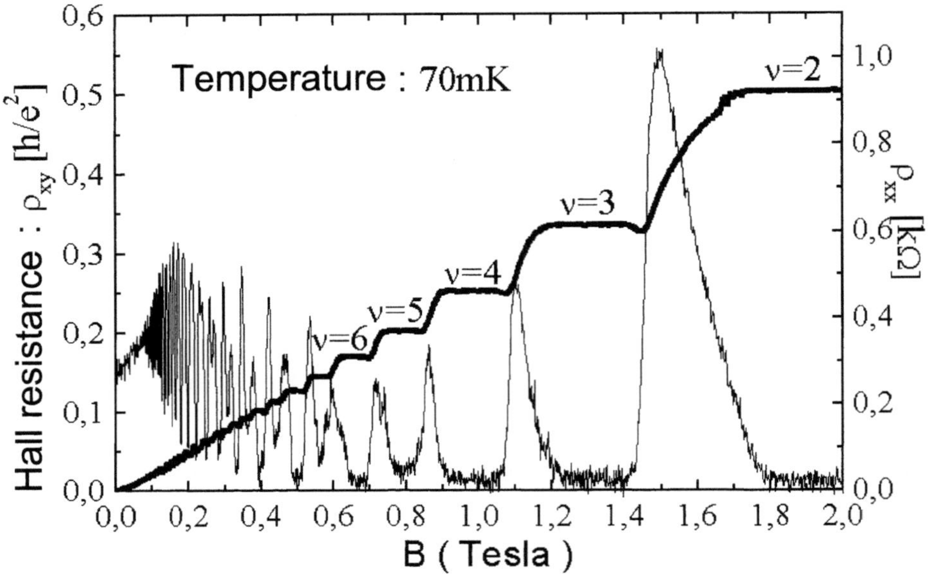

Fig. 3. Longitudinal and Hall resistance observed at T=70mK at low magnetic fields. Hall plateaus at h/pe^2 for $\nu \simeq p$ are well defined. Simultaneously the longitudinal resistance vanishes

2 Realization of Clean 2D Electron Gas

A clean 2DEG can be realized by confining electrons to a plane in a region free of positive compensating charges (for a review see [15,16]). The positive background can be either static ionized donors, randomly distributed in the host material far from the 2DEG, or the polarization charge of a positively biased metallic gate. This is different from the 3D case where the ionized impurities scatter the electrons with a dramatic decrease of the conductivity and of the electron-electron interactions effects. The electrostatics law requires another force to maintain electrons far from the donor plane. This is provided by a hard wall potential using quantum mechanical properties of the host material: the band structure discontinuity. The wall can be produced either using the interface between a semiconductor and an insulator or using two semiconductors with different conduction band energies. The first type is found in silicon devices using Si/SiO_2 interface. It is called a MOSFET. It is widely used in the micro-electronic industry for micro-processors and random access memories. The technology, used to fabricate Si-MOSFET, is not difficult and well mastered. The second type is a semiconductor heterojunction usually $GaAs/Ga_xAl_{1-x}As$, but other III-V materials are possible. It is used in the industry for making high speed switches and transistors for telecommunications. The technology is much more sophisticated and uses epitaxial growth. The purest systems need Molecular Beam Epitaxy (MBE) growth while standard heterojunctions can be realized using Chemical Vapor Deposition. Epitaxial growth provides atomically flat interface and a mean to modulate the doping during the growth. The good atomic interface is due to the very weak lattice mismatch of GaAs and Ga(Al)As.

In both systems, the conduction band is bent toward the potential wall in order to realize a triangular potential confining the charges close to the interface. Ionization of donors, randomly distributed in a region preferentially well away from the interface, is the source for the conduction band electrons which will be trapped at the interface to realize the 2DEG. In MOSFET, the electric field which bends the conduction band is provided by a positively polarized gate. The gate is deposited on the few nm thick SiO_2 insulator while the bulk semiconductor is p-doped (2D hole gas can also be formed with negative bias and n-doped material). In GaAs/Ga(Al)As heterojunctions, electrons are trapped in the GaAs side, the Ga(Al)As region with larger gap provides the potential wall. The electric field is usually not provided by a gate on the sample surface. Instead, it is provided by donors situated in the Ga(Al)As region. Such configuration is possible because the MBE technique allows to modulate the doping at will. Some of the electrons provided by the donors reach the potential well to form the mobile 2DEG, while most electrons are usually trapped at the surface of the sample by surface states and do not participate to the conduction. Fig. 4 shows schematically the two systems and the band structure.

The Integer QHE is well observed in Si-MOSFET samples, while the Fractional QHE is better observed in GaAs/Ga(Al)As heterojunctions. This is due to fundamental differences between the two materials. The effective mass to be considered in Si is 0.19 while in GaAs it is 0.068. For a similar concentration this make GaAs electrons more quantum. The typical length scale and size of the random potential experienced by the 2D electrons is also different. In a Si-

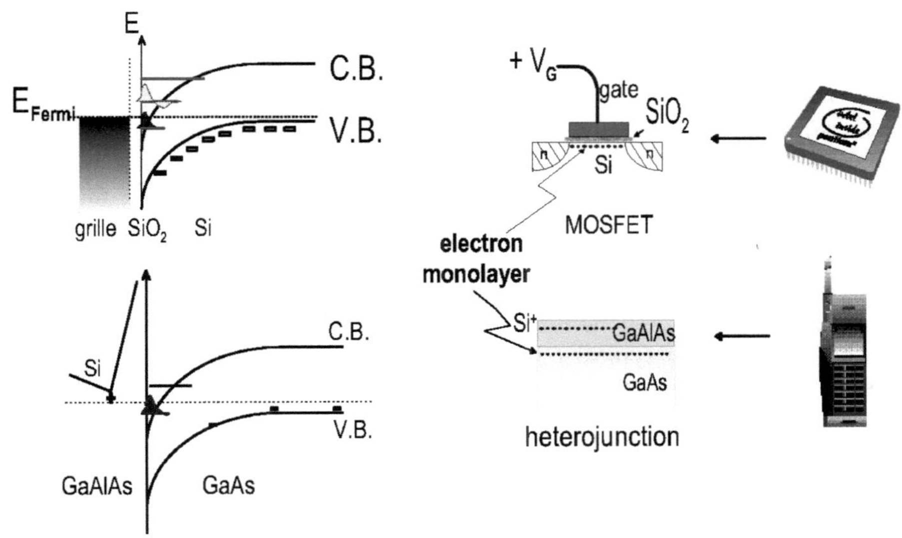

Fig. 4. Two possible realizations of a 2DEG: silicon MOSFET and GaAs/GaAlAs heterojonction

MOSFET, the SiO_2/Si interface roughness provides a wide spectrum of spatial fluctuations with a large intensity. In GaAs/Ga(Al)As the interface is atomically flat and disorder is long range and of small amplitude. Indeed, an undoped spacer layer of thickness d separates the randomly ionized donors from the 2D electrons. This filters the spatial frequencies of the random potential which are larger than d^{-1}. Also this provides spatial averaging of the potential fluctuations over a disk of diameter d, thus reducing their amplitude. As a result, zero field electron mobility hardly approaches 500000 $cm^2V^{-1}s^{-1}$ in silicon while mobility of 10^7 can be obtained in GaAs for similar density.

The Hall plateaus of the integer QHE are wider in case of larger disorder (provided the disorder strength does not overcome the Landau level separation energy). This is why IQHE is well observed in silicon. The fractional QHE requires small integer plateaus (in order to reveal the fractions in between). This implies small disorder with potential fluctuations much smaller than the FQHE gap (which is one order smaller than the Landau level separation as we will see below). Clean GaAs/Ga(Al)As heterojunctions are thus favored (but some FQHE fractions can be observed in specially grown silicon samples).

The zero field mobility at 4.2 K is a good qualitative measure of disorder. The highest mobilities have been obtained in GaAs. The improvement came in two steps: the modulation doping and, at the end of the eighties, the planar doping [11]. Further continuous improvements are now mostly due to the increase of the purity during MBE (better vacuum). Indeed, for ultra high mobility samples, residual acceptor impurities remain the dominant scattering mechanism in the GaAs side of the heterojunction, where electrons are confined.

I would not end this paragraph without talking about two new systems. The first one is MBE grown Si/SiGe heterojunctions. Interest for this material appeared recently and the growth technique improves regularly. The sample quality can reach the level of good quality GaAs/AlGaAs heterojunctions. With a different effective mass, a different valley degeneracy and a different dielectric constant Si/SiGe heterojunctions provide a set of parameters different from GaAs/Ga(Al)As heterojunctions to study the QHE [17]. The second system is more recent. The 2D electrons gas is realized in epitaxially grown molecular crystals such as Tetracene or Pentacene [18]. To confine the electrons a scheme similar to Si-MOSFET samples is used. A metallic gate, separated from the surface of the molecular crystal by an oxide layer, is used to create the 2D electrons (or the 2D holes because the molecular crystals are ambipolar). Holes seem to give the best mobility, typically $10^5 cm^2V^{-1}s^{-1}$. According to the large effective mass $m^* = 1.5m_e$, this corresponds to scattering times $\approx 10^{-10}$s comparable to the best GaAs/Ga(Al)As heterojunctions. The Fractional QHE has been indeed observed in these new molecular systems.

How far do we know the physical parameters of a 2DEG in a real sample? Indeed the basic ingredients behind the physics of the QHE are well known: the Coulomb interaction, the Fermi statistics and the quantization of the kinetic energy in magnetic field. Also the effective mass, the dielectric constant and the g factor are well known for each materials. A direct comparison between experi-

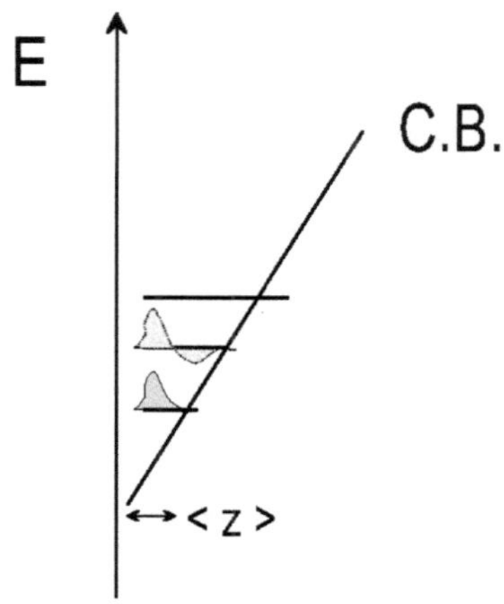

Fig. 5. Triangular well potential confining the electrons at the semiconductor interface

ments and theory should be possible, free of adjustable parameters. In fact this is not completely true. A first adjustable parameter is the strength of the disorder coming from ionized donors or interface roughness. However, it can be reduced to a very small amount in the purest structures. A second parameter comes from the width of the wavefunction in the $\widehat{z}$ direction which weakens the Coulomb interaction. Because of this, electrons are not point like but become rods. Assuming a triangular potential, as shown in Fig. 5 one can estimate the spreading $\langle z \rangle$ of the wavefunction. A crude estimation is obtained by minimizing the sum of the z kinetic energy and of the potential energy $\hbar^2/2m^* \langle z \rangle^2 + eE_z \langle z \rangle$. This yields $\langle z \rangle \sim \left(\hbar^2/m^* eE_z\right)^{1/3}$, where E_z is the electric field bending the conduction band. Typically $E_z \sim en_s/\varepsilon\varepsilon_0$ which yields $\langle z \rangle \sim (a^2.a_{\mathrm{B}})^{1/3} = a/(r_s)^{1/3}$. The length $a = (\pi n_s)^{-1/2}$ measures the typical distance between electrons and $r_s = a/a_{\mathrm{B}}$ is the correlation parameter, as in 3D metals (quantitative estimations of $\langle z \rangle$ can be found in [15]). In GaAs heterojunctions, the Bohr radius is $a_{\mathrm{B}} \simeq 10$nm and r_s is a few units (as in ordinary 3D metals). Thus the $\langle z \rangle$ extension is of the order of a and produces a significant reduction of the correlation energy. Its knowledge is important for comparison with theoretical estimations of the FQHE gaps as discussed in Shankar's chapter.

3 Quantum Hall Effect: Bulk Macroscopic Transport

The transport properties of the bulk 2DEG has been studied for a long time. They are many good reviews on this subject. In the following sections I will only present some important basic aspects of the Integer and of the Fractional Quantum Hall regime. Some of them will be illustrated by experimental results borrowed from the work of leading groups in the field. I apology for any important work which will not be quoted. Because of the broad spectrum of the present readership the aim is mostly to give a feeling for what the Quantum Hall Effect is.

3.1 The Integer Quantum Hall Effect

In presence of a magnetic field $\mathbf{B} = B\widehat{\mathbf{z}}$ the kinetic energy $K = \frac{(\mathbf{p}+e\mathbf{A})^2}{2m^*}$ of an electron moving freely in a plane is equivalent to the problem of an electron moving in a harmonic potential. The set of eigenvalues for the kinetic energy, called Landau levels, is:

$$E_n = (n + \frac{1}{2})\hbar\omega_c \tag{9}$$

where $\omega_c = eB/m^*$ is the cyclotron pulsation. Typically $\omega_c/2\pi$ =400 GHz/Tesla and $\hbar\omega_c$ =20 Kelvin/Tesla in GaAs/GaAsAl samples. The Landau levels are already well separated at liquid Helium temperature in field $\gtrsim$ Tesla. The eigen energies are function of a single quantum number n. As there are two degrees of freedom (the electron moves in 2D) there is a missing quantum number indicating a huge degeneracy. Using the cylindrical gauge $\mathbf{A} = (-By/2, Bx/2, 0)$, the meaning of the degeneracy appears clearly if one represents the conjugate pairs of electron coordinates $[x, p_x]$ and $[y, p_y]$ by a new set of conjugate pairs, see Fig.6

$$(x, y) = (X + \xi, Y + \eta) \tag{10}$$

where

$$[\xi, \eta] = [v_y/\omega_c, -v_x/\omega_c] = -i\hbar/eB \tag{11}$$

$$[X, Y] = i\hbar/eB \tag{12}$$

The Hamiltonian becomes $\mathcal{H} = \frac{1}{2}m\omega_c^2(\xi^2 + \eta^2)$ and does not depend on (X, Y). The first pair of conjugate coordinates represents the fast cyclotron motion with radius

$$r_n = \langle n|\,\xi^2 + \eta^2\,|n\rangle^{1/2} = (n + \frac{1}{2})^{1/2} l_c \tag{13}$$

The cyclotron radius increases with the orbital Landau level index n and scales as $l_c = (\hbar/eB)^{1/2}$, the magnetic length. The second pair is the center of the cyclotron orbit $\mathbf{R} = (X, Y)$. This is precisely the freedom to choose the center

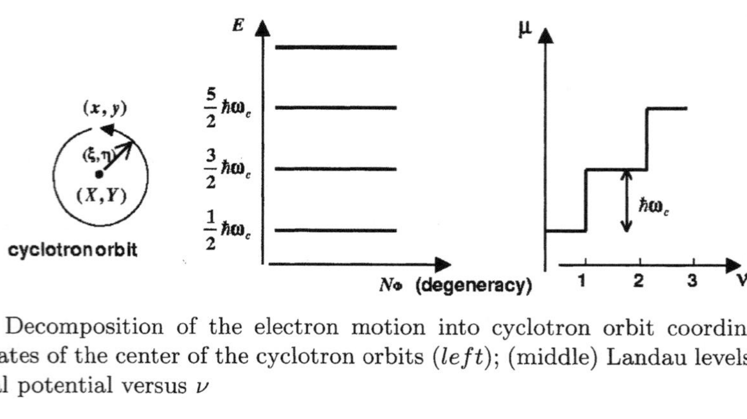

Fig. 6. Decomposition of the electron motion into cyclotron orbit coordinates and coordinates of the center of the cyclotron orbits (*left*); (middle) Landau levels; (*right*) chemical potential versus ν

of cyclotron orbits which encode the degeneracy of the Landau levels. The degeneracy is equal to the number of way to put the cyclotron orbit centers in the plane. It is equal to the number $N_\Phi = n_\Phi S = eBS/h$ of flux quanta $\Phi_0 = h/e$ in the surface S. This a direct consequence of the non commutation of coordinates (X, Y). We can also say that the 2D electron plane is analogous to the phase space $[P, Q]$ of a one-dimensional system. It is known that the area of a quantum state is h. In our 2D system with quantizing magnetic field the flux quantum h/e stands for the Planck constant h.

Assuming first fully spin polarized electrons, it is easy to understand why there is a vanishing longitudinal resistivity ρ_l and conductivity $\sigma_\mathrm{l} = \frac{\rho_\mathrm{l}}{\rho_\mathrm{l}^2+\rho_\mathrm{t}^2}$ when the Hall resistance R_H is precisely h/pe^2, p integer. Because of the Pauli principle, when the number of electrons $N = n_s S$ is exactly $pN_\Phi = pn_\Phi S$, the ground state corresponds to the complete filling of the p lowest Landau levels, while the next Landau levels are empty. Thus, there is an energy gap $\hbar\omega_c$ to create an electron-hole pair excitation. The 2DEG becomes a perfect insulator for longitudinal currents ($\sigma_\mathrm{l} = 0$), while the current parallel to equipotential line is exclusively made of dissipationless transverse currents ($\rho_\mathrm{l} = 0$). At finite temperature, longitudinal transport and dissipation are recovered via thermally activated inelastic process with energy $\geq \hbar\omega_c$.

The central role played by the filling of the Landau levels has lead to define the filling factor $v = n_s/n_\Phi$ characterizing the QHE. The IQHE occurs for $\nu = p$. Regarding thermodynamics, there is a jump $\hbar\omega_c$ of the chemical potential for the smallest change of the filling factor around the integer value p: removing a single flux quantum by lowering the field or adding just a single electron. The system is *incompressible.*

In the previous view of a clean ideal 2DEG, the striking QHE features are expected only for an infinitely small width of the parameter ν around an integer. However experiments show a remarkable and accurately quantized Hall plateau *with finite width.* In the narrow mesoscopic samples considered later, the explanation is simple: we will see that a finite plateau can be explained using a Landauer-Büttiker approach of 1D chiral modes at the boundary of the sample. In macroscopic systems, where boundary effects are negligible, the explanation is different. It requires disorder and the associated zero temperature quantum phase transition called Localization. Note that this is one of the rare cases where disorder helps for the observation of a very fundamental phenomenon. This is what we are going to discuss now.

To get some physical insight, let us assume for simplicity that the random potential $V(x, y)$ acting on the electrons is smooth such that no appreciable Landau level mixing occurs. This means no strong variations on the lengthscale l_c or $|l_c \nabla V| \ll \hbar\omega_c$. The effect of disorder is therefore to lift the degeneracy of each Landau level n separately. If we restrict the Hilbert space to the subspace of a single Landau level n, the resulting Hamiltonian becomes:

$$\mathcal{H}_n = (n + 1/2)\hbar\omega_c + V^{(n)}(X, Y) \tag{14}$$

where $V^{(n)}$ is the potential V averaged over the fast cyclotron motion (i.e. projected over n). Using $[X, Y] = i\hbar/eB$ and it is easy to show that the coordinates $\mathbf{R} = (X, Y)$ of the centre of the cyclotron orbits obey the semiclassical equation of motion:

$$d\mathbf{R}/dt = (-1/eB)(\widehat{\mathbf{z}} \times \nabla V^{(n)}) \tag{15}$$

Electrons move along the equipotential lines. Their velocity is such that the Lorenz force compensates the local electric field. As schematically shown in Fig. 7 disorder leads to localized states. Electron running along paths of localized states do not participate to the macroscopic longitudinal conductivity and if tunneling between neighboring localized states is negligible σ_{l} remains zero. Localized states are mostly found at energies in the middle between the unperturbed Landau Level energies. When varying the magnetic field the localized states provide a way to pin the Fermi level and a quantized Hall plateau associated with zero σ_{l} occurs on a finite parameter range. This is only when the equipotential lines at the Fermi level percolate through the whole sample that a macroscopic zero finite conductivity is recovered and the Quantum Hall effect breaks down. This occurs for Fermi level close to the unperturbed Landau level energies. At the separation between the two regimes, when the chemical potential is equal to the so-called mobility-edge energy, there is an abrupt transition from a conducting to localized insulating regime. Localization is a zero temperature quantum phase transition and this is the reason why the phenomenon is robust and universal. When the Fermi level is pinned in the localized region, electrons participating to the transverse current have a density still equal to pn_{Φ} and, as shown in the

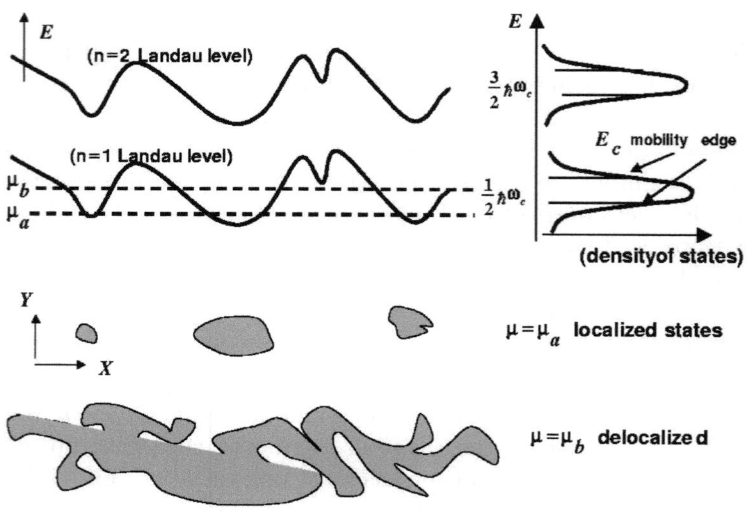

Fig. 7. Semiclassical percolation picture of the localized states of the QHE. In the schematic figure, μ_b is below the mobility edge E_C of the first Landau level: states are localized and σ_{l} vanishes. μ_a is close to the energy of the Landau level and states are delocalized

classical model of the introduction, the Hall resistance is unchanged and equal to $R_{\mathrm{H}} = (1/p)(h/e^2)$.

This picture can be extended to larger disorder, with appreciable Landau level mixing. One can show that the global result is unchanged. For more details on the localization approach see Ref. [10].

When spin is included, an even filling factor $\nu = 2p$ corresponds to the situation described above but with the $n \leq p$ orbital Landau levels doubly occupied. An odd filling factor $\nu = 2p - 1$ corresponds to $n \leq p - 1$ doubly occupied Landau level and the last Landau level $n = p$ fully polarized. There is also a gap, but the origin is now the Zeeman energy $g^* \mu_{\mathrm{B}} B$. With the g-factor of the host material ($g = -0.44$ for GaAs) the gap is expected much smaller than the cyclotron gap. Odd filling factor would be difficult to observe. However we will see later that interactions via the exchange energy enhance the gap to values close to $\hbar\omega_c$ and the effective g^* is much larger than g. This is why Hall plateaus at odd filling factor are easily observed. What is important for the occurrence of the quantum Hall effect is not the origin but the existence of the gap. States localized by disorder in the gap will pin the Fermi level and prevent long range transport in the macroscopic sample.

As pointed out by K.v.Klitzing, G. Dorda, and M. Pepper in their pioneering paper, the accuracy of the plateaus is metrological. This has lead to a resistance standard based on the fundamental constants h and e. The exactness of the quantization is theoretically supported by the gauge argument of R. Laughlin [19] and also by the concept of topological invariant developed by Thouless (see theoretical chapters in this book) [20]. Already at the beginning of the 80's experimentalists were able to produce and measure samples with longitudinal resistivity ρ_{l} as low as 10^{-10} Ohm. The Hall plateau is found very flat with $R_{\mathrm{H}}(T) - R_{\mathrm{H}}(0) = -s.\rho_{\mathrm{l}}$, s being an empirical parameter found around 0.01-0.05. Experiments show that the resistivity obeys an Arrhenius law $\rho_{\mathrm{l}} \sim \exp(-E_g/2k_{\mathrm{B}}T)$ at moderately low temperature characteristic of a conduction gap E_g. This is followed at lower temperature by a variable range hopping law $\rho_{\mathrm{l}} \sim \exp(-(T_0/T)^{1/2})$ characteristics of assisted tunneling through localized states. In the context of localization E_g is called the mobility gap and measures the distance between the critical energies separating conducting from insulating states. For even filling factors E_g is thus expected to be smaller than $\hbar\omega_c$ the separation between two consecutive Landau levels. In Si-MOSFET E_g has been found 85%-95% of the cyclotron gap. In clean GaAs/GaAlGa sample it is found 10%-25% higher for $\nu = 2$ and 4. A value larger than $\hbar\omega_c$ indicates the effect of the interaction not included in a simple description of the Integer QHE. Indeed a correlation energy of order $\sim e^2/4\pi\varepsilon\varepsilon_0 l_c$ has to be added to the cyclotron gap.

The localization theory has been a key to understand the bulk transport properties of the QHE. Also, important theoretical and experimental work has been done to test this theory. There have been numerous attempts to determine the critical exponent of the localization length ξ, the typical size of localized states which diverges at the mobility edge. The exponent enters in the temperature dependence of the transition width separating two Hall plateaus and in the temperature dependence of the height and width of the associated resistance peaks. One expects $\xi \sim |E - E_c|^{-\nu}$ where E_c is the mobility edge energy. The exponent, not to be confused with the filling factor, is $\nu = 4/3$ in a classical percolation picture [21] and $\nu = 7/3$ if quantum tunneling between close localized states is included [22]. The first experiments measured the width $\Delta B(T) \sim T^{\mu}$ and the magnetic field derivative of the conductance peaks $d\rho_l(T)/dB \sim T^{\mu}$ [23]. The experiments were not able to give a direct access to the critical exponent ν because the exponent measured $\mu = p/2\nu$ involves the exponent p of the temperature dependence of the inelastic length, which is not universal. Later, experiments based on a comparison of ξ with the sample size [24] found $\nu = 2.3$ compatible with the quantum percolation prediction as shown in Fig. 8. More recently, predictions relating the frequency dependence of the longitudinal conductivity $\sigma_l(\omega) = (2\pi/3)\varepsilon\varepsilon_0\xi\omega$ [25] have been checked experimentally and provide also a critical exponent 2.3 even for surprisingly short values of ξ [26]. Although some refinements have to be made from the theoretical side (inclusion of interactions, etc. ...) and from the experimental side, there is a common agreement to believe that Localization is the key ingredient to observe the QHE at macroscopic scale.

In the previous presentation of the integer QHE, interactions have been crudely neglected. When samples become cleaner and cleaner this is no longer legitimate. Recently new features in the bulk resistivity have been observed as an anisotropy related to the possible formation of stripes at larger Landau levels. This is due to the Coulomb interactions which lead to natural instability toward charge density formation. It is beyond the scope of this presentation to

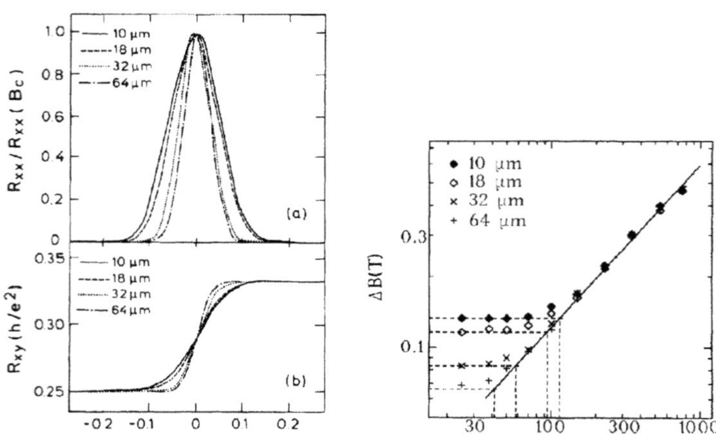

Fig. 8. (*left*) Transport data in Landau level $n = 1 \downarrow$ at T=25mK. (**a**) normalized longitudinal resistance and (**b**) Hall resistance as a function the $B - B_c$ around the percolation critical field B_c; (*right*) half witdh $\Delta B(T)$ obtained for different sample size. Information of the temperature variation and its saturation with sample size allows determination of the critical exponent ν (after Ref. [24])

tell more about this and I refer the reader to the chapter by Fogler in this book. Instead I will describe a regime now much more understood, where interactions are important and which concerns partially filled Landau levels: the Fractional Quantum Hall Effect.

3.2 The Fractional Quantum Hall Effect

We will first give a brief theoretical picture of the fractional QHE. For simpler presentation we will assume a spin polarized system and the first orbital Landau level partially filled: $N < N_{\Phi}$ or $\nu < 1$. We have seen that a Landau level is highly degenerate. For partial filling, there is a large freedom to occupy the quantum states, i.e. to fill the plane with electrons. However electrons interact and the Coulomb repulsion will reduce our freedom to distribute electrons in the plane. Let us first consider the limit of infinite magnetic field when the filling factor goes to zero. The Gaussian wavefunction describing the cyclotron motion in the first Landau level shrinks to zero. The N electrons being like point charges behave classically (no overlap between quantum states) and minimize their energy to form a crystalline state (analogous to the one observed in dilute 2D classical electron systems in zero field). In the present case, for weaker magnetic field, ν not too small, the wavefunctions overlap. Electron are not localized to a lattice but instead form a correlated quantum liquid. However interactions still break the Landau level degeneracy and, for some magic filling factors, a *unique* collective wavefunction minimizes the energy. The magic filling factors are found to be odd denominator fractions: $\nu = 1/3$,1/5, 2/3, 2/5, 3/5, 2/7, ... [27]. This is the Fractional QHE. Historically these remarkable states were not anticipated by the theory and were accidentally discovered by experimentalists doing transport measurements and searching for a crystalline state at higher field.

The Laughlin States: The unique wavefunction corresponds to a ground state separated from a continuum of excitations by a gap Δ. For $\nu = 1/(2s+1)$, s integer, Laughlin proposed a trial wavefunction for the ground state which was found very accurate. The wavefunction is built from single particle states in the cylindrical vector potential gauge. Using a representation of electron coordinates as $z = x + iy$ in unit of magnetic length l_c, the single particle states in the first Landau level are:

$$\varphi_m = \frac{1}{\sqrt{2\pi 2^m m!}} z^m \exp(-|z|^2) \tag{16}$$

It is instructive to look first at the Slater determinant of electrons at filling factor 1 which is a Van der Monde determinant. Its factorization gives the following wavefunction, up to a normalization constant:

$$\Psi_1 = \prod_{i<j\leq N} (z_i - z_j) \exp(-\sum_{i=1,N} |z_i|^2) \tag{17}$$

The polynomial part ensures a uniform distribution of electrons in the plane with on average one state, or equivalently one flux quantum, for each electron. The zeros at $z_i = z_j$ reflects the Pauli principle and their multiplicity 1 the Fermi statistics. At filling factor $1/(2s+1)$ the polynomial for each z_i should be of degree of $(2s+1)(N-1)$ such that all electrons are also uniformly distributed on the $(2s+1)N$ states available. A uniform distribution of electrons in the plane requires a very symmetrical polynomial. Also Laughlin proposed the simple polynomial form [3]:

$$\Psi_{1/2s+1} = \prod_{i<j\leq N} (z_i - z_j)^{2s+1} \exp(-\sum_{i=1,N} |z_i|^2) \tag{18}$$

The zeros of multiplicity $2s+1$ ensures that electrons keep away from each other. This minimizes efficiently the correlation energy. By exchanging two electrons the wavefunction is multiplied by $(-1)^{(2s+1)}$ and satisfies the requirement that the bare electrons obey Fermi-Dirac statistics. But there is more: the extra $(-1)^{2s}$ expresses the fact that moving two electrons around each other adds an extra phase. This phase can be viewed as the Aharanov-Bohm flux of two fictive flux quanta bound to each electron. This is at the origin of the composite Fermion picture (see section below). One can also say that electrons obey a super exclusion principle where each particle occupies $2s+1$ quantum sates (i.e. the area of $2s+1$ flux quanta) so minimizing the interaction (there are deep connections with the concepts of exclusonic statistics and anyonic statistics [4]).

One can show that the excitations above the ground state present a gap. The meaning of the excitations is particularly clear in the case of the best known state occurring at $\nu = 1/3$. The ground state corresponds to uniform distribution of electrons, one electron per area occupied by three flux quanta. The unique wavefunction cannot be continuously deformed and the only way to decrease the density is to empty a single particle quantum state, i.e. to create a hole having the area occupied by a single flux quanta, see Fig. 9. This can be realized by multiplying the Laughlin wavefunction by $\prod_{i=1,N}(z_i - z_h)$ where z_h is the position of

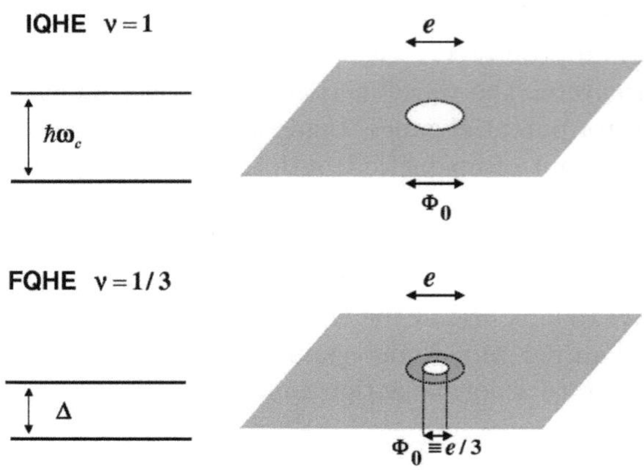

Fig. 9. In the fractional regime an elementary excitation of the hole type is created by adding a quantum flux. The hole left in the wavefunction has a size equal to a flux quantum and carries a charge $e/3$ at $\nu = 1/3$

the hole. The so called quasi-hole carry a charge $e^* = -e/3$. The energy cost Δ_h can be approximated by the energy required to create a disc of size Φ_0/B and charge $e/3$: $(4\sqrt{2}/3\pi)(e/3)^2/4\pi\varepsilon\varepsilon_0 l_c$. Similarly quasi-electron excitations with charge $e/3$ are possible and correspond to removing a flux quantum to locally increase the electronic density with an energy Δ_e. Quasi-electron or hole excitations with charge $\pm e/q$ can be generalized for other filling factor $\nu = p/q$ with q odd. The excitation gap for quasi-electron quasi-hole pairs has been numerically estimated and calculations agree with a value $\Delta \simeq 0.092e^2/4\pi\varepsilon\varepsilon_0 l_c$ for $q = 3$ and does not depend on p (as far as spin polarized electrons are considered). An interesting theoretical issue is the statistics associated with the excitations. It can be shown that when moving adiabatically two quasi-holes around each other and exchanging their positions, the collective wavefunction picks up a Berry's phase factor $\exp(i\pi(2s+1))$. The excitations are not bosons nor fermions but obey a so-called anyonic statistics, a concept first introduced in the context of particle physics.

The simple picture of the FQHE presented here is valid for spin polarized electrons. The chapter of Prof. Shankar in this book gives theoretical details when incomplete polarization plays an important role. Now I will concentrate on some experimental features of the transport properties of the FQHE.

Experimental Determinations of the Gap: Although the underlying physics is different from the physics of the Integer QHE, the basic experimental manifestations are similar. The gap associated with the incompressibility leads to a quantization of the Hall resistance at the magic filling factors and to vanishing longitudinal resistivity and conductivity.

The accuracy of the Hall plateau quantization can not be tested on a metrological level as for the Integer QHE. The reason is the lower gap. Less voltage and so less current can be passed through the sample and the signal to noise ratio is lower. However there is no reason to believe to some deviation from the fundamental resistance quantum.

The magnitude of the gap corresponds to the correlation energy anticipated by Laughlin. Attempts to measure the gap using the thermally activated dependence of the resistivity has been done by various groups. Exact comparison with theory is not that straightforward however. As for the Integer QHE, disorder leads to a mobility edge. The activation energy is then lower than the fundamental gap of a pure system. Disorder can also be view as random fluctuations of the filling factor at large length scale which, on average, reduces the effective gap. In addition, the extension of the electron wavefunction in the direction perpendicular to the plane softens the Coulomb interaction. This could be included in the computation of the gap but the extension of the wavefunction is not known accurately. Finally calculations have to include the Landau level mixing. Indeed, the gap calculated using the Laughlin wavefunction describe interactions within the first Landau level. When the correlation energy $e^2/(4\pi\varepsilon\varepsilon_0 l_c)$ is comparable or larger than $\hbar\omega_c$ the projection in the first Landau is no longer valid. This is the case in real sample where the correlation parameter $r_s \gtrsim 1$. There is a huge literature on experimental determination of the FQHE gaps using thermal

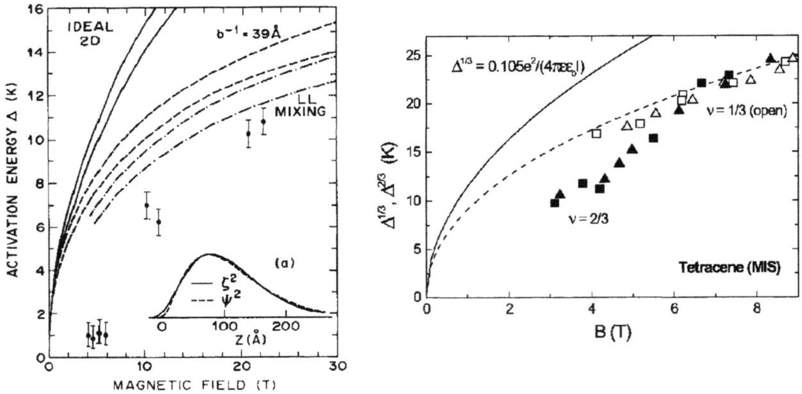

Fig. 10. The energy gap deduced from the activation energy of the longitudinal resistance is compared with theoretical values including Landau-Level mixing effect. *Left*: GaAs heterojunction (after Ref. [28]) $\nu = 1/3$, $2/3$, $4/3$ and $5/3$. *Right*: 2D hole gas confined in a Tetracene molecular crystal, $\nu = 1/3$ and $2/3$ (after Ref. [18])

activation energy. Fig. 10 shows a typical attempt to compare data with theory on two different samples. The first one is a GaAs/Ga(Al)As heterojunction [28] while the second one is a new system based on a hole inversion layer in a molecular crystal, here Tetracene [18]. In both systems the effect of the finite thickness has been taken into account. In the first one Landau level mixing has been considered and all corrections give a reasonable agreement. In the molecular system, there is a puzzling good agreement without inclusion of Landau level mixing. This is surprising regarding the larger effective mass ($1.3m_e$) and smaller dielectric constant ($\varepsilon \simeq 4$) implying $r_s \gg 1$. This shows that our understanding is far from being complete.

The intriguing excitations proposed by Laughlin have not been convincingly probed using bulk transport measurement [29]. We will see later that the mesoscopic transport properties of edge states offers a mean for unambiguous determination of their charge. It is important to remark that the fractional quantization of the Hall conductance is not a signature of the fractional excitations. It is a property of the ground state and a consequence of the fractional filling of the single particle states of the correlated electrons. This is not a signature of fractionally charge excitations (although fractional filling is probably a necessary requirement to get fractional excitations).

Composite Fermions: Following the work of Jain [7], a hierarchy of the fractional filling factors can be made using the concept of Composite Fermions as a guide. This hierarchy followed more pioneering work made by Halperin [30] using a different approach to built higher order fractions from the basic $1/(2s+1)$ states. Because this physics is also discussed elsewhere in the book, I will not talk here about the long theoretical improvements of Jain's first proposal which have lead to the present theoretical understanding. Instead I will show how the CF concept has been a guide and a source of inspiration for experimentalists and how it manifests experimentally. The concept is first based on another concept which is the statistical transmutation of electrons in 2D (or 1D). Topological considerations show that purely 2D particles are not necessarily bosons or fermions but may have any intermediate statistics (an example is the Laughlin quasiparticles). For the same reason, it is easy to "manipulate" the statistics of 3D particles such as electrons which are Fermions provided they are forced to live in 2D (or 1D). This can be done by attaching an integer number of fictitious flux quanta to each electron. The price to pay is a redefinition of the wavefunction and of the Hamiltonian. An even number of flux quanta will transform Fermions into Fermions while an odd number Fermions into Bosons. In the first case we have Composite Fermions while in the second case Composite Bosons (CB). Both approaches have been used in the FQHE context. Both have their own merit and a bridge between them is possible. CF are believed to be appropriate for high order fractions and to describe the remarkable non Quantum Hall electronic state found at $\nu = 1/2$. CB make an interesting correspondence between the $\nu = 1/(2s + 1)$ states and superfluidity [25].

By attaching $2s$ flux quanta to each electron with a sign opposite to the external magnetic field flux, the resulting CF experiences reduced mean field. A mapping can then be done between FQHE states and IQHE states. As an example, for $s = 1$, the mean field attached to the "new electrons", the composite Fermions, is equivalent and opposite to the magnetic field $B_{1/2}$ at $\nu = 1/2$. The field experienced by the CF is thus $B_{\mathrm{CF}} = B - B_{1/2}$. A filling factor $\nu = 1/3$ for electrons corresponds to a CF filling factor $\nu_{\mathrm{CF}} = 1$. Similarly $\nu = p/(2p + 1)$ becomes $\nu_{\mathrm{CF}} = p$. This describes a series of fractions observed between 1/2 and 1/3. For fields lower than $B_{1/2}$, $\nu = p/(2p - 1)$ also becomes $\nu_{\mathrm{CF}} = (-)p$ and this describes fractions from 1 to 1/2. In general attaching $2s$ flux quanta to electrons describe the fractions $\nu = p/(p.2s \pm 1)$. The following table shows the correspondence for $2s = 2$:

ν	1/3	2/5	3/7	...	1/2	...	3/5	2/3	1
ν_{CF}	1	2	3	...	∞	...	3	2	1

The composite fermion picture is supported by experimental observations. Indeed, the symmetric variations of the Shubnikov-de Haas oscillations around $B_{1/2}$ are very similar to that observed around $B = 0$. I should emphasize that this is not a real cancellation of the external field, as the Meissner effect in superconductivity is, but the phenomenon is a pure orbital effect due to the $2s$ flux attachment. Convincing experiments have shown that the quasiparticles at $\nu \simeq 1/2$ behave very similarly to the quasiparticles at zero field. Fig 11 compares the magnetic focusing experiments realized at zero field and around $B_{1/2}$ (from Ref.[32]; see also [5]). The radius of the trajectories of the quasiparticles bent by the magnetic field is $R = v_{\mathrm{F}}/\omega_c = \hbar k_{\mathrm{F}}/eB$ around zero field and $R_{\mathrm{CF}} = \hbar k_{\mathrm{F_{CF}}}/eB_{\mathrm{CF}}$, where $k_{\mathrm{F}} = (2\pi n_s)^{1/2}$, $k_{\mathrm{F_{CF}}} = (4\pi n_s)^{1/2}$ assuming spin polarized CF fermions, and $B_{\mathrm{CF}} = B - B_{1/2}$. The results of the experiment displayed here show that the ratio of the period in magnetic field of the resistance oscillations verify the relation $\Delta B_{\mathrm{CF}} = \sqrt{2}\Delta B$.

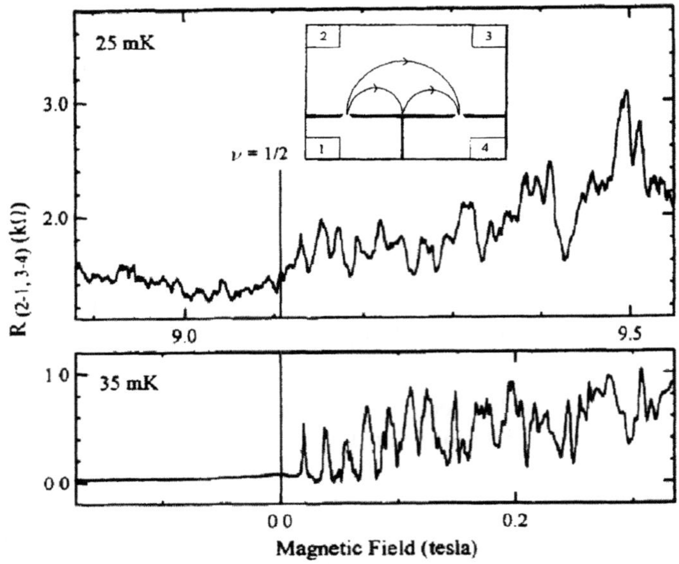

Fig. 11. Magnetic focusing for zero magnetic field (bottom trace) and for composite Fermions near $\nu = 1/2$ (top trace) for 9T. The distance between the injector and collector point contact is 5.3 μm. A peak is obtained each time a multiple integer of the cyclotron orbit diameter matches the distance (adapted from Ref.[32])

Surface acoustic waves (SAW) experiments have found also such relation. A SAW couples weakly to a 2DEG via the piezoelectric interaction. The minute change in the SAW propagation velocity resulting from this interaction is measurable and bring information about the electrons. In particular, SAW experiments probes the commensurability between the wavevector $q = \omega/v_s$ of the surface wave and the cyclotron radius of the composite Fermions. Here ω is the frequency of the SAW and v_s the propagation velocity. An unexpected finding of this experiment was the observation of a very peculiar dependance of the longitudinal CF conductivity with the wavevector: $\sigma_l(q) \sim q$. This is due to the fact that the CF do not form an ordinary 2D Fermi liquid but keep very special correlations resulting from the interactions [32]. Another interesting quantity which has been studied experimentally is the effective mass of the CF. If the correspondence between ν and ν_{CF} is exploited further one can introduce an effective mass $m_{\mathrm{CF}}(p)$ such that the FQHE gap for the $p/(2p \pm 1)$ series is defined as $\Delta_p = \hbar e B_{\mathrm{CF}}/m_{\mathrm{CF}}(p)$. The gap is determined by the activation energy of the longitudinal resistance. There is an agreement for a value $m_{\mathrm{CF}}(p \to \infty) \simeq 0.52$ at $B_{1/2}$ and for a $1/p$ or B_{CF} dependance of $m_{\mathrm{CF}} \simeq 0.52 + 0.074 B_{\mathrm{CF}}$. Also the same value is found on both side of $B_{1/2}$ (for example $\nu = 2/5$ and $2/3$, $\nu = 3/5$ and $3/7$,...) [35].

So far, we have not discussed the effect of spin in the quantum Hall regime. We have always considered the case of full polarization by the strong magnetic field. Spin effects lead to a very rich physics. Often neglected spin effects have been the subject of a growing interest the last ten years. As these effects are discussed in great details in the chapter of Prof. Shankar and in several contributions in this book, we will mentioned only a few basic points for the consistency of this review.

In GaAs, the gap between spin-split Landau level should be $\simeq 0.3$K/Tesla. However the gap observed at odd filling factors 1, 3, 5,... is much larger and even comparable to $\hbar\omega_c$ for $\nu=1$. The first explanation proposed was the enhancement due to the exchange interaction. Ando *et al.* found $\Delta = g\mu_B B + \frac{n_\downarrow - n_\uparrow}{n_\downarrow + n_\uparrow} e^2/(4\pi\varepsilon\varepsilon_0 l_m)$ [15]. Here $(n_\downarrow - n_\uparrow)/(n_\downarrow + n_\uparrow)$ denotes the relative difference of the total population of spin down and spin up electrons. For even filling factor this is always zero and the enhancement vanishes. For odd filling factor $\nu = 2k+1$, the effect decreases as $1/(2k+1)$. Indeed, the enhancement experimentally observed increases with magnetic field. Note that the correlation energy scale is the same as the one of the FQHE. Because of the spin degree of freedom, the possible excitations are not only charged excitations (electron-hole pairs), but also collective spin excitations or spin waves. In the simplest model for $\nu = 1$ the dispersion relation of spin waves is $\varepsilon_{SW}(k) = g\mu_B B$ for $k \longrightarrow 0$ and $\varepsilon_{SW}(k) = g\mu_B B + \sqrt{\frac{\pi}{2}} e^2/(4\pi\varepsilon\varepsilon_0 l_m)$ when $k \simeq 1/l_m$. The excitations have been theoretically reconsidered with the introduction of skyrmions [7][8]. It has been proposed that topological excitations should be more favorable than a single spin flip. The so-called skyrmion excitation corresponds to a spin texture in which the central flipped spin is accompanied by a gradual spin flip of the surrounding electrons so decreasing the exchange energy. This excitation carries a unit charge. Charge and spin transport can thus be coupled.

Transport measurements are not able to provide direct information on the degree of spin polarization of the 2DEG. Measurements of the transport activation energy have been made to study the competition between the Zeeman energy and the spontaneous polarization due to interaction[36]. Experiments using pressure to change the g^* factor have been made and have given indication of skyrmions [37]. Instead specific measurements are needed such as magneto-optics (see the review by Kukushkin in this book), nuclear specific heat measurements [38], optically pumped nuclear magnetic resonance [40][41] and nuclear magnetic resonance [42]. These measurements have shown that, in general, filling factors with odd numerators are well polarized while even numerators show less or zero spin polarization depending on Zeeman energy. Regarding skyrmions most experiments agree with this picture but there are still many unsettled issues.

4 Quantum Hall Effect: Mesoscopic Transport

In the following we will concentrate on transport properties at the edge of a finite quantum Hall conductor. We will see that transport is mediated by a finite number of nearly ideal one-dimensional modes representing electron running around the edges. These modes are called edge channels. They are chiral and have very nice properties which make them a prototype of quantum conductor to which the concept of mesoscopic transport can be applied. In the fractional

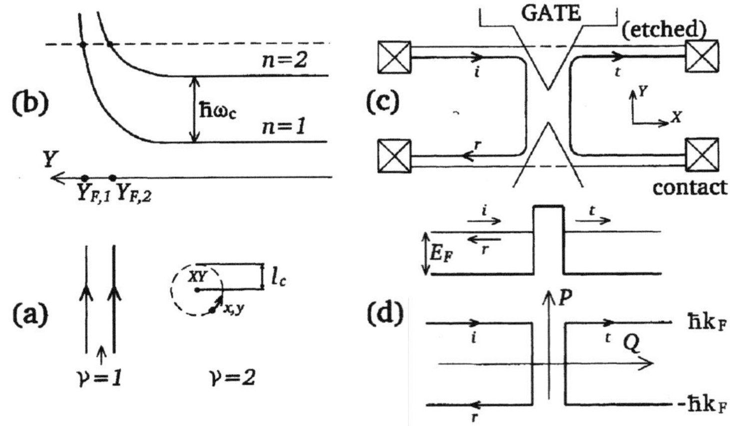

Fig. 12. (**a**) Schematic representation of edge states and of (**b**) the Landau level bending. (**c**) reflection of an edge state by a controlled artificial impurity called Quantum Point Contact. (**d**) analogy with 1D semiclassical trajectories in the phase space

regime these modes form a set of chiral Luttinger liquids, i.e. their physics is similar to that of one-dimensional interacting electron systems called Tomonaga-Luttinger liquids.

4.1 Edge Channels in the Integer Quantum Hall Regime

We have seen that the quantum Hall effect is due to the existence of a gap arising at special filling factors. However, in a finite system, gapless excitations appear at the periphery of the quantum Hall conductor. For better understanding, it is easier to consider first the case of fully spin polarized non-interacting electrons in the integer QHE regime where the physics is dominated by Landau level formation and Fermi statistics. We will also neglect coupling between Landau levels, an assumption justified in most experimental situations. Within the nth Landau level, the dynamics of electrons is described by the projected Hamiltonian:

$$\mathcal{H}_n = (n + 1/2)\hbar\omega_c + V^{(n)}(X, Y) \tag{19}$$

where $V^{(n)}(X, Y)$ is the potential which confines electrons at the boundary of the 2D conductor. The coordinates (X, Y) describing the center of a cyclotron orbit are conjugated: $[X, Y] = i\hbar/eB$. If for example, the electric field due to confinement is along the $\widehat{y}$ direction, electrons drift along the boundary, here the $\widehat{x}$ direction, with a velocity $V_x = (1/eB)\partial V^{(n)}/\partial Y$. This defines the so-called edge sates. When populated by electrons, edge states give rise to a chiral persistent current concentrated at the edges. Simultaneously the nth Landau level is bend by the confining potential, see Fig. 12. In the integer quantum Hall regime, for a filling factor $\nu = p$, there are p Landau levels which are fully occupied in the bulk of the sample. Because of bending, the filled Landau levels will cross the Fermi energy E_{F} and lead to gapless excitations at the edges. This gives rise to p lines of gapless excitations parallel to the boundary. They are located at the coordinate $Y = Y_{F,n}$, $n \leq p$, given by:

$$E_{\mathrm{F}} = (n + 1/2)\hbar\omega_c + V^{(n)}(Y_{F,n}) \tag{20}$$

The line of gapless excitations and the associated persistent currents carried by edge states form what is called an edge channel. At the opposite boundary, not shown in the figure, there are also p similar gapless excitations. Here electrons drift with the opposite direction as the confinement electric field is of reversed sign. Thus for a filling factor $\nu = p$, there are p pairs of opposite edge channels associated with the p filled Landau levels in the bulk.

Edge channels are ideal one dimensional (chiral) conductors: the physical separation between pairs of opposite edge channels prevents backscattering and electrons propagate elastically over huge distances (mm at low temperature) because only one dimensional forward scattering by phonons is permitted. These properties have make them a convenient tool to test the generalized Landauer-Büttiker relations [43] in the context of the mesoscopic quantum transport. In order to do that it is necessary to induce intentionally elastic backscattering in a controllable way. The tool used is a Quantum Point Contact as shown in Fig. 12. A negative potential applied on a metallic gate evaporated on top of the sample depletes electrons to realize a narrow constriction in the 2DEG. This allows a controllable modification of the boundaries of the sample. The separation between opposite pairs of edges channels of a given Landau level can be made so small that the overlap between wavefunctions lead to backscattering from one edge to the other. Indeed, the QPC creates a saddle shape potential and when the potential at the saddle point is close but below the value $V_{F,n} = E_{\mathrm{F}} - (n + \frac{1}{2})\hbar\omega_c$, electrons emitted from the upper left edge channel start to be reflected into the lower edge channel with probability $R \ll 1$ but are still mostly transmitted with probability $T = 1 - R$. When the saddle point potential is above $V_{F,n}$ electrons are mostly reflected and rarely transmitted $T \ll 1$. Fig. 12 shows the semi classical analogy between the real space coordinates (Y, X) of the 2D Hall conductor and the (P, Q) phase space coordinate of a real 1D conductor, as suggested by the non-commutation relations $[Y, X] = i\hbar/\left|e\right|B$ and $[P, Q] = i\hbar$. The physics of tunneling between opposite edge channels is clearly equivalent to that of the tunneling in a 1D system. However the 2D topology allows us to inject or detect electrons *at the four corners of the phase space*, something impossible with 1D systems.

Four point conductance measurements in combination with gates or QPCs have been used to probe edge states [44]. The experiments use the Büttiker-Landauer's multiterminal formula for quantum transport which relates the current at contact probe α to the chemical potential μ_β of contact β:

$$I_\alpha = \frac{e}{h}(M_\alpha - R_\alpha)\mu_\alpha - \frac{e}{h}\sum_{\beta \neq \alpha} T_{\beta\alpha}\mu_\beta \tag{21}$$

M_α is the number of (edge) channels emitted from contact α , chemical potential μ_α, and R_α the sum of their reflection probabilities. $T_{\alpha\beta}$ is the sum of transmission probabilities of the channels emitting electrons from contact β to contact

α. Fig. 13 shows an example of such measurement in the IQHE regime. If we denotes $R_{\alpha\beta,\gamma\delta}$ the resistance measured by injecting current between contacts α and β while measuring the voltage between contacts γ and δ, the following longitudinal and diagonal resistances are respectively found:

$$R_{13,12} = R_{13,43} = \frac{h}{e^2}(\frac{1}{\nu_G} - \frac{1}{\nu}) \tag{22}$$

$$R_{13,13} = \frac{h}{e^2}\frac{1}{\nu_G} \tag{23}$$

$$R_{13,24} = \frac{h}{e^2}(\frac{2}{\nu} - \frac{1}{\nu_G}) \tag{24}$$

Here ν is the filling factor in the lead and $\nu - \nu_G$ edge channels are reflected (ν_G edge channels transmitted). The two terminal conductance $G = \nu_G(e^2/h)$ is equal to the number of channels fully transmitted. One can also says that this corresponds to the filling factor at the center of the QPC. In the example shown in Fig. 13 the filling factor is $\nu = 8$ in the 2D leads. For $\nu_G = 8$ to 2 one can check that plateaus in the longitudinal resistance are found at 0, 1/56, 1/24, 3/40, 1/8, 5/24, 3/8, in quantum resistance units, as expected.

So far we have considered the integer quantum Hall regime. Note that the above picture of edge channels generalizes immediately if we include spin split Landau levels. Now the number of pairs of opposite edge channels in a given sample is given by the number of occupied spin split Landau level, i.e. the filling factor. In fact in Fig. 13 the plateaus observed for $\nu_G = 7$, 5, and 3, do correspond to spin split edge channels. In the data, the odd plateaus are clearly less well defined because the spin gap, even enhanced by exchange interactions, is less than the cyclotron gap.

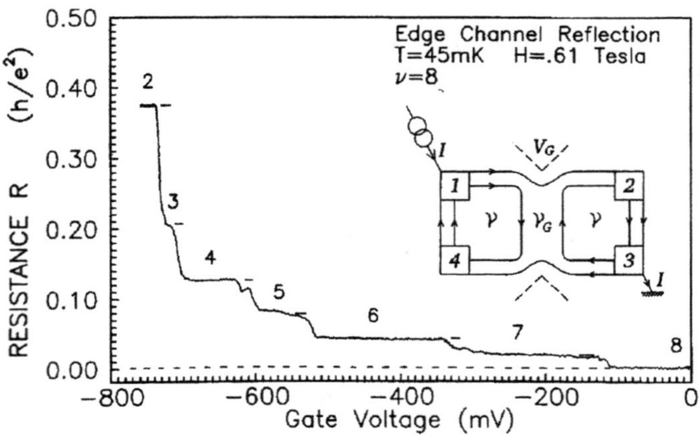

Fig. 13. Longitudinal resistance versus QPC gate voltage showing the reflection of the six first edge states for $\nu = 8$, at T=45 mK. The values of ν_G are indicated on the resistance plateaus

4.2 An Interacting Picture of IQHE Edge States

At short distance the Coulomb repulsion between electrons is responsible for correlations which eventually give rise to exchange effects and to the fractional QHE. But the Coulomb interaction is very special because it is long range. We are now going to focus on the long range part of this interaction which is responsible for screening. In a first approximation it can be treated using electrostatics and capacitive models in a Thomas-Fermi approach. This approximation is believed qualitatively correct for the smooth edge confining potentials which exist in real samples.

Screening means that electrons change their density to build up a charge distribution which in turn creates an induced electric field compensating an external force. The gap of the quantum Hall effect is however responsible for incompressibility which prevents screening. The competition between both effects leads to a strong modification of the edge channels as schematically shown in Fig. 14. In this example going from the bulk to the edge, one starts with an electron density pinned at $n_s = 2n_\Phi$ ($\nu = 2$). The confining potential is unscreened because of the energy gap $\simeq \frac{1}{2}\hbar\omega_c$ to remove one electron. Here a change of the local density is impossible. Thus, the Landau levels bend and rise. When the second Landau level crosses the Fermi energy, the energy gap vanishes. Now, it costs no energy to depopulate the second Landau level and screening is as effective as in a perfect metal. The total potential is constant and the Landau level energies are locally flat. When the second Landau level is empty, $n_s = n_\Phi$ ($\nu = 1$) the density is again pinned until the first Landau level reaches the Fermi energy, and so on. The edges can be viewed as a series of incompressible quantized region of well defined filling factors, the quantum Hall regions, which alternate with "metallic"

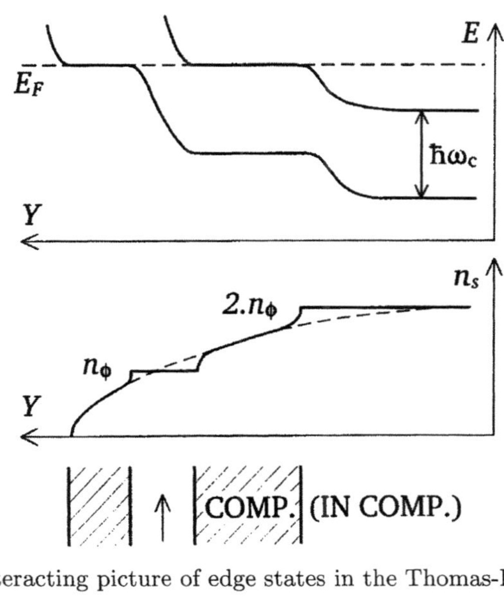

Fig. 14. Interacting picture of edge states in the Thomas-Fermi picture

strip of compressible regions, the edge channels. From simple electrostatics, the width of the incompressible strips is found of the order of $(a_{\text{Bohr}}d)^{1/2}$ where d is the lengthscale of the confining potential [45].

Regarding quantum transport, the longitudinal and diagonal resistance are expected to be the same as in the non interacting picture of edge states. Transport can not distinguish alone between the two pictures. The rules for calculating the current are the following: 1) the compressible strips form equipotential equal to the electrochemical potential μ_α of the contact α emitting electrons; 2) an incompressible strip with filling factor ν_i between two compressible strips attached to contact α and β carries a excess Hall current $(\nu_i e^2/h)(\mu_\alpha - \mu_\beta)$. Applying these rules give the same expressions as those given in Eqs. (22) to (24).

This picture can be extended to the fractional quantum Hall regime. Now the gap is that of FQHE and the ν_i's can be fractional [46]. Fractional edge states can be detected using the reflection induced by a QPCs in a similar manner as integer edge states as shown in Fig. 15 . Here the filling factor in the leads is $\nu = 2/3$ while the filling factor at the QPC ν_G is varied from 2/3 to $\lesssim 1/3$. The longitudinal resistance $(h/e^2)(\frac{1}{\nu_G} - \frac{1}{\nu})$ shows accurate quantization with plateaus at 1 and 3/2. From these values, we can infer the existence of the 2/5 and 1/3 fractional edge states respectively.

The picture described here is expected to apply to smooth edge, as it is the case in ordinary samples. Another approach has been proposed for hard wall confinement in Ref.[47].

4.3 Luttinger Liquid Properties of Fractional Edge Channels

Wen's Approach of Fractional Edge States: Wen [49] has first shown the deep connection between fractional edge channels and the concept of Tomonaga-Luttinger liquids [48]. We will here repeat the phenomenological hydrodynamic approach of Wen in the simple case of a Laughlin state in the bulk, filling factor $\nu = 1/(2s+1)$. We will start with a classical approach and keep only incompressibility as a quantum ingredient.

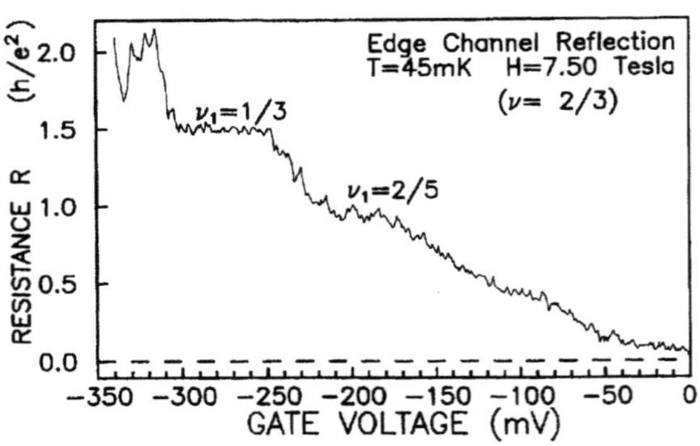

Fig. 15. Fractional edge channel reflection observed for $\nu = 2/3$. The longitudinal resistance quantization indicates $\nu_G = 2/5$ and 1/3 fractional channels

The only possible excitations are periphery deformations of the 2D quantum Hall conductor which preserves the total area (like a 2D droplet of an ordinary liquid) as shown schematically in Fig. 16 . If we denote $y_+(X,t)$ and $y_-(X,t)$ the deformation of the upper and lower boundaries which are respectively located at the positions $Y = Y_{\rm F}$ and $Y = -Y_{\rm F}$, the time varying electron density is given by:

$$n(X,Y,t) = n_s\Theta(Y - Y_{\rm F} - y_+(X,t)) - n_s\Theta(Y + Y_{\rm F} + y_-(X,t)) \tag{25}$$

where $n_s = \nu eB/h$ and $\Theta(x)$ is the Heaviside function. We wish to find the equations of motion for $y_\pm$. To do that we have to remind that, within the first Landau level, the single particle motion is given by the reduced Hamiltionian $\mathcal{H}_1 = (1/2)\hbar\omega_c + V^{(1)}(X,Y)$. X and Y being conjugate, the classical Poisson's bracket is $\{X,Y\} = h/eB$. Using the equation of motion for the 2D density $\partial n/\partial t + \{\mathcal{H}_1, n\} = 0$ and assuming translation invariance along the X axis for $V^{(1)}$, leads to the two decoupled equations describing the chiral propagation of the shape deformations y_+ and y_- along each boundary $Y = Y_{\rm F}$ and $Y = -Y_{\rm F}$ respectively:

$$\partial y_+/\partial t + v_D \partial y_+/\partial X = 0 \tag{26}$$

$$\partial y_-/\partial t - v_D \partial y_-/\partial X = 0 \tag{27}$$

where $v_D = \frac{1}{eB}\left|\partial V^{(1)}/\partial Y\rfloor_{Y=\pm Y_{\rm F}}\right|$ is the drift or local Fermi velocity. For simplicity we have assumed a similar confining potential at $Y = \pm Y_{\rm F}$ leading to equal absolute value of the drift velocity. Lets now concentrate on the upper edge. The potential energy associated with the deformation is

$$U_+ = \frac{1}{2}\int dX n_s y_+^2 \frac{\partial V}{\partial Y} = \frac{\hbar v_D}{\nu\pi}\int dX\,(\pi n_s y_+)^2 \tag{28}$$

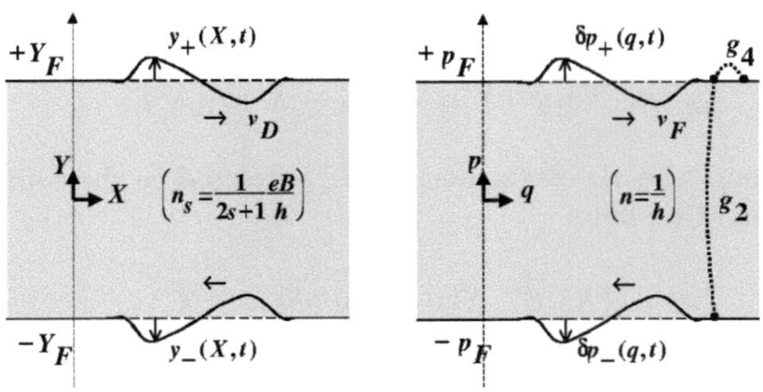

Fig. 16. *Left*: Wen's approach for the Chiral Luttinger liquid picture of a fractional edge channel at $\nu = 1(2s+1)$. *Right*: a similar picture can be used for the (zero magnetic field) 1D interacting electron system called Luttinger liquids. Here, the semiclassical phase space (p,q) replaces the 2D real space (Y,X)

If we define the charge variation integrated on the upper edge in units of π in the following way

$$\frac{\widetilde{\phi_+}}{\pi} = \int_{-\infty}^{X} n_s y_+ dX \text{or} n_s y_+ = \widetilde{\rho_+}(X) = \frac{1}{\pi}\frac{\partial \widetilde{\phi_+}}{\partial X}$$

the action

$$S = -\frac{\hbar}{\pi\nu}\int dX dt \frac{\partial \widetilde{\phi_+}}{\partial X}\left(\frac{\partial \widetilde{\phi_+}}{\partial t} + v_D \frac{\partial \widetilde{\phi_+}}{\partial X}\right) \tag{29}$$

gives the correct propagation equation for y_+ and leads to the Hamiltonian $\mathcal{H}_+ = U_+$. We finally arrive to the total hamiltonian for the two decoupled upper and lower edge channels:

$$\mathcal{H}_+ = \tfrac{\hbar v_D}{\nu\pi}\int dX\left[\left(\tfrac{\partial \widetilde{\phi_+}}{\partial X}\right)^2 + \left(\tfrac{\partial \widetilde{\phi_-}}{\partial X}\right)^2\right] \tag{30}$$

$$= \tfrac{h v_D}{2\nu}\int dX\left[\widetilde{\rho_+}^2 + \widetilde{\rho_-}^2\right] \tag{31}$$

So far the model is purely classical. We are going now to quantize the collective modes found above. From the Lagrangian, one can define momentum conjugate of $\widetilde{\phi_\pm}$ as $\widetilde{\pi_\pm} = \mp\frac{\hbar}{\pi\nu}\partial\widetilde{\phi_\pm}/\partial X$ and quantizes the bosonic field:

$$\left[\widetilde{\pi_\pm}(X), \widetilde{\phi_\pm}(X')\right] = i\hbar\delta\left(X - X'\right) \tag{32}$$

The physics of these bosonic modes describing the periphery deformations would not be interesting unless one have to consider the transfer of a bare electron or a Laughlin quasiparticle form one edge to the other. Indeed, the non trivial physics arises from the fact that adding an electron on a given edge involves an infinite number of bosonic modes. This is at the origin of strong non-linearities in the transport properties, a property not shared by ordinary Fermi liquids.

How to define the creation operator $\psi^\dagger$ for one electron on the upper edge? By definition, it satisfies:

$$\left[\rho_+(X), \psi^\dagger(X')\right] = \delta(X - X')\psi^\dagger(X') \tag{33}$$

On the other hand, the 1D excess density $\widetilde{\rho_+}$ is related to the conjugate of $\widetilde{\phi_+}$ by $\widetilde{\pi_\pm} = -\frac{\hbar}{v}\widetilde{\rho_+}$ and we have:

$$\left[\widetilde{\rho_+}(X), \widetilde{\phi_+}(X')\right] = -i\nu\hbar\delta\left(X - X'\right) \tag{34}$$

which immediately implies:

$$\psi^\dagger \propto \exp(i\widetilde{\phi_+}/\nu) \tag{35}$$

$\psi^\dagger$ creates a unit charge at X but it is not an electron operator unless it satisfies Fermi statistics. Exchanging two electrons at position X and X' gives

$\psi^{\dagger}(X')\psi^{\dagger}(X) = \exp(-i\frac{\pi}{\nu}sgn(X-X'))\psi^{\dagger}(X)\psi^{\dagger}(X')$. The requirement that the bare particles are Fermions implies

$$\nu = 1/(2k+1) \tag{36}$$

The beauty of Wen's approach is that the Laughlin filling factors appear naturally as a consequence of incompressibility and Fermi statistics. To obtain other fractional filling factors as Jain's filling factors, a generalization of this approach is possible. One must introduce additional bosonic modes at the periphery (for example: p modes for $p/(2ps+1)$ filling factors).

Finally, one can define similarly the quasiparticle operator which creates a charge $1/(2s+1)$ on the edge:

$$[\widetilde{\rho_+}(X), \psi^{\dagger}_{qp}(X')] = \nu\delta(X-X')\psi^{\dagger}_{qp}(X') \tag{37}$$

which writes as

$$\psi^{\dagger}_{qp} \propto \exp(i\phi_+) \tag{38}$$

It shows *fractional* statistics: $\psi^{\dagger}_{qp}(X')\psi^{\dagger}_{qp}(X) = e^{(-i\pi\nu sgn(X-X'))}\psi^{\dagger}_{qp}(X)\psi^{\dagger}_{qp}(X')$. This establishes a direct correspondance between quasi-particles on the edge and Laughlin quasiparticles in the bulk.

The above set of equations for the electron operator $\psi^{\dagger}$ and for the bosonic modes are characteristics of those of a Luttinger liquid. Because of the decoupling between upper and lower modes it is called a Chiral Luttinger Liquid. The conductance $\nu e^2/h$ correspond to the conductance ge^2/h of a Luttinger Liquid and one usually identifies $g = \nu$. As for Luttinger liquids there is an algebraic decay of the correlation functions. One can show that the single particle Green's function decreases as $(X - v_D t)^{-1/\nu}$. The Tunneling Density of State for an electron at energy ε is $\sim |\varepsilon - E_{\mathrm{F}}|^{(1/\nu-1)}$. This means that, for electrons tunneling between a Fermi liquid and a chiral Luttinger fractional edge channel, the finite temperature conductance $G(T)$ and the zero temperature differential conductance dI/dV show the following power laws:

$$G(T) \sim (T/T_B)^{\gamma} \tag{39}$$

$$\frac{dI}{dV} \sim (V/V_B)^{\gamma} \tag{40}$$

$$\gamma = \tfrac{1}{\nu} - 1 \tag{41}$$

where T_B and V_B are related to the coupling energy of the tunnel barrier. The power laws characterize a Luttinger liquid and have been experimentally observed (see below). The chirality leads to some differences with ordinary Luttinger Liquids for which $1/\nu$ is to be replaced by $(g^{-1}+g)/2$.

Connection with the 1D Luttinger Liquids: Because some (attendants of the School) come from the field of organic conductors and the physics of Luttinger liquids is also relevant for these systems for some energy scale, it may be useful to present the connection, and also the differences, between fractional edge channels and 1D interacting systems. To do that I will give a phenomenological derivation of Luttinger liquids, borrowing Wen's approach. I will apply this approach to the semi-classical phase space of the 1D Fermi sea (in zero magnetic field). For those not familiar in this field, I must emphasize that rigorous fully quantum mechanical microscopic derivations exist (see Ref.[50]). Here, the idea is to avoid heavy mathematics and give some physical intuition of what is going on.

In the following, the (q, p) coordinates will play the role of (X, Y). According to the Pauli principle, *the phase space is incompressible* with an electron density $n(p, q)$ pinned to the value $1/h$ for $-p_{\mathrm{F}} < p < p_{\mathrm{F}}$ in the ground state. Because it is known that for 1D interacting electrons collective excitations are more relevant than single particle excitations, the dynamics of electrons will be taken into account by considering *deformations* $\delta p^{\pm}(q, t)$ of the periphery of the Fermi sea, as shown in Fig. 16, with:

$$n(p, q, t) = \frac{1}{h}\Theta(p - p_{\mathrm{F}} - \delta p_{+}(q, t)) - \frac{1}{h}\Theta(p + p_{\mathrm{F}} + \delta p_{-}(q, t)) \tag{42}$$

A collective excitation corresponds to redistribution of the population of single particles near $\pm p_{\mathrm{F}}$. The Hamiltonian describing the dynamics of the single particles occupying the Fermi sea is given by:

$$\mathcal{H}(p, q) = \frac{1}{2m}p^2 + V(q) \tag{43}$$

We will proceed like in the previous section and find the equation of motion of the Fermi sea deformation, using the Poisson's bracket $\partial n/\partial t + \{H\mathcal{H}, n\} = 0$. In the limit $\delta p^{\pm}(q, t) \ll p_{\mathrm{F}}$ and using $\partial n/\partial p = \frac{1}{h}\delta(p \pm p_{\mathrm{F}})$ the $\pm$ deformations satisfy the following equations:

$$\left(\frac{\partial}{\partial t} + v_{\mathrm{F}}\frac{\partial}{\partial q}\right)\delta p_{+} + \frac{\partial V}{\partial q}\rfloor_{p_{\mathrm{F}}} = 0 \tag{44}$$

$$\left(\frac{\partial}{\partial t} - v_{\mathrm{F}}\frac{\partial}{\partial q}\right)\delta p_{-} - \frac{\partial V}{\partial q}\rfloor_{-p_{\mathrm{F}}} = 0 \tag{45}$$

We will assume a short range local interaction between electrons such that the potential energy experienced by each single particle occupying the Fermi sea is proportional to the linear density $\int dpn(p, q, t) = 2\frac{p_{\mathrm{F}}}{h} + \frac{1}{h}(\delta p_{\mathrm{F}}^{+} + \delta p_{\mathrm{F}}^{-}) = \rho_0 + \rho(q, t)$, i.e. $V(q) = \alpha\rho(q, t)$. This gives

$$\left(\frac{\partial}{\partial t} + (v_{\mathrm{F}} + 2\frac{\alpha}{h})\frac{\partial}{\partial q}\right)\delta p_{+} + 2\frac{\alpha}{h}\delta p_{-} = 0 \tag{46}$$

$$\left(\frac{\partial}{\partial t} - (v_{\mathrm{F}} + 2\frac{\alpha}{h})\frac{\partial}{\partial q}\right)\delta p_{-} - 2\frac{\alpha}{h}\delta p_{+} = 0 \tag{47}$$

The solution are collectives modes mixing $+p_{\mathrm{F}}$ and $-p_{\mathrm{F}}$ deformations of the Fermi sea. The eigenvalues for the propagation velocity are:

$$u_{\pm} = \pm u \quad \text{with } u = \sqrt{v_{\mathrm{F}}(v_{\mathrm{F}} + 2\frac{\alpha}{h})} \tag{48}$$

In standard notations of Luttinger liquid models, interactions between electrons moving in the same direction are characterized by an interaction parameter g_4 while interactions between electrons moving in opposite direction are characterized by the interaction parameter g_2 and the general solution is $u_\pm = \pm\sqrt{(v_\mathrm{F} + \frac{g_4}{2\pi})^2 - (\frac{g_2}{2\pi})^2}$. The interaction chosen here is such that $g_2 = g_4 = 2\pi\alpha/h$, a common situation. We will use g_2 instead of α in the following. To make better connection with standard notations, we introduce, in the so-called phase representation, the fields $\frac{1}{\pi}\phi(q) = \rho(q)/\rho_0 = \frac{1}{h\rho_0}(\delta p_+ + \delta p_-)$ and $\frac{1}{\pi}\theta(q) = -\frac{1}{h\rho_0}(\delta p_+ - \delta p_-)$ which gives

$$\frac{\partial\phi}{\partial t} - v_\mathrm{F}\frac{\partial\theta}{\partial q} = 0 \tag{49}$$

$$-\frac{\partial\theta}{\partial t} + (v_\mathrm{F} + 2\frac{g_2}{2\pi})\frac{\partial\phi}{\partial q} = 0 \tag{50}$$

From the Lagrangian density in appropriate units $\frac{\hbar}{2\pi v_\mathrm{F}}\left[\left(\frac{\partial\varphi}{\partial t}\right)^2 - u^2\left(\frac{\partial\varphi}{\partial q}\right)^2\right]$, one can define the momentum $\Pi(q) = \frac{\hbar}{\pi v_\mathrm{F}}\frac{\partial\varphi}{\partial t} = \frac{\hbar}{\pi}\frac{\partial\theta}{\partial q}$ conjugate to the field φ:

$$[\Pi(q), \varphi(q')] = -i\hbar\delta(q - q') \tag{51}$$

and one finally arrives to the well known Luttinger liquid Hamiltonian for the bosonic modes:

$$\mathcal{H} = \frac{\hbar u}{2\pi}\int dq\left(g\left(\frac{\partial\theta}{\partial q}\right)^2 + g^{-1}\left(\frac{\partial\varphi}{\partial q}\right)^2\right) \tag{52}$$

where

$$g = \sqrt{\frac{v_\mathrm{F}}{v_\mathrm{F} + 2\frac{g_2}{2\pi}}} \quad \text{with } u.g = v_\mathrm{F} \text{ and } \frac{u}{g} = v_\mathrm{F} + 2\frac{g_2}{2\pi} \tag{53}$$

The parameter g is called the Luttinger parameter. It plays the role of the filling factor ν of the fractional quantum Hall effect. One can show that the conductance without impurity and for an infinite long system is:

$$G_\mathrm{LL} = g.\frac{e^2}{h} \tag{54}$$

For repulsive interactions, $g < 1$.

As in the previous section, using the commutation relations given by (51), it is found that the operator creating a right (left) moving electron is $\psi^\dagger \propto \exp(i(\theta \mp \phi))$. One can then calculate various useful correlation functions. For example, the tunneling density of state is found to vanish at the Fermi energy as TDOS $\sim |\varepsilon - E_\mathrm{F}|^{(g+1/g-2)/2}$, a hallmark of Luttinger liquids. Finally in order to make direct comparison with the Chiral Luttinger Liquid model of fractional edge channels, let us represent the collective bosonic excitations in term of the eigenmodes solution of (46, 47)

$$\widetilde{\phi\pm} = g\theta \mp \varphi \tag{55}$$

associated with the linear densities

$$\widetilde{\rho_\pm} = \frac{1}{2\pi}\frac{\partial\widetilde{\phi\pm}}{\partial q} \tag{56}$$

The Hamiltonian simply becomes

$$\frac{hu}{2g}\int dq\left[\widetilde{\rho_+}^2 + \widetilde{\rho_-}^2\right] \tag{57}$$

and it is similar to the one obtained in Eq. (31) for the case of the fractional quantum Hall effect.

2D Chiral Luttinger Liquid	1D Luttinger Liquid
$g = \nu$ fractional,	g: any values, results from
built by interaction in the bulk	coupling between $\pm p_{\mathrm{F}}$ excitations
$Gk_{\mathrm{CLL}} = \nu.\frac{e^2}{h}$	$G_{\mathrm{LL}} = g.\frac{e^2}{h}$
TDOS $\sim \lvert\varepsilon - E_{\mathrm{F}}\rvert^{(1/\nu-1)}$	TDOS $\sim \lvert\varepsilon - E_{\mathrm{F}}\rvert^{(g+1/g-2)/2}$
right (left) bosonic modes	right (left) bosonic modes
do not mix upper and lower excitations	mix $\pm p_{\mathrm{F}}$excitations

Comparison Between Chiral and non Chiral Luttinger Liquid: The physics of both systems is very similar. Using the analogy between the 2D space and the 1D phase space, the difference in the microscopic origin of the bosonic modes is clearly established. In some experimental situations, when the 2D electron Hall bar is narrow, width w, and when the energy ε is such that $v_D h/\varepsilon > w$, the long range Coulomb interaction can couple the upper and lower modes. This introduces a parameter g_2 (here wavevector dependent) similar to that defined for non chiral Luttinger liquid. The effective Luttinger parameter is no longer $g = \nu$ but $g(\nu, g_2) \neq \nu$. In particular, in the limit $w \to 0$ the qualitative difference between both systems becomes very small.

There are also important differences. A *continuous variation* of the $g = \nu$ parameter is a priori not expected in the FQHE regime because the magnetic field stabilizes special fractional values of ν in the bulk (as the Jain's series). A generalization of Wen's approach for $\nu = p/(2sp+1)$ shows that one must have p branches of bosonic modes. These branches interact together and give a relation between the TDOS exponent not simply related as in (41). This is not the case for 1D systems for which g varies continuously and the relation between the tunneling exponent and g is expected to always hold: $\gamma = (g + g^{-1} - 2)$. For the simplest series of Jain's filling factor $p/(2p \pm 1)$ between 1 and 1/3, the exponent γ is expected to be

$$\gamma = \frac{2p+1}{p} - \frac{1}{|p|} \tag{58}$$

i.e. constant ($\gamma = 2$) between filling factors 1/2 and 1/3 and decreasing linearly ($\gamma = 1/\nu \sim B$) form 2 to 1 between filling factors 1/2 and 1.

It is also interesting to compare the practical realizations of 1D and fractional quantum Hall Luttinger liquids. Practically, it is rather difficult to built a clean single 1D quantum wire. The $\pm p_{\mathrm{F}}$ electronic modes being spatially located at the same place, any inhomogeneity will produce efficient scattering. The upper and lower fractional quantum Hall edge channels are usually well separated spatially and no unintentional backscattering is possible. This is why fractional quantum Hall systems have been the only ideal realization of Luttinger liquids found so far. Also, in 1D systems such as quantum wires [51–53], carbon nanotubes [54,55], blue-bronze or quasi-1D organic conductors it is difficult to probe power laws over more that 1 to 1.5 decades, while in fractional quantum Hall systems the TDOS has been probed over 3 decades giving good determination of the tunneling exponents and a clear evidence of Luttinger liquids properties [56,57,61].

Experimental Evidence of Chiral Luttinger Liquids: The best evidence for Luttinger liquid properties is obtained by probing the tunneling density of states. To do that, the measurement has to be *non-invasive*, i.e. a weak tunnel coupling is required. Indeed, a tunneling experiment measures the TDOS only if higher order tunneling processes are negligible, which means small transmission and small energy (as the current $\sim |\varepsilon - E_{\mathrm{F}}|^{\gamma+1}$ and, so the effective coupling, will increase with voltage $\varepsilon \equiv eV$ or temperature $\varepsilon \equiv k_{\mathrm{B}}T$).

Convincing experiments have been performed by the group of A.M. Chang [56,57,61]. The tunnel contact is realized using the cleaved edge overgrowth technique. By epitaxial growth on the lateral side of a 2DEG, a large tunnel barrier is first defined followed by a metallic contact realized using heavy doped semiconductor. The advantage is a weak coupling, a barrier strong enough to apply large voltage without significant change in the (bare) tunnel coupling, and a rather well define edge for the 2D electron system. Also, probably important is the fact that the metallic contact close to the edge provides screening of the long range Coulomb interaction (short range is needed for having power laws).

Fig. 17 shows example of IVC: a power law of the current with applied voltage $I \sim V^{\alpha}$, $\alpha = \gamma + 1$, is well defined over several current decades for $\nu = 1/3$. The exponent α found is 2.7-2.65 close to the value $\alpha = 3$ predicted by the theory. For other filling factors similar algebraic variations are also observed. In the same figure, the tunneling exponent deduced from a series of I-V curves is shown as a function of the magnetic field or $1/\nu$. For filling factor $1/2 < \nu < 1/3$ the constant exponent predicted [58–60] is not observed and instead the exponent varies rather linearly with field or $1/\nu$ (however, recent experiments made with cleaner samples have shown signs of a plateau in the exponent in a narrow filling factor value near 1/3).

Theoretical attempts to explain quantitatively the discrepancies have been made. Taking into account the long range Coulomb interaction slightly lowers the exponent. The modified Luttinger liquid theories can also include the finite conductivity in the bulk for non fractional filling factors. Indeed a finite conductivity modifies the dispersion relation of the bosonic chiral modes and so the exponents. With reasonable parameters these modifications are not yet able to fully reproduce the data [59,60]. The discrepancy between experiments and

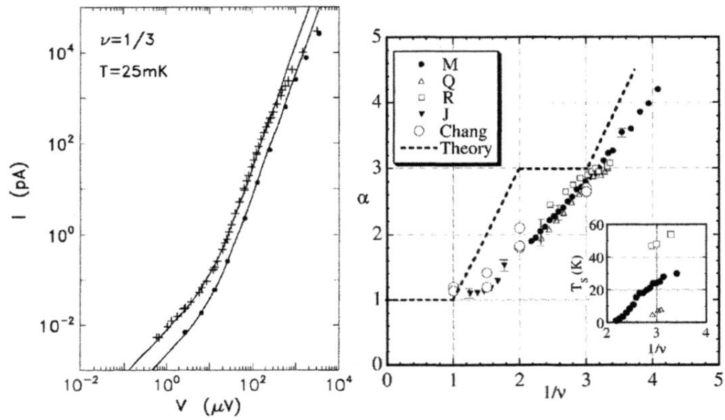

Fig. 17. *Left*: Log-Log plot of an I-V curve at $\nu = 1/3$ clearly shows the algebraic variation of current with voltage which characterizes a Luttinger liquid. *Right*: the exponent α is ploted verus $1/\nu$ and compared with theoretical predictions (adapted from Ref[56,57])

predictions may be due to the reconstruction of the edge. Wen's model assumes a sharp density variation at the edge of the 2D sample. In real samples, the density decreases smoothly and some additional edge states corresponding to filling factor lower than the bulk filling factor may also strongly modify the exponents.

Another possibility to search for chiral Luttinger liquid evidence is to make transport experiments between two fractional edges. This can be done using an artificial impurity, a Quantum Point Contact. This strategy however is less reliable than the previous one as it is likely that, when a voltage difference V_{ds} is applied between both side of the QPC, the microscopic potential which results from the gate of the QPC and the electric field associated with V_{ds} will change. This will modify the transmission in a trivial manner and will make identification of power laws difficult. In the case of a high barrier with tunneling regime established for all energies and no possibilities to reach the regime of weak backscattering, the result will be similar to that of A.M. Chang's experiments. However, because the dI/dV_{ds} characteristics will be proportional to the square of the TDOS, the exponents for the conductance will be doubled. This approach has been used by several groups. A difficulty is that sample inhomogeneities around the QPC leads to transmission resonances which are difficult to control. Exploiting Luttinger predictions for tunneling through such a resonant state has been used in [62]. A further difficulty is the high value of the power law for the conductance with temperature or voltage. For $\nu = 1/3$, $\gamma = 4$ (i.e. $2(\nu^{-1} - 1)$). It has been realized recently, with the help of exact finite temperature calculations of I-V characteristics, that the exponent 4 can be measured only if the conductance is smaller than $10^{-4}e^2/3h$ [63]. Up to now, no experimental group have tried to do measurements in this limit. For higher conductance instead, an effective exponent, much weaker than 4, will be measured. To understand this, one have to discuss the physics of Luttinger liquids with an impurity and understand how this impurity affects the conductance.

4.4 Fractional Edge Channels with an Artificial Impurity

What we are going to discuss in the following is also valid for ordinary Luttinger liquids. Indeed we have seen that the chiral Luttinger liquids and ordinary Luttinger liquids are described by the same Hamiltonian. Without loss of generality we will take the case of the special filling factor $\nu = 1/3$ as the physics is better understood and experiments are cleaner.

In the fractional regime, a controllable artificial impurity is easily realizable using a QPC. The gate potential can be used to change the strength of the impurity. We will discuss here the case of a *weak impurity*. Without Luttinger liquid effects such impurity would lead to a weak coupling between upper and lower edges. Practically, the QPC gently pushes the upper edge close to the lower edge to induce a quantum transfer of particles from one edge to the other. Also the QPC potential is weak enough to do not make appreciable change of the local filling factor. This is different from the so-called pinch-off regime where the potential is so strong that a local electron depletion sets in and a tunnel barrier forms. Nevertheless, the Luttinger liquid theory predicts that in the limit of small energy ε ($k_B T$ or $eV_{ds} \to 0$) the strength of the weak artificial impurity is *strongly renormalized* by the interaction. The system flows to an insulating state and the conductance displays the same power law with T or V_{ds} than the one expected for a strong impurity potential (tunnel barrier)

$$G \sim \frac{e^2}{3h}\left(\frac{\varepsilon}{T_B}\right)^{2(\frac{1}{\nu}-1)} \to 0 \quad \text{for } \varepsilon \ll T_B \tag{59}$$

where T_B is an energy scale related to the impurity potential. On the other hand, as the impurity is very weak one expects that at high energy a conductance close to, but smaller than the quantum of conductance $\frac{e^2}{3h}$ is recovered. Indeed the Luttinger liquid theory predicts

$$G = \frac{e^2}{3h} - G_B \quad \text{with } G_B = \frac{e^2}{3h}\left(\frac{\varepsilon}{T_B}\right)^{2(\nu-1)} \to 0 \quad \text{for } \varepsilon \gg T_B \tag{60}$$

G_B is called the backscattering conductance. Indeed if we call I the forward current, $I_0 = \frac{e^2}{3h}V_{ds}$ the current without impurity, the current associated with particles backscattered by the impurity is $I_B = I_0 - I$ from which one can define $G_B = I_B/V$. The above formulae correspond to *strong* and *weak* backscattering limits. In the first case there is a weak tunneling of particles between the left and right side, while in the second case there is a weak quantum transfer of particles between the upper edge and the lower edge. There is an interesting *duality* with $\nu \longleftrightarrow 1/\nu$.

Most experiments at $\nu = 1/3$ have been performed in an intermediate regime. First, to reach the strong backscattering regime one needs at least one order of magnitude of variation of T with the condition $T \ll T_B$. This implies working with conductances as low as $10^{-2(\frac{1}{\nu}-1)}\nu e^2/h$ to obtain conductance values. Second, if this condition is not respected, theory shows that inclusion of higher order processes gives lower effective exponents. Indeed the exponent above can be viewed as the scaling exponent of a renormalization theory. It is valid only close to the critical point (here $G = 0$).

What are the higher order tunneling terms? The tunneling Hamiltonian coupling the upper and lower edge is

$$\mathcal{H}_{\text{imp}} = \frac{1}{2}\sum_n \lambda_n \left(\exp(in\widetilde{\varphi_+(0)})\exp(-in\widetilde{\varphi_-(0)}) + \exp(in\widetilde{\varphi_-(0)})\exp(-in\widetilde{\varphi_+(0)})\right) \tag{61}$$

where λ_n is the Fourier transform of the potential at $2nk_{\text{F}}$ and the nth term correspond to the transfer of n particles. Each contribution can be calculated perturbatively using a renormalization approach. In the two limiting cases strong backscattering, for energies $\varepsilon \to 0$, it is found [58]

$$G(\varepsilon) = \nu\frac{e^2}{h}\sum_n \varepsilon_n \left|\lambda_n\right|^2 \varepsilon^{2(\frac{n^2}{\nu}-1)} \tag{62}$$

were the coefficient ε_n are non universal. The powers in the series increase rather rapidly with the order n considered. For $\nu = 1/3$ for example the second term has a power 22 while the first one has a power 4. When the energy is below some energy scale only the first term is relevant and is usually kept. When keeping only the first $\cos(\widetilde{\varphi_+(0)} - \widetilde{\varphi_-(0)})$ term in the Hamilltonian, a recent work based on conformal field theory has shown that the problem of a Luttinger liquid with one impurity is fully integrable for all energies [65,66]. The so-called FLS theory exploits the charge conservation when the particles are scattered by the impurity centered in $X = 0$. Using the $\widetilde{\rho_\pm}$ charge densities associated with the fields $\widetilde{\phi_\pm}$ considered previously, it is possible to define the even and odd charges (and corresponding fields)

$$\widetilde{\rho^e}(X,t) = \tfrac{1}{\sqrt{2}}\left(\widetilde{\rho_+}(X,t) + \widetilde{\rho_-}(-X,t)\right) \tag{63}$$

$$\widetilde{\rho^o}(X,t) = \tfrac{-1}{\sqrt{2}}\left(\widetilde{\rho_+}(X,t) - \widetilde{\rho_-}(-X,t)\right) \tag{64}$$

now define on a semi-infinite line $X \leq 0$. Clearly $\widetilde{\rho^e}$ is conserved during scattering and the physics is contained in the odd quantities. The Hamiltonian to consider is thus:

$$\frac{hv_D}{2\nu}\int_{-\infty}^{0} dX \left[\widetilde{\rho^o}(X,t)^2 + +\lambda_1 \cos\left(\sqrt{2}\widetilde{\varphi^o}(0)\right)\right] \tag{65}$$

Now remind that $\widetilde{\rho^o}$ is solution of a free propagation equation (velocity v_D) and the equation is very similar to a Sine-Gordon equation (SG) but with the SG term only at the boundary. Classically, it is easy to show that this boundary SG equation admits solutions using a combination of the natural kink and anti-kink of the ordinary SG equation. By definition, a kink (or a soliton) in the field (of the charge density) which is solution of the ordinary SG equation propagates without deformation and is also solution of the free propagation equation for $X < 0$. By linearity, superpositions of kink and anti-kinks are also solutions. The effect of the boundary term is mainly to convert kink into antikink. Physically, the effect of scattering is that a positive pulse of charge can be reflected as a negative pulse. The step from classical to quantum intregrability is beyond the skill of the author, details can be found in [65,66]. One can show that applying a voltage bias V_{ds} between reservoirs emitting electrons in the upper and lower edges is, in the convenient basis for interacting electrons, equivalent to send a regular flow of kink which are randomly transformed into antikink. Kink and antikink respectively contribute to the forward and backscattered current. The Landauer formula adapted to this approach gives the backscattering current which expresses simply as:

$$I_B(V_{ds}, T_B) = ev_D \int_{-\infty}^{A(V_{ds})} d\alpha \rho_+(\alpha) \left|S_{+-}(\alpha - \alpha_B)\right|^2 \quad \text{and} \quad I = \nu \frac{e^2}{h} V_{ds} - I_B \tag{66}$$

where $\rho_+(\alpha)$ (not to be confused with previous notations) is the density of incoming kink at energy parametrized by e^{α}, and

$$\left|S_{+-}(\alpha - \alpha_B)\right|^2 = \frac{1}{1 + \exp\left[2(1-\nu)(\alpha - \alpha_B)/\nu\right]} \tag{67}$$

is the probability for kink to anti-kink conversion (the scattering probability) with α_B related to the impurity strength T_B. A series expansion in T_B/V_{ds} and V_{ds}/T_B for respectively weal and strong backscattering gives the current where all coefficients are known analytically

$$I_B = \nu \frac{e^2}{h} V_{ds} \sum_n \nu a_n(\nu) \left(\frac{V_{ds}}{T_B}\right)^{2n(\nu-1)} \quad \text{for} T_B/V_{ds} < 1 \tag{68}$$

$$I = \nu \frac{e^2}{h} V_{ds} \sum_n a_n(\frac{1}{\nu}) \left(\frac{V_{ds}}{T_B}\right)^{2n(\frac{1}{\nu}-1)} \quad \text{for} V_{ds}/T_B < 1 \tag{69}$$

The curve $I(V_{ds})$ describing the whole transition from strong to weak backscattering can be calculated. Note again the duality $\nu \longleftrightarrow 1/\nu$. For finite temperature numerical solutions are also available giving the whole information $I(V_{ds}, T)$. It is important to realize that the solution is exact as long it is justified to neglect the non-relevant terms $\cos n(\widetilde{\varphi_+(0)} - \widetilde{\varphi_-(0)})$ for $n > 1$. Including these terms is probably very difficult, except the case $g = 1/2$ for which analytical expression with the $n = 2$ contribution have been obtained [67]. Experimentally it is possible to find situation where only the $2k_{\rm F}$ term is important such that the exact FLS theory is of precious help for quantitative comparison.

For finite temperature, one can show that the conductance is a function of the reduced variable T/T_B and $eV_{ds}/2\pi k_{\rm B}T$:

$$G(T, V_{ds}) = \frac{e^2}{3h} f\left(\frac{T}{T_B}, \frac{eV_{ds}}{2\pi k_{\rm B} T}\right) \tag{70}$$

and measuring $G(T,0)$ fixes the only parameter T_B. The V_{ds}/T scaling law also can be tested very accurately. Fig. 18 shows theoretical calculations of the differential conductance for various values of the parameter T_B (the results of Ref. [65,66] have been used). We can see that, increasing the energy (the voltage or the temperature), leads to a progressive transition from the strong backscattering regime to the weak backscattering regime.

Fig. 19 shows recent data obtained in our group for the conductance in the strong backscattering regime for $\nu = 1/3$. Experiments are made in the intermediate regime (i.e. $G > 10^{-4}e^2/3h$). The effective exponent deduced from a series of dI/dV_{ds} curves for different impurity strengths is compared with the effective one calculated using the finite temperature exact solution. The agreement is rather good. The theoretical graph in inset of the figure shows that the asymptotic scaling exponent $2(\nu^{-1} - 1) = 4$ is not expected except for conductance lower than 10^{-4}, a regime experimentally difficult to obtain. It is important to say that there are no adjustable parameters.

While one can say that the Luttinger liquid model describes qualitatively well the experiments, there are still many problems to solve for an accurate quantitative description of conductance measurements. Long range interactions are one of this. One can show that the dispersion relation of the bosonic chiral edge modes which usually varies linearly with the wavenumber k get a contribution $k\ln(k)$. Such contribution is known from edge magneto-plasmon radiofrequency experiments realized in classical or in quantum Hall 2D electron systems where the neutral collective modes are excited and resonantly detected [71]. When the energy is low enough such that the associated wavelength is larger than the width of the sample, the Coulomb interaction is able to couple the edges

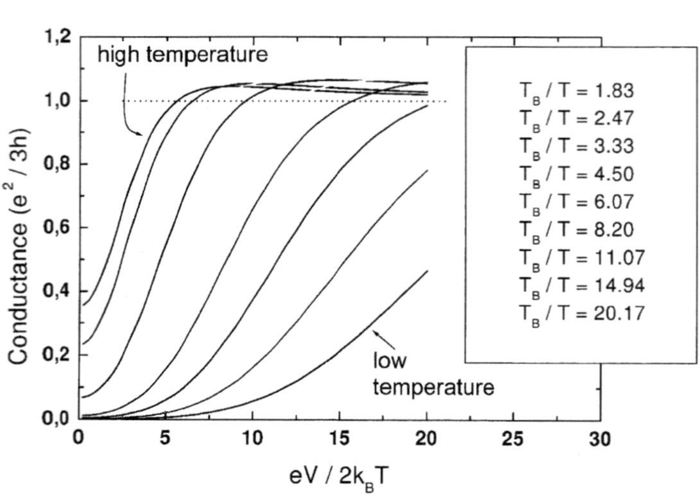

Fig. 18. Theoretical curves for the differential conductance versus voltage calculated for different values of the ratio T/T_B. The numerical exact solution of Ref.[65,66] is used. The conductance is a universal function of the variable T/T_B and $eV_{ds}/2\pi k_B T$ (adapted from [63,64])

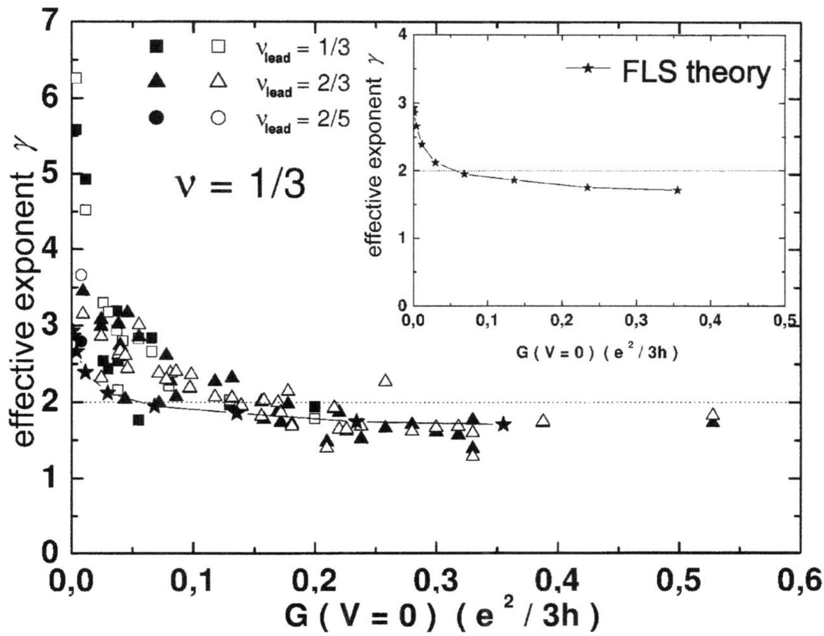

Fig. 19. Exponent of the algebraic variation of the differential conductance with voltage measured in the strong backscattering regime versus the zero bias conductance normalized to $e^2/3h$. The solide line is a comparison with the FLS predictions. The scaling exponent $\alpha = 2(\nu^{-1} - 1) = 4$ is only expected in a regime of extremely low conductance (10^{-4})

(like the g_2 parameter of 1D Luttinger liquids). Because of the $\ln(k)$ term, the power law of the TDOS is lost. The TDOS is expected to vary with energy like $\exp(const \times (\ln \varepsilon)^{3/2})$ [68–70]. In general, for a good description of transport, all geometrical details of the sample should be taken into account. A comparison with an ideal situation giving universal exponents has to be used with caution.

4.5 Shot Noise of Fractional Channels Coupled by an Artificial Impurity. Detection of Fractional Charges

Recently, the study of shot noise in quantum conductors has been of increasing importance in mesoscopic physics. This followed a series of theoretical papers [72] and experiments [73,74] showing that the quantum noise of electrons in conductors is naturally sub-Poissonian (a result opposite to the one known for photons in quantum optics). In quantum conductors, for which interactions can be neglected, the reduction of noise comes from the statistical interaction between electrons. Indeed, the Pauli principle naturally regulates the electron flow. Another source of reduction can arise when the repulsive interaction between electrons is important because of correlations. It was thus interesting to consider the noise properties of an interacting system such as a Luttinger liquid. Another stronger motivation arise when considering the fractional quantum Hall effect. Indeed shot noise is proportional to the carrier charge. Thus noise investigations could provide an unambiguous evidence of the fractionally charged Laughlin excitations.

Let us recall what shot noise in a conductor is. According to Schottky [75], the random transfer of charge q across a conductor generates an average current $\overline{I}$ but also finite temporal fluctuations of the current ΔI around the mean value when observed during a finite time τ. The current is related to the average number of transferred electrons $\overline{N}$ via $\overline{I} = q\overline{N}/\tau$, while the square of the current fluctuations are $\overline{(\Delta I)^2} = q\overline{I}/\tau\frac{\overline{(\Delta N)^2}}{\overline{N}}$. If the statistics of transfer events is Poissonian $\overline{(\Delta N)^2} = \overline{N}$ the well known Schottky formula is obtained:

$$\overline{(\Delta I)^2} = 2q\overline{I}\Delta f = S_I \Delta f \tag{71}$$

where we have introduce the effective frequency bandwidth $\Delta f = 2/\tau$ (Nyquist theorem) and the current noise power S_I. The noise power is directly proportional to the carrier charge q which manifests the fact that noise is a direct consequence of charge granularity. As an application, the simultaneous measure of the average current and its fluctuations gives a simple direct measurement of q, free of geometrical parameters.

Determination of the Laughlin quasiparticle charge using shot noise has been suggested by experimentalists rather early [10]. This is a non-equilibrium experiment. Non equilibrium is necessary condition to probe quasiparticles which are excitations *above* the ground state. A theoretical proposition taking into account the specific Luttinger liquid dynamics was given in Ref. [76]. An artificial impurity, as the one discussed in previous sections, couples the two opposite fractional edges. It induces tunneling in the strong backscattering regime or quantum transfer of quasiparticles in the weak backscattering regime. The results, obtained for both limits, are respectively:

$$S_I \simeq 2eI\coth(eV/2k_B\theta) \qquad I \ll I_0 = \frac{1}{3}\frac{e^2}{h}V \tag{72}$$

$$S_I \simeq 2(\nu e)I_B\coth((\nu e)V/2k_B\theta) \qquad I_B = I_0 - I \ll I_0 \tag{73}$$

Here, the Johnson-Nyquist thermal noise contribution $4Gk_BT$ has been subtracted for clarity. For strong backscattering, low conductance, the current I is small compared to I_0. The rate of tunneling of the charges is small compared to that of incoming charges. The statistics is expected to be Poissonian and it is natural to recover the (finite temperature) formula of Schottky. The charge found is that of electrons. In this limit only electrons can tunnel. This is intuitively expected. The system being insulating, correlations between the left and the right side of the conductor are lost and fractional excitations can not survive. For weak backscattering, conductance close to $e^2/3h$, this is the backscattering current which is now small compared to the incoming current. and the statistics of quantum transfer of charge from the upper to the lower edge is expected to be

Poissonian. The Schottky formula is also recovered. I_B replaces I. The process is also Poissonian but the effective charge is νe. Indeed, fractional excitations may be expected inside the fractional region virtually not perturbed by the weak impurity. In this limit, the noise provides a *direct* way to measure the fractional Laughlin charge νe.

A fractional νe charge is also found in the coth function. However the meaning is different. The cross-over from thermal to shot noise corresponds to electrochemical potential difference $\Delta\mu = \nu eV$ comparable to $k_B\theta$. However, this is not a measure of the fractional quasiparticle charge. This is just a measure of the fractional filling of the quantum state at equilibrium, like the conductance $\nu e^2/h$ is. Only the shot noise $S_I \simeq 2(\nu e)I_B$ really measures the quasiparticle charge. Nevertheless observation of a three times larger voltage for the thermal cross-over in noise experiments has been an important confirmation of Eq. (73).

The zero temperature limit of expressions (72) and (73) have been also derived in [77] using Luttinger liquid approaching the perturbative limit. The exact solution of the FLS model presented in the previous section allows not only to calculate the current in all regime but also to calculate the noise [65]. To obtain the noise, one can mimic the wavepacket approach used by T. Martin and R. Landauer for the noise of non interacting Fermions Ref.[78] . The incoming kinks of the field $\widetilde{\phi^o}$ correspond to a regular flow of solitons in $\widetilde{\rho^o}$. The regular flow is noiseless but the random scattering of kinks into anti-kinks produces noise in the outgoing current. When $|S_{+-}(\alpha-\alpha_B)|^2 \ll 1$ this a Poissonian processus while if $|S_{+-}(\alpha-\alpha_B)|^2$ is not negligible, the statistics is binomial and the fluctuations are proportional to $|S_{+-}(\alpha-\alpha_B)|^2\,(1-|S_{+-}(\alpha-\alpha_B)|^2)$. The expression for the noise is thus simply

$$S_I(V) = 2e^2 v \int_{-\infty}^{A(V_{ds})} d\alpha \rho_+(\alpha)\, |S_{+-}(\alpha-\alpha_B)|^2\,(1-|S_{+-}(\alpha-\alpha_B)|^2) \qquad (74)$$

The exact expression and technical mathematical details can be found in Ref. [65]. The special simple form of $|S_{+-}(\alpha-\alpha_B)|$ leads to a relation between current and noise where $S_I = \frac{v}{1-v}(V\frac{dI}{dV} - I) = \frac{v}{1-v}(I_B - V\frac{dI_B}{dV})$. From it, using the weak and strong backscattering limits of the Luttinger theory, we can easily check that $S_I \to 2(\nu e I_B)$ and $2eI$ respectively in agreement with the zero temperature limit of (72) and (73). Finite temperature predictions can also be found in Ref. [66].

4.6 Measurement of the Fractional Charge Using Noise

A difficulty of shot noise measurements in the FQH effect is that the extremely low shot noise has to be extracted from the background of relatively large amplifiers noise. Shot noise levels are extremely small both due to the smaller charge and the small available current. The latter is restricted by the fact that the FQH effect breaks down when the applied voltage is larger than the excitation gap. This excitation gap, in turn, depends crucially on the quality of the material in which the 2DEG resides. The state of the art technology currently yields samples with an excitation gap of the order of a few 100 μeV, leading to shot noise levels in the 10^{-29} A^2/Hz range.

These measurements have been performed by two groups [79][80]. A QPC is used in order to realize a local and controllable coupling between two $\nu = 1/3$ fractional edges to partially reflect the incoming current. The experiments are designed to have a best sensitivity for the weak coupling limit where Poissonian noise of the $e/3$ Laughlin quasiparticles is expected. In the experiment of Ref. [79], a cross correlation technique detects, at low frequency, the anti-correlated noise of the transmitted current I and the reflected current I_B, i.e. $S_{I,I_B} = \langle \Delta I \Delta I_B \rangle / \Delta f \simeq -2(e/3)I_B$. The magnetic field corresponds to a filling factor 2/3 in the bulk of the sample and a small region of filling factor 1/3 is created close to the QPC using the depletion effect of the gates. The size of the 1/3 region is estimated about 150 ϕ_0, sufficient to establish FQHE correlations. The advantage of doing this is that the coupling between edges occurs on a shorter scale and the controllable QPC potential is larger than the potential fluctuations inherent of sample fabrication. In the two samples measured, the combination of QPC and random potential lead to two dominant paths for backscattering. The coherent interference between paths gives rise to nearly perfect resonant tunneling peaks in the conductance. Careful measurements of the conductance resonance showed that tunnelling was coherent. This was an important check for the quasiparticle charge measurement because this ruled out the possibility of noise suppression due to multiple uncorrelated steps, similar to the 1/3 noise reduction factor in zero field diffusive conductors. Also the resonant conductance showed non-linear dependence on bias voltage consistent with Luttinger liquid model provided the filling factor of the bulk is used. The other group [80] used a high frequency technique in order to increase the signal bandwidth and measured the autocorrelation of the transmitted current. Here the magnetic field corresponded to a filling factor 1/3 throughout the sample. They found few non-linearities in the conductance, in contrast with the Luttinger liquid predictions, and this allowed them to define a bias voltage independent transmission.

In the Poissonian limit $I_B \ll I_0$, the two experiments give the same conclusion (see Fig. 20) that near filling factor 1/3, shot noise is threefold suppressed. This is the most direct evidence that the current can be carried by quasiparticle with a fraction of e and that Laughlin conjecture was correct. In addition, the data showed a cross-over from thermal noise to shot noise when the applied voltage satisfies the inequality $eV/3 > 2k\theta$ (rather than $eV > 2k\theta$), indicating that the potential energy of the quasiparticles is threefold smaller as well as predicted in Eq. (11).This experiment has been now reproduced many time with different sample and measurement conditions in both laboratories.

Is it possible to go further and probe different fractional charges for less simple filling factor? Recently measurements close to $\nu = 2/5$ have given indications that the $e/5$ quasiparticles are the relevant excitations in this regime [81]. This last result has been analyzed in a model of non-interacting composite Fermions where Luttinger effects are neglected [82]. More experiments and a better the-

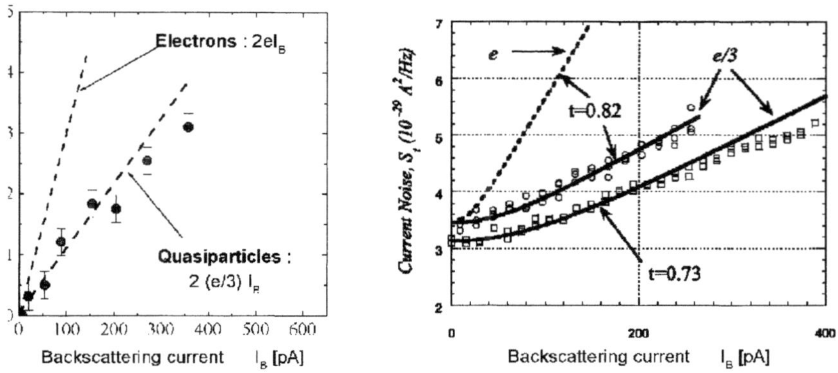

Fig. 20. Experimental Poissonian noise of the fractionally charged excitations of the FQHE, from Ref. [79] (*left*) and Ref. [80] (*right*)

oretical understanding of the noise for Jain's filling factors are certainly needed and will be done in the future.

References

1. K. von Klitzing, G. Dorda, M. Pepper: Phys. Rev. Lett. **45**, 494 (1980)
2. D.Tsui, H.Stormer, A.Gossard: Phys. Rev. Lett. **48**, 1599 (1982) (FQHE)
3. R.B. Laughlin: Phys. Rev. Lett. **50**, 1395 (1983)
4. J.M. Leynaas, L. Myrheim: Nuovo Cimento B **37**, 1 (1977) S.R. Renn, D.P. Arovas: Phys. Rev. B **51**, 16832 (1995) S. Ouvry: Phys. Rev. D **50**, 5296 (1994) S. Wu: Phys. Rev. Lett. **73**, 922 (1994)
5. J. Jain: Phys. Rev. Lett. **63**, 199 (1989)
6. S.C. Zhang, H.Hansson, S.Kivelson: Phys. Rev. Lett. **62**, 82 (1989) D.-H. Lee, S.-C. Zhang: Phys. Rev. Lett. **66**, 1220 (1991) S.C. Zhang: Int. J. Mod. Phys. **B6**, 25 (1992)
7. S.L. Sondhi, A. Karlhede, S.A. Kivelson, E.H. Rezayi: Phys. Rev. B **47**, 16419 (1993)
8. S.M. Girvin: in *'Topological Aspects of Low Dimensional Systems'*, NATO ASI, Les Houches Summer School, A. Comtet, T. Jolicoeur, S. Ouvry, F. David eds., Springer, p.551 (1999)
9. F. Calogero: J. Math. Phys. **10**, 2191 (1969) B. Sutherland: J. Math. Phys. **12**, 246 (1971) see for example D. Serban, F. Lesage, V. Pasquier: Nucl. Phys. B**466**, 499 (1996) and references therein
10. see *'The Quantum Hall Effect'*, R.E. Prange, S.M. Girvin eds., Springer-Verlag, New York (1987)
11. B. Etienne, E. Paris: J. Phys. France **48**, 2049 (1987)

12. H. Störmer: Rev. Mod. Phys. **71**, 875 (1999)
13. A. H. Mac Donald: in *'Mesoscopic Quantum Physics'*, E. Akkermans, G. Montambaux, J.L. Pichard, J. Zinn-Justin eds., Elsevier Science, Amsterdam (1994) *'Perspectives in Quantum Hall Effects'*, S. Das Sarma, A. Pinczuk eds., (Wiley, New York, 1997) *'Topological Aspects of Low Dimensional Systems'*, NATO ASI, Les Houches Summer School, A. Comtet, T. Jolicoeur, S. Ouvry, F. David eds., Springer, p.551 (1999)
14. T. Ando, Y. Matsumoto, Y. Uemura: J. Phys. Soc. Jpn. **39**, 279 (1975) P. Streda, L. Smrcka: Phys. Stat. Solidi B **70**, 537 (1975)
15. T. Ando, A.B. Fowler, F. Stern: Rev. Mod. Phys. **54**, 437 (1982)
16. A. Cho ed.: *'Molecular Beam Epitaxy'* AIP Press, Woodbury, New York 1994
17. R.B. Dunford et al.: Surf. Sci. **361/362**, 550 (1996) S.I. Dorozhkin et al.: Surf. Sci. **361/362**, 933 (1996) P.T. Coleridge et al.: Solid State Commun. **102**, 755 (1996)
18. Battlog et al.: J. Phys. Condens. Matter **13**, L163 (2001)
19. R.B. Laughlin: Phys. Rev. B **23**, 5632 (1981)
20. D.J. Thouless, M. Kohmoto, P. Nightingale, M. den Nijs: Phys. Rev. Lett. **49**, 405 (1982)
21. A.A.M. Pruisken: Phys. Rev. B **31**, 416 (1985)
22. J.T. Chalker, P.D. Coddington: J. Phys. C **21**, 2665 (1988)
23. H.P. Wei, D.C. Tsui, M.A. Paalanaen, A.M.M. Pruisken: Phys. Rev. Lett. **61**, 1294 (1988)
24. S. Koch et al.: Phys. Rev. Lett. **67**, 883 (1991)
25. D.G. Polyakov, B.I. Shklovskii: Phys. Rev. B **48**(15), 11167 (1993) Phys. Rev. Lett. **70**(24), 3796 (1993)
26. F. Hohls, U. Zeitler, R.J. Haug: Phys. Rev. Lett. **88**, 036802 (2002)
27. this is only true for the first orbital Landau level. For $\nu > 2$ fractions with even denominator, such as 5/2, may occur. The quantum states are believed to be related to non a abelian permutation symetry of the electrons, a fascinating subject not treated here.
28. R.L. Willet et al.: Phys. Rev. B **37**, 8476 (1988)
29. R.G. Clark et al.: Phys. Rev. Lett. **60**, 1747 (1988) Y. Katayama, D.C. Tsui, M. Shayegan: Phys. Rev. B **49**, 7400 (1994) S.I. Dorozhkin, R.J. Haug, K. von Klitzing, H. Ploog: Phys. Rev. B **51**, 14729 (1995)
30. B.I. Halperin: Phys. Rev. Lett. **52**, 1583 (1984)
31. V. Pasquier, F.D.M.Haldane: Nucl. Phys. **516**, 719 (1998) B.I. Halperin, P.A. Lee, N. Read: Phys. Rev. B **47**, 7312 (1993)
32. V.J. Goldman, B. Su, J.K. Jain: Phys. Rev. Lett. **72**, 2065 (1994)
33. J.H. Smet et al.: Physica B **249-251**, 15 (1998)
34. R.L. Willet et al.: Phys. Rev. Lett. **65**, 112 (1990) Phys. Rev. Lett. **71**, 3848 (1993)
35. D.R. Leadley et al.: Phys. Rev. Lett. **72**, 1906 (1994) RR. Du et al.: Phys. Rev. Lett. **73**, 3274 (1994)
36. R.G. Clark et al.: Phys. Rev. Lett. **62**, 1536 (1989)
37. D.K. Maude et al.: Phys. Rev. Lett. **77**, 4604 (1996)
38. V. Bayot et al.: Phys. Rev. Lett. **76**, 4584 (1996)
39. S. Melinte et al.: Phys. Rev. Lett. **76**, 4584 (1999)
40. S.E. Barrett et al.: Phys. Rev. Lett. **74**, 5112 (1995)
41. R. Tycko et al.: Science **268**, 1460 (1995)
42. S. Melinte, N. Freytag, M. Horvatic, C. Berthier, L.P. Levy, V. Bayot, M. Shayegan: Phys. Rev. Lett. **84**, 354 (2000)

43. M. Buettiker: Phys. Rev. B **38**, 9375 (1988)
44. R.J. Haug et al.: Phys. Rev. Lett. **61**, 2797 (1988)
45. D.B. Chklovskii, B.I. Shklovskii, L.I. Glazman: Phys. Rev. B **46**, 4026 (1992)
46. C.W.J. Beenakker: Phys. Rev Lett. **64**, 216 (1990)
47. A.H. Mac Donald: Phys. Rev. Lett. **64**, 229 (1990)
48. J.M. Luttinger: J. Math. Phys. **4**, 1154 (1963) S. Tomonaga: Prog. Theor. Phys. (Kyoto) **5**, 544 (1950)
49. X.G. Wen: Phys. Rev. B **43**, 11025 (1991) Phys. Rev. Lett. **64**, 2206 (1990) Phys. Rev. B **44**, 5708 (1991) Int. Jour. Mod. Phys. B **6**, 1711 (1992)
50. H. Schulz: *'Fermi liquids and non Fermi liquids'* in 'Mesoscopic Quantum Physics', Proceedings of the Les Houches Summer School of Theoretical Physics, Session LXI, 1994, E. Akkermans, G. Montambaux, J.L. Pichard, J. Zinn-Justin eds. (Springer-Verlag)
51. S. Tarucha, T. Honda, T. Saku: Solid State Comm. **94**, 413 (1995)
52. A. Yacoby, H.L. Stormer, N.S. Wingreen, L.N. Pfeiffer, K.W. Baldwin, K.W. West: Phys. Rev. Lett. **77**, 4612 (1996)
53. O.M. Auslaender, A. Yacoby, R. de Picciotto, K.W. Baldwin, L.N.Pfeiffer, K.W. West: Phys. Rev. Lett. **84**, 1764 (2000)
54. M. Bockrath, D.H. Cobden, J. Lu, A.G. Rinzler, R.E. Smalley, T. Balents, P.L. McEuen: Nature **397**, 598 (1999)
55. Z. Yao, H.W.C. Postma, L. Balents, C. Dekker: Nature **402**, 273 (1999)
56. A.M. Chang, L.N. Pfeiffer, K.W. West: Phys. Rev. Lett. **77**, 2538 (1996)
57. M. Grayson, D.C. Tsui, L.N. Pfeiffer, K.W. West, A.M. Chang: Phys. Rev. Lett. **80**, 1062 (1998)
58. C.L. Kane, M.P.A. Fisher: Phys. Rev. B **46**, 15233 (1992) Phys. Rev. Lett. **68**, 1220 (1992) Phys. Rev. B **51**, 13449 (1995)
59. A.V. Shytov, L.S. Levitov, B.I. Halperin: Phys. Rev. Lett. **80**, 141 (1998)
60. L.S. Levitov, A.V. Shytov, B.I. Halperin: Phys. Rev. B **6407 (7)**, 5322 (2001)
61. A.M. Chang, M.K. Wu, J.C.C. Chi, L.N. Pfeiffer, K.W. West: Phys. Rev. Lett. **87**, 2538 (2001)
62. F.P. Milliken, C.P. Umbach, R.A. Webb: Solid State Commun. **97**, 309 (1996)
63. P. Roche et al.: preprint.
64. V. Rodriguez: Thesis, Université Pierre et Marie-Curie Paris 6 Novembre 2001
65. P. Fendley, A.W.W. Ludwig, H. Saleur: Phys. Rev. Lett. **74**, 3005 (1995)
66. P. Fendley, H. Saleur: Phys. Rev. B **54**, 10845 (1996)
67. R. Egger, A. Komnik, H. Saleur: Phys. Rev. B **60**, R5113 (1999)
68. K. Moon, S.M. Girvin: Phys. Rev. B **54**, 4448 (1996)
69. K. Imura, N. Nagaosa: Solid State Commun. **103**, 663 (1997)
70. L.P. Pryadko, E. Shimshoni, A. Auerbach: Phys. Rev. B **61**, 10929 (200)
71. E.Y Andrei, D.C. Glattli, F.I.B. Williams, M. Heiblum: Surf. Sci. **196**, 501 (1988) D.B. Mast, A.J. Dahm, A.L. Fetter: Phys. Rev. Lett. **54**, 1706 (1985) D.C. Glattli et al.: Phys. Rev. Lett. **54**, 1710 (1985)
72. G.B. Lesovik: Pis'ma Zh. Eksp. Teor. Fiz. **49**, 513 (1989) [JETP Lett. **49**, 592 (1989)]
73. M. Reznikov et al.: Phys. Rev. Lett. **18**, 3340 (1995)
74. A. Kumar et al.: Phys. Rev. Lett. **76**, 2778 (1996)
75. W. Schottky: Ann. Phys. (Leipzig) **57**, 541 (1918)
76. C.L. Kane, M.P.A. Fisher: Phys. Rev. Lett. **72**, 724 (1994)
77. C. de C. Chamon, D.E. Freed, X.G. Wen: Phys. Rev. B **51**, 2363 (1995)
78. Th. Martin, R. Landauer: Phys. Rev. B **45**, 1742 (1992) Physica B **175**, 167 (1991)

79. L. Saminadayar, D.C. Glattli, Y. Jin, B. Etienne: Phys. Rev. Lett. **79**, 2526 (1997) cond-mat/9706307
80. R. de-Picciotto et al.: Nature **389**, 162 (1997) cond-mat/9707289
81. M. Reznikov, R. de-Picciotto, T.G. Griffiths, M. Heiblum: Nature **399**, 238 (1999)
82. R. de-Picciotto, cond-mat/980221

Theories of the Fractional Quantum Hall Effect

R. Shankar

Yale University, New Haven CT 06520, USA

Abstract. This is an introduction to the microscopic theories of the FQHE. After a brief description of experiments, trial wavefunctions and the physics they contain are discussed. This is followed by a description of the hamiltonian approach, wherein one goes from the electrons to the composite fermions by a series of transformations. The theory is then compared to other theoretical approaches and to experiment.

1 Introduction

In this chapter I will provide an introduction to some of the theories of the Fractional Quantum Hall Effect (FQHE), for a non-FQHE person who is familiar with graduate quantum mechanics and a little bit of second quantization. I will discuss only the simplest phenomena to illustrate the basic ideas and also focus on the parts of the subject I am most familiar with, since many excellent reviews exist [1], [2], [3]. A rule I will bear in mind is that things should be made as simple as possible but no simpler.

2 The Problem

In the Hall effect, one takes a system of electrons in the $x - y$ plane, subject to a magnetic field in the z-direction. A current density j_x is driven in the x-direction and the voltage V_y is measured in the y-direction. At equilibrium, an electric field E_y is set up in the y-direction which balances the Lorentz force $v_x B$, and the Hall conductance is

$$\sigma_{xy} = \frac{j_x}{E_y} = \frac{nev_x}{v_x B} = \frac{ne}{B} \tag{1}$$

where n is the number density. Thus we expect that σ_{xy} will be linear in ne/B. Instead we find steps at some special values as in Fig. 1.

$$\sigma_{xy} = \frac{e^2}{2\pi\hbar}\,\frac{p}{2ps+1} \equiv \frac{e^2}{2\pi\hbar}\nu \tag{2}$$

$$p = 1, 2 \ldots \quad s = 0, 1, 2 \ldots \tag{3}$$

The case $s = 0$, $\nu = p$ corresponds to the Integer Quantum Hall Effect (IQHE) discovered by von Klitzing et al. [4]. The case $s > 0$ corresponding to

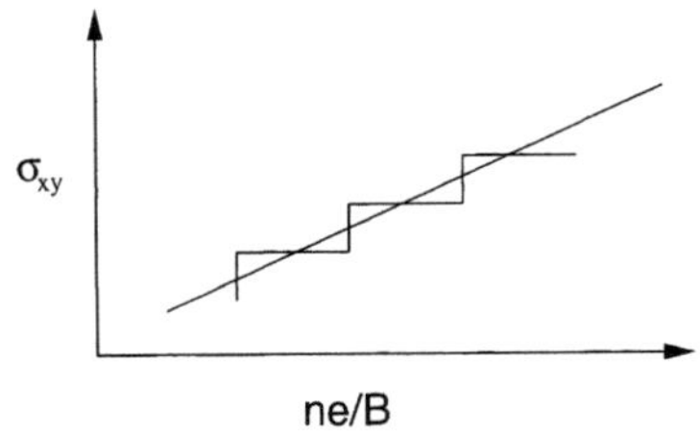

Fig. 1. Schematic form of the Hall conductance: the straight line is expected, and the steps are seen. Professor Glattli's lectures discuss the data in great detail.

the FQHE was discovered by Tsui *et al.* [5]. The case $p = 1,\ \nu = 1/(2s+1)$ was explained by Laughlin [6] with his famous trial state. The more general case $s \geq 0,\ p \geq 1, \nu = \frac{p}{2ps+1}$ was explained by Jain using the idea of composite fermions (CF) [7]. (The case $\nu = \frac{p}{2ps-1}$ is a trivial extension.)

2.1 What Is Special About These Values of σ_{xy}?

Note that where the straight line meets the plateaus,

$$\sigma_{xy} = \frac{e^2}{2\pi\hbar}\frac{p}{2ps+1} = \frac{ne}{B} \tag{4}$$

$$\frac{eBL^2}{2\pi\hbar nL^2} = 2s + \frac{1}{p} \tag{5}$$

$$\frac{\left[BL^2/(\frac{2\pi\hbar}{e})\right]}{nL^2} = \frac{\Phi/\Phi_0}{N} = 2s + \frac{1}{p} \tag{6}$$

where $\Phi_0 = \frac{2\pi\hbar}{e}$ is a quantum of flux. *At these points*, there are $\nu^{-1} = 2s + \frac{1}{p}$ flux quanta per particle. Thus IQHE has $\frac{1}{p}$ flux quanta per electron. We will see that this makes life easy, in terms of coming up with a good wavefunction for the ground state. The FQHE has $2s + \frac{1}{p} > 1$ flux quanta per electron. This will be seen to frustrate the search for a ground state.

2.2 What Makes FQHE So Hard?

The usual approach of considering interactions perturbatively fails when the number of flux quanta per particle > 1. Consider one free electron.

$$\mathcal{H}_0 = \frac{(\mathbf{p} + e\mathbf{A})^2}{2m} \quad m \text{ is the bare mass} \tag{7}$$

$$e\mathbf{A} = \frac{eB}{2}(y, -x) = -\frac{1}{2l^2}\hat{z} \times \mathbf{r} \tag{8}$$

$$l^2 = \frac{\hbar}{eB} \quad l \text{ magnetic length} \tag{9}$$

$$\nabla \times e\mathbf{A} = -eB \tag{10}$$

Note B points down the z-axis. If you do not like this complain to the person who defined the electron charge to be $-e$ or $z = x + iy$. Setting $\hbar = c = 1$,

$$\mathcal{H}_0 = \frac{\eta^2}{2ml^4} \tag{11}$$

$$\eta = \frac{1}{2}\mathbf{r} + l^2 \hat{z} \times \mathbf{p} \tag{12}$$

$$[\eta_x , \eta_y] = il^2 \tag{13}$$

$$E = (n + \frac{1}{2})\frac{eB}{m}. \tag{14}$$

One calls $\boldsymbol{\eta}$ the *cyclotron coordinate*, even though its two components form a conjugate pair. This is why the hamiltonian describes a harmonic oscillator whose energy levels are called Landau Levels (LL). Of special interest is the Lowest Landau Level (LLL) which has $n = 0$. In this state $\langle \eta^2 \rangle_{n=0} = l^2$.

Now a one-dimensional oscillator spectrum for a two dimensional problem implies the LL's must be degenerate. The degeneracy is due to another canonical pair that does not enter the hamiltonian, called the *guiding center coordinates* $\mathbf{R}$ defined by

$$\mathbf{R} = \frac{1}{2}\mathbf{r} - l^2 \hat{z} \times \mathbf{p} \tag{15}$$

$$[R_x , R_y] = -il^2 \tag{16}$$

Note that the role of $\hbar$ is played by l^2 and that

$$\mathbf{r} = \mathbf{R} + \eta. \tag{17}$$

which plays the role of phase space and the degeneracy of the LLL is

$$D = \frac{L^2}{\text{“}h\text{”}} = \frac{L^2}{2\pi l^2} = \frac{eBL^2}{2\pi\hbar} = \frac{\Phi}{\Phi_0} \tag{18}$$

where Φ_0 is the flux quantum. The **inverse filling factor** is

$$\nu^{-1} = \text{flux quanta or LLL states per electron} = \frac{eB}{2\pi n\hbar} \tag{19}$$

When $\nu^{-1} > 1$ the noninteracting problem is macroscopically degenerate since there are more LLL states than electrons. However when $\nu^{-1} > 1$ has the special values where the steps cross the straight line, we know excellent approximations to ground states and low lying excitations. At these points

$$\sigma_{xy} = \frac{ne}{B} = \frac{e^2}{2\pi\hbar}\frac{p}{2ps+1} \tag{20}$$

which means that at these special points

$$\frac{eB}{2\pi n\hbar} = \frac{\text{flux quanta density}}{\text{particle density}} = \frac{\text{LLL states}}{\text{particles}} = \frac{2ps+1}{p} \tag{21}$$

$$\textbf{flux quanta per particle} = \textbf{states per particle} = 2s + \frac{1}{p} \tag{22}$$

Thus for example at $\nu = 2/5$ there are $5/2$ states per particle in the LLL and the $V = 0$ problem is macroscopically degenerate. While this is also true when there are 2.500001 states per electron, we will see that at these special values, a single ground state built out of LLL states will emerge in accordance with the expectation that *as $m \to 0$ and $eB/m \to \infty$, a sensible low energy theory built out of the LLL states must exist!* Let us see how this happens and what happens as we move along the plateaus.

Getting to Know the LLL. The LLL condition $a_\eta|LLL\rangle = 0$ gives

$$\psi = e^{-|z|^2/4l^2} f(z) \tag{23}$$

where $z = x + iy$. A basis for ψ is

$$\psi_m(z) = z^m \, e^{-|z|^2/4l^2} \qquad m = 0, 1, \ldots \tag{24}$$

The Gaussian is usually suppressed. The state has $L_z = m$.

If $\nu = 1$, (one electron per LLL state) there is a unique noninteracting ground state (which may then be perturbed by standard means)

$$\chi_1 = \prod_{i<j}(z_i - z_j) \cdot Gaussian = Det \begin{vmatrix} z_1^0 & z_1^1 & z_1^2 & \ldots \\ z_2^0 & z_2^1 & z_2^2 & \ldots \\ \ldots & \ldots & \ldots & \ldots \end{vmatrix} \cdot Gaussian \tag{25}$$

For $\nu < 1$, we want to work entirely within the LLL. If in $H = T + V$ we set T equal to a constant ($eB/2m$ per particle), all the action is in V. Why is this a problem if V is a function of just coordinates? In terms of the density

$$\rho(\mathbf{r}) = \sum_j \delta(\mathbf{r} - \mathbf{r}_j) \tag{26}$$

$$V = \frac{1}{2}\int d^2r \int d^2r' \rho(\mathbf{r}) v(\mathbf{r} - \mathbf{r}') \rho(\mathbf{r}') \tag{27}$$

$$= \frac{1}{2}\sum_{\mathbf{q}} \rho(\mathbf{q}) v(q) \rho(-\mathbf{q}) \tag{28}$$

$$\rho(q) = \sum_j e^{-i\mathbf{q}\cdot\mathbf{r}_j} \tag{29}$$

The point is that if one sets $T = eB/2m$, the LLL value, one must project the operator $\mathbf{r}$ to the LLL. The coordinates x and y which commute in the full Hilbert space, no longer commute in the LLL.

Projection to the LLL. Let $\mathcal{P}$ denote projection to the LLL. Then

$$\mathcal{P}: \ \mathbf{r} = \mathbf{R} + \boldsymbol{\eta} \Rightarrow \mathbf{R}. \tag{30}$$

The projected components do not commute:

$$[R_x \ , R_y] = -il^2 \quad \text{or} \qquad [z \ , \bar{z}] = -2l^2 \quad \text{in the LLL} \tag{31}$$

As for the densities

$$\mathcal{P}: \ e^{-i\mathbf{q}\cdot\mathbf{r}} \Rightarrow \langle e^{-i\mathbf{q}\cdot\boldsymbol{\eta}}\rangle_{LLL} e^{-i\mathbf{q}\cdot\mathbf{R}} = e^{-q^2l^2/4} e^{-i\mathbf{q}\cdot\mathbf{R}} \tag{32}$$

Thus the projected problem is defined by

$$\bar{\bar{\mathcal{H}}} = \frac{1}{2}\sum_{\mathbf{q}} e^{-q^2l^2/2}\bar{\bar{\rho}}(\mathbf{q})v(q)\bar{\bar{\rho}}(-\mathbf{q}) \tag{33}$$

$$\bar{\bar{\rho}}(\mathbf{q}) = \sum_j e^{-i\mathbf{q}\cdot\mathbf{R}_j} \tag{34}$$

I shall refer to $\bar{\bar{\rho}}$ as the projected density, although it is actually the Magnetic Translation Operator and differs from the density by a factor $e^{-q^2l^2/4}$. (This explains the gaussian in (33)). The $\bar{\bar{\rho}}$ obey the Girvin-MacDonald-Platzman (GMP) [23] algebra:

$$[\bar{\bar{\rho}}(\mathbf{q}), \bar{\bar{\rho}}(\mathbf{q}')] = 2i\sin\left[\frac{(\mathbf{q}\times\mathbf{q}')\ l^2}{2}\right]\bar{\bar{\rho}}(\mathbf{q}+\mathbf{q}') \tag{35}$$

There is no small parameter in $\bar{\bar{\mathcal{H}}}$ and the overall energy scale is set by $v(q)$. How is one to find the ground state?

2.3 Laughlin's Answer

For $\nu = 1/(2s+1)$ Laughlin proposed a trial state:

$$\Psi_{\frac{1}{(2s+1)}} = \prod_{i=2}^{N}\prod_{j<i}(z_i - z_j)^{2s+1}\exp(-\sum_i |z_i|^2/4l^2) \tag{36}$$

Let us now contemplate its many virtues.

- It lies in the LLL. (Analytic function times gaussian)
- It has definite angular momentum (homogeneous in z's)
- It obeys the Pauli principle ($2s+1$ is odd, spin is assumed fully polarized)
- It has the right filling.
  Proof 1: Let us refer to the wavefunction as a function of z_N with all the others fixed as $\Psi(z_N)$. Consider a finite sample and drag z_N around it.

$$\#\text{ of zeros of }\Psi(z_N) = \frac{\text{Aharonov Bohm phase change}}{2\pi} = \frac{\Phi}{\Phi_0} \tag{37}$$

But we know

$$\frac{\Phi}{\Phi_0} = N\nu^{-1} \tag{38}$$

Thus $\Psi_{1/2s+1}(z_N)$ must be a polynomial of degree $N(2s+1)$ in z_N and it is.
Proof 2: Again we consider a finite droplet. Consider any single particle wavefunction in the LLL with angular momentum m.

$$|\psi_m|^2 = |z|^{2m} e^{-|z|^2/2l^2} \tag{39}$$

$$\pi z_{max}^2 = 2\pi m l^2 \tag{40}$$

$$\frac{\Phi \text{ enclosed in } \pi z_{max}^2}{\Phi_0} = \frac{2\pi m l^2 B}{(2\pi/e)} = m \tag{41}$$

Thus the single particle state of $L_z = m$ has a size (which is sharply defined at large m) such that it encloses m flux quanta. Now the largest value of m for any coordinate in

$$\Psi_{\frac{1}{(2s+1)}} = \prod_{i=2}^{N}\prod_{j<i}(z_i - z_j)^{2s+1}\exp(-\sum_i |z_i|^2/4l^2)$$

is $N(2s+1)$. Thus the droplet encloses $2s+1$ flux quanta per particle.

- Has great correlations and no wasted zeros. Halperin observed [9] that the Laughlin function has no wasted zeros in the following sense. Given that $\psi(z_N)$ has to have $N(2s+1)$ zeros per particle (given analyticity and the filling fraction) only N of these had to lie on other electrons by the Pauli principle. But they all lie on other electrons, thereby keeping them away from each other very effectively, producing a low potential energy.
- It is not enough to know what Ψ is, we need to know what kind of state it represents. Laughlin gave the following answer: it is an *incompressible fluid*, which means a fluid that abhors density changes. Whereas a Fermi gas would increase (decrease) its density globally when compressed (decompressed), an incompressible fluid is wedded to a certain density and would first show no response and then suddenly nucleate a localized region of different density. (Just the way a Type II superconductor, in which magnetic field is not welcome, will allow it to enter in quantized units in a region that turns normal.) Laughlin deduced this result based on the following plasma analogy. Note that all averages in the ground state are found using the measure

$$\langle \Omega \rangle = \int \prod_i d^2 z_i \Psi^* \Omega \Psi \Big/ \int \prod_i d^2 z_i \Psi^* \Psi \tag{42}$$

$$\Psi^*\Psi = e^{-\text{“}E/kT\text{”}} \tag{43}$$

$$\frac{1}{\text{“}kT\text{”}} = 2(2s+1) \tag{44}$$

$$\text{“}E\text{”} = \sum_{i<j} -\ln|z_i - z_j| + \sum_i \frac{\pi n}{2}|z_i|^2 \tag{45}$$

where I have used $1/l^2 = eB = 2\pi n/\nu$. This describes a plasma with "$1/kT$" $= 2(2s+1)$, particles of unit charge interacting with a log potential and in a neutralizing background given by second term. The plasma is known to hate density changes away from that of the background because of the log potential.

Laughlin also provided the wavefunction for a state with a quasihole, a state with a charge deficit. If one imagines inserting a tiny flux tube at a point z_0 and slowly increasing the flux to one quantum, one must, by gauge invariance, return to an eigenstate of H, and each particle must undergo a 2π phase shift as it goes around z_0. This and analyticity imply the ansatz

$$\Psi_{qh} = \prod_i (z_i - z_0)\Psi_{2s+1} \tag{46}$$

This is a *quasihole*. (There is a more complicated state with a quasiparticle.) The prefactor is a *vortex* at z_0, which is an analytic zero at z_0 for every coordinate. It denotes a hole near z_0 with charge deficit $1/(2s+1)$ (in electronic units) as can be seen using Laughlin's wavefunction and the plasma analogy in which there is now an extra term in energy $\frac{1}{2s+1}\sum_i \ln|z_i - z_0|$. This positive charge embedded in the plasma is immediately screened by an equal negative quasihole charge.

A second proof of quasihole charge depends only on the state being gapped and incompressible. As the flux quantum Φ_0, is inserted, the charge driven out to infinity is given by integrating the radial current density j produced by the Hall response to the induced azimuthal E field:

$$\begin{aligned} Q &= -\int j(r,t)2\pi r dt = -\sigma_{xy}\int E\ 2\pi r\ dt \\ &= -\sigma_{xy}\int \frac{d\Phi}{dt}dt = -\Phi_0\sigma_{xy} && (47) \\ &= -\frac{2\pi\hbar}{e}\ \frac{e^2}{2\pi\hbar}\ \nu = -e\nu && (48) \end{aligned}$$

Thus fractional charge is due to fractional σ_{xy} and not *vice versa*.

Finally the plateaux are produced when you go off the magic fractions because the change in density appears in the form of quasiparticles (holes) which get localized, as per standard localization mythology in $d = 2$.

While a fairly comprehensive understanding of the fundamentals FQHE was provided by Laughlin, the subsequent years brought many phenomena that needed explanation. For example Girvin, MacDonald and Platzman [23] worked out the dispersion relation for a magnetoexciton in which has the quantum numbers of a quasiparticle and hole which are however not infinitely separated. Their work showed a roton minimum just as in superfluids.

We can ask how there is a Hall current at $\nu = 1/2$ if $e^* = 0$, i.e., the quasiparticles are neutral.

But our primary question next is this: what are the wavefunctions for fractions of the form $\frac{p}{2ps+1}$? The corresponding Laughlin wavefunction with $2s+1 \rightarrow$

$2s + 1/p$ does not obey the Pauli principle. Jain gave us trial states for these fractions and also explained why they are natural, in terms of objects called *composite fermions* (CF).

3 Composite Fermions

It was pointed out by Leinaas and Myrheim [10] that in $d = 2$ one could have particles (dubbed anyons by Wilczek [10]) that suffered a phase change $e^{i\theta}$ upon exchange, with $\theta = 0$, π corresponding to bosons and fermions respectively. To do this one takes a fermion and drives through its center a point flux tube. If this contains an even/odd number of flux quanta, the composite particle one gets is a fermion/boson. (One may ask how adding a flux quantum is going to make any difference since it will not show up as one goes fully around it. Consider two particles, A at the origin and B at $x = R$. To exchange them we rotate B by π around A and shift both to the right by R.)

Jain exploited this idea as follows. Consider $\nu^{-1} = 2s + 1/p$, where each particle sees $2s + 1/p$ flux quanta of external B per particle. Let us now convert our electrons into CF by attaching $2s$ point flux quanta pointing opposite to B. Now we invoke the idea used by Laughlin et al. [11] and argue that *on average* the CF's see $1/p$ flux quanta per particle and fill up exactly p LL's. This gives the following trail state at mean field level:

$$\Psi_{p/2ps+1} = \prod_{i<j} \left| \frac{(z_i - z_j)}{|z_i - z_j|} \right|^{2s} \cdot \chi_p(z, \bar{z}) \tag{49}$$

where χ_p is the CF wavefunction with p-filled LL's and the prefactor takes you back to electrons. Jain however improved this ansatz in two ways and proposed:

$$\psi_{p/2ps+1} = \mathcal{P} \prod_{i<j} (z_i - z_j)^{2s} \cdot \chi_p(z, \bar{z})) \tag{50}$$

- He replaces flux tubes by vortex attachment: $\prod_{i<j}(z_i - z_j)^{2s}$.
- He does a projection to LLL using $\mathcal{P}$:

$$\mathcal{P} : \bar{z} \to 2l^2 \frac{\partial}{\partial z} \qquad \text{(Recall } [z, \bar{z}] = -2l^2.\text{)} \tag{51}$$

- At $p = 1$, $\chi(z, \bar{z}) = \prod(z_i - z_j)$, and we do not need $\mathcal{P}$ to get back Laughlin's answer. At $p > 1$ we have concrete expression for Ψ in terms of electron coordinates.

The Jastrow factor

$$J(2s) = \prod_{i<j} (z_i - z_j)^{2s} \tag{52}$$

describes $2s$-fold vortices on particles. This has two effects.

First, when one CF goes around a loop, it effectively sees $\nu^{-1*} = 2s + \frac{1}{p} - 2s = \frac{1}{p}$ flux quanta per enclosed particle since the phase change due to encircling vortices attached to CF's neutralizes $2s$ of the flux quanta per particle of the Aharanov - Bohm phase of the external field.

Thus while degeneracy of the noninteracting problem is present for any $\nu < 1$, at the Jain fractions one can beat it by thinking in terms of composite fermions. As we move off the Jain fractions, the incremental CF (particles or holes) get localized, giving the plateaus.

Next, the vortices reduce e down to

$$e^* = 1 - \frac{2ps}{2ps+1} = \frac{1}{2ps+1} \tag{53}$$

The idea that electrons bind to vortices, first pointed by Halperin[9], has been greatly emphasized by Read [12] who also give a physical picture to explain this binding: unlike the flux tube, the vortex is a physical LLL excitation, one which is charged and hence attracted to the electron.

It is however important to note that *$2s$-fold vortices are associated with electrons only in the Laughlin states. In the Jain states this is only before projection by $\mathcal{P}$.* The reason is that upon projection, zeros can move off the electrons (as will be explained shortly) and also get destroyed: for example at $\nu = 2/5$, after projection to the LLL (i.e., to analytic functions) there can only be 5/2 zeros per electron; one must lie on other electron as per the Pauli's principle and only 3/2 are left to form vortices. Clearly they cannot all be on electrons or be associated with them uniquely.

This brings us to the following question. The quasiparticle charge, $e^* = 1 - \frac{2ps}{2ps+1}$ is robust under $\mathcal{P}$ since e^* is tied to σ_{xy} which is presumed robust under projection. However the CF can no longer be viewed as an electron-vortex complex. *What, if any, is the particle that binds to the electron to bring e down to e^*?* The hamiltonian theory will provide an answer.

$\nu = 1/2$. The wavefunction, also called the Rezayi-Read wavefunction [13] is

$$\Psi = \mathcal{P} \prod_{i<j} (z_i - z_j)^2 Det \left| e^{i\mathbf{k}_i \cdot \mathbf{r}_j} \right| . \tag{54}$$

At $\nu = 1/2$, $e^* = 0$ since the double vortex has charge $= -1$. Read [12] gave the following argument for the dipole moment of this neutral CF. Consider

$$e^{i\mathbf{k}\cdot\mathbf{r}} = \exp \frac{i}{2}(k\bar{z} + \bar{k}z) \quad k = k_x + ik_y \tag{55}$$

Since $e^{il^2 k \frac{\partial}{\partial z}}$ causes $z \to z + ikl^2$ in the Jastrow factor producing the dipole moment $d^* = kl^2$. The energy to separate the vortex from electron (coulomb attraction) becomes kinetic energy of the CF. The leading term is quadratic in separation i.e., k, and defines an m^*.

While this picture has many merits and will be regained in an operator version later on, there are some subtleties often neglected.

- *Every z_i* gets translated so that $(z_i - z_j) \to (z_i - z_j + ik_i - ik_j)$.

- All this is for one assignment of k's. Upon antisymmetrization, we can't say where the zeros will be and there need not be a simple relation between the location of the electrons and the zeros in $\Psi_{\rm LLL}$.
- Upon projection, vortices not only get moved around, they are reduced in number: at $\nu = 1/2$ there are 2 vortices per electron (before projection there were three zeros per particle: two in the Jastrow factor and one in the determinant). One is on electrons due to the Pauli principle and the other is a *single* vortex.
- The value of dipole moment d^* is sensitive to wavefunction unlike σ_{xy}: it is zero before projection (since vortices sit on the electrons) and reaches some value d^* in the LLL, presumably $d^* = kl^2$. Where is one to look for this moment and what does it really mean? The hamiltonian approach will show how the dipole is described in an operator treatment.

The enormous success of the trial wavefunctions notwithstanding, the following challenges exist and are addressed by the hamiltonian approach.

- Separate high and low energies at the cyclotron scale and LLL.
- For the latter, obtain a limit independent of m, and an energy scale set by $v(q)$, especially for $1/m^*$.
- Find the partner to the electron that turns it into a CF.
- Obtain the right $e^* = 1/(2ps+1)$, d^*, and $\mu^* = e/2m$, the magnetic moment predicted by Simon, Stern and Halperin [14].
- Explain who carries Hall current if CF don't. (At $\nu = 1/2$, $e^* = 0$.)
- Analyze unequal time correlations, $T > 0$ correlations and disorder.

4 Hamiltonian Theory I. The Chern-Simons Approach

The aim here is to start with the electronic hamiltonian and try to reach, by a sequence of transformations, exact or approximate, fair or foul, a description of the final quasiparticles, the CF's. So we begin with

$$\mathcal{H} = \sum_i \frac{(\mathbf{p}_i + e\mathbf{A}(\mathbf{r}_i))^2}{2m} + V \tag{56}$$

The next step is to attach flux tubes to electrons by the following Chern-Simons (CS) gauge transformation [15], introduced into the FQHE work by Lopez and Fradkin [16] in the functional approach. In the hamiltonian version one trades the electronic wavefunction Ψ_e for Ψ_{CS} defined as follows :

$$\Psi_e = \prod_{i<j} \frac{(z_i - z_j)^{2s}}{|z_i - z_j|^{2s}} \Psi_{\rm CS} \equiv \exp(2is \sum_{i<j} \phi_{ij})\, \Psi_{\rm CS}. \tag{57}$$

$$\mathcal{H}_{\rm CS} = \sum_i \frac{(\mathbf{p}_i + e\mathbf{A}(\mathbf{r}_i) + \mathbf{a}_{\rm cs}(\mathbf{r}_i))^2}{2m} + V \tag{58}$$

In $\mathcal{H}_{\text{CS}}$ there appears a CS gauge field, $\mathbf{a}_{\text{cs}}$, that comes from the action of $\mathbf{p}$ on the prefactor, which is the phase of the Jastrow factor:

$$\mathbf{a}_{\text{cs}}(\mathbf{r}_i) = 2s\nabla \sum_{j\neq i} \phi_{ij} \tag{59}$$

$$\oint \mathbf{a}_{\text{cs}}(\mathbf{r}_i)\cdot d\mathbf{r}_i = 2s \oint \sum_{j\neq i} \nabla\phi_{ij}\cdot d\mathbf{r}_i \tag{60}$$

$$= 4\pi s\#\ \text{enclosed} \tag{61}$$

$$\nabla\times\mathbf{a}_{\text{cs}} = 4\pi s\rho \tag{62}$$

The above equations show that even though $\mathbf{a}_{\text{cs}}$ is a gradient, it has a curl since ϕ_{ij} is multivalued. The curl is readily calculated by the use of Stokes' theorem. The idea of flux attachment was to cancel part of the applied field on average. To this end we separate $\mathbf{a}_{\text{cs}}$ and ρ into average and fluctuating parts:

$$\nabla\times\langle\mathbf{a}_{\text{cs}}\rangle + \nabla\times : \mathbf{a}_{\text{cs}} := 4\pi sn + 4\pi s : \rho : \tag{63}$$

This gives

$$\mathcal{H}_{CS} = \sum_i \frac{(\mathbf{p}+e\mathbf{A}+\langle a_{\text{cs}}\rangle + : \mathbf{a}_{\text{cs}} :)_i^2}{2m} + V$$

$$= \sum_i \frac{(\boldsymbol{\Pi} + : \mathbf{a}_{\text{cs}} :)_i^2}{2m} + V \tag{64}$$

$$\boldsymbol{\Pi} = \mathbf{p} + e\mathbf{A} + \langle\mathbf{a}_{\text{cs}}\rangle \tag{65}$$

$$\nabla\times(e\mathbf{A}+\langle\mathbf{a}_{\text{cs}}\rangle) = -eB + 4\pi ns \tag{66}$$

$$= -\frac{eB}{2ps+1} \equiv -eB^* \qquad (A^* = \frac{A}{2ps+1}) \tag{67}$$

Here is the good news from above the picture derived by Fradkin and Lopez.

- If we ignore $: \mathbf{a}_{\text{cs}} :$ and V, the composite fermions see $1/p$ flux quanta each (since $2s + 1/p \to 1/p$) and have a unique ground state χ_p of p filled LL's. One can go on to include the neglected terms perturbatively. Excitations are given by pushing fermions into higher CF LL's.
- At mean-field, the CF wavefunction χ_p, transformed back to electrons is

$$\Psi_e = \prod_{i<j}\left(\frac{z_i - z_j}{|z_i - z_j|}\right)^{2s} \chi_p(z,\bar{z}) \tag{68}$$

  (See Rajaraman and Sondhi [17] for a way to get the entire Jastrow factor at the mean-field level, by introducing a *complex* gauge field.)
- Fluctuations at one loop give the square of the wavefunction and get rid of the $|z_i - z_j|$ but only for Laughlin like states.

- The cyclotron mode appears with the right residue.

But there is also some bad news.

- If we excite a fermion from level p to $p+1$, the energy cost (activation gap) of the particle-hole pair is $\Delta = eB^*/m$ plus corrections due to neglected terms. The dependence on m is not good. We want $\Delta \simeq e^2/\varepsilon l$.
  Jain does not have this problem: he does not use H_{CS} or χ_p or its excitations directly. For him the CS picture is a step towards getting electronic wavefunctions for the ground and excited states by attaching the Jastrow factor and projecting. The energy gap is computed as the difference in $\langle \bar{\bar{V}} \rangle$ between the ground and excited electronic wavefunctions.
- Between cyclotron mode (eB/m) and the LLL excitations ($\mathcal{O}(e^2/\varepsilon l)$) there are many spurious modes once again due to the presence of m.
- So far we have attached flux tubes and not vortices. Consequently the CF have electric charge unity and not $1/(2ps+1)$.

The Case of $\nu = 1/2$. Some remarkable predictions were made and verified at $\nu = 1/2$. Kalmeyer and Zhang [18] discussed it briefly and a very exhaustive study was made by Halperin, Lee and Read (HLR) [1].

At $\nu = 1/2$, $p = \infty$, $e^* = 0$ and $\boldsymbol{\Pi} \to \mathbf{p}$ and

$$\mathcal{H}_{\mathrm{CS}} = \sum_i \frac{(\mathbf{p} + :\mathbf{a}_{\mathrm{cs}}:)_i^2}{2m} + V \tag{69}$$

HLR used the Random Phase Approximation (RPA) and made the following predictions that *did not depend on m*.

- A fermi surface should exist with k_F determined by n.
- At $B^* = B - B_{1/2}$ CF should exhibit a bending with radius $R = \frac{k_F}{eB^*}$. This was verified by Kang et al. [20], Goldman et al. [21] and Smet et al. [22].
- When an acoustic wave is coupled the electronic system, it must undergo a velocity shift and an attentuation described by

$$\frac{\delta v_s}{v_s} - \frac{i\kappa}{q} = \frac{\alpha^2/2}{1 + i\sigma_{xx}(q)/\sigma_m} \tag{70}$$

  where α is a piezoelectric constant, v_s is the sound velocity, κ describes the attenuation, $\sigma_{xx} = \frac{e^2}{8\pi\hbar}\frac{q}{k_{\mathrm{F}}}$ and $\sigma_m = \frac{v_s\varepsilon}{2\pi}$. Theory fits the experiments of Willett *at al* [23] with a σ_m that is about five times larger than expected.
- They predicted divergences in m^*. These do not affect bosonic correlations (e.g., density-density) as shown by Kim et al. [24].

Bosonic CS Theory. There is an appealing CS theory of Zhang,Hansson and Kivelson [25] for Laughlin fractions, obtained by converting electrons to *composite bosons* in zero (mean) field upon attaching $2s+1$ flux tubes. This has many attractions: e.g., a Landau Ginsburg theory (see also Girvin, MacDonald [26], Read [27]) with $FQHE \leftrightarrow$ superfluidity. However it too has problems due to the unwanted dependence on m. Finally, unlike CF, which could be free in a first approximation, composite bosons had to be interacting for a stable ground state. Lack of time and space prevent further discussion of this alluring alternative.

5 Hamiltonian Theory II

I now turn to the extension of the CS theory that Murthy and I made in a series of papers [28], in order to cure it of some of its problems. We began by modifying the CS hamiltonian as follows.

$$\mathcal{H} = \sum_i \frac{(\boldsymbol{\Pi}_i + :\mathbf{a}_{\mathrm{cs}}: + \mathbf{a})^2}{2m} + V \tag{71}$$

$$0 = a|physical\rangle \tag{72}$$

where $\mathbf{a}$ is a transverse vector field. Since $d = 2$, it has only one component at each $\mathbf{q}$ which we denote by $a(\mathbf{q})$. We introduce a conjugate variable $P(\mathbf{q})$:

$$[a(\mathbf{q}), P(-\mathbf{q}')] = i\delta_{\mathbf{q}\mathbf{q}'} \tag{73}$$

and define a longitudinal vector field $\mathbf{P}(\mathbf{q})$ in terms of it. What is going on? First note that we are now operating in a bigger Hilbert space. On vectors that are annihilated by a, $\mathcal{H}$ is the same as $\mathcal{H}_{\mathrm{CS}}$. Since a commutes with $\mathcal{H}$, such eigenstates of $\mathcal{H}$ and a will exist and we focus on only those with zero eigenvalue for a. This is a trick adapted from the work of Bohm and Pines [29].

Why do we do this? We do this to fight fire with fire: $\mathbf{a}_{\mathrm{cs}}$ is a transverse gauge field, but it is a complicated function of the particle coordinates. On the other hand $\mathbf{a}$ is an independent transverse field. *Let us shift* $\mathbf{a}$ *by* $-\mathbf{a}_{\mathrm{cs}}$ *so as to eliminate* $\mathbf{a}_{\mathrm{cs}}$ from $\mathcal{H}$. To shift a we use the exponential of P:

$$U = \exp\left[\sum_q P(q)\frac{4\pi i s}{q}\rho(-q)\right]. \tag{74}$$

The resulting hamiltonian and constraint on physical states are

$$\mathcal{H} = \sum_i \frac{(\boldsymbol{\Pi}_i + \mathbf{a}(\mathbf{r}_i) + 4\pi s P(\mathbf{r}_i))^2}{2m} + V \tag{75}$$

$$0 = (q\, a(q) - 4\pi s\rho(q))|physical\rangle \tag{76}$$

The presence of $\mathbf{P}$ in $\mathcal{H}$ is due to the fact that U affects the particle sector as well. The nice thing is that $\mathcal{H}$ is written in terms of *independent* Fermi and Bose fields. Let us therefore focus on the quadratic part of $\mathcal{H}$:

$$\mathcal{H} = \sum_i \frac{\mathbf{\Pi}_i^2}{2m} + \underbrace{\frac{n}{2m}\sum_q (a^2(q) + (4\pi s P(q))^2)}_{\sum_q A^\dagger A \omega_0 \frac{2ps}{2ps+1}} + \mathrm{V} + :\rho: \mathrm{term} + \mathbf{j}\cdot\mathbf{A}\ \mathrm{term} \tag{77}$$

$$0 = (qa - 4\pi s\rho)|physical\rangle \tag{78}$$

For those who want to know, here is how one gets the second term.

$$\sum_i (\mathbf{a}(\mathbf{r}_i) + 4\pi s\mathbf{P}(\mathbf{r}_i))^2 =$$

$$\int d^2r\rho(\mathbf{r})(\mathbf{a}(\mathbf{r}) + 4\pi s\mathbf{P}(\mathbf{r}))^2 = n\int d^2r(\mathbf{a}(\mathbf{r}) + 4\pi sP(\mathbf{r}))^2 + \mathcal{O} : \rho :$$

$$= n\sum_{\mathbf{q}}(\mathbf{a^2(q)} + (\mathbf{4\pi sP(q)})^2 + \mathcal{O} : \rho :$$

How does this help? Focus on first two terms in (77). The ground state is a product: the fermions are in χ_p, the state with p -filled CF Landau levels, the oscillators, whose frequency is close to ω_0, are in their ground state and

$$\Psi_{CS} = e^{-C\sum_q a^2}\chi_p \qquad C \text{ is some constant} \tag{79}$$

$$= \exp[-C'\sum_q : \rho(q) : \frac{1}{q^2} : \rho(-q) :]\chi_p = |Jastrow|\chi_p$$

$$\Psi_e = \text{Jastrow} \cdot \chi_p \tag{80}$$

where $C'$is another constant and we have set $a(q) \simeq \rho(q)/q$ so that

$$C'\sum_q : \rho(q) : \frac{1}{q^2} : \rho(-q) :\simeq$$

$$\int d^2rd^2r'\sum_i s(\delta(r-r_i)-n)\ln|r-r'|(\sum_j \delta(r'-r_j)-n)$$

$$= \sum_{i<j} 2s\ln|z_i - z_j| - \sum_i \frac{|z_i|^2}{4l_v^2} + \text{constant}$$

$$\frac{1}{l_v^2} = \frac{1}{l^2}\frac{2ps}{2ps+1}$$

Going from the CS to the electron basis, we put in the phase of the Jastrow factor, and using $1/l_v^2 + 1/l^{*2}$ (fromχ_p) $= 1/l^2$, we even get the right gaussian.

Thus we get the Laughlin and unprojected Jain wavefunctions. However, if we had not known how good these wavefunctions were, we would not have bet much money on them since they came from an uncontrolled approximation. However, given what we know, it appears we are on the right track.

But we still have complaints.

- The oscillators are at $\omega_0\ \frac{2ps}{2ps+1}$ and violate Kohn's Theorem.
- The kinetic energy scale for fermions is still set by $1/m$. The potential energy V has played no role, instead of dominating the LLL physics.
- High and low energy degrees of freedom are still mixed up.
- No sign of e^*,d^*, or the moment $\mu^* = e/2m$ predicted by Simon et al. [14].

The solution is to decouple oscillators and fermions. They are coupled by the "$j \cdot A$" term. Since we have an $A^\dagger A$ term, we want to shift the oscillators to complete the squares using (schematically)

$$U = \exp\left[C\sum_q j(q)A^\dagger(q) + h.c.\right] \tag{81}$$

The transformation due to U is implemented with two approximations:

- Infrared limit: $ql \ll 1$
- RPA: $\sum_j e^{-i(\mathbf{q}-\mathbf{q}')\cdot\mathbf{r}_j} \simeq n\delta(\mathbf{q}-\mathbf{q}')$

Upon decoupling, the theory takes the following form (for $s=1$) [28]:

$$\begin{aligned}\mathcal{H} \quad &= \sum_{\mathbf{q}} \omega_0 A^\dagger(\mathbf{q})A(\mathbf{q}) + \sum_i \frac{e\delta B(\mathbf{r}_i)}{2m} \\ &+ \sum_j \frac{|\Pi_j|^2}{2m} - \frac{1}{2mn}\sum_i\sum_j \Pi_i^- e^{-i\mathbf{q}\cdot(\mathbf{r}_i-\mathbf{r}_j)}\Pi_j^+ \\ &+ V(\rho)\end{aligned} \tag{82}$$

$$J^+(\mathbf{q}) = \frac{e(q_x+iq_y)}{q\sqrt{2\pi}}\omega_0 c A(\mathbf{q}) \quad J^\pm = J_x \pm iJ_y \tag{83}$$

$$c^2 = \frac{2p}{2p+1} \tag{84}$$

$$\rho(\mathbf{q}) = \sum_j e^{-i\mathbf{q}\cdot\mathbf{r}_j}\left(1 - il^2\frac{\mathbf{q}\times\boldsymbol{\Pi}_i}{1+c} + ..\right) \tag{85}$$

$$+ \frac{cq}{\sqrt{8\pi}}(A(\mathbf{q}) + A^\dagger(\mathbf{q})) \tag{86}$$

$$\chi(\mathbf{q}) = \sum_j e^{-i\mathbf{q}\cdot\mathbf{r}_j}\left(1 + il^2\frac{\mathbf{q}\times\boldsymbol{\Pi}_j}{c(1+c)} + ..\right) \tag{87}$$

$$0 = \chi(\mathbf{q})|\text{physical}\rangle \tag{88}$$

Let us digest the various terms. First consider H_{os}, the oscillator part, to which we will add a coupling between the external potential Φ_{ext} and the oscillator part of the charge density:

$$\rho(\mathbf{q}) = \frac{cq}{\sqrt{8\pi}}(A(\mathbf{q}) + A^\dagger(\mathbf{q})) \tag{89}$$

$$\mathcal{H}_{os} = \sum_{\mathbf{q}} \omega_0 A^\dagger(\mathbf{q})A(\mathbf{q}) + e\Phi_{ext}(\mathbf{q})\frac{cq}{\sqrt{8\pi}}(A(\mathbf{q}) + A^\dagger(\mathbf{q})) \tag{90}$$

Note that the current depends only on A:

$$J^+(\mathbf{q}) = \frac{eq_+}{q\sqrt{2\pi}}\omega_0 c A(\mathbf{q}) \tag{91}$$

- Kohn's theorem is obeyed upon decoupling.
- The constraint χ makes no reference to A, which is thus truly decoupled. We have separated the high and low energy physics in the infrared limit.
- Both J and σ_{xy} are due to A alone. We find

$$\langle A\rangle = \frac{ecq}{\omega_0\sqrt{8\pi}}\Phi_{ext} \tag{92}$$

Plugging it into $J(A)$ and dividing by E we get

$$\sigma_{xy} = \frac{e^2}{4\pi}c^2 = \frac{e^2}{2\pi}\frac{p}{2p+1} \tag{93}$$

Thus the oscillators carry the Hall current, known to be non LLL effect:

$$\sigma_{xy} \simeq \underbrace{\frac{1}{m}}_{\text{from } J} \cdot \underbrace{\frac{1}{(eB/m)}}_{\text{energy denom}} \simeq m^0 \tag{94}$$

- The matrix elements of the oscillator part of ρ saturates the sum rule

$$\int_0^\infty S(q,\omega)\omega d\omega = \frac{q^2 n}{2m} \tag{95}$$

$$S(q,\omega) = \sum_N |\langle N|\rho(q)|0\rangle|^2\delta(\omega - E_N) \tag{96}$$

where

$$\rho(q) = \bar{\bar{\rho}}(q) + \text{“}A + A^\dagger\text{”part} \tag{97}$$

The LLL (CF) part (due to $\bar{\bar{\rho}}(q)$) must therefore go as q^4. Now we turn to the particle sector.

- The particles have the magnetic moment $\frac{e}{2m}$ deduced by Simon et al. [14].
- The kinetic term becomes

$$\sum_j \frac{|\Pi_j|^2}{2m} - \frac{1}{2mn}\sum_i\sum_j \Pi_i^- e^{-i\mathbf{q}\cdot(\mathbf{r}_i-\mathbf{r}_j)}\Pi_j^+ \tag{98}$$

The $i=j$ term when combined with the first, renormalizes the mass

$$\frac{1}{m^*} = \frac{1}{m}(1 - \frac{\sum_q}{n}) \tag{99}$$

As oscillators are added and decoupled, $1/m^*$ decreases. We can make it go away if we choose $\sum_q$ = number of oscillators = n, the number of particles. Let us then make this choice.

The $i \neq j$ terms in 98) can be shown to be convertible to a short range piece [28]. The infrared low energy hamiltonian and constraint become

$$\mathcal{H} = \bar{V} = \frac{1}{2}\sum_{q} \rho(q)v(q)\rho(-q) \tag{100}$$

$$\begin{aligned}
\rho &= \underbrace{\sum_j e^{-i\mathbf{q}\cdot\mathbf{r}_j}\left[1 - \frac{il^2}{(1+c)}\mathbf{q}\times\mathbf{\Pi}_j\right]}_{\bar{\rho}} + (\text{“}A + A^{\dagger}\text{”})\\
\bar{\chi} &= \sum_j e^{-i\mathbf{q}\cdot\mathbf{r}_j}\left[1 + \frac{il^2}{c(1+c)}\mathbf{q}\times\mathbf{\Pi}_j + \ldots\right]\\
c^2 &= \frac{2ps}{2ps+1}
\end{aligned}$$

Hereafter we will set oscillators in their ground state and $A = A^{\dagger} = 0$.

Murthy and I used this low q theory to study gaps [30] for potentials that vanished quickly at large q. There were however some conceptual problems that bothered me. For example the constraint algebra does not close: $[\chi_\alpha, \chi_\beta] \neq f_{\alpha\beta\gamma}\chi_\gamma$. The error was of course due to higher order terms in q. It did not make sense that two constraints each annihilated the physical states but their commutator did not. (The charge algebra did not close either, but this at least did not pose the same problem.) Secondly, it was not clear what the constraints stood for. I proposed the following remedy [31]. *Let us assume that the power series for charge and constraint mark the beginning of exponential series.* Thus we assume that they extend to:

$$\begin{aligned}
\bar{\bar{\rho}} &= \sum_j \exp(-i\mathbf{q}\cdot(\mathbf{r}_j - \frac{l^2}{1+c}\hat{\mathbf{z}}\times\mathbf{\Pi}_j)) \equiv \sum_j e^{-i\mathbf{q}\cdot\mathbf{R}_{ej}}\\
\bar{\bar{\chi}} &= \sum_j \exp(-i\mathbf{q}\cdot(\mathbf{r}_j + \frac{l^2}{c+c^2}\hat{\mathbf{z}}\times\mathbf{\Pi}_j)) \equiv \sum_j e^{-i\mathbf{q}\cdot\mathbf{R}_{vj}}
\end{aligned}$$

Note that $\mathbf{R}_e$ *and* $\mathbf{R}_v$ *were fully determined by the small* ql *theory.* Suddenly a lot of things fall into place.

- Consider $\mathbf{R}_e$ and its commutation relations

$$\mathbf{R}_e = \mathbf{r} - \frac{l^2}{(1+c)}\hat{\mathbf{z}}\times\mathbf{\Pi}, \tag{101}$$

$$[R_{ex}, R_{ey}] = -il^2, \tag{102}$$

which correspond to the electron guiding center, *but represented in the CF basis.* In particular, it is written in terms of the velocity operator $\boldsymbol{\Pi}$, which sees a weaker field A^* and will lead to a nondegenerate ground state with p filled CF Landau levels in Hartree-Fock calculations.

- The commutation relations of

$$\mathbf{R}_v \qquad = \mathbf{r} + \frac{l^2}{c(1+c)}\hat{\mathbf{z}} \times \mathbf{\Pi} \tag{103}$$

$$[R_{vx}\ , R_{vy}] = il^2/c^2, \tag{104}$$

correspond to the guiding center coordinates of a particle of charge $-c^2 = -2ps/(2ps+1)$. This is the right charge for an object that should bind with the electron to form the CF. We shall refer to it as the vortex even though, as we have seen, in the Jain case there aren't enough zeros to associate a $2s$-fold vortex with each electron.
- The two sets of coordinates commute:

$$[\mathbf{R}_e\ , \mathbf{R}_v] = 0. \tag{105}$$

- Will the vortex pair with the electron? Figure 2 shows that the separation between the two is of order Π. Since there will be terms in $\mathcal{H}$ that associate an energy with Π, the two will bind. Finally since $\mathcal{H}$ is just V in the new basis, the binding has its origin in the electrostatic interaction between electrons, as envisaged by Read.
- The constraint $\bar{\bar{\chi}} = 0$ states that the vortices will have no density fluctuations. In other words, one member of every CF (the vortex) obeys a collective constraint on density.
- The density and constraint obey the following algebras:

$$\begin{aligned}[]
[\bar{\bar{\rho}}(q), \bar{\bar{\rho}}(q')] &= 2i \sin\left[\frac{l^2(\mathbf{q}\times\mathbf{q}')}{2}\right] \bar{\bar{\rho}}(q+q') \\
[\bar{\bar{\chi}}(q), \bar{\bar{\chi}}(q')] &= -2i \sin\left[\frac{l^2(\mathbf{q}\times\mathbf{q}')}{2c^2}\right] \bar{\bar{\chi}}(q+q') \\
[\bar{\bar{\chi}}(q), \bar{\bar{\rho}}(q')] &= 0
\end{aligned} \tag{106}$$

Pasquier and Haldane [32], working in the LLL, studied bosons at $\nu = 1$ and obtained an algebra that coincides with the above for the case $c = 1$.

Here is where we stand:

$$\mathcal{H} = \frac{1}{2}\sum \bar{\bar{\rho}}(q)\ v(q) e^{-(ql)^2/2}\ \bar{\bar{\rho}}(q) \tag{107}$$

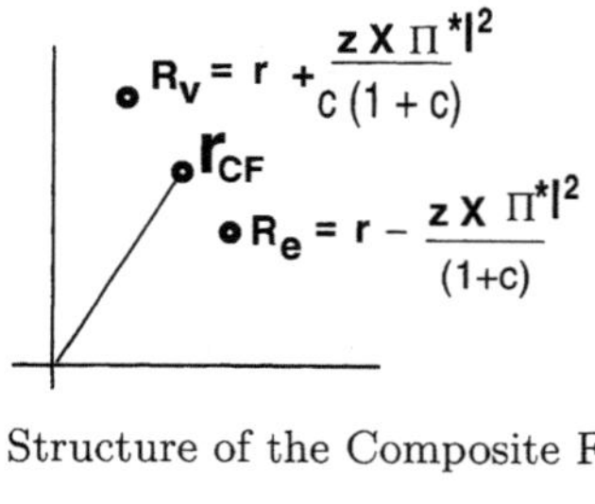

Fig. 2. Structure of the Composite Fermion

$$0 = [\mathcal{H}, \bar{\bar{\chi}}] \tag{108}$$

$$\bar{\bar{\chi}} = 0 \tag{109}$$

As in Yang-Mills, the constraints form an algebra that commutes with $\mathcal{H}$.

Could we perhaps ignore the constraint $\bar{\bar{\chi}} = 0$? Consider $\nu = 1/2$ where

$$\bar{\bar{\rho}}(\mathbf{q}) = \sum_j e^{-i\mathbf{q}\cdot\mathbf{r}_j}\left[1 - \frac{il^2}{2}\mathbf{q}\times\mathbf{p}_j + \cdots\right] \tag{110}$$

The fermion has unit charge and half the expected dipole moment. Presumably $\bar{\bar{\chi}}$ should send $e \to e^*$ and $d \to d^*$. Read [33] showed in a conserving approximation that $\bar{\bar{\chi}}$ generates a gauge field whose longitudinal part makes $e^* = 0$ and the surviving dipoles interact via the transverse field whose propagator is peaked at $\omega \simeq q^3$ and thus expected to matter only at $T = 0$ and at the Fermi surface.

Murthy [34] computed magnetoexciton dispersion in a conserving approximation. The results obeyed Kohn's theorem ($S(q) \simeq q^4$).

A shortcut to χ in the safe region ($T > 0$, and/or ω not $\lll q^3$) was provided by Murthy and myself[28,30]. We used the *preferred combination*

$$\bar{\bar{\rho}}^p = \bar{\bar{\rho}} - c^2\bar{\bar{\chi}} \tag{111}$$

equivalent to $\bar{\bar{\rho}}$ and *weakly gauge invariant*:

$$[\bar{\bar{\chi}}\ , \bar{\bar{\rho}}^p] \simeq \bar{\bar{\chi}}. \tag{112}$$

Clearly $\mathcal{H}(\bar{\bar{\rho}}^p)$ is also weakly gauge invariant. Consider the series

$$\bar{\bar{\rho}}^p = \sum_j e^{-i\mathbf{q}\cdot\mathbf{r}_j}\left(\frac{1}{2ps+1} - il^2 q \times \mathbf{\Pi}_j + 0\cdot(q\times\mathbf{\Pi}_j)^2 + \cdots\right) \tag{113}$$

Amazingly this single combination yields the correct e^* and d^* and can be verified to have q^2 matrix elements between Hartree-Fock states, in compliance with Kohn's theorem. It also makes physical sense: $\bar{\bar{\rho}}^p = \bar{\bar{\rho}} - c^2\bar{\bar{\chi}}$ adds the charge of electrons to that of correlation hole, namely $-c^2 = -\frac{2ps}{2ps+1}$ and describes the correlated entity, the CF.

Henceforth we will work with $\mathcal{H}(\bar{\bar{\rho}}^p)$. Its significance is the following. In the constrained space $\bar{\bar{\chi}} = 0$, there are many equivalent hamiltonians. In the HF approximation, these are not equivalent and $\mathcal{H}(\bar{\bar{\rho}}^p)$ best approximates (within HF states and in the infrared) the true hamiltonian between true eigenstates. In contrast to a variational calculation where one searches among states for an optimal one, here the HF states are the same for a class of hamiltonians (where $\bar{\bar{\chi}}$ is introduced into $\mathcal{H}$ in any fashion as long as rotational invariance holds) and we seek the best hamiltonian: $\bar{\bar{\rho}}^p$ encodes the fact that every electron is accompanied by a correlation hole of some sort which leads to a certain e^*, d^* and obeys Kohn's theorem (q^2 matrix element in the LLL).

However when gauge invariance (constraints) are crucial, one must not use $\mathcal{H}(\bar{\bar{\rho}}^p)$ but revert to the conserving approximation. Here is an example.

Compressibility at $\nu = 1/2$. A question that came up when the dipolar form of the charge operator was derived was this. If $\bar{\bar{\rho}}^p$ is dipolar, is the system compressible? Early approximate calculations suggested that compressibility vanishes as q^2 due to the extra q in the dipole moment.

Halperin and Stern [35] showed through an example that dipolar fermions can be compressible if $\mathcal{H}$ has the right symmetries. These are part of the symmetries generated by $\bar{\bar{\chi}}$. In particular as $q \to 0$, $\bar{\bar{\chi}}(\mathbf{q}) \simeq \sum_j e^{-i\mathbf{q}\cdot\mathbf{r}_j}$. Its action is

$$\delta \mathbf{r}_j = [\bar{\bar{\chi}}, \mathbf{r}_j] = 0 \tag{114}$$

$$\delta \mathbf{p}_r = [\bar{\bar{\chi}}(\mathbf{q}_0), \mathbf{p}_j] \simeq \mathbf{q}_0 \quad \text{as } \mathbf{q}_0 \to 0 \tag{115}$$

This symmetry of $\mathcal{H}$ is called K-invariance and had been noticed earlier by Haldane. It makes the Fermi surface very squishy and produces a response whose singularity cancels the q^2. The detailed work of Stern et al. [35], showed that finite compressibility is ensured if the CF has a Landau parameter $F_1 = -1$.

Compressibility at $\nu = 1/2$ was also confirmed by Read [33] within the conserving approximation and by D.H.Lee [36].

That the correct compressibility obtains only if the constraint $\bar{\bar{\chi}} = 0$ is correctly imposed makes physical sense: if one end of every dipolar fermion (the vortex) is actually part of an inert system with no density fluctuations (for this is what $\bar{\bar{\chi}} = 0$ means) the other, electronic, end (with charge unity) must responds with a nonvanishing compressibility.

5.1 My Final Answer

For the rest of this chapter I will use

$$\mathcal{H}^p = \frac{1}{2}\sum_q \bar{\bar{\rho}}^p(q)\check{v}(q)\bar{\bar{\rho}}^p(-q) \tag{116}$$

$$\bar{\bar{\rho}}^p(q) = \sum_j e^{-i\mathbf{q}\cdot\mathbf{r}_\mathrm{j}}\left[\frac{1}{2ps+1} - il^2\mathbf{q}\times\mathbf{\Pi}_\mathrm{j} + \text{known series}\right]$$

$$\check{v}(q) = v(q)\, e^{-q^2l^2/2} = \frac{2\pi e^2 e^{-ql\lambda}}{q} e^{-q^2l^2/2} \qquad ZDS \tag{117}$$

Note that $v(q)$ is the Zhang-Das Sarma potential [37]. I use it just to illustrate the formalism: it has a free parameter λ which allows one to suppress large q. Roughly speaking, λ is a measure of sample thickness in units of l.

I illustrate the rather unusual form of $\mathcal{H}$ we have been led to by considering the simplest case of $\nu = 1/2$. When we square $\bar{\bar{\rho}}^p$ we get a double sum over particles whose diagonal part is the one particle (free field) term:

$$\mathcal{H}^0_{\nu=1/2} = 2\sum_j \int \frac{d^2q}{4\pi^2}\sin^2\left[\frac{\mathbf{q}\times\mathbf{k}_j l^2}{2}\right]\check{v}(q) \tag{118}$$

This is not a hamiltonian of the form $k^2/2m^*$. However if the potential is peaked at very small q, we can expand the sin and read off an approximate $1/m^*$

$$\frac{1}{m^*} = \int \frac{qdqd\theta}{4\pi^2} \left[(\sin^2\theta)\ (ql)^2\right] \check{v}(q) \tag{119}$$

which has its origin in electron-electron interactions. However we can do more: we have full $\mathcal{H}_0$ as well as the interactions. The point I want to emphasize is that $\mathcal{H}$ is not of the traditional form and that there is no reason it had to be.

6 Computation of Gaps

The formalism will be illustrated with some examples, starting with activation gaps for a fully polarized sample. The expression for the gap is

$$\Delta_a = \langle \mathbf{p} + PH|\mathcal{H}|\mathbf{p} + PH\rangle - \langle \mathbf{p}|\mathcal{H}|\mathbf{p}\rangle \tag{120}$$

where $|\mathbf{p}\rangle$ stands for a HF ground state with p-filled LL's and $|\mathbf{p} + PH\rangle$ for the state with a widely separated particle-hole pair. The hamiltonian is $\mathcal{H} = \mathcal{H}^p$. It turns out all the matrix elements can be analytically evaluated. Figure 3 compares the numbers so obtained with those obtained by Park, Meskini, and Jain [38] using Jain wavefunctions. The gap formula looks the same for them, except that the states are not the simple ones (like χ_p) mentioned above, but these multiplied by the Jastrow factor and projected down to the LLL. On the other hand $\mathcal{H} = V(\rho)$ is very simple in the electronic basis with $\rho(q) = \sum_j e^{-i\mathbf{q}\cdot\mathbf{r}_j}$.

Note that for $p = 3, 4$, the HF answer is not necessarily above the PMJ results, perhaps because the PMJ results may not be ideal benchmarks. Indeed

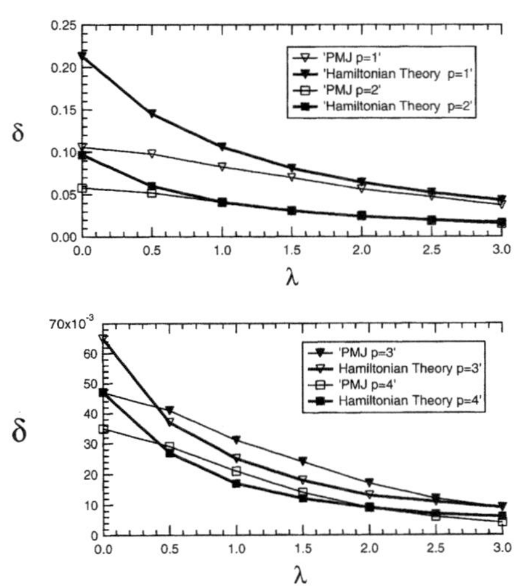

Fig. 3. Dimensionless activation gaps $\delta = \Delta_a/(e^2/\varepsilon l)$ compared the work of PMJ.

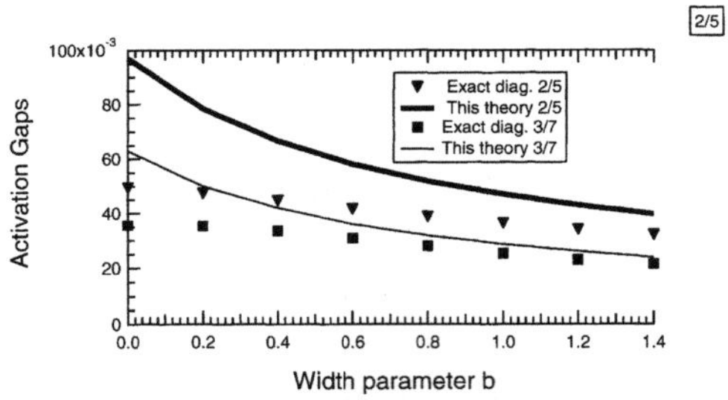

Fig. 4. Dimensionless gaps compared to exact diagonalization by Morf et al.

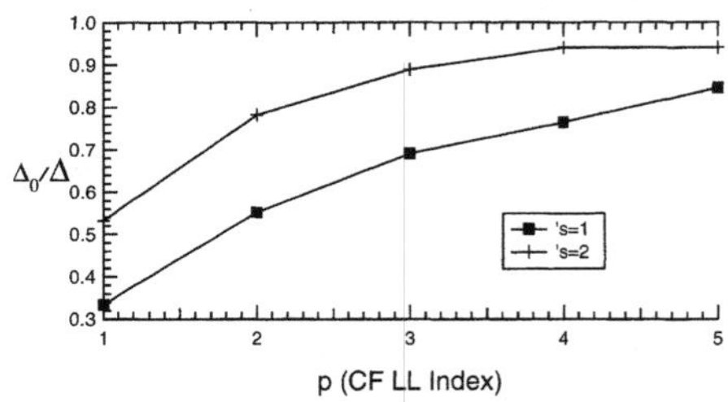

Fig. 5. Effect of turning off the interaction on gaps.

when I compare my results to the exact diagonalization work of Morf et al. [39], I find that the HF numbers are systematically above, as shown in Fig. 4. (The b parameter is like λ and defines a very similar potential.)

I also compared the numbers to PMJ for a gaussian potential and found almost perfect agreement except for 1/3. In general it appears the theory works best for potentials that are soft at large q, say for $\lambda > 1$.

Are CF weakly interacting? Given that two different mass scales control activation and polarization processes, one expects the answer to be negative, though only the hamiltonian formalism, with a concrete $\mathcal{H}$, allows us to ask this in a meaningful way. In Fig. 5, I compare the effect of turning off the interaction on gaps. Note that interactions seem less important for $\nu = 1/4$ and systematically get less important as p increases. I do not understand this.

I next compare the theory to the experiments of Du et al. [40] in Table 1. This is very tricky since we do not know the exact form of interactions and have not included disorder. The idea is to see what sort of λ fit the data.

Table 1. Comparison to Du et al..

ν	$B(T)$	$\Delta_a^{exp}(^oK)$	$\Delta_a^{theo}(^oK)$	λ
1/3	13.9	8.2	$5.3\sqrt{B(T)}/\lambda$	2.4
2/5	11.6	3	$2.08\sqrt{B(T)}/\lambda$	2.4
3/7	10.8	2	$1.23\sqrt{B(T)}/\lambda$	2.0

The value of $\lambda \simeq 2$ is roughly double what one expects for a pure system. Additionally, the gaps, when extrapolated to $\nu = 1/2$ have a negative intercept of a few Kelvin. These suggest that disorder is very strong and that it is not possible to describe a disordered system by an effective λ.

Consider now the results of Pan et al. [41] who found that

$$m_a^{nor} = \frac{m_a}{m_e\sqrt{B(T)}} \simeq .25 - .35. \tag{121}$$

near $\nu = 1/2$ and $1/4$. How does this rough equality of normalized masses of fermions with two and four vortices fit in the present theory? I find

$$m_a^{nor} = .163\lambda \quad (s = 1) \tag{122}$$

$$= .175\lambda^{5/4} \quad (s = 2) \tag{123}$$

where the powers of λ are approximate. The theory does explain the near equality of these masses, but clearly does not attribute any fundamental significance to it since the answer is sensitive to the potential (λ).

6.1 Magnetic Phenomena

So far we have assumed the spin to be fully polarized along the applied field. Thus in the fraction $p/(2ps + 1)$, the CF fill p-LL with spins up. This however costs a lot of kinetic energy, which would favor filling spin-up and down LL's equally. If one could vary the Zeeman coupling (by placing the sample in a tilted field whose normal part remains fixed) one could drive the system through various transitions. If $E(p - r, r)$ is the energy of the state with $p - r$ up and r down LL's (not including Zeeman energy) then the transition $r \to r + 1$ will take place at a field B^c given by

$$E(p - r, r) - E(p - r - 1, r + 1) = g\frac{eB^c}{2m_e}\frac{n}{p} \tag{124}$$

When the energies are computed, a strange fact, previously seen by Park and Jain, emerges: even though the CF are not free, the energy differences behave as if CF were free and occupied LL with a (polarization) gap Δ_p. That is, we find

$$E(p-r, r) - E(p-r-1, r+1) = \frac{n(p - 2r - 1)}{p}\Delta_p \tag{125}$$

(This would be the relation in a free theory since (n/p) spin-up fermions of energy $(p - r - 1 + \frac{1}{2})\Delta_p$ drop to the spin-down level with energy $(r + \frac{1}{2})\Delta_p$).

In the gapless case, the polarization is determined by

$$\mathcal{E}_+(k_{+\mathrm{F}}) - \mathcal{E}_-(k_{-\mathrm{F}}) = g\frac{e}{2m_e}\frac{B_\perp}{\cos\theta} \tag{126}$$

where $k_{\pm\mathrm{F}}$ are the Fermi momenta of up and down fermions and $\mathcal{E}_{\pm\mathrm{F}}$ the corresponding Fermi energies. Calculations show once again that the Fermi energy differences may be fit by

$$\mathcal{E}_+(k_{+\mathrm{F}}) - \mathcal{E}_-(k_{-\mathrm{F}}) = \frac{k_{+\mathrm{F}}^2 - k_{-\mathrm{F}}^2}{2m_p} \tag{127}$$

where m_p is the polarization mass.

But we know CF cannot be not free because the activation gap $\Delta_a = eB^*/m_a$ and polarization gap $\Delta_p = eB^*/m_p$ are very different. *Here is my argument [42] that this free field behaviour for magnetic phenomena is accidental and reflects $d = 2$ and rotational invariance.* Let us assume that the energy as a function of total spin S has the form

$$E(S) = E(0) + \frac{\alpha}{2} S^2 \tag{128}$$

where α is the inverse linear static susceptibility. Consider the gapless case for simplicity. When dn particles go from spin-down to spin-up,

$$dE = \alpha\ S\ dS = \alpha\ S\ (2\ dn) \tag{129}$$

$$= \alpha \frac{k_{+\mathrm{F}}^2 - k_{-\mathrm{F}}^2}{4\pi} (2\ dn) \tag{130}$$

using the areas of the circular Fermi seas. We see that dE has precisely the form of the kinetic energy difference of particles of mass m_p given by

$$\frac{1}{m_p} = \frac{\alpha}{\pi}. \tag{131}$$

Thus m_p is essentially the static susceptibility, which happens to have dimensions of mass in $d = 2$. With this understanding I give the calculated values

$$\frac{1}{m_p^{(2)}} = \frac{e^2 l}{\varepsilon} C_p^{(2)}(\lambda) \qquad C_p^{(2)}(\lambda) = \frac{.087}{\lambda^{7/4}} \tag{132}$$

$$\frac{1}{m_p^{(4)}} = \frac{e^2 l}{\varepsilon} C_p^{(4)}(\lambda) \qquad C_p^{(4)}(\lambda) = \frac{.120}{\lambda^{7/4}} \tag{133}$$

where the superscripts on C refer to the number of vortices attached and the exponent $7/4$ is approximate.

Comparison to $T = 0$ Experiments. Since magnetic transitions are controlled by ground state energies, perhaps disorder can be incorporated via an effective potential. *I have shown [43] that under certain restrictive conditions this is possible and will try to fit theory to experiment by using one data point to extract an effective λ and use it to make predictions for other measurements made on that sample (using simple scaling laws for λ if needed).*

Kukushkin et al. [44] vary both n and a perpendicular B to drive the system through various $T = 0$ (by extrapolation) transitions. I will compare the hamiltonian theory to these experiments by computing the B^c's at which the systems at $1/4, 2/5, 3/7, 4/9$, and $1/2$ lose full polarization ($r = 0$ for gapped cases, saturation for the gapless cases) and, for $4/9$, also the $r = 1$ transition, $|\mathbf{3},\mathbf{1}\rangle \rightarrow |\mathbf{2},\mathbf{2}\rangle$. I fit λ to the $\nu = 3/7$ transition $|\mathbf{3},\mathbf{0}\rangle \rightarrow |\mathbf{2},\mathbf{1}\rangle$ at $B^c = 4.5T$. The results are in Table 2.

Table 2. Comparison to Kukushkin et al. Critical fields based on a fit at 3/7.The rows are ordered by the last column which measures density.

ν	comment	B^c (exp)	B^c (theo)	νB^c (exp)
4/9	$(3,1) \rightarrow (2,2)$	2.7 T	1.6 T	1.2
2/5	$(2,0) \rightarrow (1,1)$	3 T	2.65 T	1.2
1/4	saturation	5.2 T	4.4 T	1.3
3/7	$(3,0) \rightarrow (2,1)$	4.5 T	4.5 T	1.93
4/9	$(4,0) \rightarrow (3,1)$	5.9 T	5.9 T	2.62
1/2	saturation	9.3 T	11.8 T	4.65

7 Physics at $T > 0$

The hamiltonian formalism is particularly suited to study physics at $T > 0$ in the infinite volume limit. Given a concrete hamiltonian, one can work out the HF energies at $T > 0$

$$\begin{aligned}\mathcal{E}^{\mathrm{Z}}_{\pm}(k) &= \\ &\mp\frac{1}{2}g\left[\frac{eB}{2m}\right] + 2\int\frac{d^2q}{4\pi^2}\tilde{v}(q)\sin^2\left[\frac{\mathbf{k}\times\mathbf{q}l^2}{2}\right] \\ &-4\int\frac{d^2k'}{4\pi^2}n^{\mathrm{F}}_{\pm}(|k'|)\tilde{v}(|\mathbf{k}-\mathbf{k}'|)\sin^2\left[\frac{\mathbf{k}'\times\mathbf{k}l^2}{2}\right]\end{aligned}$$

where the superscript on $\mathcal{E}^{\mathrm{Z}}_{\pm}$ reminds us it is the total energy including the Zeeman part, the Fermi functions

$$n^{\mathrm{F}}_{\pm}(|k|) = \frac{1}{\exp\left[(\mathcal{E}^{\mathrm{Z}}_{\pm}(k)-\mu)/kT\right]+1} \tag{134}$$

depend on the energies $\mathcal{E}^{\mathrm{Z}}_{\pm}(k)$ and the chemical potential μ. At each T, one must choose a μ, solve for $\mathcal{E}^{\mathrm{Z}}_{\pm}(k)$ till a self-consistent answer with the right total particle density n is obtained. From this one may obtain the polarization by taking the difference of up and down densities.

The computation of $1/T_1$ is more involved. The theory predicts [43]

$$\frac{1}{T_1} = -16\pi^3 k_B T \left(\frac{K_\nu^{max}}{n}\right)^2 \times \int_{E_0}^{\infty} dE \left(\frac{dn^F(E)}{dE}\right) \rho_+(E)\rho_-(E)F(k_+,k_-) \tag{135}$$

$$F = e^{-(k_+^2+k_-^2)l^2/2} I_0(k_+k_-l^2) \tag{136}$$

$$\rho_\pm(E) = \int \frac{kdk}{2\pi}\delta(E-\mathcal{E}_\pm^Z(k)) \tag{137}$$

where E_0 is the lowest possible energy for up spin fermions,$k_\pm$ are defined by $\mathcal{E}_\pm^Z(k_\pm) = E$ and K_ν^{max} is the measured maximum Knight shift.

Comparison to Experiment. We now compare to some experiments at $\nu = 1/2$ and $T > 0$. Consider first Dementyev et al. [45]. From their data point $P = .75$ for $B = B_\perp = 5.52T$ at 300 mK I deduce $\lambda = 1.75$. I have once again chosen to match my HF results with the above data point, and see to what extent a *sole* parameter λ, can describe P and $1/T_1$ for the given sample at a given $B_\perp$, but various temperatures and tilts. The results are shown in Fig. 6. Dementyev *et al* had pointed out that a two parameter fit (mass m and interaction J), led to four disjoint sets for these four curves. Given that $\mathcal{H}$ is neither free nor of the standard form $(p^2/2m + V(x))$ this is to be expected. By contrast, a single λ is able to describe the data since H has the right functional form.

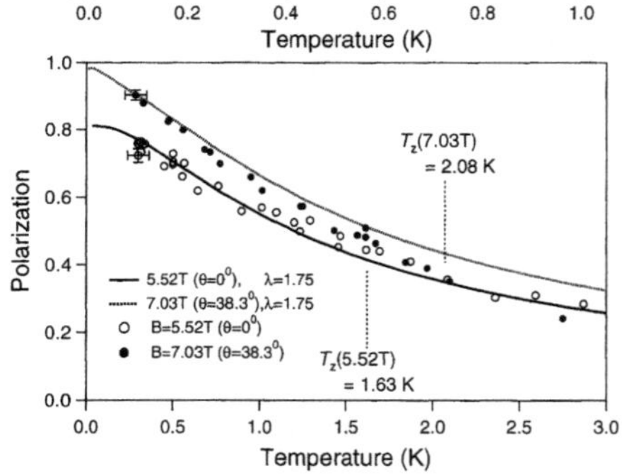

Fig. 6. Comparison to the work of Dementyev et al.. The value of λ is fit to P at 300 mK, $B_\perp = 5.52$ T and the rest follows from the theory.

Next I compare the results to that of Melinte et al. [46] for polarization and Freytag et al. [46] for $1/T_1$. A value of $\lambda = 1.6$ was extracted from one polarization point for sample M280. Note that a factor of two difference in $1/T_1$. I do not have an explanation for this. If the theory is correct, the figure implies that there are other means of relaxation besides the one introduced into the calculation.

8 Summary

Electrons in a magnetic field form degenerate Landau Levels and for the cases we were interested in ($\nu < 1$), all the electrons could be fit into the Lowest Landau Level (LLL) with room to spare. The macroscopic degeneracy of the noninteracting ground state ruled out perturbative treatments. Interactions were expected to lift the degeneracy and produce a unique ground state and tower of excited states with a scale set by the interactions. How was one to find them?

One way out was to write down inspired trial wavefunctions in the LLL that had all the right properties, a trail blazed by Laughlin for $\nu = 1/(2s+1)$. From his work we learnt also that the system was an incompressible fluid that allowed only localized density deviations. For example a vortex at z_0 described a quasihole with a charge deficit of $1/(2s+1)$.

Jain extended the wavefunctions to fractions $\nu = p/(2ps+1)$. By trading an electron for CF which carried $2s$ units of flux that, on average, canceled $2s$ of the $2s + 1/p$ flux quanta per particle, Jain obtained a particle that saw $1/p$ flux quanta per particle, which was just right to fill p LL's in the noninteracting case, a state we called χ_p. The electronic wavefunction was obtained by undoing the flux attachment. In practice, the flux tubes were replaced by vortices and a projection was done to the LLL.

From these wavefunctions we learnt [7],[12] that in the Laughlin case electrons were bound to $2s$-fold vortices to form CF with a charge e^*. This was true in the Jain case only before projection, which kills many zeros and also moves them off the particles. It was however true in both cases that the CF had a charge e^*.

We then asked how all these could be derived directly from the hamiltonian. We saw how flux could be attached by the Linneas and Myrrheim or Chern-

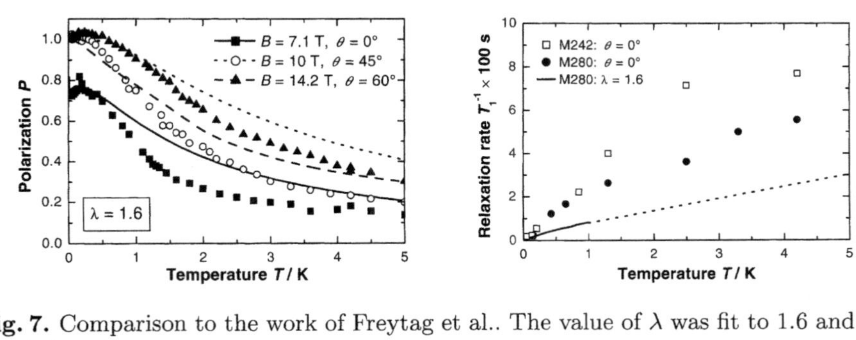

Fig. 7. Comparison to the work of Freytag et al.. The value of λ was fit to 1.6 and the rest follows from the theory.

Simons transformation following Fradkin and Lopez. Then the work of Murthy and myself was described. Here one introduces additional (oscillator) degrees of freedom at the cyclotron scale and a corresponding number of constraints χ. Placing the oscillators in the ground states gave us the modulus of the Jastrow factor which combined with the phase in the CS transformation to give the Laughlin and unprojected Jain wavefunctions. These oscillators were then decoupled from the fermions in the infrared-RPA approximation. The oscillators, now at exactly the cyclotron frequency, were seen to carry all the Hall current. The particles' kinetic energy could be quenched if the number of oscillators was chosen to equal the number of particles. In the low energy sector we were then left with just the potential energy V written in terms of the projected density $\bar{\rho}$ (in the new basis) and a set of constraints $\bar{\chi}$.

We then turned to my extension of these expressions for all ql (denoted by $\bar{\bar{\rho}}$ and $\bar{\bar{\chi}}$) which not only made the problem mathematically more alluring, but captured much of the known CF physics in operator form. In particular, one could see a CF made of the electron and another object which had the charge of the $2s$-fold vortex (called vortex for simplicity). The potential energy of electrons was seen to bind the two and to give the CF its charge and internal structure, as well as kinetic and interaction energies exactly as desired. The Hilbert space of the CF was seen to spawn two guiding center coordinates, one for the electron and one for the vortex. The constraints prevented the latter from having density fluctuations. Hence the proper implementation of constraints was needed to recover the HLR result of finite compressibility at $\nu = 1/2$: with one end of every CF frozen, only the electronic end responded to any applied potential as a unit charge object.

There were two approaches to solving the theory defined by $\mathcal{H} = V(\bar{\bar{\rho}})$ and subject to the constraints $\bar{\bar{\chi}} = 0$. Since the $[\bar{\bar{\chi}}, \mathcal{H}] = 0$ one could find an approximation that respected the constraint algebra, as was done by Read for the gapless case or Murthy [34] for magnetoplasmons. This is ideal when gauge invariance is important, typically at $T = 0$, in the gapless cases, at the Fermi surface. For the other cases, (emphasized here), we use the one proposed by Murthy and myself: make the replacement $\bar{\bar{\rho}} \to \bar{\bar{\rho}} - c^2\bar{\bar{\chi}} \equiv \bar{\bar{\rho}}^p$ which incorporates many of the effects of the constraints in a HF calculation and respects Kohn's theorem.

We focused on a few illustrative examples of this formalism. We saw that gaps could be computed for the polarized case in HF in closed form and that they were within about 10% of the wavefunction based results or exact diagonalization results, for potentials that were soft for large q, which meant roughly that $\lambda > 1$, λ being the parameter in the Zhang-Das Sarma potential.

We could see that CF were not free by turning off the interaction term and observing sizeable changes in gaps. (Without a hamiltonian such a question did not even have a meaning.) We could understand the rough equality of the renormalized masses observed by Pan et al..

Turning to magnetic phenomena, we understood how polarization phenomena could be mimicked by a free theory with a LL gap Δ_p due a conspiracy involving $d = 2$ and rotational invariance. Magnetic transitions at $T = 0$ observed by Kukushkin et al. were described by the theory given one data point from which an effective λ was deduced. (Heuristic arguments suggesting that magnetic phenomena in a disordered system could be described by an effective λ was demonstrated elsewhere [43].)

We saw how it was possible to compute polarization and relaxation at $\nu = 1/2$ as a function of T using just one data point to extract λ. The power of this approach was evident in the comparison to the data of Dementyev et al: whereas a single $\lambda = 1.75$ gave a very good description of two polarization and two relaxation rate graphs, a fit to the data with a canonical mass plus interaction term required four disjoint sets of values. The theory was also in fair agreement with the Melinte et al. and Freytag et al. data.

In summary, we understand the FQHE in two complementary fronts: trial wavefunctions and hamiltonian approaches. The former give excellent numbers where applicable and the latter provide many interpolating steps, insight, and facilitate otherwise impossible computations such as unequal-time correlations, coupling to disorder or relaxation rates at $T > 0$, all in the infinite volume limit.

After this brief introduction, inevitably limited and idiosyncratic, you should be ready to pursue variants of the simple FQHE system studied here. Some are reviewed in Ref. 2 by Eisenstein (experiments on double layer systems), Girvin and MacDonald (theory of the same), and Kane and Fisher (edge physics). Other possible areas that might interest you are areas are drag [47] and skyrmions [48].

References

1. For a review see *The Quantum Hall Effect*, Edited by R.E.Prange and S.M Girvin, Springer-Verlag, 1990, T. Chakraborty and P. Pietiäïnen, *The Fractional Quantum Hall Effect: Properties of an incompressible quantum fluid*, Springer Series in Solid State Sciences, **85**, Springer-Verlag New York, 1988, A.H. MacDonald ed., *Quantum Hall Effect: A Perspective*, Kluwer, Boston, 1989. A. Karlhede, S. A. Kivelson and S. L. Sondhi, "The Quantum Hall Effect: The Article", in V.J. Emery ed.,"Correlated Electron Systems", World Scientific, Singapore (1993)
2. *Perspectives in Quantum Hall Effects*, Edited by Sankar Das Sarma and Aron Pinczuk (Wiley, New York, 1997).
3. *Composite Fermions*, Edited by Olle Heinonen, World Scientific, Singapore, (1998).
4. K. von Klitzing, G. Dorda and M. Pepper, *Phys. Rev. Lett.*, **45**, 494, (1980)
5. D.Tsui, H.Stromer and A.Gossard, Phys. Rev. Lett., **48**, 1599, (1982). (FQHE).
6. R. B. Laughlin, Phys. Rev. Lett, **50**, 1395, (1983). Needs no explanation.
7. J. Jain, Phys. Rev. Lett., **63**, 199, (1989). For the latest summary see J.K. Jain and R. Kamilla, in Ref(3).
8. S.M.Girvin, A.H. MacDonald and P. Platzman, Phys. Rev. **B33**, 2481, (1986).
9. B.I. Halperin, Helv. Phys. Acta **56**, 75, (1983). B.I. Halperin, Phys. Rev. Lett., **52**, 1583, (1984).
10. J.M.Leinaas and J.Myrheim, *Nuovo Cimento* **37B**, 1 (1977). F. Wilczek, Phys. Rev. Lett. **48**, 1144, (1982).
11. R. L. Laughlin, Phys. Rev. Lett., **60**, 2677 , (1988). See also A.L.Fetter, C.B. Hanna, and R.B. Laughlin, Phys. Rev. **B39**, 9679, (1989), *ibid* **43**, 309, (1991).

12. N.Read Semi. Cond. Sci. Tech., **9**, 1859, (1994), and Surface Science, **361/362**, 7, (1995).
13. E. Rezayi and N.Read, Phys. Rev. Lett,**72**, 900, (1994).
14. A.Stern, and B.I.Halperin, Phys. Rev. **B54**, 11114 (1996). See the review on modifications of RPA by S.Simon, J. Phys. Cond. Mat., **48**, 10127 (1996).
15. S. Deser, R. Jackiw and S. Templeton, Phys. Lett., B139, 371, *Field theories in condensed matter physics*, E. Fradkin, Addison-Wesley, Reading MA 1991.
16. A. Lopez and E.Fradkin, Phys. Rev. **B 44**, 5246, (1991), *ibid* **47**, 7080, (1993), Phys. Rev. Lett., **69**, 2126, (1992).
17. R. Rajaraman and S. L. Sondhi, Mod. Phys. Lett. B **8**, 1065 (1994).
18. V.Kalmeyer and S. C. Zhang, Phys. Rev.**B46**, 9889, (1992).
19. B. I. Halperin, P.A.Lee and N.Read, Phys. Rev. **B47**, 7312, (1993).
20. W.Kang, H.L. Störmer, L.N. Pfeiffer, K.W. Baldwin and K.W. West, *Phys. Rev. Lett.*, **71**, 3850, (1993).
21. V.J. Goldman, B. Su and J. K. Jain, *Phys. Rev. Lett.*, **72**, 2065, (1994).
22. J.H.Smet, D.Weiss, R.H. Blick, G. Lutjering and K. von Klizting, *Phys. Rev. Lett.*, **77**, 2272, (1996).
23. R.L. Willett, M .A. Paalanen, R.R.Ruel, K.W.West, L.N.Pfeiffer, and D.J.Bishop, Phys. Rev. Lett. **54**, 112 (1990) For a review see R. Willett, Advances in Physics., **46**, 447, (1997).
24. Y.B. Kim, A. Furusaki and P.A. Lee,Phys. Rev.**B50**, 17917, (1994).
25. S.C. Zhang, H.Hansson and S.Kivelson, Phys. Rev. Lett., **62**, 82, (1989). See also D.-H. Lee and S.-C. Zhang, Phys. Rev. Lett. **66**, 1220 (1991), S.C. Zhang, Int. J. Mod. Phys., **B6**, 25, (1992) and C. L. Kane, S. Kivelson, D.H. Lee and S.C.Zhang, Phys. Rev. **B 43** , 3255 (1991).
26. S.M. Girvin and A.H. MacDonald, Phys. Rev. Lett. **58**, 1252, (1987). See also S. M. Girvin in Ref. 1 for work pointing to Chern-Simons theories.
27. N. Read, Phys. Rev. Lett. **62**, 86, (1989).
28. R.Shankar and G. Murthy, Phys. Rev. Lett., **79**, 4437, (1997). For details see "Field Theory of the FQHE", in Heinonen's book (Ref. (3) or cond-mat 9802244.
29. D. Bohm and D. Pines, Phys. Rev. **92**, 609, (1953).
30. G. Murthy and R.Shankar,Phys. Rev. **B 59** , 12260, (1999), G.Murthy, R.Shankar, K.Park, and J.K.Jain, Phys. Rev. **B58**, 13263, (1998).
31. R.Shankar, Phys. Rev. Lett., **83**, 2382, (1999) cond-mat 9903064.
32. V. Pasquier and F.D.M.Haldane Nucl. Phys. **516**, 719,(1998).
33. N. Read, Phys. Rev. **B58**, 16262, (1998).
34. G. Murthy, cond mat. 9903187.
35. B.I.Halperin and A.Stern, Phys. Rev. Lett., **80**, .5457 (1998); G. Murthy and R.Shankar, *ibid* , 5458 (1998). A. Stern, B.I. Halperin, F. von Oppen and S.Simon, Phys. Rev., **59**,12547, (1999). cond-mat 9812135.
36. D.H. Lee, Phys. Rev. Lett., **80**, 4745 (1998).
37. F.C. Zhang and S. Das Sarma, Phys. Rev. B 33, 2908, (1986).
38. K. Park, N.Meskini, and J.K. Jain, J. Phys. Condensed Matter, **11**, 7283, (1999). Has a response to R.Morf, Phys. Rev. Lett. **83**, 1485, (1999).
39. R.H. Morf, N. d'Ambrumenil and S. Das Sarma (to be published).
40. R.R.Du, A.S. Yeh, H.L. Störmer, D.C. Tsui, L. N. Pfeiffer, K.W.Baldwin and K.W. West, Phys. Rev. Lett., **70**, 2944, (1993).
41. W. Pan, H.L. Störmer, D.C.Tsui, L.N. Pfeiffer, K.W. Baldwin and K.W. West, Phys. Rev., **B61**, R5101, (2000).
42. R.Shankar, Phys. Rev. Lett., **84**,3946, (2000) cond-mat 9911288.

43. R.Shankar, Phys. Rev **B 63**, 085322, (2001), cond-mat 0009361.
44. I.V.Kukushkin, K. v. Klitzing and K. Eberl, Phys. Rev. Lett. **82**, 3665, (1999.)
45. A.E. Dementyev, N.N. Kuzma, P. Khandelwal, S.E. Barrett, L.N. Pfeiffer, and K. W. West, Phys. Rev. Lett., **83**, 5074, (1999). cond-mat/9907280.
46. S.Melinte, N.Freytag, M. Horvatic, C. Berthier, L.P. Levy, V. Bayot and M. Shayegan, Phys. Rev. Lett., **84**, 354, (2000). cond-mat/9908098 are the source of $P(\theta = 0)$. The rest are from N. Freytag, L. Levy, M. Horvatic, C. Berthier, and M. Shayegan, unpublished.
47. For a recent review see A.G. Rojo, Electron-drag effects in coupled electron systems J. Phys. Cond. Mat., **11 (5)**: R31-R52, (1999). For some very early work see G.L. Zheng and A.H. Macdonald, Phys. Rev **B 48**, 8203, (1993). For work on FQHE see for example I. Ussishkin I and A. Stern Phys. rev. , **56 (7)**, 4013, (1997).
48. For a review see S. M. Girvin, 'The Quantum Hall Effect: Novel Excitations and Broken Symmetries,' 120 pp. Les Houches Lecture Notes, in Topological Aspects of Low Dimensional Systems, ed. by Alain Comtet, Thierry Jolicoeur, Stephane Ouvry and Francois David, (Springer-Verlag, Berlin and Les Editions de Physique, Les Ulis, 2000, ISBN: 3-540-66909-4). cond-mat/9907002, The first papers were S. L. Sondhi, A. Karlhede, S. A. Kivelson, E. H. Rezayi, Phys. Rev. B **47**, 16419 (1993). H.A. Fertig, L. Brey, R. Côté, A.H. MacDonald, A. Karlhede, and S.L. Sondhi, Phys. Rev. B **55**, 10671 (1997), and S. E. Barrett et al., Phys. Rev. Lett., **74**, 5112 (1995).

Magneto-optics of Composite Fermions

I.V. Kukushkin

Max-Planck-Institut für Festkörperforschung,
Heisenbergstr. 1, 70569 Stuttgart, Germany
and
Institute of Solid State Physics, RAS, Chernogolovka, 142432 Russia

Abstract. From the degree of circular polarization of the time-resolved radiative recombination of 2D-electrons with photo-excited holes bound to acceptors we have measured the magnetic field dependence of the electron spin polarization for various fractional and Composite Fermions (CF) states. The Fermi energies of these CF states are measured for the first time and the corresponding value of the CF density of states mass at $\nu = 1/2$ is found to be about 4 times heavier than the previously reported values of the "activation" mass. We demonstrate that the Zeeman splitting of CFs is enhanced by a factor of 2.5 due to the interaction between CFs. The latter is very sensitive to the finite width of the 2D-channel. The spin polarization at $\nu = 1/3$ and $\nu = 2/3$ displays an activated behavior and the derived spin-wave gaps are compared with simultaneously measured transport values.

The existence of the microwave (12-40 GHz) cyclotron resonance of Composite Fermions has been demonstrated in a 2D-electron system with an artificial density modulation using a sensitive optical detection scheme. This allows one to measure the cyclotron mass of Composite Fermions. This cyclotron mass increases from 0.68 m_0 to 1.21 m_0 when the electron density varies between 0.6×10^{11} cm^{-2} and 1.5×10^{11} cm^{-2}.

1 Introduction

It is well known, in condensed matter physics, that no single theory can exactly describe real experiments due to a huge number of interacting particles involved in the process. Nevertheless one expects a "good" theory to provide original concepts giving a clear physical picture and explaining the essential experimental results. The new theory of Composite Fermions (CFs) belongs to this class: it explains the Fractional Quantum Hall Effect in terms of a novel composite particle consisting of an electron to which a given number of magnetic flux quanta are attached. In general such composite particles behave like boson or fermion when the number of flux attached is odd or even respectively. CFs are electrons dressed with two magnetic flux quanta. Attaching flux quanta to an electron minimizes the energy of the system, since this vortex expels other electrons which results in a decrease of the repulsion interaction between 2D-electrons. The main consequence of the theory of CFs is that, at half filling of the lowest Landau level ($\nu = 1/2$), an external magnetic field is effectively compensated by two flux quanta and a metallic state of these composite particles is formed. This state can be characterized by a Fermi wave-vector and a Fermi-energy [1]. The deviation of the magnetic field from exact half filling results in the appearance of a non-zero effective magnetic field, that quantizes the CF-motion and discretizes their energy spectrum into Landau levels. In this model, the FQHE is a manifestation of the Landau quantization of CFs and a rich variety of experimental observations can be straightforwardly understood in terms of nearly independent CFs [7,3]. Recent experiments [4–6] support the validity of this theoretical concept and moreover demonstrate the semiclassical behavior of these composite particles. In this chapter, I will describe results obtained from a magneto-optical study of Composite Fermions which allows to measure their spin polarization and to detect the cyclotron resonance of composite particles.

2 Spin Polarization of Composite Fermions

As mentioned above there exists very strong experimental support of the concept of CFs, but it remains an apparent inconsistency within this simple picture. In the case of normal 2D-electron gas (2DEG), their Fermi energy is equal to the cyclotron gap at filling factor $\nu = 2$. If this holds for CFs, their Fermi energy, E_F^{CF}, can be taken as the value of the energy gap measured for $\nu = 1/3$ (for this state, the CF filling factor is $\nu_{CF} = 1$). Since for the magnetic field $B \approx 10$ T this gap is typically 8-12 K [7,8], E_F^{CF} can be estimated as 10 K (the same value can also be directly obtained from the published CF mass $m_{CF} \approx 0.6 m_0$ [7,9] at $B \approx 10$ T). Due to the small value of the g-factor of electrons in GaAs, the Zeeman energy (E_Z) at $B = 10$ T is only about 3 K, which means that $E_F^{CF} \gg E_Z$ and, therefore, one may expect that the Fermi gas of CF is practically spin unpolarized at $B \approx 10$ T. In contrast, it follows from the reported experiments, in which the CF wave-vector was measured, that the system of CFs is completely spin polarized at $B \approx 10$ T. This apparent discrepancy was our main motivation to measure the spin polarization of CFs using a well established optical method [10].

In this section we will consider the magnetic field dependence of the spin polarization of a 2D-system in the $\nu = 1/2$, $3/2$ and $1/4$ CF states. The CF Fermi energies for each of these states have been determined from the value of the critical magnetic field B_c beyond which a system of CFs becomes completely spin polarized ($E_F^{CF} = E_Z$ at $B = B_c$). The CF density of states mass is found to be about 4 times heavier than the previously reported "activation" mass [7,9] and closer to the value obtained from the temperature dependence of CFs scattering [7]. Another achievement of the present work concerns the phase transitions between differently spin polarized ground states of the FQHE. Several spin transitions between unpolarized, partly polarized and fully polarized spin states, which are accompanied by abrupt changes in the electron spin polarization, γ_e, are detected for the $\nu = 2/3$, $3/5$, $4/7$, $2/5$, $3/7$, $4/9$, $8/5$, $4/3$ and $7/5$ FQHE states and absolute values of γ_e are determined for these partly spin polarized FQHE ground states.

2.1 Samples and Method

We studied several low-density ($n_s = (0.36 - 2.4) \times 10^{11}$ cm^{-2}) and high-quality (electron mobility $\mu = (0.9 - 3) \times 10^6$ cm^2/Vs) GaAs/Al$_x$Ga$_{1-x}$As single heterojunctions with a δ-doped monolayer of Be acceptors ($n_A = 2 \times 10^9$ cm^{-2}) located in the wide ($1\mu m$) GaAs buffer layer at a distance of 30 nm from the interface [11]. In all samples a variation of the 2D-electron concentration was possible by using a top gate. For photoexcitation, we used pulses from a tunable Ti-sapphire laser (the wavelength was close to 780 nm) with a duration of 20 ns, a peak power of $10^{-4} - 10^{-2}$ W/cm^2, and a repetition rate of $10^4 - 10^6$ Hz. Luminescence spectra were detected by a gatable photon counting system with a spectral resolution of 0.03 meV. To analyze the circular polarization of the luminescence signal at low temperatures (down to 100 mK), we used an optical fiber system with a quarter wave plate and a linear polarizer placed in liquid helium just nearby the sample. This set-up allows the detection of σ^-/σ^+ signal ratios up to 1000 providing an accuracy of about 1 percent for the polarization measurements.

The spin polarization of the ground state of a 2DEG is determined by the competition between the Zeeman and Coulomb (E_C) energies. This was well established both theoretically [12,13] and experimentally [14,5,16] for the ground states of different FQHE states in the extreme quantum limit. In our previous work [10], we demonstrated unambiguously that such spin transitions can be directly detected from the degree of circular polarization γ_L of time-resolved radiative recombination of 2DEG with photoexcited holes bound to acceptors ($\gamma_L = (I_- - I_+)/(I_- + I_+)$, I_- and I_+ are integrated intensities of the luminescence measured in the σ^- and σ^+ polarizations, respectively). There are two independent reasons for the luminescence to be circularly polarized. One reason is the spin polarization of the hole system due to the Zeeman effect, which depends on the magnetic field and the temperature of the photoexcited holes (i.e. on the population of the different spin-split levels). The other reason is the spin polarization of the 2DEG, which depends on filling factor, temperature and magnetic field. It is possible to derive the contribution of the holes to the circular polarization of the luminescence separately by investigating the emission from fully occupied electron Landau levels (at $\nu = 2, 4, 6...$ and also well below the Fermi surface), because in this case both spin up and down states of electrons are completely occupied and the circular polarization of the luminescence is determined only by the spin polarization of the holes. It has been demonstrated experimentally (and in agreement with calculations [10]), that the spin polarization of the hole system is defined by the ratio B/T, so that it can be subtracted out from the polarization of the luminescence for a fixed value of B/T and a direct correspondence between γ_L and γ_e can be established. Alternatively time resolved techniques have also been used to ensure the complete relaxation of the hole system in its fundamental state, which allows to avoid or minimize the corrections [17].

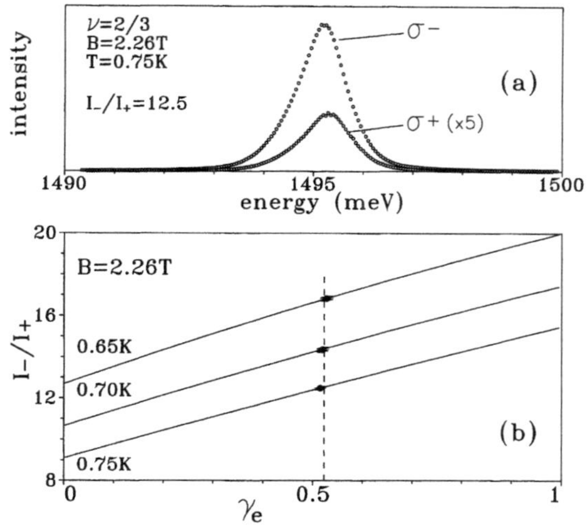

Fig. 1. (**a**) Typical luminescence spectra measured in the σ^- and σ^+ polarizations for $\nu = 2/3$, $B = 2.26$ T, $T = 0.75$ K and (**b**) calibration dependence of the ratio I_-/I_+ on the electron spin polarization γ_e, obtained for $B/T = 3.01$ ($T = 0.75$ K), $B/T = 3.23$ ($T = 0.70$ K), $B/T = 3.48$ ($T = 0.65$ K). I_- and I_+ are the integrated intensities of the luminescence, measured in the σ^- and σ^+ polarizations, respectively.

2.2 Spin Transitions

Typical luminescence spectra, recorded for both σ^- and σ^+ polarizations and the calibration curves, obtained for different temperatures for a given magnetic field are shown in Fig. 1. For a fixed value of B/T, the spin polarization of the holes determines the interval of variation of γ_L. For $B/T = 3$ this interval corresponds to the ($\gamma_e = 0$) low boundary of the ratio $I_-/I_+ = 9.1$ and to the high ($\gamma_e = 1$) boundary $I_-/I_+ = 15.3$. Within this interval, the ratio I_-/I_+ is an almost linear function of γ_e. It was established that γ_e increases slightly with decreasing temperature, but saturates at low temperatures (see Fig. 1b). The value of γ_e found from the extrapolation of $\gamma_e(T)$ to $T \to 0$ was taken as the correct one. The magnetic field dependence of the electron spin polarization $\gamma_e(B)$ measured for two different 2DEG concentrations are plotted in Fig. 2. It shows pronounced minima and maxima at several FQHE states. Both fundamental sequences of fractional states $n/(2n+1)$ ($\nu = 2/5$, 3/7, 4/9) and $n/(2n-1)$ ($\nu = 2/3$, 3/5, 4/7) are present in these data. In addition, the series of fractions 5/3, 8/5, 7/5 and 4/3 around $\nu = 3/2$ is clearly visible in $\gamma_e(B)$ (insets in Fig.2). However, it is rather inconvenient to analyze the magnetic field dependence of γ_e at a fixed n_s since, depending on n_s, the features at fractional filling factors sometimes show up as minima or maxima or even disappear. Instead a much more clear behavior of the dependence $\gamma_e(B)$ is obtained for a fixed filling factor value of ν.

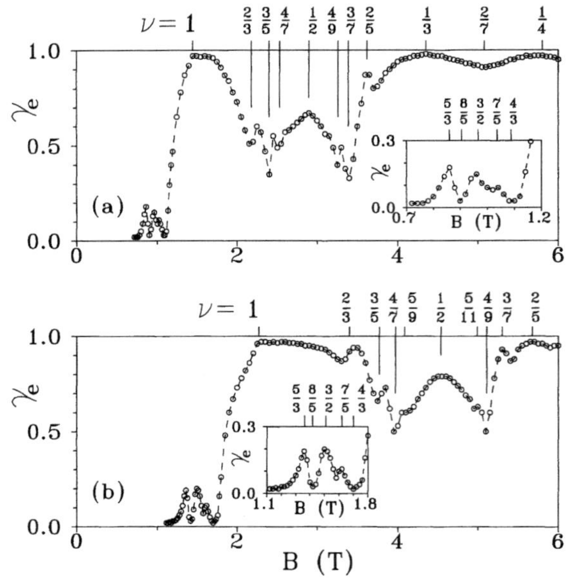

Fig. 2. Magnetic field dependence of the electron spin polarization γ_e measured for 2 different 2D-electron concentrations: (a) $n_S = 3.5 \times 10^{10}$ cm^{-2} and (b) $n_S = 5.5 \times 10^{10}$ cm^{-2}. Insets show the details around $\nu = 3/2$

In Fig. 3a we plot the magnetic field dependence of the electron spin polarization measured for several FQHE states from the sequence $\nu = n/(2n-1)$. To obtain these data, we investigated 6 different samples with top gates, which allows to vary n_s, so that a rather broad interval of magnetic fields was accessible for each fraction. Data measured for different samples are shown by different symbols (for $\nu = 2/3$) and illustrate the consistency of the results. Well defined spin transitions between various different spin-polarized ground states, governed by the magnetic field are clearly visible in Fig.3a for all fractions under study. For $\nu = 2/3$ only one transition (if one neglects the weakly developed feature around $B = 2.3$ T) from spin unpolarized to completely spin polarized state was observed at $B \approx 2$ T. In contrast, the magnetic field dependence of γ_e obtained for other FQHE states, such as 3/5 and 4/7, clearly indicates a broad region of stability of the partly polarized spin states (γ_e is close to 1/3 for $\nu = 3/5$ and it is about 1/2 for ν=4/7). Additional transitions between unpolarized, partially polarized and completely spin polarized ground states were detected (for $\nu = 4/7$ such transitions take place at $B \approx 1.8$ T and $B \approx 4.8$ T). Interestingly, a similar sequence of spin transitions with nearly identical absolute values of γ_e was observed also for FQHE states from the other sequence, $\nu = n/(2n+1)$ (see Fig. 3b). A correspondence between 2/3, 3/5 and 3/7 states from one side and 2/5, 3/7 and 4/9 states from the other side is apparent from a comparison

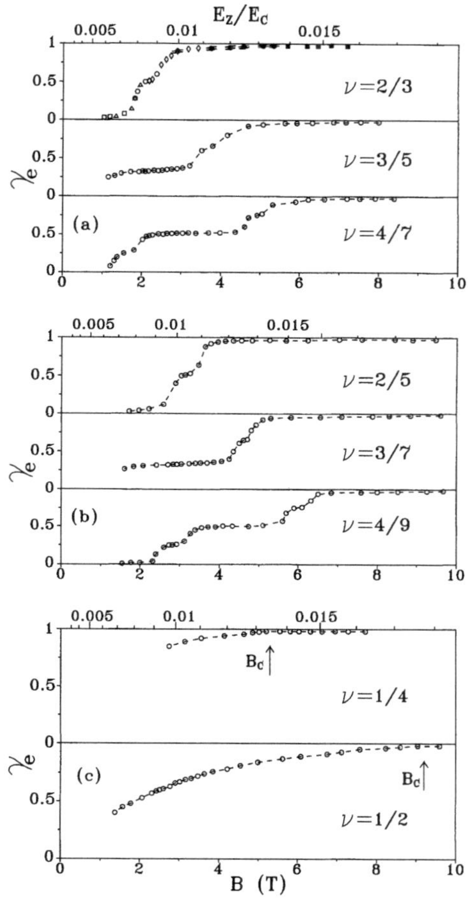

Fig. 3. Magnetic field dependence of the electron spin polarization γ_e measured for different FQHE and CF states (**a**) sequence $n/(2n-1)$, (**b**) sequence $n/(2n+1)$ and (**c**) $\nu = 1/2$ and $1/4$. The results shown for $\nu = 2/3$ by different symbols correspond to the data obtained for different gated samples

of Figs. 3a and 3b. Such similarity is consistent with the model of Composite Fermions.

Figure 3c shows the magnetic field dependence of the electron spin polarization measured for the CF state at $\nu = 1/2$. It is clear from the data that the system of CFs becomes fully spin polarized at $B_c = 9.3 \pm 0.5$ T. This means that at $B = B_c$ the CF Fermi energy is equal to the Zeeman splitting, which is equal to 2.8 K at 9.3 T (the CF g-factor is the same than that of electrons $g_{CF} = g_e = -0.44$). Note that the Zeeman energy of the CFs can be enhanced by interaction between them. However, this effect is expected for odd CF filling factors (for example, 1/3 FQHE state), but not for the $\nu = 1/2$ state. Assuming a parabolic dispersion law for the CFs at $\nu = 1/2$, the value of their density of state mass $m_{\rm DS}$ (or "polarization" mass [18]) can be estimated from the values of $E_{\rm F}^{\rm CF}$ and density ($n_{\rm CF} = n_{\rm S} = 1.26 \times 10^{11}$ cm^{-2} for $B = 9.3$ T and $\nu = 1/2$), which yields $m_{\rm DS}$ =2.27 m_0 (m_0 is the free electron mass in vacuum). The deduced mass is several times heavier than the previously reported values, measured from the activation energies. However, these masses have a different meaning [18]. Indeed, the filling factor dependence of the ground state energy has a cusp at fractional ν and the activation energy is sensitive to the sharpness of this cusp (slopes), but not to the absolute value of the minimum corresponding to the ground state energy. In contrast, the competition between different spin-polarized ground states, driven by the Zeeman energy, is sensitive only to the energy of this minimum and, therefore, the density of states mass is related to the energy splitting between the states with different $\gamma_{\rm e}$. Since there is no energy gap (cusp) at $\nu = 1/2$ the only relevant CF parameters are their Fermi energy and the density of state mass. It is important to notice that, if one assume that CFs are 2D-particles with a parabolic dispersion law (at effective magnetic field $B^* = 0$ at $\nu = 1/2$), the magnetic field dependence of $\gamma_{\rm e}$ should be a linear function of the parameter $E_{\rm Z}/E_{\rm F}$ ($\gamma_{\rm e} = E_{\rm Z}/2E_{\rm F}$), so that $\gamma_{\rm e} \propto B^{1/2}$ (due to $E_{\rm F} \propto n_{\rm s} \propto B$ and the expected increase in the CF mass, as $\sqrt{B}$ [18]). The dependence $\gamma_{\rm e}(\sqrt{B})$ measured for $\nu = 1/2$ was found to be a slightly sub-linear function of $\sqrt{B}$, which is mainly due to the temperature smearing of the Fermi surface. The dependence of $\gamma_{\rm e}(B)$ measured for $\nu = 1/4$ is shown in Fig.3c. Using the value of $B_{\rm c} = 5.2 \pm 0.2$ T we derived that for $\nu = 1/4$ and $n_{\rm S} = 0.32 \times 10^{11}$ cm^{-2}, $m_{\rm DS} = 1.13 m_0$. Note that the values of the CF mass obtained for $\nu = 1/2$ and 1/4 are simply $2\nu m_0/g_{\rm e}$, which follows from equaling the Zeeman and Fermi energies.

In Fig. 4a we plot the magnetic field dependence of $\gamma_{\rm e}(B)$ measured for several FQHE states from the sequences $2 - [n/(2n \pm 1)]$ ($\nu = 8/5$, 7/5 and 4/3) and also for $\nu = 3/2$ (the dashed line $\gamma_{\rm e} = 1/3$ shown on this figure corresponds to the complete spin polarization of the CFs at $\nu = 3/2$). It is clear from these figures that various spin transitions are also available for these FQHE states. However, in contrast to the fractional states at $\nu < 1$, these transitions are shifted to much higher magnetic fields. Such a breaking of the electron-hole symmetry is in agreement with previously reported transport data [7,14]. The analysis of the dependence $\gamma_{\rm e}(B)$ measured for $\nu = 3/2$ shows a very strong sub-linearity of the dependence $\gamma_{\rm e}(\sqrt{B})$, indicating a considerable nonparabolicity of the CFs dispersion at $\nu = 3/2$ (in contrast to the ν=1/2 state, we find that the CF dispersion law for the $\nu = 3/2$ state is not parabolic, but closer to a linear dependence). In Fig. 4b we plot the phase diagram. It illustrates at which magnetic fields (and at what value of $E_{\rm Z}/E_{\rm C} \propto \sqrt{B}$) the spin-transition to a fully polarized state takes place for different fractions (open circles) and also for various CF states $\nu = 1/2$ and 1/4 (filled circles). The importance of the parameter $E_{\rm Z}/E_{\rm C}$ for this problem was already mentioned (also demonstrated in a recent theory [18]) and explains the choice of the coordinates in Fig. 4b. The dashed lines on this figure correspond to a linear dependence plotted through

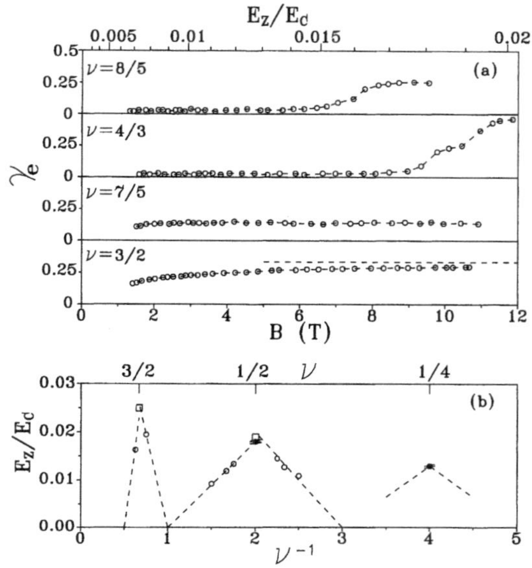

Fig. 4. (**a**) Magnetic field dependence of the electron spin polarization γ_e measured for the $\nu = 8/5, 4/3$ and $7/5$ FQHE states and for the $\nu = 3/2$ CF state. (**b**) Phase diagram $E_Z/E_C(\nu^{-1})$ illustrating at which magnetic fields the spin-transition into fully polarized states takes place for different fractions (open circles) and also for various CF states $\nu = 3/2$, $1/2$ and $1/4$ (filled circles). Other details are explained in the text

the experimental points and also through the points at $\nu = 1$, $1/3$ and $1/5$ since at these filling factors the 2DEG is expected to be fully spin polarized even for vanishing Zeeman energy. Empty squares correspond to the extrapolation of these lines to $\nu = 1/2$ and $3/2$ (note that for $\nu = 1/2$ the result obtained from the extrapolation is in a good agreement with the directly measured value). The threshold value of E_Z/E_C obtained for $\nu = 3/2$ from the extrapolation procedure corresponds to $B_c = 17.5 \pm 0.5$ T, which yields $m_{CF} = 2.27\ m_0$ at $n_S = 6.3 \times 10^{11}$ cm^{-2} (for $\nu = 3/2$, $n_{CF} = n_s/3 = 2.1 \times 10^{11}$ cm^{-2}).

2.3 Spin-Wave Excitations in the FQHE States and Interaction Between Composite Fermions

The attractiveness of the CF-picture is based on the assertion that dressing the electrons with two flux quanta constitutes the main effect of the interaction between the 2D-electrons. The remaining interaction between CFs is weak, so that in many cases the system can be considered as a nearly ideal Fermi-gas of composite particles [19]. The residual interaction between CFs does however exist and plays an essential role in the dispersion of the neutral excitations at FQHE states [20]. No interaction between CFs would imply no dispersion of the neutral excitations, yet it is well known [20,21] that the dispersion of the CFexciton is a rather complicated function of the excitonic momentum. Two types of intra-Landau-level neutral excitations for FQHE states below $\nu = 1$ have been recognized. They are neutral charge density (CD) excitations and neutral spin density (SD) excitations, associated with changes of the charge and spin degrees of freedom, respectively. The collective CD-mode has a finite gap at zero wave vector ($k = 0$) and displays a characteristic "magnetoroton" minimum at the inverse magnetic length $k = 1/l_B$. Its energy approaches in the limit of large k the FQHE energy gap, i.e. the energy to create infinitely separated quasiparticle-quasielectron pairs [20,23]. The other branch of collective excitations, the SD-mode, takes on the Zeeman energy at zero wave-vector (due to Larmor's theorem) and increases monotonically as a function of momentum until it approaches the exchange energy gap in the limit of infinite wave-vector [22,21]. The theoretically calculated dispersion laws for the CFs indicate that the interaction between CFs is indeed about an order of magnitude weaker than the interelectron interaction [20,22,23].

In order to measure the interaction energy directly, one thus needs to study the dispersion law of the CD-excitations or, alternatively, measure the spin-wave energy gap at infinite momentum. The number of experimental possibilities are very limited. Some results were obtained from inelastic light scattering [24,25] and NMR [26] investigations, however complementary measurements using alternative methods are highly desirable. In the present report we investigate the temperature dependence of the degree of electron spin polarization of the 1/2-CF state. This allows to determine the exchange interaction. Analogue experiments at the 1/3- and 2/3-FQHE-states allow to extract the spin wave energy gap. These results equally illustrate the importance of exchange interaction between CFs.

Fig. 5a shows the temperature dependence of the electron spin polarization (γ_e), measured at different magnetic fields but at the fixed filling factor $\nu = 1/2$. Below and above the critical magnetic field $B_c = 9.3$ T [17], for which the CF-system becomes fully spin polarized, the electron spin polarization saturates at low temperatures. In contrast, at $B = B_c$ a well defined linear dependence of γ_e on T is observed in the low T limit. Very similar behaviors at $B = B_c$, $B > B_c$ and $B < B_c$ were predicted theoretically for non-interacting CFs [27] and it was demonstrated that the linear dependence at low temperatures at $B = B_c$ results from the Fermi statistics (this linear term is expected to be stable for the case of weakly interacting particles). The slope of this dependence is determined by a single parameter, the Fermi energy of CFs: $\gamma_e(T) = 1 - (2T/E_F) \cdot ln((1+\sqrt{5})/2)$. The linear extrapolation of the low temperature portion of $\gamma_e(T)$ at $B = B_c$ yields a CF Fermi energy of 6.9 K. At $B = B_c$ the Fermy energy equals the Zeeman energy, and one must conclude that the spin splitting of CFs is strongly (by a factor of 2.5) enhanced in comparison with the bare Zeeman energy (2.8 K at 9.3 T).

Interaction phenomena are quite sensitive to the width of the 2D-channel. They are gradually suppressed upon increasing the channel width. In order to vary this width, we applied a substrate-bias voltage (V_{SB}), while maintaining

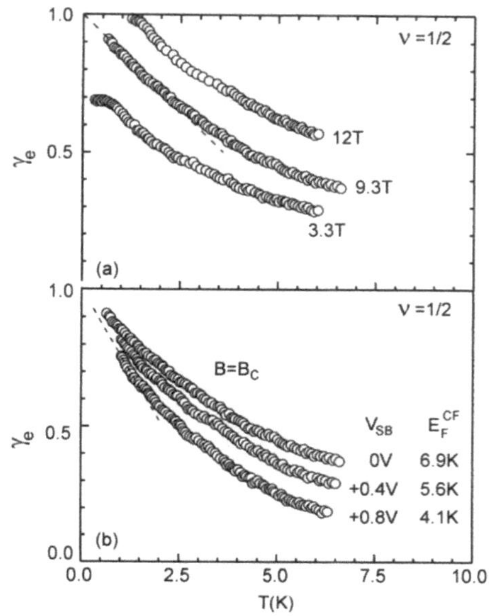

Fig. 5. (**a**) The temperature dependence of the electron spin polarization measured for $\nu = 1/2$ and $V_{\mathrm{SB}} = 0$ V for different magnetic fields. The dashed line corresponds to the linear extrapolation of the experimental data taken at $B = B_{\mathrm{c}} = 9.3$ T. From its slope the Fermi energy of CF metal was determined. (**b**) Influence of the substrate-bias voltage on the temperature dependence of the electron spin polarization at $\nu = 1/2$ and $B = B_{\mathrm{c}}$

a fixed carrier concentration through a simultaneous change of the top gate bias. The intersubband splitting E_{10} provides a measure for the finite width w of the 2D-channel and can be obtained directly from luminescence spectra at small delays after the laser pulse, when recombination from both subbands is observable. Alternatively, numerical simulations in the Hartree approximation of the eigenvalues and wave-functions of the ground and first excited subband can serve this purpose. The effective width w can be estimated from a fit of the calculated wave-function to the Fang-Howard function $\psi(z) = z \cdot \exp(-bz/2)$, with $w = 1/b$ [28]. Fig. 5b depicts the temperature dependence of the electron spin polarization at $B = B_{\mathrm{c}}$ for different V_{SB}. Fig. 6a shows the intersubband splitting and the critical magnetic field B_{c} as a function V_{SB} at a fixed density of $n_{\mathrm{s}} = 1.1 \times 10^{11}$ cm^{-2}. The latter depends only weakly on the substrate bias. On the top axis calculated values for the channel width are indicated for several bias voltages. Fig. 6b displays the Zeeman splitting (and Fermi energy) as extracted from the temperature dependent data. Positive (negative) substrate bias voltages correspond to a decrease (increase) of E_{10} and to an increase (decrease) of the channel width. The enhanced spin splitting of CFs is considerably suppressed for

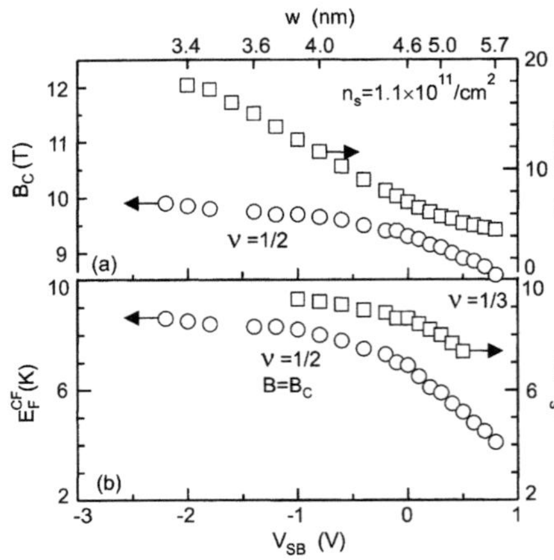

Fig. 6. (a) The dependencies of the intersubband energy E_{10} (squares, $n_s = 1.1 \times 10^{11}$ cm^{-2}) and the critical magnetic field B_c (circles, $\nu = 1/2$)) on the substrate-bias voltage V_{SB}. On the top axis the width of wavefunction w, as obtained from a comparison of numerical simulations with the Fang-Howard function is indicated for several values of V_{SB}. (b) The enhanced spin splitting of CFs (circles, $\nu = 1/2$, $B = B_c$) and of the spin-flip gap at $\nu = 1/3$ (squares) as a function of the substrate-bias voltage V_{SB}

positive V_{SB}, which means that the exchange interaction between CFs drastically drops for wide channels ($w > 6$ nm).

Since the concept of CFs is quite successful (both in numerical calculations and in experiment) in explaining many properties of the FQHE-states, we can proceed in a similar manner and estimate the interaction energy between CFs from the temperature dependence of the electron spin polarization at the 1/3- and 2/3-FQHE states. The analysis of our magnetooptical measurements is very similar to that used for activated magnetotransport investigations. However, the deduced energy gap is quite different since it is selectively sensitive to the spin degree of freedom. It therefore provides information about the intra-CF-Landau-level SD-excitation (or CF-spin-waves, CFSW), whereas magnetotransport data delivers the inter-CF-Landau-level CD-excitation gap (or CF-magnetoplasmon (CFMP)).

In Fig. 7a we show the temperature dependence of γ_e, measured at different magnetic fields, for the $\nu = 1/3$ FQHE state. The presence of a gap in the CFSW mode makes an Ahrenius-type analysis, as shown in Fig. 7b, reasonable. At low temperatures, the deviation of γ_e from 1 as a function of temperature is well described by a single exponential dependence: $\gamma_e = 1 - 2 \times \exp(-\Delta/2k_BT)$. Such a dependence is expected theoretically as described in Ref. [29]. Two possible values for Δ were discussed in this reference, because of the presence of two different gaps: Zeeman gap (Δ_Z) and CFSW-gap ($\Delta^{SW}_{1/3}$). In the former case the activated gap Δ equals $2\Delta_Z$ (Ref. [29]), whereas for the latter case Δ equals

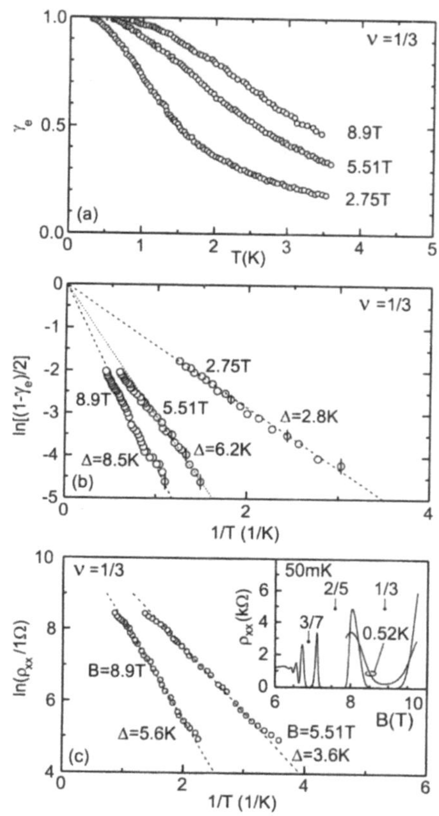

Fig. 7. The temperature dependence of the electron spin polarization (a) and Ahrenius plots (b) $ln((1-\gamma_e)/2)$ vs $1/T$ measured at $\nu = 1/3$ and $V_{SB} = 0$ V for different magnetic fields. (c) Activation behavior of ρ_{xx}, measured at $B = 5.51$ T and $B = 8.9$ T. The inset depicts the magnetic field dependence of ρ_{xx} near $\nu = 1/3$ at two different temperatures

$\Delta_{1/3}^{SW}$. The infinite density of states associated with spin wave transitions at large wave-vectors makes it plausible that $\Delta_{1/3}^{SW}$ determines the activated behavior. Indeed, the measured values of Δ are considerably larger than $2\Delta_Z$ and exhibit a nonlinear magnetic field dependence. A comparison of the measured CFSW-gap with gaps derived from activated transport under the same conditions revealed that the gap derived from $\gamma_e(T)$ is systematically larger than the transport gaps as shown in Fig. 7b and 7c. We therefore can conclude that in transport a different, smaller, gap, the CFMP gap, is measured. The dependence of the CFSW-gap measured for the 1/3-FQHE state on V_{SB} (and thus the channel width) is illustrated in Fig. 7b. The interaction energies between CFs, measured

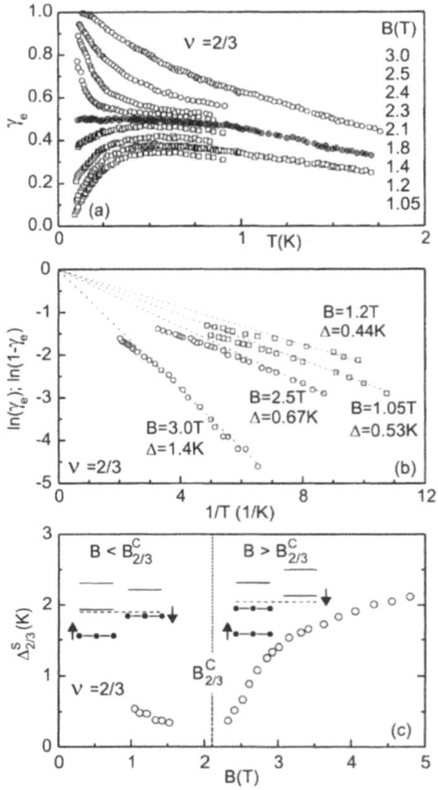

Fig. 8. (**a**) The temperature dependence of the electron spin polarization at $\nu = 2/3$ and V_{SB}=0 V for different magnetic field values in the vicinity of the spin-transition $B^{C}_{2/3} = 2.1$ T. (**b**) Ahrenius plots $ln(\gamma_e)$ vs $1/T$ (for $B < B^{c}_{2/3}$, squares) and $ln(1-\gamma_e)$ vs $1/T$ (for $B > B^{c}_{2/3}$, circles) at several magnetic fields. (**c**) Magnetic field dependence of the spin-flip activation energy for $\nu = 2/3$ around $B^{C}_{2/3}$. In the inset the CF spin-split Landau level diagrams are presented

for $\nu = 1/2$ and $\nu = 1/3$, as well as their dependence on the channel width are in good agreement with each other.

The $\nu = 2/3$ FQHE state is known for its spin transition at $B = B^{C}_{2/3} = 2.1$ T [17] from an unpolarized to a completely spin polarized ground state. Because of the close relationship between the exchange interaction energy of CFs and the spin polarization of the system, the spin transition at $B = B^{C}_{2/3}$ may be an excellent illustration of the CF-interaction. The temperature dependence of the electron spin polarization at $\nu = 2/3$ for different magnetic fields both below and above $B^{C}_{2/3}$ is shown in Fig. 8a. The behavior of $\gamma_e(T)$ at low temperatures is qualitatively different for $B > B^{C}_{2/3}$ (circles) and $B < B^{C}_{2/3}$ (squares), even though the temperature dependence in both cases is well described by a single exponential at low temperatures: $\gamma_e = 1-\exp(-\Delta/2k_BT)$ (for $B > B^C_{2/3}$) and $\gamma_e = \exp(-\Delta/2k_BT)$ (for $B < B^C_{2/3}$). These equations can be easily obtained with the aid of a CF spin-split Landau level chart as shown in the inset of Fig. 8c for $B > B^C_{2/3}$ and $B < B^C_{2/3}$. A best fit to data (Ahrenius plot analysis) with a line passing through the origin yields the CFSW energy gaps $\Delta^S_{2/3}$ given in Fig. 8b. The magnetic field dependence of the $\Delta^S_{2/3}$ in the vicinity of $B^C_{2/3}$ is presented in Fig. 8c. The abrupt enhancement of the CFSW gap at $B > B^C_{2/3}$ is obvious and demonstrates once more the importance of the exchange interaction between CFs.

3 Cyclotron Resonance of Composite Fermions

Since the kinetic energy of electrons is entirely quenched when applying a magnetic field, the CF cyclotron mass is not a renormalized version of the electron conduction band mass, but must originate entirely from electron-electron interactions [1]. The search for the CF cyclotron resonance requires substantial sophistication over conventional methods used to detect the electron cyclotron resonance, since Kohn's theorem [30] must be outwitted. It states that in a translationally invariant system, the radiation can only couple to the center-of-mass coordinate and can not excite other internal degrees of freedom. Phenomena originating from electron-electron interactions will thus not be reflected in the absorption spectrum. An elegant way to bypass this theorem is to impose a periodic density modulation to break the translational invariance. The non-zero wavevectors defined by an appropriate modulation may then offer access to the cyclotron transitions of CFs, even though they are likely to remain very weak. Therefore, the development of an optical detection scheme, that boosts the sensitivity to resonant microwave absorption by up to two orders of magnitude in comparison with traditional techniques, was a prerequisite for these studies. Furthermore, we took benefit of the accidental discovery that microwaves, already incident on the sample, set up a periodic modulation through the excitation of surface acoustic waves (SAW).

The 2DEG, patterned into a disk with 1 mm diameter [31], is placed near the end of a 16- or 8-mm short-circuited waveguide in the electric field maximum of the microwave excitation inside a He3-cryostat. At a fixed B-field, luminescence spectra with and without microwave excitation were recorded consecutively. The differential luminescence spectrum is obtained by subtracting both spectra. To improve the signal-to-noise ratio, the same procedure was repeated N times ($N = 2-20$). Subsequently, we integrated the absolute value of the averaged differential spectrum over the entire spectral range and hereafter refer to the value of this integral as the microwave absorption amplitude. The same procedure is then repeated for different values of B. To establish the trustworthiness in this unconventional procedure, we applied it to the well-known case of the electron cyclotron resonance $\omega_{cr} = eB/m^*$ (Fig. 9), m^* being the effective mass of

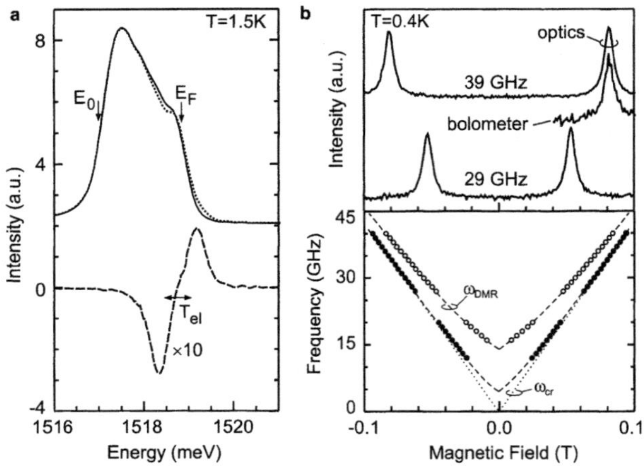

Fig. 9. (**a**) Luminescence spectrum in the presence of (dotted line) and without (solid line) a 0.05 mW microwave excitation of 18 GHz obtained on a disk-shaped 2DEG with a diameter of 1 mm and carrier density $n_S = 5.8 \times 10^{10}/\text{cm}^2$ at a magnetic field $B =$ 22 mT. (**b**) Top panel: the microwave absorption amplitude at 29 GHz and 39 GHz as a function of B-field by recording differential luminescence spectra as in (a) for 1 mT field increments at $n_S = 1.09 \times 10^{11}/\text{cm}^2$. The inset shows a conventional bolometer measurement. Bottom panel: resonance position for $n_S = 1.09 \times 10^{11}$ (open circles) and $1.1 \times 10^{10}/\text{cm}^2$ (solid circles) as a function of incident microwave frequency. The dashed lines represent the theoretical dependence of the hybrid dimensional magnetoplasma-cyclotron resonance. The dotted line corresponds to the cyclotron mode only

GaAs ($0.067\ m_0$). Due to its limited size, the sample also supports a dimensional plasma mode with a frequency ω_p, that depends on both the density n_S and diameter d of the sample, according to $\omega_p^2 = 3\ \pi^2 n_S e^2/(2m^* \epsilon_{\text{eff}} d)$. The plasma and cyclotron modes hybridize and the resulting resonance frequency of the upper dimensional magnetoplasma-cyclotron mode ω_{DMR} equals $\omega_{\text{cr}}/2 + \text{sqrt}[\omega_p^2 + (\omega_{\text{cr}}/2)^2]$ [31,32].

The optical method indeed records this mode. A comparison with the theoretical expression for ω_{DMR} yields excellent agreement. No fitting is required, since the density can be independently extracted from the luminescence at higher B-fields, where Landau levels can be resolved. At sufficiently low density, the influence of ω_p on the hybrid mode drops and one recovers at large enough B the anticipated $\omega_{\text{cr}} = eB/m^*$-dependence. Further details of the electron cyclotron resonance are discussed elsewhere [31]. Additional support for the validity of the detection method comes from a comparison with measurements based on the conventional approach using a bolometer (inset Fig. 9b). One finds not only the same resonance position, but also the same line shape. The only difference is the improved signal to noise ratio (30–100 times) for the optical detection scheme, that enables to observe the electron cyclotron resonance at microwave levels below 10 nW.

Disorder and the finite dimensions of the sample are in principle sufficient to break translational invariance as attested by the interaction of the cyclotron and dimensional plasma mode. However, they give access to internal degrees of freedom, other than the center-of-mass motion of the electrons, either at poorly defined wavevectors or too small a wavevector for practical sample sizes. Therefore, the imposition of an additional periodic density modulation, that introduces larger and well-defined wavevectors to circumvent Kohn's theorem, is desirable. Transport experiments in the Hall bar geometry disclosed that additional processing is not required, since the microwaves, already incident on the sample, concomitantly induces a periodic modulation at sufficiently high power. A clear signature is the appearance of commensurability oscillations in the magnetoresistance due to the interplay between the B-dependent cyclotron radius of electrons and the length scale imposed by the modulation [33]. Examples are displayed in Fig. 10. We understand these oscillations with the following scenario: owing to the piezoelectric properties of the AlGaAs-crystal, the radiation is partly transformed into SAW with opposite momentum, so that both energy and momentum are conserved. Reflection from cleaved boundaries of the sample then produces a standing wave with a periodicity determined by the sound wavelength. Photoexcitation creates a very poorly conducting parallel 3D-layer in the Si-doped portion of the AlGaAs-barrier and may enhance the influence of the standing acoustic waves. Carriers are collected in the nodes and affect the local density of the 2DEG. The involvement of sound waves, in this configuration, can be deduced from transport data, since from the minima we expect the modulation period to be approximately 200 and 250 nm for frequencies of 17 and 12 GHz respectively. The ratio of this period to the sound wavelength at these frequencies is 1.12 and 1.15.

Fig.11a depicts the microwave-absorption amplitude as a function of field. Apart from the strong dimensional magnetoplasma-cyclotron resonance signal at low B-field discussed above, several peaks, that scale with a variation of the density, emerge near filling factors 1, 1/2 and 1/3. Those peak positions associated with $\nu = 1$ and 1/3 remain fixed when tuning the microwave frequency and are ascribed to the heating induced by non-resonant absorption of microwave power. In contrast, the weak maxima surrounding filling factor 1/2 readily respond to a change in frequency as illustrated in Fig. 11b. They are symmetrically arranged around half filling and their splitting grows with frequency. The B-dependence is summarized in Fig.11c for two densities. To underline the symmetry, the effective CF- field B^* is chosen as the abscissa. The linear relationship between frequency and field extrapolates to zero at vanishing B^*. We do not expect a deviation at small B^* due to a plasma-like contribution as in Fig. 9c. Excitations for the 1/3, 2/5, 3/7 and other fractional quantum Hall states exhibit in numerical simulations no magnetoplasmon-like linear contribution to the dispersion at small values of k [19]. We conclude that the resonance in Fig. 11 features the long searched signature for cyclotron resonance of CFs.

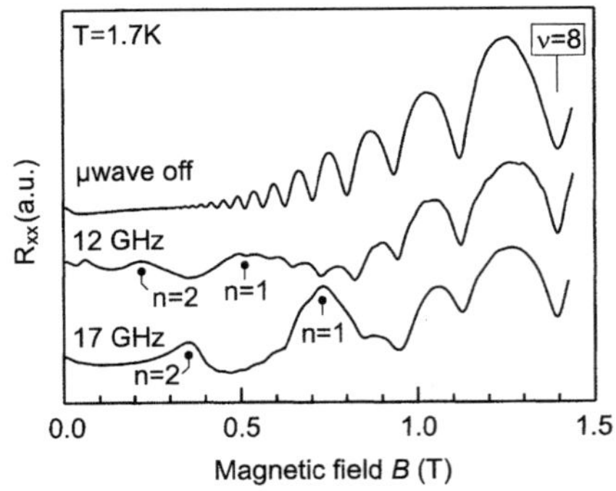

Fig. 10. Magnetotransport data without (top curve) and under 0.1 mW of microwave radiation at 12 GHz (middle curve) and 17 GHz (bottom curve). Curves are offset for clarity. Besides the well-known Shubnikov-de Haas oscillations, additional magnetoresistance oscillations appear under microwave radiation

Geometric resonances (GR), as they occur in transport at low fields due to the density modulation (Fig. 10), are excluded as an alternative interpretation for the observed features on the following grounds: (i) in the optical data, only the electron cyclotron resonance peak is observed. Contrary to optical quantities, transport is also sensitive to semi-classical phenomena non related to changes in the density of states; (ii) even if the 2DEG condenses in a FQHE-state and the chemical potential is located in a gap, the resonance peaks surrounding $\nu = 1/2$ occur (Fig. 11c). Commensurability effects are not observable in this regime; (iii) the observation of GRs requires that the density modulation is temporally static on the time scale for the CFs to execute their cyclotron orbit. For electrons at low fields this condition is met and accordingly transport displays GRs. For the anticipated enhanced mass of CFs, this condition is violated; (iv) Analogous resonance peaks were also detected for the higher order CFs around $\nu = 1/4$. Since this CF metallic state is characterized by the same Fermi wavevector GRs would show up at the same distance from $\nu = 1/4$ as they do at $\nu = 1/2$. The observed peaks are located at different positions rendering a commensurability picture untenable (see below Fig. 12b).

In contrast to electron cyclotron resonance, the intensity of the CF cyclotron resonance is a strong non-linear function of microwave-power (Fig. 12a). Moreover, its observability at only high power correlates with the first appearance of commensurability oscillations. The drop in intensity at even higher power is most probably due to heating. The intensity decreases to zero at temperatures above 0.7 K, whereas the electron cyclotron resonance persists up to $T > 2$ K. The slope of the CF cyclotron frequency as a function of B^* in Fig. 11c defines the cyclotron mass $m_{\mathrm{cr}}^{\mathrm{cf}}$. This mass is set by the electron-electron interaction scale, so that a square root behavior on density or B-field is fore-

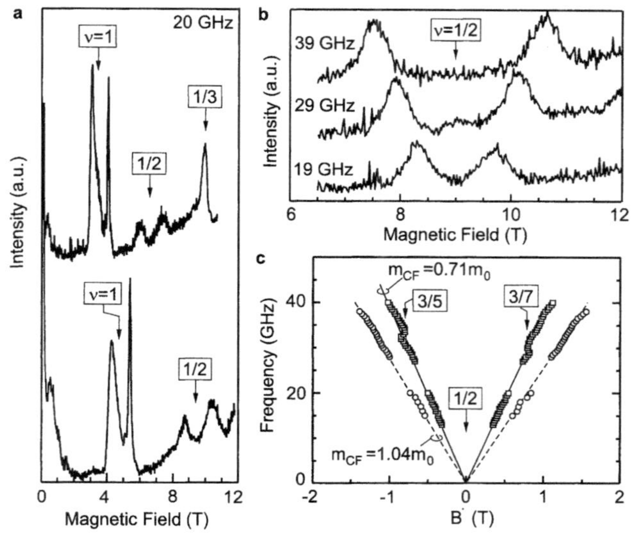

Fig. 11. (**a**) Microwave absorption amplitude measured at high magnetic fields for $n_S = 0.81 \times 10^{11}$ and $1.15 \times 10^{11}/\mathrm{cm}^2$ and frequency of 20 GHz ($T = 0.4$ K). (**b**) Microwave absorption amplitude in the vicinity of $\nu = 1/2$ measured at three different frequencies and $n_S = 1.09 \times 10^{11}/\mathrm{cm}^2$. (**c**) Position of the CF cyclotron mode as a function of the effective magnetic field B^* measured for $n_S = 1.09 \times 10^{11}/\mathrm{cm}^2$ (circles) and $n_S = 0.59 \times 10^{11}/\mathrm{cm}^2$ (squares). The CF effective mass equals 1.04 m_0 and 0.71 m_0 respectively

casted from a straightforward dimensional analysis [1]. Numerical calculations predict $m_{\mathrm{cr}}^{\mathrm{cf}}/m_0 = 0.079 \times \sqrt{(}B[\mathrm{T}])$ for an ideal 2DEG, without including Landau level mixing or finite width contributions [18]. The data, shown in Fig. 12b, confirm qualitatively the strong enhancement in comparison with the electron mass (more than 10 times). However a fit to the square root dependence requires a prefactor that is four times larger. Previously reported mass values based on activation energy gap measurements [7,37] must be distinguished from the cyclotron mass. The former corresponds to the limit of infinite momentum, whereas here k approaches zero. Moreover, activation gaps can only be extracted at well-developed fractional quantum Hall states and their accurate determination suffers from disorder induced broadening. These and other limitations have been discussed in Ref. [37] for example. The technique discussed here can be applied at arbitrary filling factor values.

In summary, the introduction of judicious fictitious entities occasionally allows to interpret, in a single particle picture, properties which are otherwise difficult to discuss in strongly interacting many-body systems. We can then take the customary physical approach, using concepts and representations which formerly could only be applied to systems with weak interactions, and yet still capture

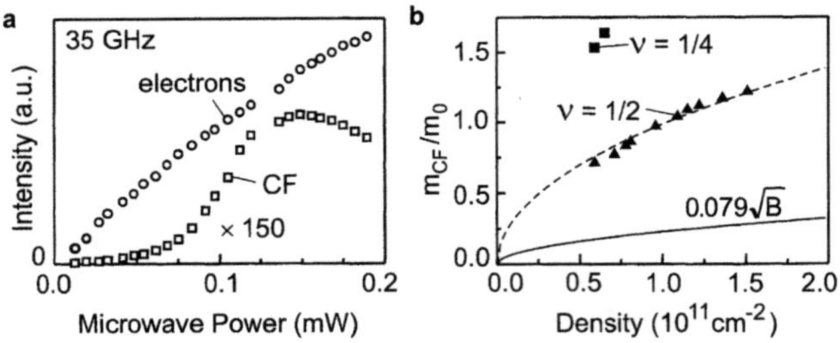

Fig. 12. (**a**) Incident microwave power dependence of the amplitude at the cyclotron resonance (circles for electrons, squares for CFs). (**b**) Dependence of the CF effective mass near $\nu = 1/2$ on the carrier density n_S (solid triangles). The dashed line is a square root fit to the data. The solid curve is the prediction from theory reported in Ref. [18]. Analogous resonance peaks have also been detected around $\nu = 1/4$. The corresponding effective mass values are indicated as solid squares for two different densities

the essential physics. A noticeable recent example of this approach occurs in the physical properties of a two dimensional electron system subjected to a strong perpendicular magnetic field B. They are governed by electron-electron interactions that bring about the fractional quantum Hall effect. Composite fermions, that do not experience the external magnetic field but a reduced effective magnetic field B^*, were identified as apposite quasi-particles that simplify our understanding of the FQHE. They precess, like electrons, along circular cyclotron orbits, with a diameter determined by B^* rather than B. The frequency of their cyclotron motion remained hitherto enigmatic, since the effective mass is no longer related to the band mass of the original electrons and entirely originates from electron-electron interactions. Here, we demonstrated the enhanced absorption of a microwave field that resonates with the frequency of their circular motion. From this cyclotron resonance, we derive a composite fermion effective mass that varies from 0.7 to 1.2 times the electron mass in vacuum as their density is tuned from $0.6 \times 10^{11}/\text{cm}^2$ to $1.5 \times 10^{11}/\text{cm}^2$.

The financial support from the Volkswagen Stiftung, Russian Fund of Fundamental Research and INTAS-grant is gratefully acknowledged.

References

1. B.I. Halperin et al.: Phys. Rev. **B47**, 7312 (1993)
2. R.R. Du et al.: Phys. Rev. Lett. **70**, 2944 (1993) Phys. Rev. Lett. **75**, 3926 (1995)
3. R.L. Willett et al.: Phys. Rev. Lett. **71**, 3846 (1993)
4. V.J. Goldman et al.: Phys. Rev. Lett. **72**, 2065 (1994)
5. J.H. Smet et al.: Phys. Rev. Lett. **77**, 2272 (1996)
6. R.L. Willett et al.: Surf. Sci. **361/362**, 38 (1996)

7. R.R. Du et al.: Phys. Rev. Lett. **70**, 2944 (1993) Phys. Rev. Lett. **75**, 3926 (1995)
8. I.V. Kukushkin et al.: Phys. Rev. Lett. **72**, 736 (1994)
9. D.R. Leadley et al.: Phys. Rev. Lett. **72**, 1906 (1994)
10. I.V. Kukushkin et al.: Phys. Rev. **B55**, 10607 (1997)
11. I.V. Kukushkin et al.: Phys. Rev. **B40**, 7788 (1989)
12. B.I. Halperin: Helv. Phys. Acta **56**, 75 (1983)
13. T. Chakraborty et al.: Phys. Rev. B **29**, 7032 (1984)
14. J.P. Eisenstein et al.: Phys. Rev. Lett. **62**, 1540 (1989)
15. R.G. Clark et al.: Phys. Rev. Lett. **62**, 1536 (1989)
16. L.W. Engel et al.: Phys. Rev. B **45**, 3418 (1992)
17. I.V. Kukushkin et al.: Phys. Rev. B **40**, 7788 (1989) Phys. Rev. B **55**, 10607 (1997) Phys. Rev. B **60**, 2554 (1999) Phys. Rev. Lett. **82**, 3665 (1999)
18. K. Park, J.K. Jain: Phys. Rev. Lett. **80**, 4237 (1998)
19. J.K. Jain: Phys. Rev. Lett. **63**, 199 (1989) Adv. Phys. **41**, 105 (1992)
20. J.K. Jain, R.K. Kamilla: in 'Composite Fermions' p.3 (1998)
21. G. Murthy: Phys. Rev. B **60**, 13702 (1999)
22. T. Nakajima, H. Aoki, Phys. Rev. Lett. **73**, 3568 (1994)
23. S.M. Girvin et al.: Phys. Rev. Lett. **54**, 581 (1985)
24. A. Pinczuk et al.: Phys. Rev. Lett. **70**, 3983 (1993) Phys. Rev. Lett. **68**, 3623 (1992)
25. H.D.M. Davies et al.: Phys. Rev. Lett. **78**, 4095 (1997)
26. P. Khandelwal et al.: Phys. Rev. Lett. **81**, 673 (1998)
27. I.V. Kukushkin et al.: JETP Lett. **70**, 722 (1999)
28. F.C. Zhang, S. Das Sarma: Phys. Rev. B **33**, 2903 (1986)
29. A.H. MacDonald, J.J. Palacios, Phys. Rev. B **58**, R10171 (1998)
30. W. Kohn: Phys. Rev. **123**, 1242 (1961)
31. S.I. Gubarev et al.: JETP Lett. **72**, 324 (2000)
32. S.J. Allen et al.: Phys. Rev. **B 28**, 4875 (1983)
33. R.R. Gerhardts et al.: Phys. Rev. Lett. **62**, 1173 (1989)
34. J.K. Jain, R.K. Kamilla, Composite Fermions, ed. O. Heinonen, World Scientific Publishing, 1 (1998)
35. K. Park, J.K. Jain, Phys. Rev. Lett. **80**, 4237 (1998)
36. R.R. Du et al.: Phys. Rev. Lett. **70**, 2944 (1993) Phys. Rev. Lett. **75**, 3926 (1995)
37. R.L. Willett: Composite Fermions, ed. O. Heinonen, World Scientific Publishing, 349 (1998)

Stripe and Bubble Phases in Quantum Hall Systems

Michael M. Fogler

Department of Physics, Massachusetts Institute of Technology,
77 Massachusetts Avenue, Cambridge, MA 02139, USA

Abstract. We present a brief survey of the charge density wave phases of a two-dimensional electron liquid in moderate to weak magnetic fields where several higher Landau levels are occupied. The review follows the chronological development of this new and emerging field: from the ideas that led to the original theoretical prediction of the novel ground states, to their dramatic experimental discovery, to the currently pursued directions and open questions.

1 Historical Background

Until recently, the quantum Hall effect research effort has been focused on the case of very high magnetic fields where electrons occupy only the lowest and perhaps, also the first excited Landau levels (LL). Investigation of the weak magnetic field regime where higher LLs are populated, was not considered a pressing matter because no particularly interesting quantum features could be discerned in the magnetotransport data. The experimental situation has changed around 1992, when extremely high purity two-dimensional (2D) electron systems became available. At low temperatures, $T < 30$ mK, such samples would routinely demonstrate very deep resistance minima at integral filling fractions down to magnetic fields of the order of a tenth of a Tesla [1]. This indicated that the quantum Hall effect could persist up to very large LL indices, such as $N \sim 100$, and called for the theoretical treatment of the high LL problem. Although the integral quantum Hall effect could be explained without invoking electron-electron interaction, two other experimental findings strongly suggested that the interaction *is* important at large N. One was the prominent enhancement of the bare electron g-factor [2] and the other was a pseudogap in the tunneling density of states [3]. Surprisingly, *no* interaction-induced fractional quantum Hall effect has ever been observed at higher N in contrast to the case of the lowest and the first excited LLs ($N = 0$ and 1). An effort to understand this puzzling set of facts led A. A. Koulakov, B. I. Shklovskii, and the present author to the theory of charge density wave phases in partially filled $N \geq 2$ LLs [4,5]. When this theory received a dramatic experimental support [6,7], a broad interest to the high LL physics has emerged. Below we give a brief review of this new and exciting field. Some of the ideas presented here are published for the first time. For previous short reviews on the subject see Refs. [8,9].

2 Landau Quantization in Weak Magnetic Fields

Consider a 2D electron system with the areal density n in the presence of a transverse magnetic field B. For the case of Coulomb interaction, at zero temperature and without disorder, the properties of such a system are determined by exactly three dimensionless parameters: r_s, ν, and $E_Z/\hbar\omega_c$. The first of these, $r_s = (\pi n a_B^2)^{-1/2}$, measures the average particle distance in units of the effective Bohr radius $a_B = \hbar^2\kappa/me^2$. The properties of the electron gas are very different at large and small r_s. In these notes will focus exclusively on the case $r_s \lesssim 10$. This is roughly the condition under which the electron gas in zero magnetic field behaves as a Fermi-liquid [10]. As we discuss below, the system is no longer a Fermi-liquid at any finite B; however, the basic structure of LLs separated by the gaps $\hbar\omega_c$, where $\omega_c = eB/mc$ is the cyclotron frequency, survives at arbitrary low B. In this situation, the second dimensionless parameter, $\nu = 2\pi l^2 n$, specifies how many LL subbands are occupied. Here $l = (\hbar c/eB)^{1/2}$ is the magnetic length. The lower LLs are fully occupied, while the topmost level is, in general, partially filled. Therefore, $\nu = 2N + \nu_N$, where the factor of two accounts for the spin degree of freedom and ν_N is the filling fraction of the topmost (Nth) LL, $0 < \nu_N < 2$ (see the sketch in Fig. 1a).

The remaining dimensionless parameter $E_Z/\hbar\omega_c$ introduced above is the ratio of the Zeeman and the cyclotron energy. It affects primarily the dynamics of the spin degree of freedom, which is beyond the scope of these notes. Suffices to say that in the ground state the topmost Nth LL is thought to be fully spin-polarized for $N > 0$ [11] with a sizeable spin gap. In the important case of GaAs, the spin gap greatly exceeds E_Z due to many-body effects.

Because of the spin and the cyclotron gaps, the low-energy physics is dominated by the electrons residing in the single spin subband of a single (topmost) LL. All the other electrons play the role of an effective dielectric medium, which merely renormalizes the interaction among the "active" electrons of the Nth LL. This elegant physical picture was first put forward in an explicit form by Aleiner and Glazman [12].

The validity of such a picture in weak magnetic fields is certainly not obvious. Naively, it seems that as B and $\hbar\omega_c$ decrease, the LL structure should eventually be washed out by the electron-electron interaction. The following reasoning shows that this does not occur (see also Ref. [12] for somewhat different arguments). Let us divide all the interactions into three groups: (a) intra-LL interaction within Nth level, (b) interaction between the electrons of Nth level and its near neighbor LLs, and (c) interaction between Nth and remote LLs (with indices N' significantly different from N). The last group of interactions is characterized by frequencies much larger than ω_c. It is not sensitive to the presence of the magnetic field and leads only to Fermi-liquid renormalizations of the quasiparticle properties. Interactions within the groups (a) and (b) have roughly the same matrix elements but the latter are suppressed because of the cyclotron gap. Hence, it is the interactions among its own quasiparticles that are the most "dangerous" for the existence of a well-defined Nth LL. It is crucial that the intra-LL interaction energy scale does not exceed the typical value of

$$E_{\text{ex}} \sim 0.1 e^2/\kappa R_{\text{c}}, \tag{1}$$

where $R_{\text{c}} = \sqrt{2N+1}\, l$ is the classical cyclotron radius [12,4]. The ratio $E_{\text{ex}}/\hbar\omega_{\text{c}} \sim 0.1 r_{\text{s}}$ is B-independent; thus, there is a good reason to think that the validity domain of the proposed single-Landau-level approximation extends down to arbitrary small B's and, in fact, is roughly the same as that of the Fermi-liquid ($r_{\text{s}} \lesssim 10$). This is certainly borne out by all available magnetoresistance data [1,2,6,7]. Henceforth we focus exclusively on the quasiparticles residing at the topmost LL.

The inequality $E_{\text{ex}} < \hbar\omega_{\text{c}}$ means that the cyclotron motion is the fastest motion in the problem, and so on the timescale at which the ground-state correlations are established, quasiparticles behave as clouds of charge smeared along their respective cyclotron orbits, see Fig. 1b. The only low-energy degrees of freedom are associated with the guiding centers of such orbits. In the ground state they must be correlated in such a way that the interaction energy is the lowest. This prompts a quasiclassical analogy between the partially filled LL and a gas of interacting "rings" with radius R_{c} and the areal density $\nu_N/(2\pi l^2)$. Note that for $\nu_N > 1/N$ the rings overlap strongly in the real space.

Strictly speaking, the guiding center can not be localized a single point, and so our analogy is not precise. However, the quantum uncertainty in its position is of the order of l. At large N, where $l \ll R_{\text{c}}$, the proposed analogy becomes accurate and useful. For example, it immediately clarifies the physical meaning of E_{ex} as a characteristic interaction energy of two overlapping rings.

Technically, the high LL problem is equivalent to the more studied $N = 0$ case if the bare Coulomb interaction $\tilde{v}_0(q)$ is replaced by the renormalized interaction

$$\tilde{v}(q) = \frac{\tilde{v}_0(q)}{\epsilon(q)} \left[L_N \left(\frac{q^2 l^2}{2} \right) \right]^2, \tag{2}$$

where $\epsilon(q)$ is the dielectric constant due to the screening by other LLs [13,12] and the bracketed expression compensates for the difference in the form-factor

$$F_N(q) = L_N \left(\frac{q^2 l^2}{2} \right) \mathrm{e}^{-q^2 l^2/4} \tag{3}$$

of the cyclotron orbit at Nth and at the lowest LLs, with $L_N(z)$ being the Laguerre polynomial. In the next section we will discuss the consequences of having such an unusual interaction.

3 Charge Density Wave Instability

The mean-field treatment of a partially filled LL amounts to the Hartree-Fock approximation, first examined in the present context by Fukuyama et al. It is worth pointing out the differences between their paper [14] and our own work [4]

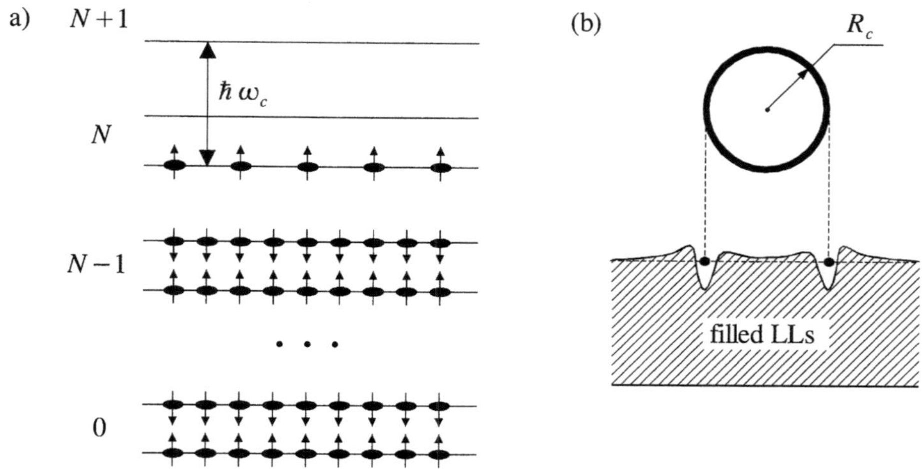

Fig. 1. (a) Landau levels. Darkened ellipses symbolize electrons and arrows — their spins (b) A quasiparticle at the Nth LL viewed as a ring-shaped object immersed into a medium formed by the filled lower LLs

reviewed below in this section. The pioneering work of Fukuyama et al. [14] appeared in 1979. A few years later, after the discoveries of integral and fractional quantum Hall effects, it became clear that for the exception of dilute limit, the Hartree-Fock approximation is manifestly incorrect for the lowest LL case [15]. By 1995 when we started to work on the high LL problem, Ref. [14] has been effectively shelved away. In contrast to Ref. [14], who did not try to assess the validity of the Hartree-Fock approximation, our theory of a partially filled high LL [4] was based on this kind of approximation because it *is* the correct tool for the job. This point is elaborated further in Sect. 5. Another important difference from Ref. [14] is a parametric dependence of the wavevector q_* of the CDW instability on the magnetic field: we find $q_* \propto B^{-1}$ instead of $\propto B^{-1/2}$ in the theory of Fukuyama et al. Finally, the physical picture of ring-shaped quasiparticles that guided our intuition is quite novel and applies only for high LLs.

Within the Hartree-Fock approximation, the free energy of the system is given by

$$\mathcal{F}_{\mathrm{HF}} = \frac{1}{4\pi l^2}\sum_q \tilde{u}_{\mathrm{HF}}(q)|\langle\tilde{\Delta}(q)\rangle|^2 + k_{\mathrm{B}}T\sum_X \langle n_X \ln n_X + (1-n_X)\ln(1-n_X)\rangle, \quad (4)$$

where $\tilde{\Delta}(q) = (2\pi l^2/L_x L_y)\sum_X \mathrm{e}^{-\mathrm{i}q_x X} a^\dagger_{X+q_y l^2/2} a_{X-q_y l^2/2}$ are guiding center density operators, L_x and L_y are system dimensions, $a^\dagger_X$ (a_X) are creation (annihilation) operators of Landau basis states $|X\rangle$, and $n_X = a^\dagger_X a_X$ are their occupation numbers subject to the constraint $\sum_X \langle n_X \rangle = L_x L_y \nu_N/(2\pi l^2)$. Here and below we assume that $0 < \nu_N < 1$ because the states with $1 < \nu_N < 2$ are the particle-hole transforms of the states with $2 - \nu_N$ and do not require a special consideration. The Hartree-Fock interaction potential is defined by $\tilde{u}_{\rm HF}(q) = (1 - \delta_{q,0})\tilde{u}_{\rm H}(q) - \tilde{u}_{\rm F}(q)$, where

$$\tilde{u}_{\rm H}(q) = \tilde{v}(q) {\rm e}^{-q^2 l^2/2}, \quad \tilde{u}_{\rm ex}(q) = 2\pi l^2 u_{\rm H}(q l^2) \tag{5}$$

are its direct and exchange components (tildes denote Fourier transforms) [4]. Their q-dependence is illustrated in Fig. 2.

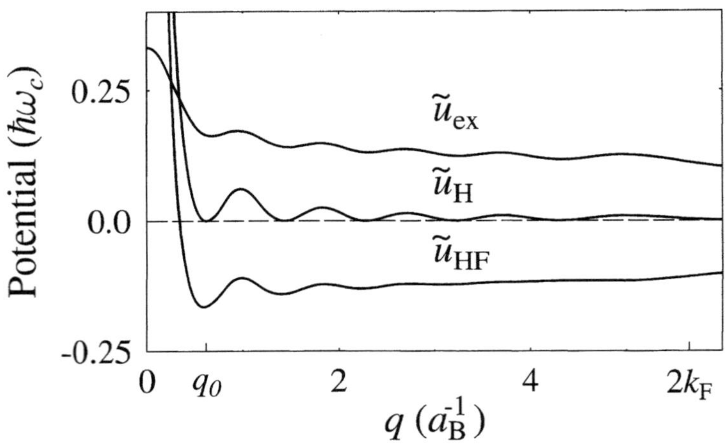

Fig. 2. Direct, exchange, and the total Hartree-Fock potentials in q-space for $N = 5$ and $r_s = 0.5$. [Reproduced with permission from Fig. 2 of Ref. [4] (a)]

The inherent feature of $\tilde{u}_{\rm HF}(q)$ is a global minimum at

$$q_* \approx 2.4/R_{\rm c}, \tag{6}$$

where it is negative, $\tilde{u}_{\rm HF}(q_*) < 0$. Within the Hartree-Fock theory, it leads to a charge density wave (CDW) formation at low enough temperatures T [14]. Indeed, at high temperatures the entropic term dominates and the equilibrium state is a uniform uncorrelated liquid with $\langle n_X \rangle = \nu_N$ and $\langle \tilde{\Delta}(q) \rangle = \delta_{q,0}\nu_N$. At lower T, it is more advantageous to forfeit some entropy but gain some interaction energy by creating a guiding center density modulation $\langle \tilde{\Delta}_q \rangle \neq 0$ with wavevector $q = q_*$.

Since the quasiparticles are extended ring-shaped objects, the actual quasiparticle density modulation in any of our states is given by the product of the amplitude $\tilde{\Delta}(q)$ of the guiding center density wave and the cyclotron orbit form-factor:

$$\tilde{\rho}(q) = \tilde{\Delta}(q) F_N(q). \tag{7}$$

The physical electric charge modulation in the system is further suppressed by the additional factor of $\epsilon(q)$ due to the screening by the lower LLs. A peculiarity of the $N \gg 1$ case is that q_* is very close to the first zero q_0 of $\tilde{u}_{\mathrm{H}}(q)$, which it inherits from the form-factor $F_N(q)$[1]. On the one hand, this means that the physical electric charge modulation is always rather small — a few percent in realistic experimental conditions. On the other hand, it explains why this instability develops in the first place. Indeed, usually the direct electrostatic interaction is repulsive and dominates over a weak attraction due to exchange. In our system $\tilde{u}_{\mathrm{H}}(q)$ vanishes at the "magic" wavevector q_0 because no charge density is induced, $\rho(q_0) = 0$. As a result, the exchange part dominates and gives rise to a range of q's around q_0 where the net effective interaction $\tilde{u}_{\mathrm{HF}}(q)$ is attractive, which leads to the instability.

The nodes of $F_N(q)$ responsible for the vanishing of $\rho(q)$ exist for a purely geometric reason that the quasiparticle orbitals are extended objects of a specific ring-like shape. The size of the orbitals is uniquely defined by the total density and the total filling factor. These two facts make the position of the global minimum q_* very insensitive to approximations contained in the Hartree-Fock approach as well as many microscopic details, e.g., the functional form of $\epsilon(q)$, thickness of the 2D layer, which affects $\tilde{v}_0(q)$, etc. At asymptotically large N where the rings are very narrow, $F_N(q)$ is closely approximated by a Bessel function $J_0(qR_{\mathrm{c}})$; hence, (6). Surprisingly, (6) is quite accurate even at $N \sim 1$.

The mean-field transition temperature $T_{\mathrm{c}}^{\mathrm{mf}}$ can be estimated as the point where the stability criterion $1/\epsilon_{\mathrm{tot}}(q) < 1$ of the uniform liquid state is first violated. Here $\epsilon_{\mathrm{tot}}(q)$ is the total dielectric function, including both the lower and the topmost LLs. A simple derivation [16] within the (time-dependent) Hartree-Fock approximation gives

$$\epsilon_{\mathrm{tot}}(q) = \epsilon(q) \left\{ 1 + \tilde{u}_{\mathrm{H}}(q) \left[\frac{2\pi l^2 k_{\mathrm{B}} T}{\nu_N (1 - \nu_N)} - \tilde{u}_{\mathrm{ex}}(q) \right]^{-1} \right\}. \tag{8}$$

The T-dependence of this function at $q = q_*$ is sketched in Fig. 3.

The stability criterion leads to the estimate

$$k_{\mathrm{B}} T_{\mathrm{c}}^{\mathrm{mf}} = \frac{\nu_N (1 - \nu_N)}{2\pi l^2} |\tilde{u}_{\mathrm{HF}}(q_*)|. \tag{9}$$

For $r_{\mathrm{s}} \sim 1$, $N \gg 1$, and $\nu_N = 1/2$, it can be approximated by

$$k_{\mathrm{B}} T_{\mathrm{c}}^{\mathrm{mf}} (\nu_N = 1/2) \approx 0.02 \hbar \omega_{\mathrm{c}}. \tag{10}$$

Compared to the expression originally given in Ref. [4](b), a certain $1/N$ term is omitted here, out of precaution that the Hartree-Fock approximation, which

[1] In contrast, in the $N = 0$ case studied by Fukuyama et al., $\tilde{u}_{\mathrm{H}}(q)$ does not have nodes.

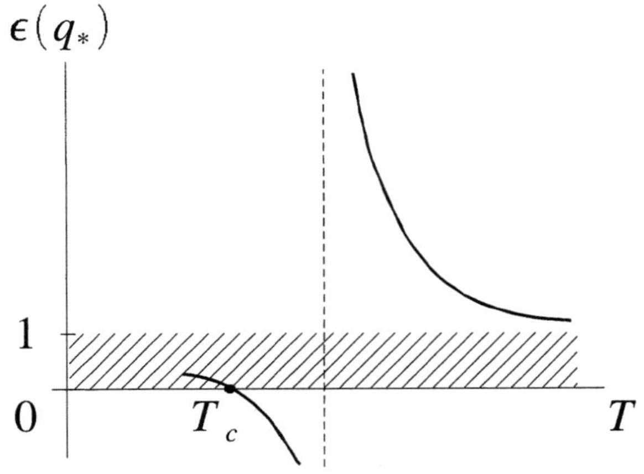

Fig. 3. T-dependence of the dielectric function of the liquid state. The unstable region is hatched

is valid in the large-N limit, may not have enough accuracy to describe $1/N$-corrections. This caveat should always be kept in mind when using numerical estimates of $k_B T_c^{\rm mf}$, such as those reported in Ref. [17]. For the typical experimental situation [6,7], (9) gives $T_c^{\rm mf} \sim 1$K.

4 Mean-Field Phase Diagram

4.1 Charge Density Wave Transition

A more systematic way to study the CDW transition is via a Landau expansion of the free energy F in powers of the order parameters $\langle \tilde{\Delta}(\mathbf{q})\rangle$ where $\mathbf{q}$ are restricted to the locus of the soft modes, $|\mathbf{q}| = q_*$:

$$\mathcal{F} = \sum_{n=2}^{\infty} a_n(T) \sum_{\mathbf{q}_1+\mathbf{q}_2+\ldots+\mathbf{q}_n=0} \prod_{i=1}^{n} \langle \tilde{\Delta}(\mathbf{q}_i)\rangle . \tag{11}$$

Note that in the quasiclassical large-N limit, the order parameter $\langle \Delta(\mathbf{r})\rangle$ is proportional to the local filling factor, $\langle \Delta(\mathbf{r})\rangle = \nu_N(\mathbf{r})/2\pi l^2$.

The above linear stability criterion is equivalent to the condition $a_2 > 0$. At $\nu_N = \frac{1}{2}$ where the cubic term vanishes by symmetry, $a_3 = 0$, the Landau theory's estimate for T_c coincides with (9). It predicts a second-order transition [14,18], which occurs by a condensation of a single pair of harmonics, whose direction is chosen spontaneously, e.g., $\mathbf{q} = \pm q_* \hat{\mathbf{x}}$. The resultant low-temperature state is a unidirectional CDW or the *stripe phase*. Away from the half-filling, $a_3 \neq 0$. Hence, at $\nu_N \neq \frac{1}{2}$ the transition is of the first order, takes place at a temperature somewhat higher than predicted by (9), and is from the uniform liquid into a CDW phase with the triangular lattice symmetry [14,18], the *bubble phase*, see Fig. 4 (left).

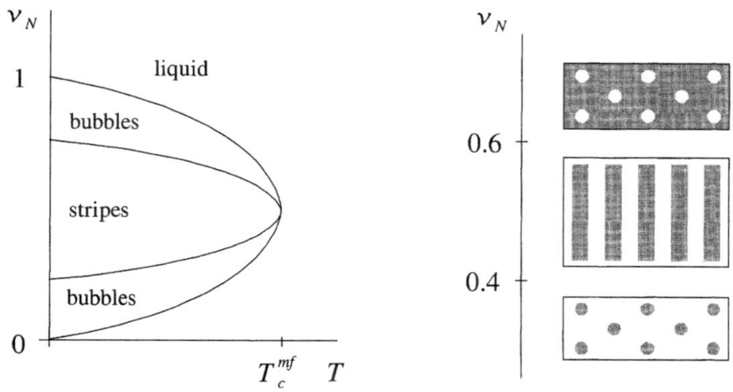

Fig. 4. Left: Mean-field phase diagram. Right: Guiding center density domain patterns at $T = 0$. Shaded and blank areas symbolize filled and empty regions, respectively

Near its onset, the CDW order brings about only a small modulation of the local filling factor, so that the topmost LL remains partially filled everywhere. As T decreases, the amplitude of the guiding center density modulation increases and eventually forces expulsion of regions with partial LL occupation. The system becomes divided into (i) depletion regions where $\langle \Delta(\mathbf{r}) \rangle = 0$ and local filling fraction is equal to $2N$, and (ii) fully occupied areas where $\langle \Delta(\mathbf{r}) \rangle = (2\pi l^2)^{-1}$ and local filling fraction is equal to $2N + 1$ (however, see Ref. [4] for a discussion of truly small r_s). At these low temperatures the *bona fide* stripe and bubble domain shapes become evident, see Fig. 4 (right).

4.2 Stripe to Bubble Transition

Near T_c, the Landau theory [18] predicts the stripe-bubble transition to be of the first order. This seems to be the case at $T = 0$ as well, at least at large N, where the this transition occurs at $\nu_N \approx 0.39$ [4], see Fig. 5a. In systems with only short-range interactions a density-driven first order transition is accompanied by a global phase separation. An example is the usual gas-liquid transition. The densities n_s and n_b of the two co-existing phases are determined by Maxwell's tangent construction, Fig. 5b.

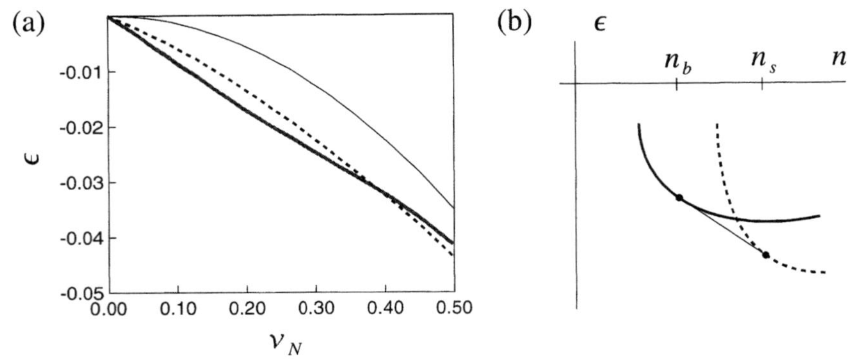

Fig. 5. (**a**) The energy density ϵ in units of $\hbar\omega_\mathrm{c}/2\pi l^2$ as a function of ν_N for the bubbles (thick solid line), the stripes (dashed line), and the uniform liquid (thin solid line) for $r_\mathrm{s} = 1$, $N \gg 1$, and $T = 0$. (**b**) Schematics of the conventional tangent construction. As explained in the main text, it is too crude to account for the specifics of the long-range Coulomb interaction. For example, for the graph on the left, such a construction would give a vanishing n_b and the maximum possible n_s, which is misleading

In the present case the long-range Coulomb interaction changes the situation drastically. The macroscopic phase separation into two phases of different charge density is forbidden by an enormous Coulomb energy penalty. Only a phase separation on a finite lengthscale, i.e., domain formation may occur. Since stripes and bubbles are two the most common domain shapes in nature [20], it opens an intriguing possibility that bubble- or stripe-shaped domains of the stripe phase inside of the bubble phase may appear, i.e., the "superbubbles" (Fig. 6) or the "superstripes". Their size would be determined by the competition between the Coulomb energy and the domain wall tension γ. The superbubbles would have a diameter

$$a \sim \frac{1}{n_\mathrm{s} - n_\mathrm{b}} \left(\frac{\gamma\kappa}{\mathrm{e}^2}\right)^{1/2} \tag{12}$$

and increase the net energy density by $\sim \gamma p/a$, p being the fraction of the minority phase. It turns out that this additional energy cost shrinks the range of the phase co-existence considerably compared to that in the conventional tangent construction. It may totally preclude the phase co-existence if

$$\frac{\mathrm{e}^2\gamma}{\kappa} > (n_\mathrm{s} - n_\mathrm{b})^2 \frac{d^2 f_\mathrm{s}}{dn^2} \frac{d^2 f_\mathrm{b}}{dn^2}\,, \tag{13}$$

where $f_{\mathrm{s(b)}}(n)$ is the free energy density of stripes (bubbles).

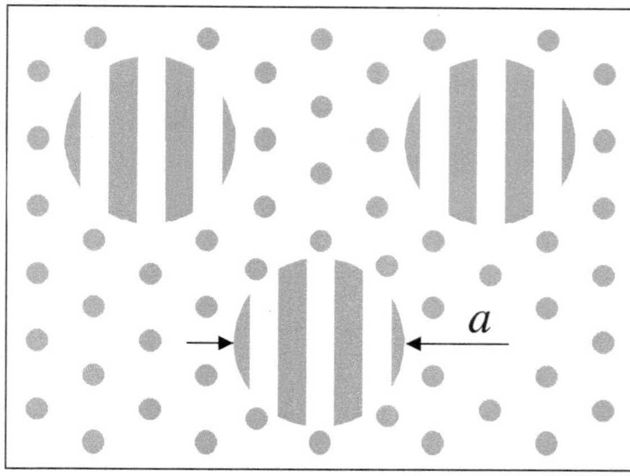

Fig. 6. A cartoon of the conjectured superbubble phase

Incidentally, these considerations are also relevant for the main transition from the uniform state into the CDW one at $T = T_c$. We can think of the primary stripe and bubble phases as examples of a frustrated phase separation. At $\nu_N = 1/2$ the conventional tangent construction would predict $k_B T_c = E_{ex}/4$, whereas the actual (mean-field) transition temperature is lower [(10)] because of the extra energy density associated with the edges of the stripes.

Returning to the case of superbubbles, we estimate $d^2 f_{s(b)}/dn^2 \sim f_{s(b)}/n^2 \sim E_{ex}/n^2$ and $\gamma \sim E_{ex} R_c n_s$ at $T = 0$. The criterion (13) becomes $(n_s - n_b)/n_s \lesssim R_c \sqrt{n_s} \sim \sqrt{N}$. Thus, we can be certain that the superstructures *do not appear* at high LLs. The cases of small N or high temperatures require further study [16].

4.3 Transitions Caused by Particle Discreteness

With the periodicity of the bubble phase set by the preferred wavevector q_*, the area of unit cell of the bubble lattice is equal to $S_0 = 2\sqrt{3}\,\pi^2/q_*^2$. The number of particles per bubble is therefore $M = S_0 \nu_N / 2\pi l^2$. Using (6) we obtain [4]

$$M \approx 3\nu_N N, \quad N \gg 1. \tag{14}$$

It is natural to ask whether this formula should be taken literally, even when it predicts a nonintegral M. Strictly speaking, CDWs with a fractional number of particles per unit cell are not ruled out. However, we choose to ignore such an exotic possibility because early Hartree-Fock studies for the lowest Landau level [19] concluded that only the phases with integer-valued M are stable. In our own numerical studies only integral M were examined. The results are reproduced in Fig. 7 (see Ref. [5] for details). We found that the ground state value of M never deviates from the prediction of (14) by more than unity at $N \leq 10$ and all ν_N. This confirms the robustness of the optimal period q_*. It also suggests a practical rule for calculating the optimal M: one should evaluate the right-hand side of (14) and round it to the nearest integer.

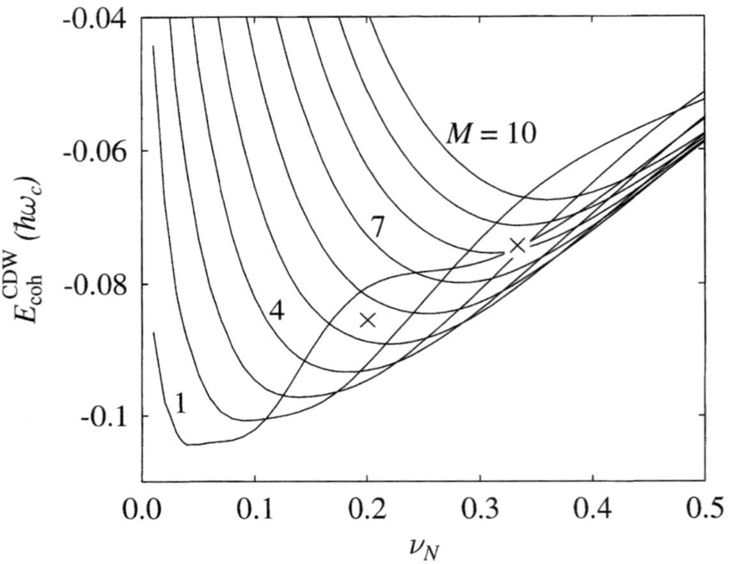

Fig. 7. Cohesive energy $E_{\rm coh}^{\rm CDW}$ in a set of bubble phases with different number of particles per bubble M. $E_{\rm coh}^{\rm CDW}$ is defined as the interaction energy per particle $E_{\rm int}$ relative to the uncorrelated liquid [where $E_{\rm int} = -(\nu_N/2)E_{\rm ex}$]. Calculation parameters: $N = 5$, $r_{\rm s} = \sqrt{2}$, $T = 0$. The crosses represent the Laughlin liquid energies. (Reproduced with permission from Fig. 2 of Ref. [5])

Under the assumptions we made, M exhibits a step-like behavior with unit jumps at a finite set of filling fractions. They correspond to the first-order transitions between distinct bubble phases. Everything said earlier about the possible phase co-existence near the first-order transitions applies here as well. For example, we do not expect any "bubbles of bubbles" at large N although at moderate N there is such a possibility. As ν_N decreases, the lattice constant of the bubble phase changes smoothly in between- and discontinuously at the transitions, but always remains close to $4\pi/\sqrt{3}\,q_* \approx 3.0R_{\rm c}$. Only in the Wigner crystal state

($M = 1$) the lattice constant is no longer tied to the cyclotron radius but is equal to $(4\pi/\sqrt{3}\,\nu_N)^{1/2}l$, which is much larger than R_c at $\nu_N \ll 1/N$.

One warning is in order here. While very useful, the Hartree–Fock approximation and the Landau theory of the phase transitions do not properly account for thermal and quantum fluctuations in 2D. A preliminary attempt to include these effects reveals additional phases and phase transitions of different order. The revised phase structure will be discussed in Sect. 7 below.

5 Validity of the Hartree–Fock Theory

It has been mentioned above that the Hartree–Fock ground state of a partially filled LL is always a CDW. In particular, for $N = 0$ it is a Wigner crystal [19]. It is well established by now that the latter prediction is in error at most filling fractions ν_N. Instead, Laughlin liquids and other fractional quantum Hall states appear [15]. The Wigner crystal is realized only when ν_N is very small (dilute particles) or very close to 1 (dilute holes). Below we present heuristic arguments and numerical calculations, which together make a convincing case that the situation at large N is *different*, so that the ground state has a CDW order at all ν_N and not just in the dilute limit.

5.1 Quantum Lindemann Criterion

A heuristic criterion of the stability of periodic lattices is the smallness of the fluctuations about the lattice sites compared to the lattice constant. Unless r_s is extremely small, the stripes and bubbles are *contiguous* completely filled (or completely depleted) regions of the topmost LL. In this situation the fluctuating objects are the edges of the stripes and bubbles. Using the suitably modified theory of the quantum Hall edge states (see Sect. 9), one can estimate their fluctuations δr to be of the order of the magnetic length, $\delta r \sim l$. Since the period Λ of the stripe and bubble lattices is at least a few $R_c = \sqrt{2N+1}\,l$, the Lindemann criterion $\delta r \ll \Lambda$ is well satisfied, and so the CDW states should be stable at any ν_N and sufficiently large N. Basically, when the amplitude of the local filling factor modulation is appreciable and the stripes (bubbles) are so wide, they are very "heavy," quasiclassical objects and quantum fluctuations are unable to induce a quantum melting of their long-range crystalline order. This is in contrast to the small N case where the quantum fluctuations in the real space are of the order of the lattice constant except in the dilute limit.

5.2 Diagrammatic Arguments

In a work [18] published soon after our original papers [4], Moessner and Chalker systematically analyzed the perturbation theory series for the partially filled high LL problem. They were able to achieve definitive results under two simplifying assumptions: (a) there is no translational symmetry breaking and (b) the range R of the quasiparticle interaction is smaller than l. [This interaction is given by the inverse Fourier transform of $\tilde{v}_0(q)/\epsilon(q)$]. Under such conditions the Hartree-Fock diagrams were shown to dominate in the $N \gg 1$ limit. This is a very important result. Even though it enables us to make controlled statements only about the high-temperature uniform state, it certainly enhances the credibility of the Hartree-Fock results at all T. Due to the screening by the lower LLs embodied in the dielectric function $\epsilon(q)$ [(2)], the effective range R of interaction turns out to be of the order of [12] $a_{\rm B} = \hbar^2\kappa/me^2 = l\sqrt{2/\nu r_{\rm s}^2}$; thus, condition (b) is satisfied at $N \gg r_{\rm s}^{-2}$, which is not too restrictive in practice.

There is still an interesting albeit academic question of what happens at truly small $r_{\rm s}$ and intermediate filling factors where $1 \ll N \ll r_{\rm s}^{-2}$, so that $R \gg l$. It will be discussed shortly below.

5.3 CDW Versus Laughlin Liquids

Obvious competitors of the CDW states at simple odd-denominator fractions $\nu_N = 1/(2k+1)$ are the Laughlin liquids. The interaction energy per particle in a Laughlin liquid can be found summing a rapidly converging series [21]

$$E_{\rm LL} = -\frac{\nu_N}{2} E_{\rm ex} + \frac{\nu_N}{\pi} \sum_{K=1}^{\infty} c_K V_K, \tag{15}$$

where V_K are Haldane's pseudopotentials

$$V_K = \frac{1}{2\pi} \int d^2 q\, \tilde{u}_{\rm H}(q)\, F_K(\sqrt{2}\, q), \tag{16}$$

$E_{\rm ex}$ (briefly introduced in Sect. 2 as a characteristic energy scale) is defined by

$$E_{\rm ex} = \tilde{u}_{\rm ex}(0)/(2\pi l^2), \tag{17}$$

and c_K are coefficients calculable by the Monte-Carlo method [21,5]. Numerically, about a dozen terms in the series (15) are needed to get an accurate value of $E_{\rm LL}$ at $\nu_N = \frac{1}{3}$ and $\nu_N = \frac{1}{5}$. In the large-N limit we can also derive an analytical estimate, guided by the asymptotic relation

$$V_K \simeq \tilde{u}_{\rm ex}(2\sqrt{K}/l)/l^2, \quad K \gg 1. \tag{18}$$

It indicates that $E_{\rm LL}$ is determined by the behavior of $\tilde{u}_{\rm ex}(q)$ at $q \sim l^{-1}$. Since the exchange and the direct interaction potentials are linked by the Fourier transform [cf. (5)], $\tilde{u}_{\rm ex}$ has the effective range of R/l^2 in the q-space [4]. Thus, two cases have to be distinguished.

1. $R \ll l$.— Physically, this corresponds to $N \gg \max\{1, r_{\rm s}^{-2}\}$ (recall that the interaction range R is of the order of $a_{\rm B}$). Provided $R \ll l$, only first few terms in the series (15) are important, leading to the estimate $E_{\rm LL} \sim \tilde{u}_{\rm ex}(1/l)/l^2$. On the other hand, the interaction energy per particle in the CDW ground state, $E_{\rm CDW} \sim -\tilde{u}_{\rm ex}(q_0)/l^2 \sim -E_{\rm ex}$, is significantly lower, i.e.,

the CDW wins. In the practical case of $r_s \sim 1$, the CDW should be lower in energy than the Laughlin liquids at $N \geq N_c$ where N_c is a small number. The numerical calculations reviewed below indicate that this "critical" number is $N_c = 2$.

2. $R \gg l$.— This regime appears in the parameter window $1 \ll N \ll r_s^{-2}$. It is mostly of *academic* interest because it can be realized only in very high density 2D systems where $r_s \ll 1$. Such systems are unavailable at present.

The theoretical analysis proceeds as follows. It turns out that (18) is correct with a relative accuracy $1/\ln N$ even for moderate K. With the same accuracy we can replace all V_K's in (15) by $2\pi E_{ex}$. Using the sum rule $\sum_K c_K = (1 - \nu_N^{-1})/4$ [21], we arrive at the asymptotic formula

$$E_{LL} \simeq -E_{ex}/2, \tag{19}$$

which is of the same order as E_{CDW}. To compare the energies of the two states, we have to exercise some care. Skipping the derivation, which relies heavily on another sum rule [19], for the Hartree-Fock states

$$\sum |\langle\tilde{\Delta}(\mathbf{q})\rangle|^2 = \nu_N, \tag{20}$$

we quote only the final result,

$$E_{CDW} \simeq -E_{ex}/2. \tag{21}$$

It signifies that the Laughlin liquid and the CDW have the same energy with a relative accuracy of $1/\ln N$. We may understand this surprising near equality as follows. Consider a system of particles interacting via a long-range two-body potential $v(r)$, which is nearly constant up to a distance R and then gradually decays to zero at larger distances. The particles are presumed to be spread over a uniform substrate "of opposite charge" with which they interact via a potential $-n_0 v(r)$. This fixes the average particle density to be n_0. It is easy to see that for any configuration of particles, which is uniform on the lenghscale of R, two potentials nearly cancel each other: (i) the total potential created at the location of a given particle by all the other particles of the system and (ii) the potential due to the substrate. The net potential is equal to $\Sigma = -v(0)$ because the particle does not interacts with itself. Due to the pairwise nature of the interaction, the interaction energy per particle (including the interaction with the substrate) is one half of Σ, i.e., $-v(0)/2$. Now we just need to recall from Sect. 2 that the energy of the self-interaction in our system is E_{ex} to recognize (19) and (21) as particular cases of this general relation.

The above discussion has several implications. First, it clarifies the physics behind the arguments of Moessner and Chalker [18] that the CDW states are likely to face a strong competition from certain uniform states when the interaction is sufficiently long-range, $R \gg l$. Second, it leaves the nature of the ground state at $1 \ll N \ll r_s^{-2}$ an open question at the moment. The CDW states seem to be favored in numerical calculations done for $2 \leq N \leq 10$ and both $r_s \sim 1$ and $r_s \ll 1$ (see below). Truly high N have not been investigated yet.

One intriguing possibility is the emergence of novel phases where the CDW ordering is not static but dynamic. One particular example is a quantum nematic phase, which can be visualized as a "soup" of fluctuating stripes. Such phases are actively discussed both in the context of the quantum Hall effect [22–27] and the high-temperature superconductivity [28–30]. We would like to reiterate that an experimental search for these exotic phases would require very special samples, e.g., with r_s much lower than presently available.

5.4 Numerical Results

Trial Wavefunctions. The analytical estimates for E_{LL} and E_{CDW} derived above become accurate only at very large N. At moderate N the comparison of trial states has to be done numerically. The procedure of calculating the energies of Laughlin liquids has been outlined above. It relies on the mapping of the Nth LL problem onto a problem at the lowest LL with the modified interaction, (2). To compute the energy of a Hartree–Fock CDW state we use the same trick: the trial state is chosen from the Hilbert space of the lowest LL, but the interaction potential is appropriately modified.

The first step is to define the wavefunction of a single bubble:

$$\Psi_0\{\mathbf{r}_k\} = \prod_{1\le i<j\le M} (z_i - z_j) \times \exp\left(-\sum_{i=1}^{M} \frac{|z_i|^2}{4l^2}\right), \qquad (22)$$

where $z_j = x_j + iy_j$ are complex coordinates of M quasiparticles that compose this bubble. A well-known property of the Vandermonde determinant indicates that Ψ_0 is in fact, a Hartree-Fock state. It is easy to see also that Ψ_0 defines the most compact arrangement of M quasiparticles allowed at the lowest LL, i.e., a circular droplet of a completely filled LL centered at the point $x = y = 0$.

The trial state we studied is composed of bubbles arranged in triangular lattice. It can be obtained by replicating the bubble (22) and translating its multiple copies to the appropriate lattice sites. This has to be followed by the antisymmetrization with respect to particle exchanges among different bubbles. The resulting wavefunction does not have a simple explicit form. Fortunately, to calculate the Hartree–Fock energy we do not need the wavefunction but only the particle density, see (4) and Ref. [19]. An excellent approximation for the latter is simply the sum of the densities of the individual bubbles. Strictly speaking, it is not a fully self-consistent solution of the nonlinear Hartree–Fock equations because of a small nonorthogonality among the wavefunctions of different bubbles. However, even the nearest bubbles are effectively so far away from each other that the deviations from the self-consistency are extremely small. If desired, the degree of self-consistency can be further improved using an iterative procedure of the relaxation type with the described density distribution as the initial guess. We have done this kind of calculations [5] and found that the iterations lower the energy of the state by less than one part in 10^6, which does not affect the comparison with the Laughlin liquid energy (known much less accurately).

On the basis of such calculations, we concluded that the CDW becomes the ground state at $\nu_N = \frac{1}{3}$ for $N \geq 2$ and at $\nu_N = \frac{1}{5}$ for $N \geq 3$. For the latter fraction and $N = 2$ the energies of the two trial states are so close that no definite conclusion could be made. Nevertheless, in practice the samples always contain some amount of disorder, which would favor the CDW state over the liquid state. Therefore, we established $N_c = 2$ as the "critical" LL index where the transition from the Laughlin liquids to CDW phases occurs. N_c turned out to be the same both for $r_s \sim 1$ and $r_s \ll 1$, and whether or not we included the effect of the finite-thickness of the 2D layer [5].

Since the fractional quantum Hall effect (FQHE) is traditionally associated with the Laughlin states while the CDW does not exhibit the FQHE, our results imply that the FQHE is restricted to the lowest and the first excited LLs, $N = 0$ and $N = 1$. *It is in principle impossible to observe the FQHE at* $N \geq 2$, and to date no one has. We are therefore led to propose the global phase diagram of the 2D electron systems shown in Fig. 8.

Exact Diagonalization of Small Systems. Strong evidence in favor of the CDW order in a partially filled $N \geq 2$ LL has been given by Rezayi, Haldane, and Yang [31], who studied systems up to 14 electrons by means of the direct numerical diagonalization of the Hamiltonian. Crucial for their success was employing periodical boundary conditions along the $\hat{\mathbf{x}}$ and $\hat{\mathbf{y}}$-directions (torus geometry). This setup avoids imposing the defects into the CDW lattice, aligns the CDW in a specific direction (which facilitates its detection), and enables to deduce the number of particles per unit cell simply from the multiplicity of the ground state manifold. Rezayi, Haldane, and Yang found no evidence of incompressible FQHE states at $2 \leq N \leq 6$ and $\nu_N = \frac{1}{4}, \frac{1}{3}, \frac{2}{5}$, etc. Instead, they reported the ground state degeneracies and quasi-Bragg peaks in the structure factor fully consistent with the formation of the stripe phase near the half-filling and a bubble phase at $\frac{1}{3}$ and $\frac{1}{4}$. The periodicity deduced from the quasi-Bragg peak positions agreed with the Hartree-Fock prediction (6) within a few percent.

Density Matrix Renormalization Group. Shibata and Yoshioka [32] studied $N = 2$ case using another powerful numerical technique, the density matrix renormalization group. Although not exact, it is presumed to be highly accurate both for the ground state energy and the ground state wavefunction. They were able to study larger systems, up to 18 electrons. Shibata and Yoshioka presented pair correlation functions unambiguously showing the stripe and bubble phases and pinpointed the transition point between them to be at $\nu_N \approx 0.38$.

6 Experimental Evidence for Stripes and Bubbles

6.1 Resistance Anisotropy

The existence of the stripe phase as a physical reality was evidenced by a conspicuous magnetoresistance anisotropy observed near half-integral fractions of

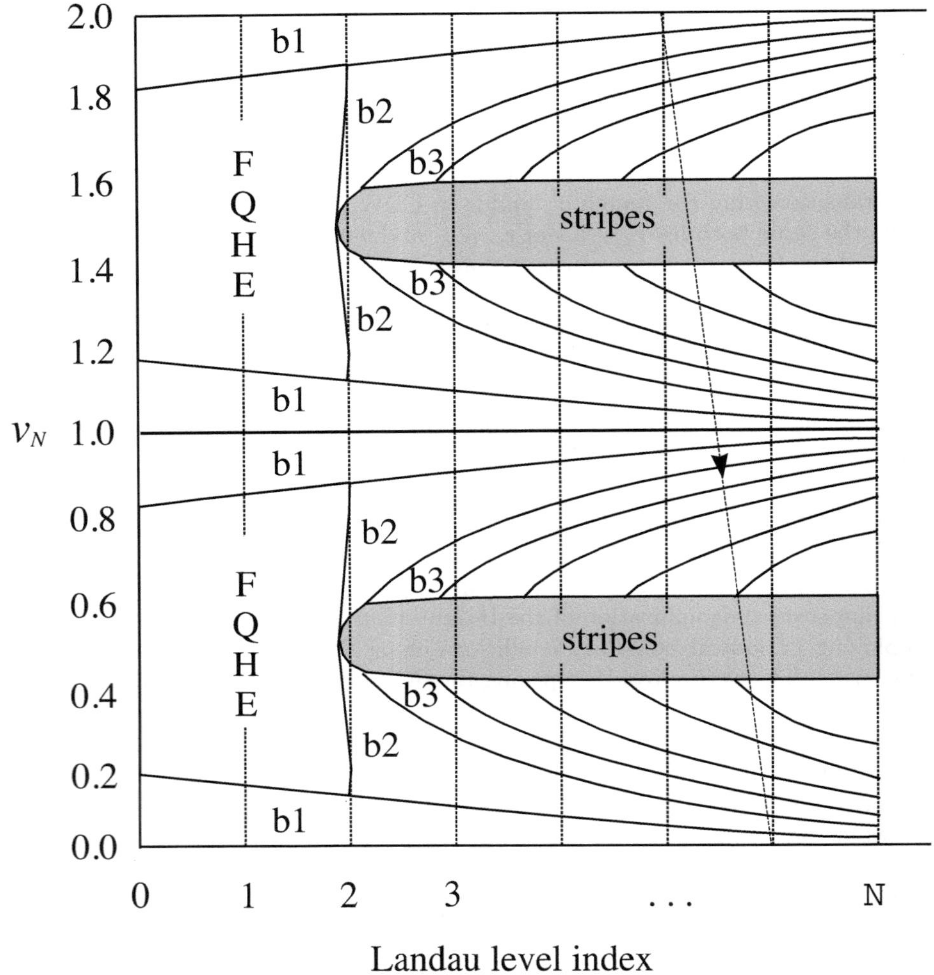

Fig. 8. The global phase diagram of a 2D electron system in the axes (N, ν_N) for the practical case $r_s \lesssim 1$ (schematically). The labels "b1", "b2", etc. denote the bubble phases with 1, 2, *etc.* particles per unit cell. As the magnetic field decreases, the system traces a zigzag path through this diagram, which starts at the lower left corner and proceeds along continuous segments $(N, 0) \to (N, 2)$ connected by discontinuous jumps $(N, 2) \to (N + 1, 0)$. One such jump is shown by the dashed line with the arrow. In principle, points away from the zigzag path can also be sampled if the functional form of the bare interaction can be modified sufficiently strongly compared to the Coulomb law

high LLs [6,7]. This anisotropy develops at low temperatures, $T \lesssim 0.1\,\mathrm{K}$, and only in very clean samples. The anisotropy is the largest at $\nu = 9/2$ ($N = 2$, $\nu_N = 1/2$) and decreases with increasing LL index. At $T = 25\,\mathrm{mK}$ it remains discernible up to $\nu \sim 11\frac{1}{2}$ whereupon it is washed out, presumably, due to residual disorder and/or finite temperature. The main anisotropy axes seem to be always oriented along the crystallographic axes of the GaAs crystal: $[1\bar{1}0]$ (the high-resistance direction) and [110] (the low-resistance one). Especially striking are the data obtained using the square samples in the van-der-Pauw measurement geometry: at $\nu = 9/2$ the resistances along the two principal directions differ by three orders of magnitude. However, one has to keep in mind that the van-der-Pauw measurements exaggerate the bare anisotropy of the resistivity tensor [33], due to current channeling along the easy (low-resistance) direction. Indeed, Hall-bar measurements, which provide a faithful representation of the resistivity tensor, show much smaller but still a very significant anisotropy up to 7:1. Once the current distribution effects are taken into account, the following picture emerges. At $T \gtrsim 0.1\,\mathrm{K}$, the transport is isotropic. As T decreases, the resistivity along the [110] direction (ρ_{yy}) decreases but only slightly. In contrast, the resistivity in the $[1\bar{1}0]$ direction (ρ_{xx}) rapidly increases, growing by almost an order of magnitude when T drops down to $25\,\mathrm{mK}$. The anisotropy is the largest at $\nu_N = 1/2$ but persists in a sizable interval $0.4 < \nu_N < 0.6$ of filling factors. Thus, as a function of ν_N, ρ_{xx} exhibits a peak resembling the quantum Hall transition peaks in dirtier samples at higher temperatures.

In contrast, the magnetotransport measurements near half-integral fillings of $N = 0$ and $N = 1$ LLs reveal no significant anisotropies and no peaks in the longitudinal resistance. Instead, the resistance exhibits a *minimum* as a function of ν_N, which deepens at T decreases [34]. Such unambiguous distinctions indicate that the electron ground states at high LLs $N \geq 2$ are qualitatively different from those in lower LLs. The $\nu = 9/2$ is the fraction that demarcates the transition to the realm of novel high LL physics.

The emergence of the anisotropy is very natural once we assume that the stripe phase forms. It is based on two concepts: pinning of stripes by disorder and the edge-state transport. Each of these topics deserves a separate discussion, which will be given (in a brief form) in Sects. 10 and 9, respectively. Here we only sketch the basic ideas. The pinning serves the purpose of preventing the global sliding of the stripes. This singles out the edge-transport as the only viable mechanism of current propagation. The edges of the stripes can be visualized as metallic rivers, along which the transport is "easy." The charge transfer among different edges, i.e., across the stripes, requires quantum tunneling and is "hard" because the stripes are effectively far away. Thus, if the stripes are preferentially oriented along the [110] direction, the sample would exhibit the anisotropy of the kind observed in the experiment.

The physical mechanism responsible for the alignment of the stripes along the definite crystallographic direction is debated at present (see Sect. 12). In the absence of external aligning fields, the stripe orientation would be chosen spontaneously. On the other hand, due to enormous collective response of the stripe-ordered phase, the stripes can be easily oriented by a tiny bare anisotropy of the medium or the substrate.

6.2 New Insulating States

The existence of the *bubble phases* at high LLs is supported by another striking experimental discovery: reentrant integral quantum Hall effect (IQHE) at $\nu \approx 4.25$ and $\nu \approx 4.75$. The Hall resistance at such filling factors is quantized at the value of the nearest IQHE plateau, while ρ_{xx} and ρ_{yy} show a deep minimum with an activated temperature dependence. The transport is isotropic in these novel insulating states, $\rho_{xx} \approx \rho_{yy}$. The current-voltage ($I$-$V$) characteristics exhibit pronounced nonlinearity, switching, and hysteresis. Such phenomena are hallmarks of the glassy behavior common for pinned crystalline lattices and conventional CDWs. Hence, these observations are fully consistent with the theoretical picture of a bubble lattice pinned by disorder. At $N = 2$ we expect only two bubble phases: with two ($M = 2$) and with one ($M = 1$) particle per bubble. Both phases are subject to pinning and should be insulating at $T = 0$. To understand the reentrancy phenomenon we have to take into account the finite-temperature effects. The $M = 2$ bubble phase is more rigid than the $M = 1$ (Wigner crystal) phase, and remains stable at temperatures where the Wigner crystal is already melted. As ν_N is varied towards the nearest integer, the insulating $M = 2$ bubbles are replaced by a conducting plasma formed in place of the Wigner crystal. Close enough to the integer ν_N, the plasma is so dilute that weakly interacting quasiparticles become localized by disorder and the conventional IQHE results.

Very recently, several more reentrant insulating states has been discovered also in the $N = 1$ LL [35]. Their nature remains to be determined but it is tantalizing to suggest that these are also the bubbles phases.

6.3 Other Experimental Findings

It was shown in a set of remarkable experiments that the anisotropy near half-integral fillings can be strongly affected by an in-plane magnetic field $B_{\|}$. When applied along the easy resistance direction, the hard and the easy anisotropy axes interchange at $B_{\|} \gtrsim 0.5\,\mathrm{T}$ [36,37]. When applied along the hard direction, the influence of $B_{\|}$ is much less pronounced and is to somewhat suppress the anisotropy. This intriguing behavior is thought to originate from the orbital effects of $B_{\|}$ in a finite-thickness 2D layer [17,38]. They can be crudely described as squeezing of the cyclotron orbits in the direction perpendicular to $B_{\|}$. For such distorted orbits the stripe phase energy depends on the orientation of the stripes with respect to the in-plane magnetic field. Calculations based on a suitably generalized Hartree-Fock theory of the previous sections show that the preferred orientation of the stripes can be both parallel and perpendicular to the in-plane magnetic field, depending on microscopic details of the real systems [17,38]. For the specific parameters believed to accurately describe the samples examined in Refs. [36,37], the perpendicular orientation is preferred (in agreement with the experiment). However, further theoretical and experimental work is needed to fully understand these issues.

The magnetotransport anisotropy at high LLs was also observed for p-type GaAs samples [39].

The higher current transport regime near $\nu = 9/2$ was investigated. Gradual increase in the differential resistance along the hard direction was reported. Compared to the strong nonlinearities at the reentrant IQHE states, it is a relatively weak effect.

Finally, the degree of anisotropy and the effect of the in-plane fields were found to be more pronounced in the lower spin subtend of the same Landau level. A possible explanation within the Hartree–Fock theory was recently suggested by Wexler and Dorsey [40].

7 Many Faces of the Stripe Phase

In the wake of the experiments, a considerable amount of work has been devoted to the stripe phase in recent years [24,43–49]. It led to the understanding that the "stripes" may appear in several distinct forms: an anisotropic crystal, a smectic, a nematic, and an isotropic liquid (Fig. 9). These phases succeed each other in the order listed as the magnitude of either quantum or thermal fluctuations increases. Thus, at small N ($N = 2, 3$) or close to T_c the phase diagrams of Fig. 4a and Fig. 8 need modifications to incorporate some (if not all) of those phases. The general structure of the revised phase diagram for the quantum ($T = 0$) case was discussed in the important paper of Fradkin and Kivelson [24]. Pinpointing the new phase boundaries in terms of the conventional parameters r_s and ν will require further analytical and numerical work. Once again, we wish to emphasize that at large N these additional phases have very narrow regions of existence, if any. Let us now give the definitions of these intriguing phases and discuss their basic properties.

Stripe Crystal. This state may in principle be understood at the Hartree–Fock level. It was shown [44] that the initially proposed Hartree-Fock solution with smooth edges and a strictly 1D periodicity [4] is not the global energy minimum. A further gain in energy is attained once the stripes acquire a periodic modulation in the longitudinal direction, in antiphase on each pair of neighboring stripes. The resultant phase breaks the translational symmetry in *both* spatial directions and thus is equivalent to a 2D crystal. The unit cell of such a crystal is very anisotropic, with the aspect ratio of the order of $N : 1$. However, there is only a single particle per unit cell; thus, this state is an anisotropic Wigner crystal. It is presumably the true ground state of the system at sufficiently large N where the Hartree-Fock is deemed to be exact.

Smectic State. The usual definition of the smectic is a "liquid with the 1D periodicity." A smectic is less ordered than a crystal because the translational symmetry is broken only in one spatial direction. The rotational symmetry is of course broken as well. The smectic stripe phase can be thought of as a descendant of the stripe crystal where the longitudinal modulations on neighboring

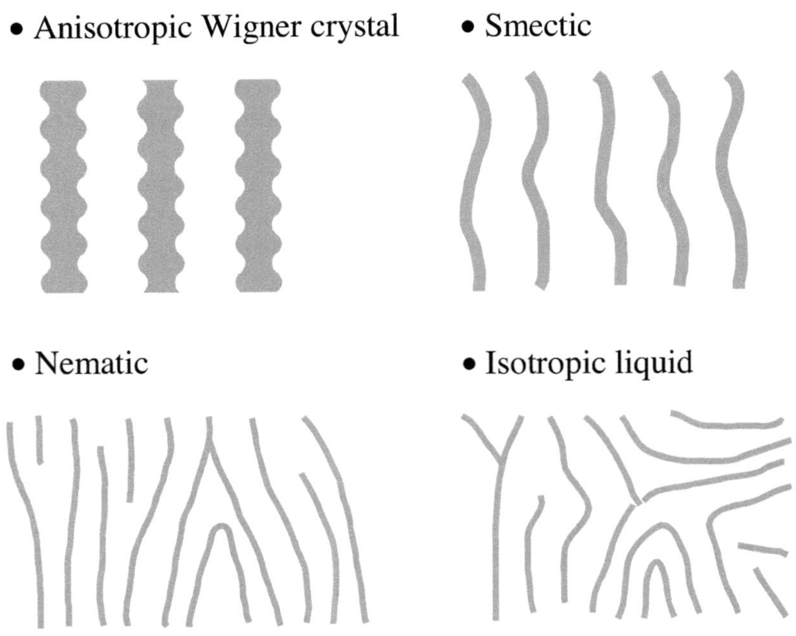

Fig. 9. Sketches of possible stripe phases

stripes persist locally, but have no long-range antiphase order because of dynamic phase slips. In other words, the neighboring stripes are *unlocked* [44]. In the thermodynamic limit this kind of state is equivalent to a stripe phase with no modulations (smooth edges), akin to the original Hartree–Fock solution [4]. The necessary condition for the smectic order is the continuity of the stripes. If the stripes are allowed to rupture, the dislocations are created. They destroy the 1D positional order and convert the smectic into the nematic.

Nematic State. By definition, the nematic is an anisotropic liquid. There is no long-range positional order. As for the orientational order, it is long-range at $T = 0$ and quasi-long-range (power-law correlations) at finite T. The nematic is riddled with dynamic dislocations. Other types of topological defects, the disclinations, may also be present but remain bound in pairs, much like vortices in the 2D X-Y model.

Isotropic Liquid. Once the disclinations in the nematic unbind, all the spatial symmetries are restored. The resultant state is an isotropic liquid with short-range stripe correlations. As the fluctuations due to temperature or quantum mechanics increase further, it gradually crosses over to the "uncorrelated liquid" where even the local stripe order is obliterated.

8 Effective Theories of the Stripe Phase

As often the case, the low-frequency long-wavelength physics of the system is governed by an effective theory involving a relatively small number of dynamical variables. The basic form of the effective theory is essentially fixed by the symmetry considerations. Let us outline how such theories are constructed for the various stripe states introduced in the previous section.

8.1 Stripe Crystal

The low-energy dynamical variables of this state are the elastic deformation $\mathbf{u} = \{u_x(\mathbf{r},t), u_y(\mathbf{r},t)\}$ of the crystalline lattice and the effective description is basically the elasticity theory. In addition, we have to account for the long-range Coulomb interaction between the density fluctuations $n(\mathbf{r},t) = -n_0\nabla\mathbf{u}$, where $n_0 = \nu_N/(2\pi l^2)$ is the average particle density at the Nth LL. The symmetry arguments identify four nonvanishing elastic moduli: c_{11}, c_{12}, c_{22}, and c_{44}, so that the effective Hamiltonian takes the form

$$\mathcal{H} = \frac{c_{11}}{2}(\partial_x u_x)^2 + \frac{c_{22}}{2}(\partial_y u_y)^2 + c_{12}\partial_x u_y \partial_y u_x + \frac{c_{44}}{8}(\partial_x u_y + \partial_y u_x)^2 + \frac{1}{2}n_0^2(\nabla\mathbf{u})u_{\mathrm{H}}(\nabla\mathbf{u}), \tag{23}$$

where u_{H} should be understood as the integral operator. The dynamics of the system is governed by the Lorentz force and can be studied with the help of the effective Lagrangean

$$\mathcal{L} = mn_0\omega_c u_y \partial_t u_x - \mathcal{H}. \tag{24}$$

Solving the corresponding equations of motion, we find the following excitation spectrum of lattice vibrations (magnetophonons):

$$\omega(\mathbf{q}) = \frac{\omega_p(q)}{\omega_c} q \left[\frac{c_{44} + (c_{11} + c_{22} - 2c_{12} - c_{44})\sin^2 2\theta}{4mn_0}\right]^{1/2}. \tag{25}$$

Here $\omega_p(q) = [n_0\tilde{u}_{\mathrm{H}}(q)q^2/m]^{1/2}$ is the plasma frequency and $\theta = \arctan(q_y/q_x)$ is the angle between the propagation direction and the $\hat{\mathbf{x}}$-axis. For Coulomb interactions $\omega_p(q) \propto \sqrt{q}$ leading to the well-known dispersion relation [50] $\omega(q) \propto q^{3/2}$. In a strongly anisotropic stripe crystal, $c_{12}, c_{44} \ll c_{11}$ causing the angular dependence $\omega(\mathbf{q}) \propto \sin 2\theta$ starting from relatively small q. All these results are valid for an idealized clean system. In reality the low-q magnetophonon modes will be profoundly affected by disorder, which will be discussed in Sect. 10.

8.2 Smectic State

As mentioned above, the smectic state is most closely related to the original Hartree–Fock solution [4]. It may be visualized as Hartree-Fock stripes slightly decorrelated by phonon-like thermal fluctuations.

Harmonic Approximation. In the smectic only u_x retains its direct meaning of the elastic displacement. On the other hand, the density fluctuations n become an independent degree of freedom. For example, in the case of incompressible stripes, n comes from the stripe width fluctuations, which are separate from the "shape" fluctuations described by u_x. The number of dynamical variables in the smectic and in the crystal is therefore the same. Moreover, the smectic can be thought of as a crystal that lost its shear rigidity because of phase slips between nearby crystalline rows. This intuitive picture enables us to deduce the effective theory for the smectic from (23) and (24) by a certain reduction. Of course, at the end we should verify that we did not miss any terms allowed by symmetry. The first step is to formally reintroduce u_y as a solution of the equation $\partial_y u_y = -n/n_0 - \partial_x u_x$ and use it to replace all instances of $\partial_y u_y$ in (23). To ensure that the stripes are free to slide with respect to each other in the $\hat{\mathbf{y}}$-direction, the terms that depend on $\partial_x u_y$ should be dropped: $c_{12} \to 0$, $c_{44} \to 0$. Yet we need to be careful and recall that the complete elasticity theory always contains higher gradients such as $(\partial_y^2 u_x)^2$ [suppressed in (23)]. Once the coefficient in front of the first-order gradient term $(\partial_y u_x)^2$ vanishes, higher gradients become dominant and must be included. The resultant effective Hamiltonian and the Lagrangean take the form

$$\mathcal{H} = \frac{Y}{2}(\partial_x u)^2 + \frac{K}{2}(\partial_y^2 u)^2 + \frac{1}{2}n(u_{\rm H} + \chi^{-1})n + Cn\partial_x u, \tag{26}$$

$$\mathcal{L} = p\partial_t u - \mathcal{H}, \quad \partial_y p = -m\omega_{\rm c}(n + n_0\partial_x u). \tag{27}$$

Here we switched to notations more natural for the smectic: u_x became u, u_y was traded for the canonical momentum p, the sum $c_{11} + c_{22}$ became Y, *etc.* The physical meaning of new phenomenological coefficients is as follows. Y and K are the compression and the bending elastic moduli, χ is the compressibility, and $C = 2\pi l^2 Y d\ln\Lambda/d\nu_N$ accounts for the dependence of the mean interstripe separation Λ on the average filling factor.

Since the number of dynamical variables in the smectic is the same as in the crystal state, the collective mode count is also unchanged. We will keep referring to them as magnetophonons. Solving the equations of motion for n and u we obtain the dispersion relation of such magnetophonons [45]:

$$\omega(\mathbf{q}) = \frac{\omega_p(q)}{\omega_{\rm c}}\frac{q_y}{q}\left[\frac{Yq_x^2 + Kq_y^4}{mn_0}\right]^{1/2}. \tag{28}$$

Unless propagate nearly parallel to the stripes, $\omega(\mathbf{q})$ is proportional to $\sin 2\theta\, q^{3/2}$. Unlike in the stripe crystal, this relation is obeyed even at $q \to 0$ (again, in the absence of disorder or orienting fields). One immediate consequence of this dispersion is that the largest velocity of propagation for the magnetophonons with a given q is achieved when $\theta = 45°$.

Thermal Fluctuations and Anharmonisms. From Eq. (26) we can readily calculate the mean-square fluctuations of the stripe positions at finite T, e.g.,

$$\langle[u(0,0)-u(0,y)]^2\rangle = 2\int \frac{d^2q}{(2\pi)^2}\frac{k_{\mathrm{B}}T}{Yq_x^2+Kq_y^4}(1-\mathrm{e}^{\mathrm{i}k_y y}) = \frac{k_{\mathrm{B}}T}{2\sqrt{YK}}|y|. \quad (29)$$

This formula is valid if y is large so that magnetophonons with wavevectors $q_y \lesssim 1/y$ can be treated classically [$\hbar\omega(\mathbf{q}) \ll k_{\mathrm{B}}T$]. As one can see, at any finite temperature magnetophonon fluctuations are growing without a bound; hence, the positional order of a 2D smectic is totally destroyed [51] at sufficiently large distances along the $\hat{\mathbf{y}}$-direction, $|y| \gg \Lambda\sqrt{YK}/k_{\mathrm{B}}T \equiv \xi_y$. Similarly, along the $\hat{\mathbf{x}}$-direction, the positional order is lost at lengthscales larger than $\xi_x = (Y/K)^{1/2}\xi_y^2$.

Another type of excitations, which decorrelate the stripe positions are the aforementioned dislocations. The dislocations in a 2D smectic have a finite energy $E_D \sim K$. At $k_{\mathrm{B}}T \ll E_D$ the density of thermally excited dislocations is of the order of $\exp(-E_D/k_{\mathrm{B}}T)$ and the average distance between dislocations is $\xi_D \sim \Lambda\exp(2k_{\mathrm{B}}T/E_D)$. At low temperatures $\xi_x, \xi_y \ll \xi_D$; therefore, the following interesting situation emerges (Fig. 10). On the lengthscales smaller than ξ_y (or ξ_x, whichever appropriate) the system behaves like a usual smectic where (26–28) apply. On the lengthscales exceeding ξ_D it behaves[2] like a nematic [52]. In between the system is a smectic but with very unusual properties. It is topologically ordered (no dislocations) but possesses enormous fluctuations. In these circumstances the harmonic elastic theory becomes inadequate and anharmonic terms must be included. The most important anharmonisms are captured in the following elastic Hamiltonian, which should be substituted in place of the first two terms in (26) [51]:

$$\mathcal{H}_{el} = \frac{Y}{2}\left[\partial_x u - \frac{1}{2}(\nabla u)^2\right]^2 + \frac{K}{2}(\partial_y^2 u)^2. \quad (30)$$

It can be easily checked that the expression inside the square brackets, which is the compressional strain, is invariant under rotations of the reference frame by an *arbitrary* angle ϕ. For example, if in the initial reference frame $u = 0$, then in the new frame $u(x,y) = (1-\cos\phi)x + \sin\phi\, y$ so that $\partial_x u - \frac{1}{2}(\nabla u)^2 = 0$.

What is the role of anharmonisms? As shown by Golubović and Wang [53], they cause power-law dependence of the parameters of the effective theory on the wavevector $\mathbf{q}$:

$$Y \sim Y_0(\xi_y q_y)^{1/2}, \quad K \sim K_0(\xi_y q_y)^{-1/2}, \quad (31)$$

for $q_x \ll \xi_x^{-1}(q_y\xi_y)^{3/2}$, $q_y \ll \xi_y^{-1}$, and

$$Y \sim Y_0(\xi_x q_x)^{1/3}, \quad K \sim K_0(\xi_x q_x)^{-1/3}, \quad (32)$$

for $q_x \ll \xi_x^{-1}$ and $q_y \ll \xi_y^{-1}(q_x\xi_x)^{2/3}$. The lengthscale dependence of the parameters of the effective theory is a common feature of fluctuation-dominated

[2] In a more precise treatment [53], the lengthscales $\xi_{Dx} \propto \xi_D^{6/5}$ and $\xi_{Dy} \propto \xi_D^{4/5}$ are introduced such that $\xi_{Dx}\xi_{Dy} = \xi_D^2$.

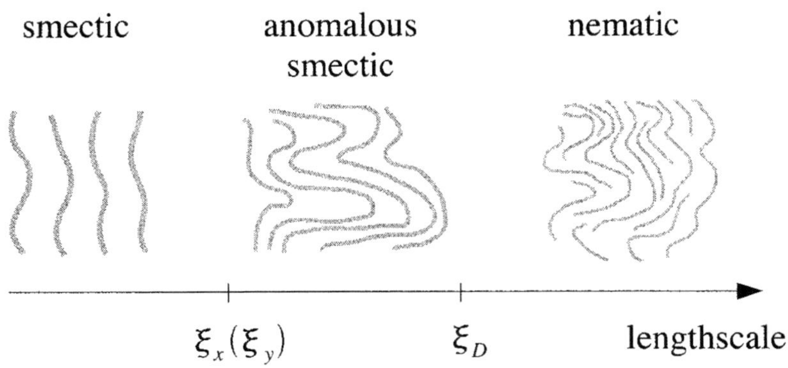

Fig. 10. Portraits of the stripe phase on different lengthscales

phenomena. Other famous examples include the criticality near phase transitions and in systems at their lower critical dimension, such as the 2D X-Y model. It should be mentioned that the lower critical dimension for the smectic order is $d = 3$ [51], so that the 2D smectic is *below* its lower critical dimension. This is the reason why the scaling behavior (31) and (32) does not persist indefinitely but eventually breaks down above the lengthscale ξ_D where the crossover to the thermodynamic limit of the nematic behavior commences.

Equations (31) and (32) indicate that the compression modulus decreases while the bending modulus increases as the lengthscale grows. This can be understood from the following qualitative reasoning. Thermal fluctuations create a lot of wiggles on the stripes. When an external compressional stress is applied, it can be relieved not just by compression but by flattening of the stripes. Since the latter involves unbending of the crumpled stripes and the bending costs less energy than the compression (q^4 instead of q^2), the apparent compressional modulus Y is smaller than its bare value Y_0. Similarly, when one attempts to bend crumpled stripes, some compression is necessarily involved, and so the bending modulus K appears larger.

The scaling shows up not only in the static properties such as Y and K but also in the dynamics. The role of anharmonisms in the dynamics of conventional 3D smectics has been investigated by Mazenko et al. [54] and also by Kats and Lebedev [55]. For the quantum Hall stripes the analysis had to be done anew because here the dynamics is totally different. It is dominated by the Lorentz force rather than a viscous relaxation in the conventional smectics. This task was accomplished in Ref. [45]. The calculation was based on the Martin-Siggia-Rose formalism combined with the ϵ-expansion below $d = 3$ dimensions. One set of results concerns the spectrum of the magnetophonon modes, which becomes

$$\omega(\mathbf{q}) \sim \sin\theta \cos^{7/6}\theta \, (\xi_x q)^{5/3} \frac{\omega_{\mathrm{p}}(\xi_x^{-1})}{\omega_{\mathrm{c}}\xi_x} \sqrt{\frac{Y_0}{m n_0}}. \tag{33}$$

Compared to the predictions of the harmonic theory, (28), the $q^{3/2}$-dispersion changes to $q^{5/3}$. Also, the maximum propagation velocity is achieved for the angle $\theta \approx 53°$ instead of $\theta = 45°$. These modifications, which take place at long wavelengths, are mainly due to the renormalization of Y in the static limit and can be obtained by combining (28) and (32). Less obvious dynamical effects peculiar to the quantum Hall smectics include a novel dynamical scaling of Y and K as a function of frequency and a specific q-dependence of the magnetophonon damping [45].

The latter issue touches on an important point. Our effective theory defined by (26) and (27) is based on the assumption that u and n are the only low-energy degrees of freedom. It is probably well justified at $T \to 0$ but becomes incorrect at higher temperatures. The point of view taken in Ref. [45] is that in the latter case thermally excited quasiparticles ("normal fluid") should appear and that they should bring dissipation into the dynamics of the magnetophonons.

Another intriguing possibility is for quasiparticles or other additional low-energy degrees of freedom to exist even at $T = 0$. Such more complicated smectic states are not ruled out and are interesting subjects for future study.

8.3 Nematic State

Much like loosing the shear rigidity due to phase slips converts a crystal to a smectic, loosing the compressional rigidity due to mobile dislocations can convert a 2D smectic into a nematic. The collective degree of freedom associated with the nematic ordering is the angle $\phi(\mathbf{r}, t)$ between the local normal to the stripes $\mathbf{N}$ and the $\hat{\mathbf{x}}$-axis orientation. Due to inversion symmetry, $\mathbf{N}$ and $-\mathbf{N}$, i.e., ϕ and $\phi + \pi$ are equivalent, that is why $\mathbf{N}$ is often referred to as the *director* rather than a vector [51]. The effective Hamiltonian for $\mathbf{N}$ is dictated by symmetry to be

$$\mathcal{H}_N = \frac{K_1}{2}(\nabla \mathbf{N})^2 + \frac{K_3}{2}|\nabla \times \mathbf{N}|^2. \tag{34}$$

The coefficients K_1 and K_3 are termed the splay and the bend Frank constants [51]. They control the cost of the two possible elementary types of director nonuniformity shown in Fig. 11. Note that in the smectic phase $\phi = -\partial_y u$. This entails the relation $K_3 \simeq K$ between the parameters of the nematic and its parent smectic. On the other hand, the value of K_1 is expected to be determined largely by the properties of the dislocations [52]. The elastic part has a particularly simple form if $K_1 = K_3$, in which case $\mathcal{H}_N = (K_3/2)(\nabla\phi)^2$ just like in the X-Y model.

Another obvious degree of freedom in the nematic are the density fluctuations $n(\mathbf{r}, t)$. A peculiar fact is that in the static limit n is totally decoupled from $\mathbf{N}$, and so it does not enter (34). However, since the nematic is less ordered

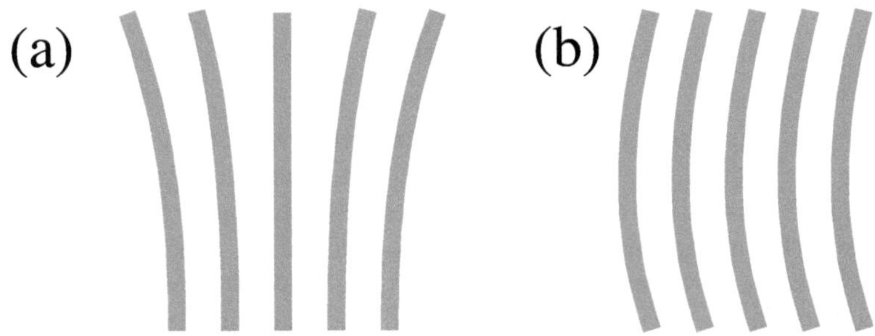

Fig. 11. Splay (**a**) and bend (**b**) distortions in a nematic

than even a smectic, the question about extra low-energy degrees of freedom or additional quasiparticles become especially relevant. We believe that different types of quantum Hall nematics are possible in nature. In the simplest case scenario $\mathbf{N}$ and n are the only low-energy degrees of freedom. This type of state has been studied by Balents [23] and recently by the present author [25]. It was essentially postulated that the effective Largangean takes the form

$$\mathcal{L} = \frac{1}{2}\gamma^{-1}(\partial_t \mathbf{N})^2 - \mathcal{H}. \tag{35}$$

(As hinted above, the full expression contains also couplings between $\partial_t \mathbf{N}$ and mass currents but they become vanishingly small in the long-wavelength limit). The collective excitations are charge-neutral fluctuations of the director. They have a linear dispersion,

$$\omega(\mathbf{q}) = q\sqrt{K_1 \gamma \cos^2\theta + K_3 \gamma \sin^2\theta}, \tag{36}$$

and resemble spinwaves in the X-Y quantum rotor model.

Also discussed in Ref. [25] was a dislocation-mediated mechanism of the smectic-nematic transition within the framework of duality transformations developed earlier by Toner and Nelson [52], Toner [56], and Fisher and Lee [57]. This theory predicts the existence of a second excitation branch with a small gap. This gapped mode can be considered a descendant of the magnetophonon mode of the parent smectic.

Very recently, Radzihovsky and Dorsey [27] formulated a different theory of the quantum Hall nematics, whose predictions disagree with our (35) and (36). Logically, there are two possibilities. Either, as mentioned above, there are several distinct kinds of nematic states possible in nature or some of the theoretical constructions advanced in Refs. [23,25,27] are incorrect. To resolve these issues it is imperative to bring the discussion from the level of effective theory to the level of quantitative calculations. One promising direction is to investigate concrete trial wavefunctions of quantum nematics, e.g., the one proposed by Musaelian and Joynt [22]:

$$\Psi = \prod_{j<k} (z_j - z_k)[(z_j - z_k)^2 - a^2] \times \exp\Big(-\sum_j |z_j|^2/4l^2\Big). \tag{37}$$

Here a is a complex parameter that determines the degree of orientational order and the direction of the stripes. This particular wavefunction corresponds to $\nu = \frac{1}{3}$. (As explained earlier, it can also be used investigate the higher Landau level states with $\nu_N = \frac{1}{3}$). Recently, the work in this direction was continued by Ciftja and Wexler [26].

Other contributions to the theory of quantum Hall nematics have been focused on a finite temperature case. They include a work of Fradkin et al. [41] who investigated the role of external anisotropy field in the 2D X-Y formulation and also a paper by Wexler and Dorsey [40], where a quantitative Hartree-Fock analysis of the Toner-Nelson disclination unbinding scenario has been done.

It is worth mentioning that in the quantum case the chain of transitions crystal → smectic → nematic → isotropic liquid may or may not be realized in full. A finite-size study by Rezayi et al. [58] suggests that the transition from the smectic to an isotropic phase, as the interaction parameters are varied away from their $N = 2$ Coulomb values, can also occur via a first-order transition, without the intermediate nematic phase. The resultant isotropic state is highly correlated and has little in common with the "uncorrelated" Hartree-Fock liquid. The natural candidates for the isotropic state include a Fermi-liquid-like state of composite fermions and a Pfaffian (Moore-Read) state [58]. Understanding all these competing quantum orders and transitions between them remains a major intellectual challenge. In contrast, the phase structure at larger N, away from the "transition point" $N = 2$, is much simpler and is adequately captured by the Hartree-Fock theory.

9 Edge State Models

Another theoretical approach to the physics of the stripe phases is based on the edge-state formalism. The advantages of the edge-state models are two-fold. First, they allow to bridge the gap between the microscopic theory and the formulations based on the elasticity theory or hydrodynamics (albeit under some crucial simplifying assumptions). Second, they enable one to calculate not only collective but also single-particle properties, such as the tunneling density of states. On the other hand, being more specialized than the hydrodynamics, the edge models have a somewhat restricted domain of validity. The models considered so far [24,43,48,49] apply when the stripes are incompressible. i.e., when they are contiguous regions of $\nu_N = 1$ separated by *sharp* boundaries from the regions with $\nu_N = 0$. In our opinion the edge state models are better suited for describing the crystal and the smectic states. In the nematic state stripes exhibit violent quantum (or thermal) fluctuations and the picture of well defined sharp edges is questionable.

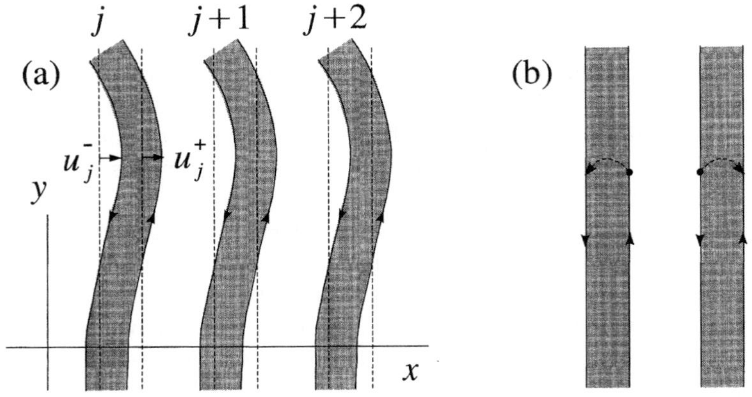

Fig. 12. (**a**) Geometrical meaning of variables in the edge-state effective theory. (**b**) Backscattering

The first step in constructing the edge theory is to identify the low-energy degrees of freedom with the deviations $u_\pm^j$ of the stripe boundaries from their equilibrium positions $x_\pm^j = (j \pm \nu_N/2)\Lambda$, see Fig. 12a. Such deviations induce extra 1D guiding center densities along the edges, $\rho_\pm^j = \pm u_\pm^j/(2\pi l^2)$. The next step is to find an effective Hamiltonian $\mathcal{H}$ that governs the interaction among $\rho_\pm^j$. To this end it is convenient to introduce the Fourier transform

$$\tilde{\rho}_\sigma(\mathbf{q}) = \Lambda \sum_j \exp(-\mathrm{i}q_y x_\sigma^j) \int dy \mathrm{e}^{-\mathrm{i}q_y y} \rho_\sigma^j(y), \tag{38}$$

and define the center-of-mass stripe displacement u_x and the net local density fluctuation per unit area,

$$\tilde{u}_x(\mathbf{q}) = 2\pi l^2 \left[\exp(\tfrac{\mathrm{i}}{2} q_x \nu_N \Lambda)\tilde{\rho}_+(\mathbf{q}) + \exp(-\tfrac{\mathrm{i}}{2} q_x \nu_N \Lambda)\tilde{\rho}_-(\mathbf{q})\right], \tag{39}$$

$$\tilde{n}(\mathbf{q}) = \Lambda^{-1} \left[\exp(\tfrac{\mathrm{i}}{2} q_x \nu_N \Lambda)\tilde{\rho}_+(\mathbf{q}) - \exp(-\tfrac{\mathrm{i}}{2} q_x \nu_N \Lambda)\tilde{\rho}_-(\mathbf{q})\right]. \tag{40}$$

The general symmetry requirements compel the long-wavelength part $\mathcal{H}_0$ of the Hamiltonian to take the form (26), reproduced here for convenience:

$$\mathcal{H}_0 = \frac{Y}{2}(\partial_x u_x)^2 + \frac{K}{2}(\partial_y^2 u_x)^2 + \frac{1}{2} n(u_{\mathrm{H}} + \chi^{-1}) n + C n \partial_x u_x. \tag{41}$$

The edge-state formalism enables one go further and obtain quantitative estimates for the phenomenological parameters Y, K, χ, and C from the microscopic Hamiltonian. It should be clarified that the exact calculation of these parameters remains out of reach for the moment. Available derivations [43,44,47,40,49] are certainly not exact. It can be shown [16] that all of them rely on one specific approximation scheme, the time-dependent Hartree–Fock approximation (TDHFA). In the spirit of the discussion in the previous sections, we expect the TDHFA to be quantitatively accurate at $N \gg 1$ but only qualitatively correct in the case of current experimental interest, $N \sim 2$. Let us therefore proceed with the exposition of the field-theoretic edge-state formalism.

Besides $\mathcal{H}_0$, the full effective Hamiltonian contains another term $\mathcal{H}_{2q_F}$, whose importance was first emphasized by Fradkin and Kivelson [24]. To clarify its origin let us recall that in the Landau gauge $A = (0, Bx, 0)$ each single-particle basis state $|X\rangle$ has a definite momentum in the $\hat{\mathbf{y}}$-direction, proportional to the mean value X of the x-coordinate in such a state, $k_y = X/l^2$. In the simplest mean-field realization of the stripes, a given state is filled if $k_y \in (x_+^j/l^2, x_-^j/l^2)$ and is empty otherwise. Thus, $k_y = x_\pm^j$ play the role of Fermi points in an equivalent quasi-1D electron system. Note that "+" ("−") stands for the sign of the Fermi velocity near a given Fermi points, in the positive (negative) $\hat{\mathbf{y}}$-direction. The density fluctuations $\rho_\pm^j$ propagate along the edges only in the direction specified by this sign. Such a property is called chirality. Thus, the stripe phase is characterized by a Manhattan grid of alternating *chiral* edge states. From this perspective, $\mathcal{H}_0$ describes the particle-particle interactions, which cause only small changes in X and thus small momentum transfer in the vicinity of the Fermi points. Another low-energy process is to scatter particles between different Fermi points, e.g., $x_+^j/l^2 \to x_-^j/l^2$ and $x_-^{j+1}/l^2 \to x_+^{j+1}/l^2$, see Fig. 12b. Such scattering acts must involve a pair of particles to conserve the total momentum: one particle gains and the other looses the same amount of momentum, $2q_F^e \equiv (x_+^j - x_-^j)/l^2 = \nu_N \Lambda/l^2$. $\mathcal{H}_{2q_F}$ accounts precisely for the processes of this type. Since they cause the reversal of the propagation direction for the particles involved, it is natural to call it backscattering. Remarkably, it is possible [42] to express H_{2q_F} in terms of density fluctuations $\rho_\pm^j$. Without going into details, we quote the final result [24,43],

$$\begin{aligned}\mathcal{H}_{2q_F} = &-\sum_j \lambda^e \cos[(\phi_+^j - \phi_-^j) + (\phi_-^{j+1} - \phi_+^{j+1})]\\ &-\sum_j \lambda^h \cos[(\phi_-^j - \phi_+^{j-1}) + (\phi_+^j - \phi_-^{j+1})]\end{aligned} \tag{42}$$

Here $\phi_\pm^j$ are auxiliary dynamical variables related to $\rho_\pm^j$ as follows:

$$\rho_\pm^j = \partial_y \phi_\pm^j/(2\pi). \tag{43}$$

The phenomenological parameters λ^a are proportional to $\tilde{u}_{\mathrm{H}}(2q_F^a)$ but in general depends on how the theory is formulated (the ultraviolet cutoff). In principle, interactions can scatter the particles not only between nearest edges but also next-nearest ones, etc. The corresponding amplitudes are proportional to $\tilde{u}_{\mathrm{H}}(q)$ where q is the appropriate momentum transfer. Since $\tilde{u}_{\mathrm{H}}$ decreases exponentially at such large q, these other processes are expected to have negligible effect. Thus, the total Hamiltonian is $\mathcal{H} = \mathcal{H}_0 + \mathcal{H}_{2q_{\mathrm{F}}}$.

The final step in constructing the effective edge theory is determining the kinetic term. Either from Wen's bosonization theory [42] of by comparing (27) and (43), one can come to the conclusion that $\phi_{\pm}^j$ and $\rho_{\pm}^j$ are canonically conjugate variables, so that the appropriate Lagrangean is

$$\mathcal{L} = \frac{\hbar}{4\pi} \sum_{j,\sigma=\pm} \sigma \partial_t \phi_\sigma^j \partial_y \phi_\sigma^j - \mathcal{H}_0 - \mathcal{H}_{2q_{\mathrm{F}}}. \tag{44}$$

Let us now explain how this theory can lead either to crystalline or to smectic behavior. The idea is to treat $\mathcal{H}_{2q_{\mathrm{F}}}$ as a small perturbation [24,43,44,47]. If this perturbation is *irrelevant*, in the long-wavelength low-frequency limit $\mathcal{H}$ reduces to $\mathcal{H}_0$ and $\mathcal{L}$ to the smectic form (27). On the other hand, if $\mathcal{H}_{2q_{\mathrm{F}}}$ is relevant, then the "+" and "−" edges of each stripe become strongly mixed so that a static $2q_{\mathrm{F}}$ density modulation in the $\hat{\mathbf{y}}$-direction appears. Its spatial period is $a = 2\pi/(2q_{\mathrm{F}}) = 2\pi l^2/\nu_N \Lambda$; hence, the number of particles on a given stripe within one modulation period is $\nu_N \Lambda \times a/(2\pi l^2) = 1$. This can be visualized as a 1D chain of particles. The neighboring chains are locked in antiphase to lower their interaction energy. This state is an anisotropic Wigner crystal with the aspect ratio of the unit cell $\Lambda : a = (\nu_N/2\pi)(\Lambda/l)^2 \sim N : 1$. Two types of stripe crystals are possible. If the first term in $\mathcal{H}_{2q_{\mathrm{F}}}$ is relevant, it is an "electron" crystal, if the second term is relevant, it is a "hole" crystal.

The elasticity theory (23) of, e.g., the electron crystal can be recovered as follows. Since the first sum in (42) is relevant, the large fluctuations of the argument of the corresponding cosines are inhibited. It is legitimate to expand these cosines to arrive at

$$\mathcal{H}_{2q_{\mathrm{F}}} \to -\sum_j \frac{\lambda_*}{2} (u_y^j - u_y^{j+1} - \Lambda \partial_y u_x)^2, \tag{45}$$

$$u_y(\mathbf{q}) \equiv \frac{l^2}{\nu_N \Lambda} [\phi_+(\mathbf{q}) \mathrm{e}^{\mathrm{i}\Lambda q_x/2} - \phi_-(\mathbf{q}) \mathrm{e}^{-\mathrm{i}\Lambda q_x/2}]. \tag{46}$$

In the long-wavelength limit it gives a term proportional to $(\partial_x u_y + \partial_y u_x)^2$, which makes the shear modulus c_{44} finite [cf. (23)]. A more careful treatment presumably recovers the remaining elastic constant c_{12}. Our choice of u_y in Eq. (46) is not arbitrary because we want to preserve the physical interpretation of u_y as the elastic displacement. This requires $\mathcal{H}_{2q_{\mathrm{F}}}$ to be invariant under the shift by a lattice constant, $u_y \to u_y + a$. Since $\mathcal{H}_{2q_{\mathrm{F}}}$ is invariant under the shift $\phi_{\pm}^j \to \phi_{\pm}^j + 2\pi$, it fixes the coefficient of proportionality in (46) to the value given.

The relevance or irrelevance of $\mathcal{H}_{2q_{\mathrm{F}}}$ depends on the stiffness of the stripes measured, roughly, by the dimensionless parameter $Yl^4/\tilde{u}_{\mathrm{H}}(\pi/\Lambda)$. If it is small enough, then high-q and ω fluctuations of $\phi_{\pm}^j$ and u_y^j (either quantum or thermal) are sufficiently strong to cause effective averaging out of the cosines in $\mathcal{H}_{2q_F}$. As a result, the coefficient λ renormalizes to zero in the low q and ω limit. In contrast, for rigid stripes, the averaging of the cosine terms is not important. The perturbative renormalization group (RG) analysis [43,44,47,49] allows to make this argument more precise. [3] However, to decide between the crystal or the smectic behavior, the RG requires accurate estimates for Y and other parameters of the edge theory as an input. Attempts to extract such parameters from the TDHFA for the case of current experimental interest, $N \sim 2$, led to contradictory statements in the literature [43,47,49]. In this regard, we wish to reiterate that at $N \sim 2$ the stripe phase is very fragile and faces a strong competition from other quantum Hall states. Therefore, the Hartree–Fock and its time-dependent extensions are not quantitatively reliable. At the present stage the controlled theoretical analysis can be envisioned only for large N, building on the work of Moessner and Chalker [18]. Note that the ratio $Yl^4/\tilde{u}_{\mathrm{H}}(\pi/\Lambda)$ becomes an N-independent number of the order of unity at $N \gg 1$ and $r_{\mathrm{s}} \sim 1$.

Concluding this section, let us discuss the tunneling into the stripe phase from the normal metal. Two setups can be imagined: tunneling from a bulk metal or from an STM tip. In the first case the electron-electron interactions are screened by the metal, in the second they remain long-range. The differential tunneling conductance $G_{\mathrm{T}} = dI/dV$ at bias voltage V is proportional to the tunneling density of states $g(eV)$, which is defined by $g(E) = (1/\pi)\Im\mathrm{m}\tilde{G}_{\mathrm{R}}(E/\hbar)$, $\tilde{G}_{\mathrm{R}}(\omega)$ being the single-electron Green's function,

$$G_{\mathrm{R}}(t) = -i\theta(t)\langle\Psi(0)\Psi^{\dagger}(t)\rangle. \tag{47}$$

Here $\Psi^{\dagger}$ and Ψ are the electron creation and annihilation operators. The crude overall behavior of $g(E)$ can be obtained at the Hartree-Fock level. In the simplest case of the smectic (unmodulated stripes), $g(E)$ is determined by the slope of the self-consistent potential $E(X)$ that defines the energies of the quasiparticle states $|X\rangle$,

$$g_{\mathrm{HF}}(E) = |4\pi l^2 \Lambda (dE/dX)|^{-1}. \tag{48}$$

The tunneling density of states calculated according to this formula is nonvanishing in a finite interval of width $\sim E_{\mathrm{ex}}$ around $E = 0$. At both ends of such an interval one finds two divergencies (van Hove peaks). The interior of the interval can be described as a shallow pseudogap, see Fig. 13. The negative-E peak corresponds to tunneling into the states with X's in the middle of the filled stripes and the positive-E one — into the middle of the empty stripes. Conversely, at low energies (low V) the tunneling can only occur into points in a vicinity of stripe edges [4]. The Hartree–Fock results for $g(E)$ are certainly not exact. However, the van Hove peaks and the pseudogap are expected to be true features and the Hartree–Fock estimate for the energy separation between the peaks should be quite reliable. The largest deviations from the Hartree–Fock predictions should

[3] These four works are in mutual agreement on how the RG procedure needs to be formulated. A different type of RG analysis suggested in Ref. [48] is believed to be in error.

occur at low E, which we now address. We will use the evolution in the imaginary time $\tau = it$ picture common in quantum tunneling problems.

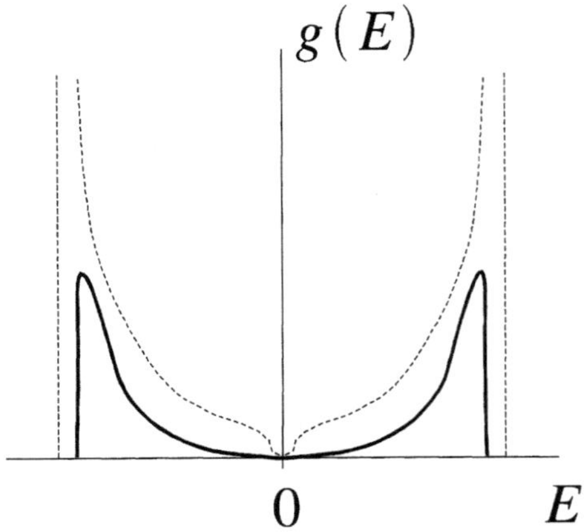

Fig. 13. The sketch of the true tunneling density of states (solid line) and its Hartree–Fock approximation (dashed line)

Since the edges are effectively far away from one another, the incoming electron is initially accommodated into the closest edge, say, j_+. It produces a compact charge disturbance with a high Coulomb energy. The charge has to gradually spread over the entire area. This takes place by the emission of the magnetophonons. (In the context of the edge-state theories they are traditionally called edge magnetoplasmons). What is missing in the Hartree–Fock picture is precisely this relaxation from a point-like disturbance into an uniform state with an added quasiparticle of the given small energy E. Its net result is a suppression of the true $g(E)$ compared to $g_{\mathrm{HF}}(E)$. To calculate the suppression factor, one can use the bosonization prescription $\Psi \propto \exp(i\phi_+^j)$ [42], which entails

$$\tilde{G}_{\mathrm{R}}(\mathrm{i}\omega_n) \propto \int_0^\infty d\tau \exp\left[\mathrm{i}\omega_n \tau - \frac{1}{2}\langle \phi_+^j(0)\phi_+^j(\tau) - \phi_+^j(0)\phi_+^j(0)\rangle\right]. \tag{49}$$

This expression can be evaluated with the help of the effective action (44) and then analytically continued to real frequencies $\mathrm{i}\omega_n \to \omega + \mathrm{i}0$ [10]. For the smectic, such a calculation has been performed by Lopatnikova it et al. [49]. They obtained $I(V) \propto \exp(-\sqrt{V_0/V})$ and $I(V) \propto \exp[-\ln^2(V_0/V)]$ for the Coulomb and short-range interaction case, respectively. The last formula was also obtained by Barci *et al.* [48]. As one can see, the tunneling conductance vanishes at zero voltage, which is a typical result for tunneling into a correlated electron system. For the crystal, the calculation is more difficult because the action is essentially nonlinear [the harmonic approximation (45) is not adequate for the tunneling problem]. It has not been reported in the literature. On the physical grounds, we expect the tunneling conductance to be exactly zero unless eV exceeds the energy needed to nucleate a phase slip $u_y \to u_y + a$ (an analog of an interstitial in conventional crystals).

10 Pinning of Stripes by Disorder

Disorder in the form of randomly positioned impurities is expected to strongly affect the degree of the positional, orientational, and topological orders in the stripe phase at low T. According to modern understanding, none of these survives in the thermodynamic limit in two dimensions. However, in the most interesting case of weak disorder, the details of how each type of order is destroyed are subtly different. Let us start the discussion with the case of a smectic. Here the positional order is limited primarily by random elastic distortions. They can be characterized by the roughness function

$$f(r) = \langle [u(\mathbf{r}') - u(\mathbf{r}' + \mathbf{r})]^2 \rangle, \tag{50}$$

so that the positional correlation length ξ_P can be defined by the condition $f(\xi_P) = \Lambda^2$. On the other hand, the correlation lengths for the topological order ξ_T and orientational order ξ_O are set by the mean distance between the disorder-generated dislocations and disclinations, respectively. Note that an appreciable magnetotransport anisotropy can be detected only if the sample size is smaller than ξ_O. This explains why samples of macroscopic dimensions $L \sim 0.25\,\mathrm{mm}$ show no anisotropy unless they are extremely pure [6,7]. In addition to weakness of disorder, the gigantic ξ_O in these experiments may have originated from a small bare anisotropy, which favored the alignment of the stripes along the [110] direction. Indeed, an external anisotropy qualitatively modifies the smectic state properties by generating a finite tilt modulus $Y_\perp$. Correspondingly, the elastic part $\mathcal{H}_{el}$ of the Hamiltonian becomes

$$\mathcal{H}_{\mathrm{el}} = \frac{Y}{2}\left[\partial_x u + \frac{1}{2}(\nabla u)^2\right]^2 + \frac{Y_\perp}{2}(\partial_y u)^2 + \frac{K}{2}(\partial_y^2 u)^2. \tag{51}$$

The full Hamiltonian $\mathcal{H} = \mathcal{H}_{\mathrm{el}} + \mathcal{H}_{\mathrm{pin}}$ also includes the interaction with disorder. It can be approximated by $\mathcal{H}_{\mathrm{pin}} = \int \exp[\mathrm{i}q_* u(\mathbf{r})] V(\mathbf{r})$ where $V(\mathbf{r})$ is a random short-range potential. The behavior of the static correlation functions, such as $f(r)$, as functions of r and disorder strength has been evaluated within this model by Scheidl and von Oppen [59] building on the theory developed earlier for other pinned systems [60,61]. Scheidl and von Oppen suggested that in the current experiments the characteristic lengthscales satisfy the chain of inequalities $\xi_\mathrm{P} < \xi_\mathrm{T} < L < \xi_\mathrm{O}$, so that the stripe phase resembles an instantaneous snapshot of a nematic with no long-range positional order and a number of frozen-in dislocations.

Besides decorrellating the stripe positions, another interesting disorder effect in the smectic phase can be inducing quenched random $2q_{\mathrm{F}}$-modulations along the stripes, as suggested by Yi et al. [47]. This is similar to the effect of impurities on a 1D "Wigner crystal" [62] (where without impurities the density is uniform because of strong quantum fluctuation). To properly describe this effect and its consequences, a comprehensive *quantum* theory of pinned smectics is required.

The pinning effects in the stripe crystal phase are qualitatively similar because H_{el} is already not too different from the elastic Hamiltonian of an anisotropic crystal. The main difference is the doubling of the number of components of the elastic displacement field $\mathbf{u} = \{u_x, u_y\}$.

The combined effects of finite temperature and disorder have not been investigated in any detail, but based on the extensive work done in other contexts [63,64] we may expect a thermal depinning at a certain "glass transition temperature."

Exceedingly interesting topic is the dynamics of pinned electron (liquid) crystals in high magnetic field. First of all, there are number of open questions in the general theory of pinned manifolds (such as CDWs, vortex lattices, magnetic domains, etc. The current interests in this subject area include dislocation dynamics and non-equilibrium states (e.g., high electric field sliding phases). In addition to that, there are interesting dynamical effects unique to quantum Hall systems. For instance, in the recent years it was experimentally discovered [65] that even the linear response of a pinned Wigner crystal in a magnetic field is dramatically different from conventional expectations [66]. Typically, one anticipates the response to be dominated by low-frequency magnetophonon modes, which in the presence of pinning become localized wavepackets. Since the disorder is different in different parts of the sample, a considerable inhomogeneous broadening is expected. Instead, a sharp resonance-like mode in the microwave absorption spectrum appears (typically around 1 GHz). Assuming that the quasiclassical description is adequate, Huse and the present author [67] explained the appearance of a narrow mode as a combined effect of the special Lorentz-force dynamics and long-range Coulomb interaction. However, the theory was done for the $T = 0$ while the experimental line width seem to be dominated by thermal broadening. In fact, the thermal broadening is much more pronounced than one would naively expect and may reflect some nontrivial physics overlooked in a formulation based on harmonic elasticity theory [67]. The similar pinning resonance should exist in a stripe crystal as well and may even persists in a smectic phase. Since at high LLs all the energy scales are smaller, the pinning mode should have a lower frequency, perhaps, of the order of 100 MHz at $\nu = 9/2$ for the kind of samples used in Refs. [6,7].

11 Magnetotransport in the Stripe Phase

The magnetotransport in a partially filled Landau level is a problem of two extremes. In a perfectly clean system, it is completely trivial and amounts to the uniform sliding motion of all the particles in the direction perpendicular to the applied electric field, leading to $\sigma_{xx} = 0$, $\sigma_{xy} = (e^2/h)\nu$. On the other hand, in the presence of disorder, the magnetotransport immediately becomes very complicated. Especially for the stripe phase, the magnetotransport theory is very incomplete at the moment. For instance, the transport in the nematic phase is a total enigma. Certain progress has been achieved regarding the crystal and the smectic phases. As discussed above, pinning by random impurities eliminates the possibility of a global sliding motion of the electron liquid under the action of an external electric field. As a result, the stripe crystal is an insulator. In the smectic, gapless edge states may exist in which case the edge transport is possible. The quasiparticles travel along the edges in the direction prescribed by the edge chiralities and also be scattered between the edges by impurities. The scattering probability depends on the distance between the edges. For a general filling fraction ν_N the widths of the filled (electron) and empty (hole) stripes are unequal, and so the corresponding mean-free paths are also different, $l_e \neq l_h$. On a large scale the quasiparticle motion resembles a random walk with unequal elementary steps: along the stripes, it is a properly weighted average of l_e and l_h, across the stripes it is the interstripe separation Λ. If we assume that (a) the stripes are continuous and are all oriented along the $\hat{\mathbf{y}}$-direction, (b) the scattering occurs only between nearest edges, and (c) quantum localization effects can be neglected, then a simple calculation leads to the following formulas for the components of the conductivity tensor [43,68]:

$$\sigma_{xx} = \frac{e^2}{h}\frac{\Lambda^{-1}}{l_e^{-1} + l_h^{-1}}, \quad \sigma_{yy} = \frac{e^2}{h}\frac{\Lambda}{l_e + l_h}, \quad \sigma_{xy} = \frac{e^2}{h}\left(N + \frac{l_e}{l_e + l_h}\right). \tag{52}$$

At the symmetric point $l_e = l_h$, the transport anisotropy ratios are given by $R = \sigma_{yy}/\sigma_{xx} = \rho_{xx}/\rho_{yy} = l_e^2/\Lambda^2$, which is large for dilute impurities. However, there are reasons to doubt that (52) directly applies to the experimental practice. Indeed, from the empirical estimate $R \sim 7$ one would conclude that $l_e \sim 2.6\Lambda \sim 0.2\,\mu$m, which seems too short for the extreme purity samples used. The most probable reason for reduction of the anisotropy ratio compared to the idealized formula (52), is the absence of high degree of the orientational and topological orders. As explained above, the pinning by disorder introduces pronounced elastic deformations as well as dislocations and disclinations into the perfect stripe pattern. Thus, in a more realistic model the above condition (a) should be abandoned. This leads to a view of the low-temperature stripe phase is a collection of elongated puddles of $\nu_N = 1$ state preferentially aligned in a certain direction (a kind of frozen nematic). The conductivity tensor can no longer be calculated in any simple way; however, it satisfies the generalized Dykhne-Ruzin semicircle law [69]

$$\sigma_{xx}\sigma_{yy} + (\sigma_{xy} - \sigma_0)^2 = (e^2/2h)^2, \quad \sigma_0 \equiv (e^2/2h)(2N + 1), \tag{53}$$

pointed out by von Oppen, Halperin, and Stern [68]. In addition, exactly at half-filling $\sigma_{xy} = \sigma_0$ and the product rule [69,43,9] applies:

$$\sigma_{xx}\sigma_{yy} = (e^2/2h)^2. \tag{54}$$

Both relations agree quite well with the low-temperature experimental data at $\nu = 9/2$ [8], which is encouraging. However, one has to be careful with drawing conclusions from such a comparison. The status of (53) and (54) in the quantum domain is poorly understood and is even somewhat compromised by the lack of agreement with the data from dirtier samples. The situation is complicated by a unconventional behavior of the resistivity peak at $\nu = 9/2$ (low-T width saturation), precisely where the product rule is supposed to apply.

At higher temperatures where the CDW amplitude is small, the stripe edges are not well defined objects. It is better to describe the system as a "compressible liquid." A new type of transport theory is needed in this regime. It is reasonable to assume, for example, that the stripes can be depinned already by a vanishingly small electric field, so that a global sliding of the electron liquid becomes possible. If disorder is sufficiently smooth, it would be similar to a flow of a viscous 2D liquid across a disordered substrate [16].

Sufficiently close to T_c the dislocations must become highly mobile so that the stripe phase should behave as a nematic on all relevant lenghtscales. A phenomenological approach to the magnetotransport in this regime was advocated by Fradkin et al. [41]. They argue on the basis of symmetry that the deviation of the resistivity ratio $R = \rho_{xx}/\rho_{yy}$ from unity is proportional to the degree of the orientational order in the nematic. Under a simplifying assumption that the Frank constants are equal, they map the problem onto a 2D X-Y model in the presence of a small orienting field. Using the Monte-Carlo results for the latter model, they fitted the experimental data for R with the help of two free parameters.

12 Other Topics

The space constraints force us to switch to a telegraphic-style overview of the remaining topics explored in connection with the CDW phases in high LLs.

One of those is the bare anisotropy that aligns the stripes in the [110] direction. Its origin remains unclear. A careful study by Cooper et al. [71] has ruled out a few earlier suggestions that included substrate morphology [70], vicinal steps from a slight miscut of the wafer, and a small anisotropy of the effective mass [72,73]. One of the remaining viable alternatives is piezoelectric effects [74]. The orienting effects of the in-plane magnetic field was further studied in wide quantum wells [78]. The combination of the empirical findings and the Hartree–Fock calculations suggest that the anisotropy is indeed tiny, $\sim 1\,\mathrm{mK}$ per electron. The corresponding analysis for the p-type GaAs [39] has also been performed [75]. It is more involved because of the band-structure anisotropy. The preferred stripe orientation was predicted to sensitively depend on various details such as the width of the quantum well.

The time-dependent Hartree–Fock calculations of the collective modes of the stripe crystal has been done by Fertig et al. [46,47]. At low q's the results are in agreement with the general structure predicted by the effective theories, Sects. 7

and 9. At higher q a new feature is revealed: a roton-like minimum at $q \approx q_*$, which may be thought of as a precursor to the stripe-bubble transition.

Stripe phases in bilayer systems have been studied by Brey and Fertig [76] and also by Demler et al. [77]. The additional layer degree of freedom leads to a rich phase diagram. Some of that physics may be operational in wide quantum wells [78].

The emergence of the CDW indicates some sort of instability of the conventional uniform FQHE states at high LLs. Such an instability was indeed found in numerical studies of Scarola et al. [79] who relied on the composite-fermion theory. It was also discussed within the phenomenological mean-field composite-boson formulation [73].

13 Conclusions

The physics of high LLs has fledged into a fast growing and vibrant field. Despite a significant progress, a multitude of open problems remains. Some of them are brand new (the existence of qualitatively novel states of matter such as quantum nematic), others are venerable quantum Hall problems with a new twist (the integral quantum Hall transition in the stripe phase). The real progress in understanding these intriguing yet difficult problems is likely to be tied with future experimental advances. It should be mentioned that the magnetotransport studies have certainly not exhausted their potential. Ongoing fantastic achievements in sample fabrication continue to deliver the ever more perfect systems, which should expedite the discovery of new phases. A fresh example is the aforementioned group of new insulating states in the $N = 1$ LL of $3.1 \times 10^7\,\mathrm{cm}^2/\mathrm{Vs}$ mobility sample [35] (Sect. 6). Besides relying on increasing sample quality, it may be interesting to experiment with different sizes of already available samples in search for the sequence of "spaghetti" phases sketched in Fig. 10. On the other hand, finite-frequency tools, such as microwaves, surface acoustic waves, and inelastic light scattering will provide other invaluable information inaccessible by the dc magnetotransport. Finally, the real-space imaging [80,81] of the stripes and bubbles would be the most definitive proof of their existence.

It is a safe bet that we will see a great number of surprises and new developments in this area.

Acknowledgements

This work is supported by the MIT Pappalardo Fellowships Program in Physics. I would like to thank A. A. Koulakov, B. I. Shklovskii, and V. M. Vinokur for previous collaboration during which the key ideas and insights surveyed in these notes have emerged. I also wish to thank B. I. Shklovskii for valuable comments on the manuscript.

References

1. P.T. Coleridge, P. Zawadzki, A. Sachrajda: Phys. Rev. B **49**, 10798 (1994)
2. L.P. Rokhinson, and V.J. Goldman, unpublished; S.A.J. Wiegers, M. Specht, L.P. Lévy, M.Y. Simmons, D.A. Ritchie, A. Cavanna, B. Etienne, G. Martinez, P. Wyder: Phys. Rev. Lett. **79**, 3238 (1997) for a discussion, see M.M. Fogler, and B.I. Shklovskii: Phys. Rev. B **52**, 17366 (1995)
3. N. Turner, J.T. Nicholls, E.H. Linfield, K.M. Brown, G.A.C. Jones, D.A. Ritchie: Phys. Rev. B **54**, 10614 (1996)
4. (a) A.A. Koulakov, M.M. Fogler, B.I. Shklovskii: Phys. Rev. Lett. **76**, 499 (1996) (b) M.M. Fogler, A.A. Koulakov, B.I. Shklovskii: Phys. Rev. B **54**, 1853 (1996)
5. M.M. Fogler, and A.A. Koulakov: Phys. Rev. B **55**, 9326 (1997)
6. M.P. Lilly, K.B. Cooper, J.P. Eisenstein, L.N. Pfeiffer, K.W. West: Phys. Rev. Lett. **82**, 394 (1999)
7. R.R. Du, D.C. Tsui, H.L. Störmer, L.N. Pfeiffer, K.W. West: Solid State Commun. **109**, 389 (1999)
8. J.P. Eisenstein, M.P. Lilly, K.B. Cooper, L.N. Pfeiffer, K.W. West: Physica E **9**, 1 (2001) J.P. Eisenstein: Solid State Commun. **117**, 132 (2001)
9. F. von Oppen, B.I. Halperin, A. Stern: Advances in Quantum Many-Body Theory, vol. 3, eds. R.F. Bishop, N.R. Walet, Y. Xian: (World Scientific, 2000) (also available as cond-mat/0002087)
10. G.D. Mahan: *Many-Particle Physics* (Plenum, New York, 1990)
11. At least for thin 2D layers, see X.-G. Wu, and S.L. Sondhi: Phys. Rev. B **51**, 14725 (1995) N.R. Cooper: Phys. Rev. B **55**, 1934 (1997)
12. I.L. Aleiner, and L.I. Glazman: Phys. Rev. B **52**, 11296 (1995)
13. I.V. Kukushkin, S.V. Meshkov, V.B. Timofeev: Usp. Fiz. Nauk **155**, 219 (1988) [Sov. Phys. Usp. **31**, 511 (1988)]
14. H. Fukuyama, P.M. Platzman, P.W. Anderson: Phys. Rev. B **19**, 5211 (1979)
15. For review, see *Quantum Hall Effect* eds. R.E. Prange, and S.M. Girvin: (Springer-Verlag, New York, 1990) C. Glattli: Chapter I in this book; R. Shankar: Chapter II *ibid.*
16. M.M. Fogler: unpublished
17. T. Stanescu, I. Martin, P. Philips: Phys. Rev. Lett. **84**, 1288 (2000)
18. R. Moessner, and J.T. Chalker: Phys. Rev. B **54**, 5006 (1996)
19. D. Yoshioka, and H. Fukuyama: J. Phys. Soc. Jpn. **47**, 394 (1979) D. Yoshioka, and P.A. Lee: Phys. Rev. B **27**, 4986 (1983)
20. M. Seul, and D. Andelman: Science, **267**, 476 (1995)
21. S.M. Girvin: Phys. Rev. B **30**, 558 (1984) S.M. Girvin, A.H. MacDonald, P.M. Platzman: Phys. Rev. B **33**, 2481 (1986)
22. K. Musaelian, and R. Joynt: J. Phys. Cond. Mat. **8**, L105 (1996)
23. L. Balents: Europhys. Lett. **33**, 291 (1996)
24. E. Fradkin, and S.A. Kivelson: Phys. Rev. B **59**, 8065 (1999)
25. M.M. Fogler: cond-mat/0107306
26. O. Ciftja, and C. Wexler: cond-mat/0108119
27. L. Radzihovsky, and A.T. Dorsey: cond-mat/0110083
28. J.M. Tranquada et. al.: Phys. Rev. Lett. **73**, 1003 (1994) J.M. Tranquada et. al.: Nature **375**, 561 (1995) S. Mori, C.H. Chen, S.-W. Cheong: Nature **392**, 473 (1998)
29. V.J. Emery, E. Fradkin, S.A. Kivelson: Nature **393**, 550 (1998)

30. J. Zaanen, O.Y. Osman, H.V. Kruis, Z. Nussinov, J. Tworzydlo: cond-mat/0102103 and references therein
31. E.H. Rezayi, F.D.M. Haldane, K. Yang: Phys. Rev. Lett. **83**, 1219 (1999) *ibid* **85**, 5396 (2000)
32. N. Shibata, and D. Yoshioka: Phys. Rev. Lett. **86**, 5755 (2001)
33. S.H. Simon: Phys. Rev. Lett. **83**, 4223 (1999)
34. At ultralow temperatures the $\nu = 5/2$ fraction exhibits a quantum Hall effect: a deep minimum in ρ_{xx} with the activated behavior and a concomitant quantum Hall plateau in ρ_{xy}, see W. Pan, J.-S. Xia, V. Shvarts, D.E. Adams, H.L. Stormer, D.C. Tsui, L.N. Pfeiffer, K.W. Baldwin, K.W. West: Phys. Rev. Lett. **83**, 3530 (1999) Recently, the quantum Hall effect was also discovered at $\nu = 7/2$ fraction, see Ref. [35]
35. J.P. Eisenstein, K.B. Cooper, L.N. Pfeiffer, K.W. West: cond-mat/0110477.
36. W. Pan, R.R. Du, H.L. Stormer, D.C. Tsui, L.N. Pfeiffer, K.W. Baldwin, K.W. West: Phys. Rev. Lett. **83**, 820 (1999)
37. M. Lilly, K.B. Cooper, J.P. Eisenstein, L.N. Pfeiffer, K.W. West: Phys. Rev. Lett. **83**, 824 (1999)
38. T. Jungwirth, A.H. MacDonald, L. Smrcka, S. Girvin: Phys. Rev. B **60**, 15574 (1999)
39. M. Shayegan, H.C. Manoharan, S.J. Papadakis, E.P. DePoortere: Physica E **6**, 40 (2000)
40. C. Wexler, and A.T. Dorsey: Phys. Rev. B **64**, 115312 (2001)
41. E. Fradkin, S.A. Kivelson, E. Manousakis, K. Nho: Phys. Rev. Lett. **84**, 1982 (2000)
42. X.-G. Wen: Int. J. Mod. Phys. B **6**, 1711 (1992)
43. A.H. MacDonald, and M.P.A. Fisher: Phys. Rev. B **61**, 5724 (2000)
44. H.A. Fertig: Phys. Rev. Lett. **82**, 3693 (1999)
45. M.M. Fogler, and V.M. Vinokur: Phys. Rev. Lett. **84**, 5828 (2000)
46. R. Côté, and H.A. Fertig: Phys. Rev. B **62**, 1993 (2000)
47. H. Yi, H.A. Fertig, R. Côté: Phys. Rev. Lett. **85**, 4156 (2000)
48. D.G. Barci, E. Fradkin, S.A. Kivelson, V. Oganesyan: cond-mat/0105448
49. A. Lopatnikova, B.I. Halperin, S.H. Simon, X.-G. Wen: Phys. Rev. B **64**, 155301 (2001)
50. L. Bonsall, and A.A. Maradudin: Phys. Rev. B **15**, 1959 (1977)
51. P.G. de Gennes, and J. Prost: *The Physics of Liquid Crystals* (Oxford University Press, New York, 1995)
52. J. Toner, and D.R. Nelson: Phys. Rev. B **23**, 316 (1981)
53. L. Golubović, and Z.-G. Wang: Phys. Rev. Lett. **69**, 2535 (1992)
54. G.F. Mazenko, S. Ramaswamy, J. Toner: Phys. Rev. Lett. **49**, 51 (1982) Phys. Rev. A **28**, 1618 (1983)
55. E.I. Kats, and V.V. Lebedev: *Fluctuational Effects in The Dynamics of Liquid Crystals* (Springer-Verlag, New York, 1994)
56. J. Toner: Phys. Rev. B **26**, 462 (1982) A.R. Day, T.C. Lubensky, A.J. McKane: Phys. Rev. A **27**, 1461 (1983)
57. M.P.A. Fisher, and D.H. Lee: Phys. Rev. B **39**, 2756 (1989)
58. E.H. Rezayi, and F.D.M. Haldane: Phys. Rev. Lett. **84**, 4685 (2000)
59. S. Scheidl, and F. von Oppen: Europhys. Lett. **55**, 260 (2001)
60. J. Villain, and J.F. Fernandez: Z. Phys. B **54**, 139 (1984)
61. L. Radzihovsky, and J. Toner: Phys. Rev. B **60**, 206 (1999)
62. L.I. Glazman, I.M. Ruzin, B.I. Shklovskii: Phys. Rev. B **45**, 8454 (1992)

63. G. Blatter, M.V. Feigelman, V.B. Geshkenbein, A.I. Larkin, V.M. Vinokur: Rev. Mod. Phys. **66**, 1125 (1994)
64. T. Nattermann, and S. Scheidl: Adv. Phys. **49**, 607 (2000)
65. C.-C. Li, L.W. Engel, D. Shahar, D.C. Tsui, M. Shayegan: Phys. Rev. Lett. **79**, 1353 (1997) A.S. Beya: Ph.D. Thesis, L'Université Paris VI (1998) C.-C. Li, J. Yoon, L.W. Engel, D. Shahar, D.C. Tsui, M. Shayegan: Phys. Rev. B **61**, 10905 (2000)
66. H. Fukuyama, and P.A. Lee: Phys. Rev. B **18**, 6245 (1978)
67. M.M. Fogler, and D.A. Huse: Phys. Rev. B **62**, 7553 (2000)
68. F. von Oppen, B.I. Halperin, A. Stern: Phys. Rev. Lett. **84**, 2937 (2000)
69. A.M. Dykhne, and I.M. Ruzin: Phys. Rev. B **50**, 2369 (2000)
70. R.L. Willett, J.W.P. Hsu, D. Natelson, K.W. West, L.N. Pfeiffer: Phys. Rev. Lett. **87**, 803 (2001)
71. K.B. Cooper, M.P. Lilly, J.P. Eisenstein, T. Jungwirth, L.N. Pfeiffer, K.W. West: Solid State Commun. **119**, 89 (2001)
72. E.E. Takhtamirov, and V.A. Volkov: cond-mat/0106162
73. B. Rosenow, and S. Scheidl: Int. J. Mod. Phys. B **15**, 1905 (2001)
74. D.V. Fil: [Sov. J. Low Temp. Phys. **26**, 581 (2000)] See also E.Ya. Sherman: Phys. Rev. B **52**, 1512 (1995)
75. T.S. Kim, S.-R.E. Yang, A.H. MacDonald: Phys. Rev. B **62**, R7747 (2000)
76. L. Brey, and H.A. Fertig: Phys. Rev. B **62**, 10268 (2000)
77. E. Demler, D.-W. Wang, S. Das Sarma, B.I. Halperin: cond-mat/0110126
78. W. Pan, T. Jungwirth, H.L. Stormer, D.C. Tsui, A.H. MacDonald, S.M. Girvin, L. Smrcka, L.N. Pfeiffer, K.W. Baldwin, K.W. West: Phys. Rev. Lett. **85**, 3257 (2000)
79. V.W. Scarola, K. Park, J.K. Jain: Phys. Rev. B **62**, R16259 (2001)
80. G. Finkelstein, P.I. Glicofridis, R.C. Ashoori, M. Shayegan: Science **289**, 90 (2000)
81. S. Ilani, A. Yacoby, D. Mahalu, H. Shtrikman: Phys. Rev. Lett. **84**, (2000)

Low Dimensional Magnets

T.M. Rice

Theoretical Physics, ETH–Hönggerberg, CH–8093 Zürich, Switzerland

Abstract. The properties of antiferromagnetically coupled $S = 1/2$ spin chains and unfrustrated spin planes have been understood for some time. When the chains are coupled together to form spin ladders, the properties do not simply interpolate between the two limits but new spin liquid groundstates appear with strictly short range correlations and with a spin gap in the excitation spectra. These in turn undergo a soft mode transition under the application of an external magnetic field or through the introduction of unfrustrated interladder coupling.

1 Introduction

Quantum effects are enhanced in magnetic systems with a low spin, e.g. $S = 1/2$, and with antiferromagnetic coupling. These quantum effects are further magnified in low dimensions. For these reasons the study of low dimensional $S = 1/2$ magnets has been a very active topic in quantum magnetism in recent times. The simple examples such as a spin chain for one dimension [1,2] and a square lattice for two dimensions [3], have been well understood for some time. However it was only relatively recently that it was realized that when spin chains are coupled together to form spin ladders, novel properties are obtained [4]. These lectures will be mainly devoted to a discussion of spin ladders with emphasis on an intuitive and phenomenological description of their properties rather than the sophisticated analytic and numerical methods that have been developed to analyze them. The recent interest in cuprates, because of their unique superconductivity, has given a strong stimulus to this field since the Cu^{2+}-ion of the parent insulators has an ideal $S = 1/2$ local moment. These local moments can be arranged in a very flexible manner to form spin ladders and other interesting structures. Stoichiometric, i.e. undoped, cuprates give us almost ideal realizations of many spin structures of theoretical interest. Of course the effects of doping the cuprates are also of great interest but that is a subject outside the scope of these lectures.

The outline of the lectures is as follows. After a brief introduction on spin chains, the spin liquid state of 2-leg ladders will be discussed. In keeping with the overall theme of this summer school, the effects of a strong magnetic field will be reviewed. Then the discussion will be generalized to wider ladders again with and without an external magnetic field. Lastly the effect of interladder coupling to form higher dimensional structures and the resulting quantum critical point at the transition to long range antiferromagnetic order will be treated.

2 Antiferromagnetically Coupled S=1/2 Spin Chains

Spin chains with $S = 1/2$ moments coupled by nearest neighbor (n.n.) antiferromagnetic interactions are most conveniently analyzed by starting with a so called XXZ-model which has differing exchange constants for spin orientations in the xy-plane ($J_{xy}(> 0)$) and perpendicular ($J_z(> 0)$) to it. In the presence of an external field, H^z, one has a Hamiltonian

$$\mathcal{H} = \sum_{j=1}^{N} \left[J_{xy} \left(S_j^x S_{j+1}^x + S_j^y S_{j+1}^y \right) + J_z S_j^z S_{j+1}^z - H^z S_j^z \right] . \tag{1}$$

The notation is standard. To analyze this model it is convenient to introduce the raising and lowering spin operators; $S_j^{\pm} = S_j^x \pm i S_j^y$ and rewrite (1) as

$$\mathcal{H} = \sum_{j=1}^{N} \left[\frac{1}{2} J_{xy} (S_j^+ S_{j+1}^- + S_j^- S_{j+1}^+) + J_z S_j^z S_{j+1}^z - H^z S_j^z \right] . \tag{2}$$

Introducing the well known Jordan-Wigner transformation [1] allows us to express H in terms of spinless fermion operators $c_j^+(c_j)$ defined by

$$\begin{aligned} S_j^+ &= c_j^+ \exp\Big(\pi i \sum_{n<j} c_n^+ c_n \Big); \; S_j^- = \exp\Big(\pi i \sum_{j<n} c_n^+ c_n \Big) c_j \\ S_j^z &= c_j^+ c_j - 1/2 \end{aligned} \tag{3}$$

leading to the spinless fermion Hamiltonian, ($n_j = c_j^+ c_j$)

$$\mathcal{H}_F = t \sum_{j=1}^{N} (c_j^+ c_{j+1} + \text{h.c.}) + +V \sum_{j=1}^{N} n_j n_{j+1} - \mu \sum_{j=1}^{N} n_j \tag{4}$$

with a hopping parameter $t = J_{xy}/2$, a n.n. interaction $V = J_z$ and a chemical potential, $\mu = J_z + H^z$.

In the limit $J_z \to 0$, only the hopping term remains and the resulting XY-model is immediately solved as free fermions with a dispersion $\varepsilon(k) = 2t \cos(k)$. In the absence of a magnetic field this band is half filled as illustrated in Fig. 1. The excitation spectrum consists of particle-hole excitations across the Fermi energy with energy spectrum, $\omega_q = \varepsilon(k+q) - \varepsilon(k)$. The result as shown in Fig. 2 is a continuum spectrum stretching down to zero energy at the points $q = 0$ and $q = \pi$.

The J_z-term introduces interactions between the fermions and transforms the free fermions into a Luttinger liquid, which is characterized by power law forms for the correlation functions [5]. The characteristic exponents are functions of the ratio, J_{xy}/J_z. The isotropic Heisenberg form, $J_{xy} = J_z = J$, is the boundary between the gapless Luttinger liquid regime and the Ising regime which occurs when $J_z > J_{xy}$.

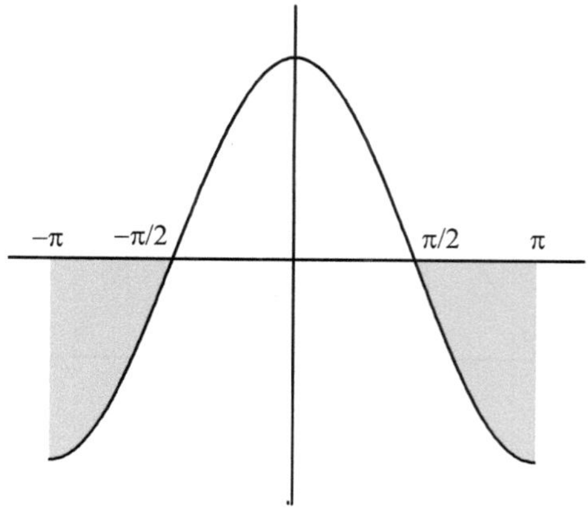

Fig. 1. The dispersion relation for the spinless fermion band which is obtained by applying a Jordan-Wigner transformation to the XY-model for a single chain

In the latter case there is a finite gap in the energy spectrum to turn over a spin. In the Heisenberg case there are logarithmic corrections and the principal properties in the low temperature regime can be obtained by renormalization group methods [5]. However the description of these methods is outside the scope of these lectures and we will just quote the main results here.

The spin-spin correlation function has a power law falloff

$$\langle \boldsymbol{S}_r \cdot \boldsymbol{S}_0 \rangle \underset{r \to \infty}{\sim} (-1)^r \ln^{1/2}(r)/r \,. \tag{5}$$

The uniform spin susceptibility goes to a constant as the temperature $T \to 0$ with a logarithmic correction of the form

$$\chi(0,T) = \pi^{-2} J^{-1}\left(1 + 1/2 \ln^{-1}(T_0/T)\right) + \dots \tag{6}$$

while the staggered susceptibility diverges as

$$\chi(\pi,T) = A \ln^{1/2}(T_0/T)/T \underset{T \to 0}{\longrightarrow} \infty \,. \tag{7}$$

The excitation spectrum is a continuum similar to that sketched in Fig. 2. In spin language this continuum can be interpreted as the decay of a $S = 1$ excitation obtained from turning over a spin into two separate excitations which are referred to as spinons. A simple intuitive picture of this process can be obtained by considering the evolution of an overturned spin in a AF ordered chain (see Fig. 3). Such an $S = 1$ configuration under the action of the spin flip terms in the Heisenberg Hamiltonian (2) will decay into two separate excitations, each with domain wall character as sketched in Fig. 3, which will propagate independently. In summary, the strong quantum fluctuations in the Heisenberg $S = 1/2$ chain prevent true long range AF order even in the groundstate but the correlation functions have a critical power law decay at large distances.

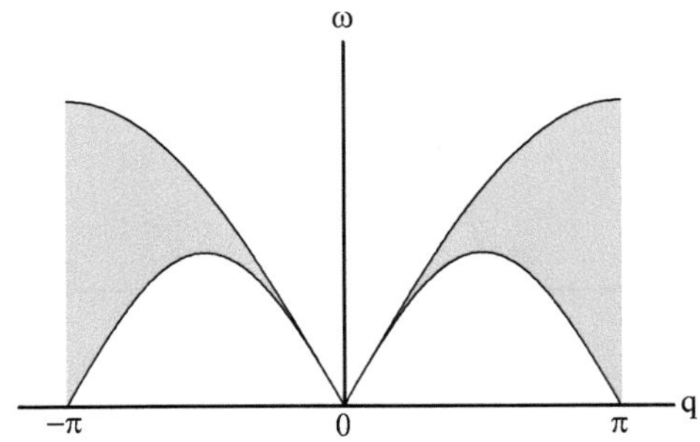

Fig. 2. The particle-hole excitation spectrum of the spinless fermion model. Note there are gapless excitations with arbitrarily small energies near to wave vectors 0 and π

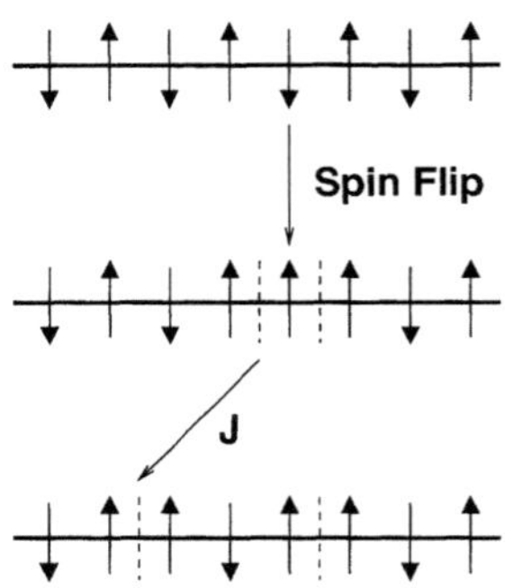

Fig. 3. The decay of an $S = 1$ excitation in an AF Heisenberg chain can be simply understood by considering the consequences of turning over a single spin to create a $S^z = 1$ configuration in an AF chain (upper line). Subsequently under the action of the spin flip terms in Eq. (2) 2 domain wall excitations are obtained, which propagate independently (lower lines)

If we define a spin liquid as a state with strictly exponentially decaying correlation functions the simple Heisenberg chain with n.n. exchange interactions is not a spin liquid. However a spin liquid can be made by introducing frustration, e.g. through the addition of a n.n.n. (next nearest neighbor) AF coupling. The Hamiltonian now has the form

$$\mathcal{H}_{\mathrm{MG}} = J\sum_j \boldsymbol{S}_j \cdot \boldsymbol{S}_{j+1} + J'\sum_j \boldsymbol{S}_j \cdot \boldsymbol{S}_{j+2} \tag{8}$$

with both $J > 0$ and $J' > 0$ which leads to frustration. This model is particularly easy to analyze in the special case, $J' = J/2$. This Majumdar-Ghosh model [6] has a two fold degenerate groundstate consisting of uncoupled dimer singlets as shown in Fig. 4. This is easily seen if we separate $\mathcal{H}_{\mathrm{MG}}$ into intradimer and interdimer terms, $\mathcal{H}'$ and $\mathcal{H}''$ with

$$\begin{aligned}
\mathcal{H}' &= J\sum_{j=1}^{N/2} \boldsymbol{S}_{2j} \cdot \boldsymbol{S}_{2j+1} \\
\mathcal{H}'' &= J\sum_{j=1}^{N/2} \boldsymbol{S}_{2j+1} \cdot \boldsymbol{S}_{2j+2} + 1/2\left(\boldsymbol{S}_{2j} \cdot \boldsymbol{S}_{2j+2} + \boldsymbol{S}_{2j+1} \cdot \boldsymbol{S}_{2j+3}\right) \\
&= J/2\sum_{j=1}^{N/2} \left(\boldsymbol{S}_{2j+1} + \boldsymbol{S}_{2j}\right) \cdot \boldsymbol{S}_{2j+2} + \boldsymbol{S}_{2j+1} \cdot \left(\boldsymbol{S}_{2j+2} + \boldsymbol{S}_{2j+3}\right) . \qquad (9)
\end{aligned}$$

We see at once that a product of singlets, $\prod_{j=1}^{N/2} \left(\uparrow_{2j}\downarrow_{2j+1} - \downarrow_{2j}\uparrow_{2j+1}\right)$, is an eigenstate of $\mathcal{H}'$ and also an eigenstate of $\mathcal{H}''$ with eigenvalue zero. Moreover in such a product of singlets the spins in neighboring singlets are totally uncorrelated so that $\langle \boldsymbol{S}_{j+r} \cdot \boldsymbol{S}_j \rangle = 0$ for $r \geq 2$. There is now an energy gap to the lowest excited state which, as shown by Shastry and Sutherland [7], is created by forming two separated domain walls in the dimerized chains and recoupling the intervening spins to form spin singlet dimers. This costs just a single singlet energy, $3J/4$, and sets the lower bound of a continuum of excited states.

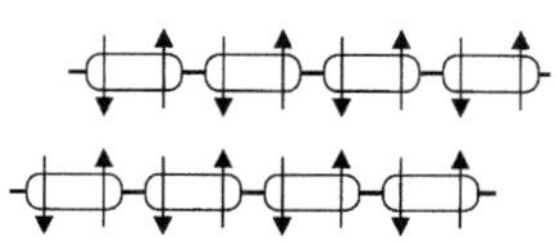

Fig. 4. The groundstate of the Majumdar-Ghosh model (Eqn. (8) with $J' = J/2$) is exactly a set of uncoupled spin singlet dimers and is two-fold degenerate

The spin liquid state at $J' = J/2$ is qualitatively different to the Heisenberg model groundstate at $J' = 0$ which is characterized by a critical power law decay of the spin-spin correlations. If the ratio J'/J is varied continuously, then one finds that the two forms of groundstate are separated by a quantum critical point so that for $J'/J < \gamma$, the system is gapless with powerlaw correlations while for $J'/J > \gamma$, there is a spin gap and exponentially decaying correlations. Numerical estimates give a value $\gamma \simeq 1/4$ [8].

3 Spin Liquid State of 2-Leg Ladders

When two chains are coupled together to form a 2-leg ladder, one might have guessed that the AF correlations would be reinforced by the coupling since there is no frustration involved. However the opposite is the case. We consider just the spin symmetric case with Heisenberg couplings of the form

$$\mathcal{H}_L = J_\perp \sum_i^N \boldsymbol{S}_{i1} \cdot \boldsymbol{S}_{i2} + J_{//} \sum_{i=1}^N \sum_{j=1}^2 \boldsymbol{S}_{i,j} \cdot \boldsymbol{S}_{i+1,j} \tag{10}$$

where the index i labels the rungs and j the sites on a rung. The simplest case occurs when the exchange coupling along the rungs, $J_\perp$, greatly exceeds that on the legs, $J_{//}$. In this case a good starting approximation is simply a product of singlets on each rung as sketched in Fig. 5. This state like the Majumdar-Ghosh state we discussed above, has clearly only short range spin correlations. There is also an energy gap to the lowest excited state — a triplet on a rung which costs an energy, $J_\perp$ (see Fig. 5). Note that in the case of the 2-leg ladder, a triplet on a single rung forms a bound state since to separate the spins one must break two rung singlets which costs more energy. This triplet can however disperse along the ladder through the action of the ladder coupling leading to a dispersion relation [9].

$$\omega(k) = J_\perp + J_{//} \cos k + \left(J_{//}^2/4J_\perp\right)(3 - \cos 2k) + \ldots \tag{11}$$

in an expansion in $J_{//}/J_\perp$. The minimum value of the excitation energy deter-

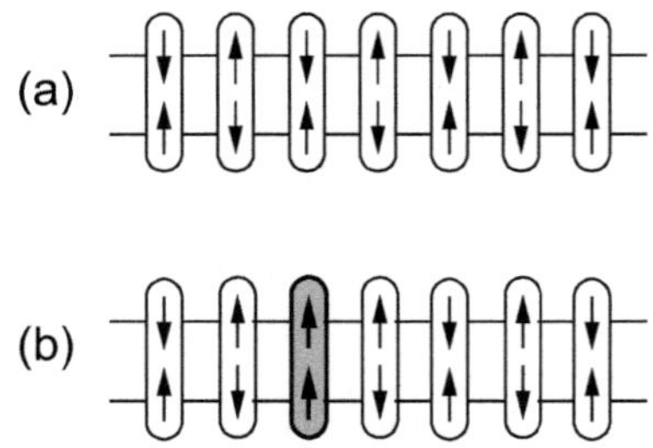

Fig. 5. In the limit that the rung coupling, $J_\perp$ is much larger than the leg coupling, $J_{//}$, the groundstate becomes simply a product of rung singlets (**a**). A triplet excitation is formed by changing a singlet to a triplet spin configuration on a single rung (**b**)

mines the spin gap, $\omega(\pi) = J_\perp - J_{//} + J_{//}^2/2J_\perp$ $(J_{//}/J_\perp \ll 1)$. The existence of a spin gap in this limit is not surprising but the result of Dagotto, Riera and Scalapino [10] that a finite spin gap persists for arbitrary values of $J_{//}/J_\perp$, was a surprise.

It is very illuminating to rewrite the Heisenberg Hamiltonian on the ladder in terms of bosonic rung operators following the approach of Sachdev and Bhatt [11]. These are defined for the singlet and the three components of the triplet on a rung as

$$\begin{aligned} s^+|0\rangle &= 2^{-1/2}\,|\uparrow\downarrow - \downarrow\uparrow\rangle\ ;\ t_x^+|0\rangle = -2^{-1/2}\,|\uparrow\uparrow - \downarrow\downarrow\rangle \\ t_y^+|0\rangle &= i\,2^{-1/2}\,|\uparrow\uparrow + \downarrow\downarrow\rangle\ ;\ t_z^+|0\rangle = 2^{-1/2}\,|\uparrow\downarrow + \downarrow\uparrow\rangle\ . \end{aligned} \tag{12}$$

The Heisenberg Hamiltonian for the ladder (10) can be rewritten in terms of these bosonic rung operators [12] as the sum of three terms

$$\mathcal{H}_L = \mathcal{H}_0 + \mathcal{H}_1 + \mathcal{H}_2 \,. \tag{13}$$

The first term $\mathcal{H}_0 = J_\perp \sum_j \big(-(3/4)\, s_j^+ s_j + (1/4)\, t_{j\alpha}^+ t_{j\alpha}\big) - \sum_j \mu_j \big(s_j^+ s_j + t_{j\alpha}^+ t_{j\alpha} - 1\big)$ describing intrarung processes. The interrung terms are divided into those processes which involve both singlet and triplet operators

$$\mathcal{H}_1 = (J_{/\!/}/2) \sum_j \big(t_{j\alpha}^+ \, t_{j+1\alpha}\, s_{j+1}^+ \, s_j + t_{j\alpha}^+ \, t_{j+1\alpha}^+ \, s_j \, s_{j+1} + \text{h.c.}\big) \tag{14}$$

and a term $\mathcal{H}_2$ involving only triplet-triplet interactions. The local Lagrange multiplier term in $\mathcal{H}_0$ is introduced to satisfy the constraint that each rung can be occupied only by the singlet or one of the triplet configurations. In this form $\mathcal{H}_L$ can be treated in a mean field approximation for a Bose condensate of singlets with site independent values $\mu_j = \mu$ and $\langle s_j \rangle = \bar{s}$, ignoring the term $\mathcal{H}_2$. The values of μ and $\bar{s}$ are determined by minimizing the energy values. This mean field approximation gives accurate results for the triplet magnon dispersion $\omega(k)$, at least for values, $1 \geq J_{/\!/}/J_\perp \geq 0$ as can be seen from Fig. 6 [13]. An advantage of this mean field approximation scheme is that it be straightforwardly extended to coupled ladder systems in higher dimensions [15].

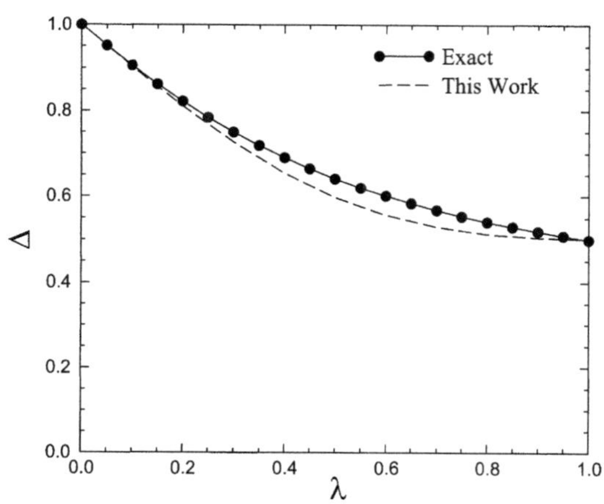

Fig. 6. The evolution of the spin gap, Δ, with the parameter $\lambda = J_{/\!/}/J_\perp$ as found in the mean field approximation to $\mathcal{H}_\mathrm{L}$ and from exact diagonalization of finite ladders following Piekarewicz and Shepard [13,14]

The existence of a finite energy gap between the quantum coherent ground-state and all excited states causes the uniform susceptibility, $\chi(T)$ to vanish exponentially at low T. In this limit Troyer et al. [16] treated the system as a dilute gas of thermally excited triplet magnons, each of which contributes a Curie term to $\chi(T)$ and arrived at the form

$$\chi(T) = 1/2\,(\pi a T)^{-1/2} \exp(-\Delta/T) \tag{15}$$

where the parameters Δ and a are determined by an expansion of $\omega(k)$ around the spin gap minimum, $\omega(k) = \Delta + a(k - \pi)^2$. At higher temperatures this approach breaks down as the local constraint on the occupation of a rung by a triplet or a singlet becomes relevant. Troyer et al. [16] proposed an approximate scheme to include the constraint and interpolate between the low temperature form and a Curie-Weiss form at high temperatures

$$\chi(T) = 1/4T - 3J/16T^2 + \ldots \quad (J_{//} = J_\perp = J)\,. \tag{16}$$

Their interpolation formula works well when compared to a numerical calculation of $\chi(T)$ which was based on a transfer matrix method. This method has the advantage that it allows a direct calculation to be carried through for an infinite ladder.

Turning now to the weak coupling region $J_\perp/J_{//} < 1$, there have also been extensive numerical investigations, particularly by Greven, Birgeneau and Wiese [17]. These authors simulated large systems down to very low temperature using the powerful Loop Algorithm in quantum Monte Carlo (QMC) calculations. Their results for the spin gap are shown in Fig. 7. The spin gap vanishes linear in the ratio $J_\perp/J_{//}$ in the limit $J_\perp/J_{//} \to 0$. This agrees with the results of field theoretic treatments by Shelton, Nersesyan and Tsvelik [18], and Hatano and Nishiyama [19]. The opening of a spin gap for arbitrarily small values of $J_\perp$ can be intuitively understood in terms of the critical state of a single chain which causes the rung coupling to be an essential perturbation on the limit of two decoupled chains at $J_\perp = 0$ [24]. If we remember the simple representation of a spinon excitation on a chain as a domain wall in an AF background then as illustrated in Fig. 3, it is easy to see that there is a linear attractive potential between spinons on neighboring legs of the ladder (see Fig. 8). This is obviously an essential perturbation which can qualitatively change the behavior of the system. As mentioned earlier, examples of antiferromagnetically coupled $S = 1/2$ systems are realized in cuprates. The Cu^{2+}-ion generally has a CuO_4-square coordination with the 4 neighboring O^{2-}-ions in cuprates. The 9 electrons in the 3d-shell of a Cu^{2+}-ion leaves a single hole (or equivalently a single electron) in the uppermost anti-bonding level which gives rise to a $S = 1/2$ local moment. The exchange interaction between n.n. Cu^{2+}-ions depends on the coordination of the CuO_4-squares (Fig. 9). Corner sharing of the CuO_4 squares results in a 180° Cu-O-Cu superexchange path between the local moments and an intermediate state with both holes on the 2p-O orbital that is σ-bonded to both Cu-ions. According to the Goodenough-Kanamori rules, this singlet intermediate state leads to a very strong AF coupling. Typical values in the cuprates are very large $J \sim 1,500 - 2,000$ K. Edge sharing of the CuO_4 squares leads to 90° Cu-O-Cu superexchange paths and an intermediate state with the holes in orthogonal 2p-orbitals and therefore weak ferromagnetic exchange, with typical

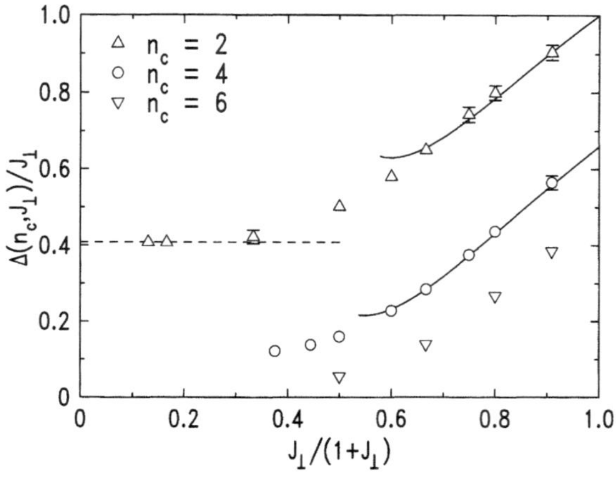

Fig. 7. The spin gap, Δ, for a 2-leg ladder (upper curve) determined from extensive numerical simulations by Greven, Birgeneau and Wiese [17] (note $J_{//}$ is set equal to 1 here). The lines at small and large $J_\perp$ are the asymptotic forms in these limits, [18,?] and [9]. Also shown are the results for 4-leg and 6-leg ladders which have much reduced values of Δ

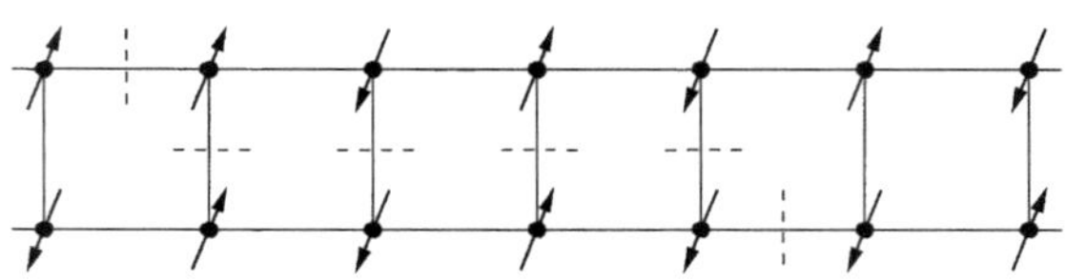

Fig. 8. Illustrating the attractive interaction between spinons introduced on opposite legs due to the presence of rungs with ferromagnetic configurations in the intervening region between the spinons

value of order $J' \sim -100$ K from the Goodenough-Kanamori rules. This difference and the flexibility of the chemistry allows different magnetic structures to be synthesized. To illustrate this flexibility in the superexchange coupling we consider two examples. In Sr_2CuO_3 the chains of corner sharing CuO_4-squares are well isolated from each other (Fig. 10) so that it can be viewed as an assembly of non-interacting AF $S = 1/2$ chains. Motoyama et al. [25] could fit their measurements of the spin susceptibility accurately to the form (6) and obtained a value for the n.n. $J = 2.2 \times 10^3$ K (Fig. 11). Weakly coupled spin ladders with strong exchange constants are realized in the planar compounds $SrCu_2O_3$ and $Sr_2Cu_3O_5$ since the intraplanar coupling of the ladders is through frustrated and weak F-exchange couplings (see Fig. 12) as was pointed out by Rice, Gopalan and Sigrist [26]. The spin susceptibility reported by Azuma et al. [27] for $SrCu_2O_3$ shows an exponential fall off at low temperatures after a small Curie term due to impurities. A more careful analysis by Johnston et al. [28] gave a good fit to an isolated two leg ladder model but with substantial anisotropy between the n.n. exchange constants along legs $J_{//}$ and rungs, $J_\perp$ for

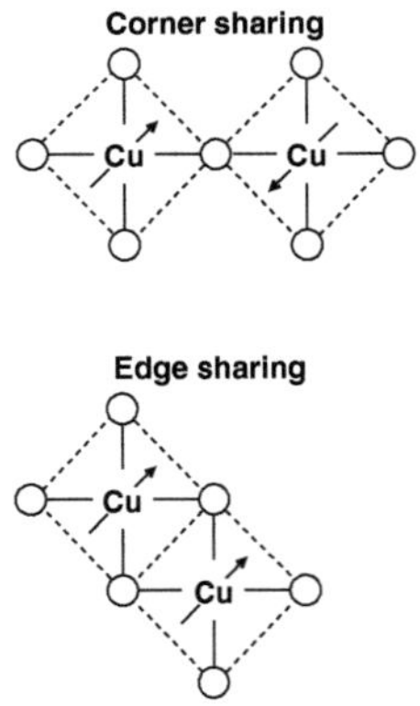

Fig. 9. Two neighboring CuO_4-squares which are arranged in a corner sharing and in an edge sharing configuration. The Cu-O-Cu bond angle is then 180° and 90° respectively

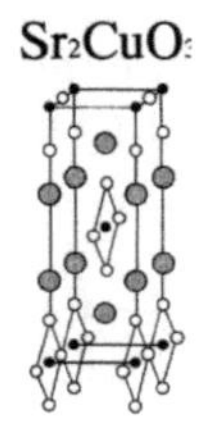

Fig. 10. The crystal structure of Sr_2CuO_3 showing the chains of corner sharing CuO_4-squares which are well isolated from each other

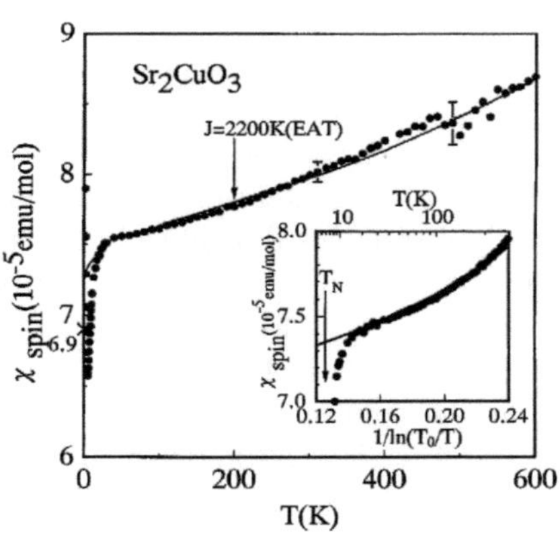

Fig. 11. The uniform spin susceptibility $\chi(T)$ of Sr_2CuO_3 and a theoretical fit to the theoretical form, Eq. (21), from 2-loop RG calculations for isolated AF Heisenberg chains with a n.n. coupling constant $J = 2,200$ K, from Motoyama et al. [25]

which they found a best fit with the value $J_\perp/J_{//} \simeq 0.5$ (see Fig. 13). Note in both cases there is a 180° Cu-O-Cu superexchange path. The analysis by Johnston et al. [28] shows that more complicated superexchange paths than those thru' the intervening O-atoms are entering the exchange constants. The results for the 3-leg ladder compound $Sr_2Cu_3O_5$ will be discussed later.

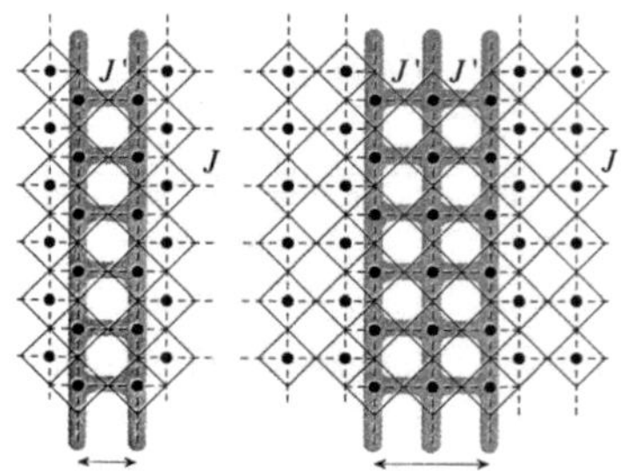

Fig. 12. Illustrating the crystal structures of the planar cuprates, $SrCu_2O_3$ and $Sr_2Cu_3O_5$. The pattern of 180° Cu-O-Cu patterns shown shaded is that of isolated 2-leg and 3-leg ladders respectively. The ladders are coupled through 90° Cu-O-Cu bonds which are weak and ferromagnetic. In addition the frustration in the interladder coupling greatly reduces its effectiveness

4 Two-Leg Ladders in an External Magnetic Field

One aspect of the study of spin liquids as exemplified by 2-leg ladders, that belongs in this school, is their behavior in an external magnetic field, H^z. This has been extensively studied both theoretically and experimentally. We shall not attempt a complete review here but concentrate on introducing the basic physics. In the presence of an external magnetic field the threefold degenerate $S = 1$ magnon splits into three separate excitations with $S^z = \pm 1$ and 0. The lowest component with S^z parallel to H^z moves lower in energy linearly in the field. There has been much interest in the quantum critical point that appears when the energy of this excitation passes through zero. In view of the relatively large exchange constants in the simple ladder cuprates discussed in the previous section, their critical magnetic field is very large. But there is another family of spin ladder systems also based on the Cu^{2+}-ion, $Cu_2(C_5H_{12}N_2)_2Cl_4$, which has weaker exchange constants, and so is amenable to study in attainable magnetic fields [25–27] (Fig. 14). The physics of this system is easiest to see again in the limit of strong rung coupling, $J_\perp \gg J_{//}$ as discussed by Mila [28]. It is also useful to generalize the ladder Hamiltonian (10) to add a n.n.n. diagonal coupling, (Fig. 15), J_2 and of course an external field H^2. Consider first an isolated rung. The energies of the singlet $(-3J_\perp/4)$ and the $S^z = +1$ component

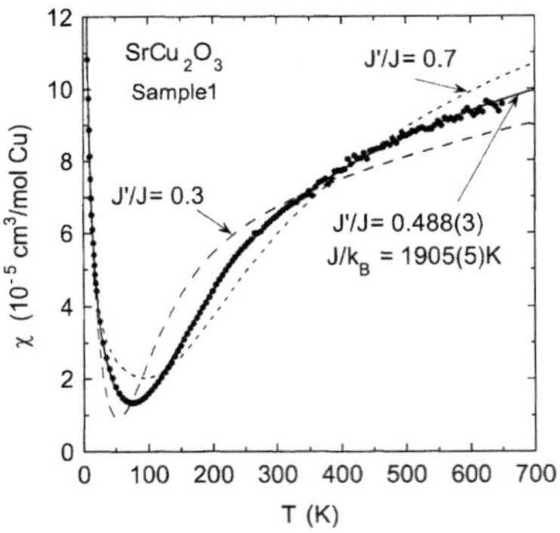

Fig. 13. The spin susceptibility $\chi(T)$ versus temperature T for $SrCu_2O_3$ and fits with the theoretical form for isolated 2-leg ladders with an impurity Curie term. The best fit is obtained with a rung coupling $J_\perp(=J')$ and a leg coupling $J_{//}(=J)$ in the ratio $J_\perp/J_{//}=0.488$ and a value $J_{//}=1905$ K, from Johnston et al. [28]

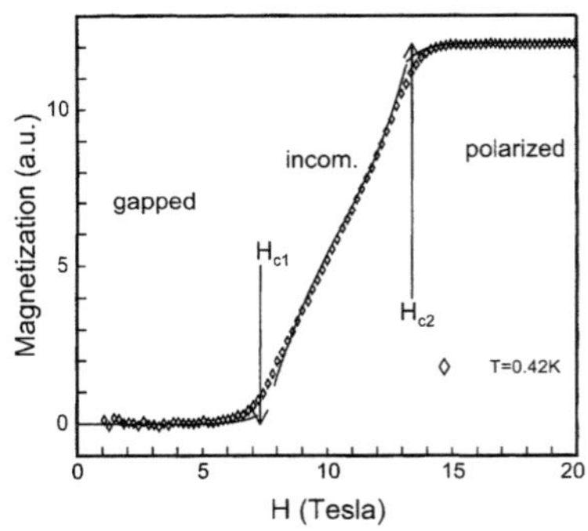

Fig. 14. The magnetization of the 2-leg ladder compound Cu versus magnetic field measured by Chaboussant et al. [27] at a low temperature. The solid line is a fit with a theory for isolated ladders

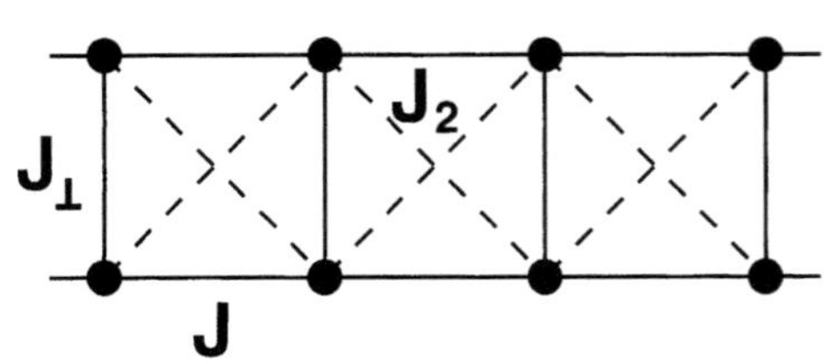

Fig. 15. The frustrated spin ladder with a diagonal n.n.n. coupling J_2 in addition to the rung and leg couplings, $J_\perp$ and $J_{//}$ respectively

of the triplet $(+J_\perp/4-H^z)$ become degenerate at a field strength $H_c^z=J_\perp$. At fields $H^z\sim H_c^z$ these two states form a nearly degenerate doublet and so it is useful to decompose the Hamiltonian by splitting off a single rung term in a magnetic field H_c^z,

$$\mathcal{H}_{\mathrm{L},0} = J_\perp \sum_{i=1}^{N} \boldsymbol{S}_{i1} \cdot \boldsymbol{S}_{i2} - J_\perp \sum_{i=1}^{N} \sum_{j=1}^{2} S_{i,j}^z . \tag{17}$$

The remaining terms including the diagonal coupling and external field are

$$\begin{aligned} \mathcal{H}_{\mathrm{L},1} = {} & J_{/\!/} \sum_{i=1}^{N} \sum_{j=1}^{2} \boldsymbol{S}_{i,j} \cdot \boldsymbol{S}_{i+1,j} + J_2 \sum_{i=1}^{N} \left(\boldsymbol{S}_{i,1} \cdot \boldsymbol{S}_{i+1,2} + \boldsymbol{S}_{i,2} \cdot \boldsymbol{S}_{i+1,1} \right) \\ & - \left(H^z - J_\perp \right) \sum_{i=1}^{N} \sum_{j=1}^{2} S_{i,j}^z . \end{aligned} \tag{18}$$

Now the lowest lying doublet of $\mathcal{H}_{\mathrm{L},0}$ is split off from the remaining two states by an energy $J_\perp$ which is large in this limit. Therefore to describe the low energy properties we can restrict our attention to this lowest lying doublet and examine how its degeneracy is lifted by $\mathcal{H}_{L,1}$. To this end it is convenient to introduce effective $S = 1/2$ operators on each rung $\{\boldsymbol{\sigma}_i\}$, following Mila [28], defined by the equations

$$\begin{aligned} \sigma_i^z \mid S; i\rangle &= -1/2 \mid S; i\rangle, \ \sigma_i^z \mid T_1; i\rangle = +1/2 \mid T_i; i\rangle \\ \sigma_i^+ \mid S; i\rangle &= \mid T_1; i\rangle, \ \sigma_i^- \mid T_1; i\rangle = \mid S; i\rangle, \ \sigma_i^- \mid S; i\rangle \\ &= \sigma_i^+ |T_1; i\rangle = 0 . \end{aligned} \tag{19}$$

Here $|S; i\rangle$ and $|T_1; i\rangle$ denote the states with a singlet and the $S^z = +1$ component of the triplet on rung, i, respectively.

The Hilbert subspace of doublets on each rung is spanned by an effective Hamiltonian which has the form of an XXZ-chain (1) with the following coupling constants; $J_{xy}^{\mathrm{eff}} = J_\| - J_2$, $J_z^{\mathrm{eff}} = 1/2(J_\| + J_2)$ and an effective field $H^{\mathrm{eff}} = H^z - J_\perp - \frac{1}{2}(J_\| + J_2)$. Again this model is most easily analyzed by using a Jordan-Wigner transformation (3) to obtain a spinless fermion model of form displayed in (4). The parameters in the spinless fermion model now are

$$t = 1/2\, J_{xy}^{\mathrm{eff}} , \quad V = J_z^{\mathrm{eff}} \ \text{and} \ \mu = H^{\mathrm{eff}} + J_z^{\mathrm{eff}} . \tag{20}$$

In analyzing this spinless fermion model we must distinguish two cases. First is the case when the frustrating interaction, J_2 is weak, then the XXZ-chain falls in the XY-universality class: $J_{xy}^{\mathrm{eff}} \geq J_z^{\mathrm{eff}}$ or in terms of the original exchange constants: $3J_2 \leq J_\|$. Consider what happens when we increase the external field from zero. The chemical potential is initially below the bottom of fermion band but then the chemical potential enters the band interacting when $\mu \geq -2t$ which corresponds to critical field $H_{c,1}^z = J_\perp - J_\| + J_2$. The fermions form a Luttinger liquid in the one dimensional band. But for H^z slightly above H_{c1}^z the fermions are dilute and therefore the interaction term can be neglected. As a result the dependence of the chemical potential on density is that of free fermions in one dimension giving in turn a magnetization $M \propto (H^z - H_{c1}^z)^{1/2}$.

Similarly there is an upper critical field H^z_{c2} which corresponds to a completely filled fermion band. Near H^z_{c2} the system is most easily analyzed by making a fermion-hole transformation leading to again a dilute hole gas of fermions for $H^z \leq H_{c2}$ which is then analogous to the dilute fermion limit. The critical field now corresponds to a chemical potential $\mu = +2t + V$ and a field $H^z_{2c} = J_\perp + 2J_\parallel$. In between these two values of the chemical potential the fermions form a Luttinger liquid and the magnetization varies continuously in the range $H_{c1} < H^z < H_{c2}$. The values of the critical fields agree quite well with values found in $Cu_2(C_5H_{12}N_2)_2Cl_4$ but for an accurate comparison to experiment various corrections e.g. interladder interactions, impurities etc. need to be considered.

When the frustrating interaction J_2 is increased a new phenomena appears. The character of the fermion model changes from XY-universality to Ising universality when $J^{\text{eff}}_{xy} > J^{\text{eff}}_z$ or $3J_2 > J_\parallel$. In this case the strength of the fermion repulsion is strong enough to stabilize an alternating density modulation when the fermion band is half-filled, i.e. $\mu = V$, and to open a gap in the energy spectrum around this value of the chemical potential. The constant fermion density, when the chemical potential lies in this gap, corresponds to a plateau in the magnetization as a function of the external magnetization at a value of one half the saturation value. In this region the fermion density modulation with period π, corresponds to an alternating modulation of the magnetization with the same period. The plateau width, ΔH increases as J^{eff}_{xy} is reduced reaching the maximum value of $\Delta H = 2J_\parallel$ when $J^{\text{eff}}_{xy} \to 0$ as $J_\parallel \to J_2$. In this case the plateau corresponds to a state with a simple alternation of singlets and triplets.

The special case of a frustrated two leg ladder with $J_2 = J_\parallel$ has been numerically simulated by Honecker, Mila and Troyer [29] using a density matrix renormalization group (DMRG) scheme. With this choice of the diagonal coupling it is possible to have another phase with spin triplets on each rung and an AF interrung coupling when $J_\perp / J_\parallel < 1.4$. This region behaves as an effective AF $S = 1$ chain and so has a singlet groundstate following Haldane [30]. The full zero temperature phase diagram in the presence of a magnetic field including an alternating singlet-triplet magnetization plateau is shown in Fig. 16.

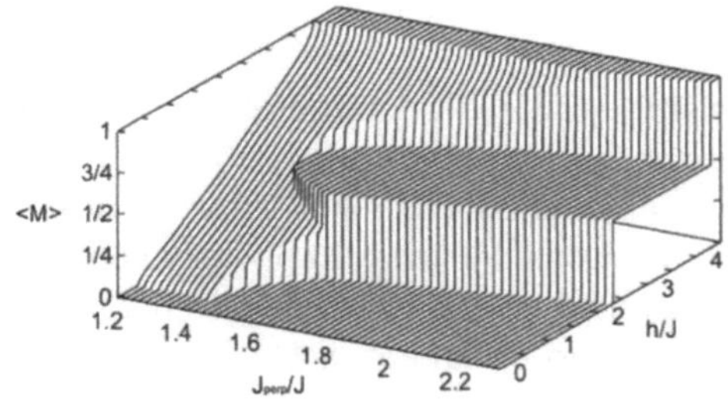

Fig. 16. The groundstate phase diagram of a frustrated 2-leg ladder with a value of the n.n.n. frustrating interaction $J_2 = J_\parallel$, in an external magnetic field as determined from numerical simulations by Honecker, Mila and Troyer [29]

5 N-Leg Spin Ladders

As the width of the spin ladders is increased, the dramatic difference that exists between the single chain ($N = 1$) and the two leg ladder ($N = 2$) repeats itself a difference between odd and even values of N. This is illustrated by the calculations in Fig. 17 which shows the temperature dependence for the uniform spin susceptibility, $\chi(T)$ from quantum Monte Carlo calculations by Frischmuth, Amman and Troyer [31], using a loop algorithm for ladders with $N = 1-6$ legs. The even N ladders ($N = 2, 4, 6$) have an exponential drop in $\chi(T) \to 0$ as $T \to 0$ and indicate the presence of a spin gap although the value of the spin gap decreases rapidly as N is increased (see also Fig. 7). The ladders with odd N, ($N = 1, 3, 5$), however all extrapolate to a finite $\chi(0)$. Curiously the value of $\chi(0)$ when expressed as the spin susceptibility per rung of N spins seems to be independent of N which implies that the low energy, low T behavior is effectively that of a single chain with a spinon velocity which is apparently independent of N. This difference between N even and N odd is easy to understand in the strong coupling limit, $J_\perp/J_\parallel \gg 1$. In this limit one solves first for the groundstate of a rung of N spins. For N odd this is a doublet with a total spin, $S = 1/2$. Introducing a finite value of $J_\parallel$ couples these effective spins together to form an effective single chain of AF coupled $S = 1/2$ spins at low energies. Frischmuth et al. [32] calculated the effective interactions in the single chain as a series in $J_\parallel/J_\perp$ for $N = 3$. The single chain develops n.n.n. and longer range interactions but interestingly these are unfrustrated with a ferromagnetic sign for the n.n.n. This means that as $J_\parallel$ increases one moves further away from the Majumdar-

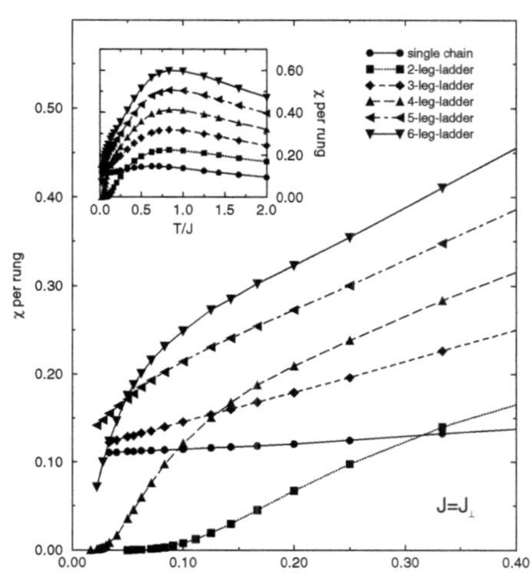

Fig. 17. The temperature dependence of the uniform spin susceptibility $\chi(T)$ per rung of N spins for a N-leg ladder simulated by Frischmuth, Amman and Troyer [31] using a loop algorithm form of QMC

Ghosh dimerized groundstate [6] towards a state with enhanced AF correlations and a longer correlation length, ξ. Qualitatively this is not a surprise since the limit of a 2D antiferromagnet is reached as $N \to \infty$ and this has an AF ordered groundstate.

A more complete analysis can be made by fitting the low temperature behavior to that of a single spin chain as calculated by Nomura [33] and Eggert et al. [34] using a RG treatment. The interaction terms due to back scattering of the Jordan-Wigner fermions give rise to marginally relevant operator and logarithmic corrections to $\chi(T)$. The 2-loop RG result is

$$\chi(T) = \frac{1}{2\pi v}\left[1 + 1/2\ln(T_0/T) - \ln\left(\ln(T_0/T) + 1/2\right)/4\ln^2(T_0/T)\right] \qquad (21)$$

where v is the spinon velocity. The characteristic temperature T_0 in the RG form was obtained by Frischmuth et al. [32] by fitting to the QMC results for $\chi(T)$ (see Fig. 18). They found a substantial decrease in T_0 between the value for a single chain $N = 1$ ($T_0 = 2.5J$) to a value $T_0 \simeq 0.25J$ for $N = 3$ in the case of isotropic coupling $J_{\|} = J_{\perp} = J$. This drop in T_0 enhances the logarithmic corrections causing $\chi(T)$ to rise more rapidly as T increases. This more rapid rise is consistent with the more rapid breakdown in the approximation which replaces the rung of N spins by just the lowest doublet.

More extensive QMC calculations have been made by Greven et al. [17] who studied the evolution of the correlation length, $\xi(T, N)$ as N is increased. Eventually as the 2D limit is reached with $\xi(T, N \to \infty)$ rising exponentially fast, $\xi \propto e^{\mathrm{J/T}}$.

The compound $Sr_2Cu_3O_5$ illustrated in Fig. 12 can be viewed as a collection of weakly coupled 3 leg ladders [26]. The susceptibility tends to a constant value as $T \to 0$ consistent with the above analysis [27]. Recently Johnston et al.

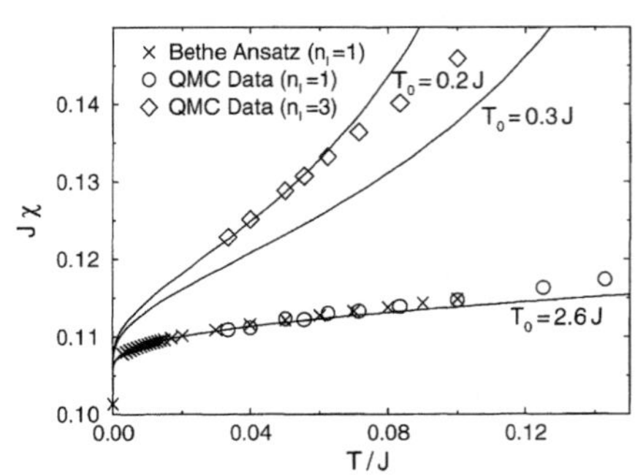

Fig. 18. The simulated values of the low temperature behavior of the uniform spin susceptibility, $\chi(T)$ for N-leg ladders with $N = 1, 3$ are fit to the 2-loop RG form, Eq. (21) for a single chain to obtain the characteristic temperatures, T_0. These fits by Frischmuth et al. [32] show a considerable reduction in the value of T_0 as N is increased from $N = 1$ to $N = 3$

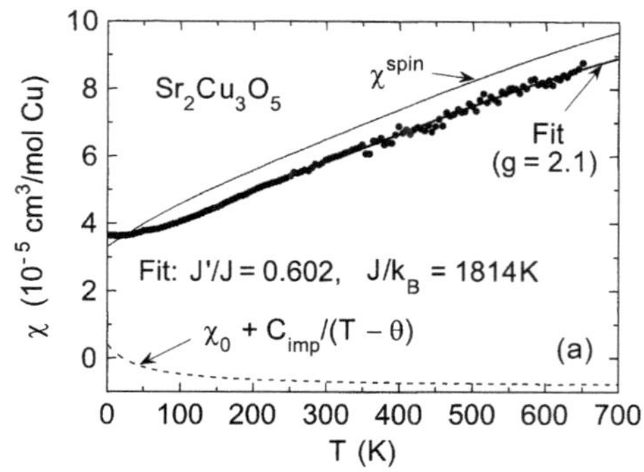

Fig. 19. The spin susceptibility $\chi(T)$ versus temperature T for $Sr_2Cu_3O_5$ and fits with the theoretical form for isolated 3-leg ladders including also an impurity term. The best fit is obtained with a rung coupling $J_\perp (= J')$ and a leg coupling $J_\parallel (= J)$ in the ratio, $J_\perp / J_\parallel = 0.602$ and a value $J_\parallel = 1814$ K, from Johnston et al. [28]

[28] made a careful fit to the susceptibility data and found parameter values of $J_\parallel = 1800$ K and $J_\perp / J_\parallel = 0.6$. Again as in the two leg ladder compound $SrCu_2O_3$ the difference between the rung and leg coupling is attributed to corrections to the shortest 180° Cu-O-Cu superexchange path through further neighbors, Cu 4s-states etc.

The application of an external magnetic field to ladders with $N = 3$ and 4 has been examined by Cabra et al. [35]. Again the strong coupling limit $J_\perp / J_\parallel \gg 1$ is the simplest case to analyze. For $N = 3$ the lowest $S = 1/2$ doublet is well separated from the next excited state on a single rung of three spins when one compares to the interrung coupling controlled by the weaker coupling, $J_\parallel$. Therefore in an applied field it is possible to saturate the magnetization of just these effective $S = 1/2$ rung spins. The result is a plateau of constant magnetization as illustrated in Fig. 20, which shows the results of exact diagonalization of finite length ladders. Note in this case the plateau corresponds to a magnetization which is uniform along the ladder and does not alternate as in the case of frustrated 2-leg ladder discussed earlier. The plateau however does not occur for arbitrary values of the ratio $J_\perp / J_\parallel$. As $J_\parallel$ is increased relative to $J_\perp$ the band of states associated with the lowest $S = 1/2$ doublet broadens relative to the separation to the higher lying rung configurations, leading to an overlap and a closing of the plateau. The calculations of Cabra et al. [35] give an estimate for the critical ratio $J_\perp / J_\parallel \simeq 0.8$ which is necessary to close the plateau. Below this critical ratio the finite spin susceptibility causes the magnetization rises continuously from zero in an external field to a single plateau that corresponds to a fully polarized state.

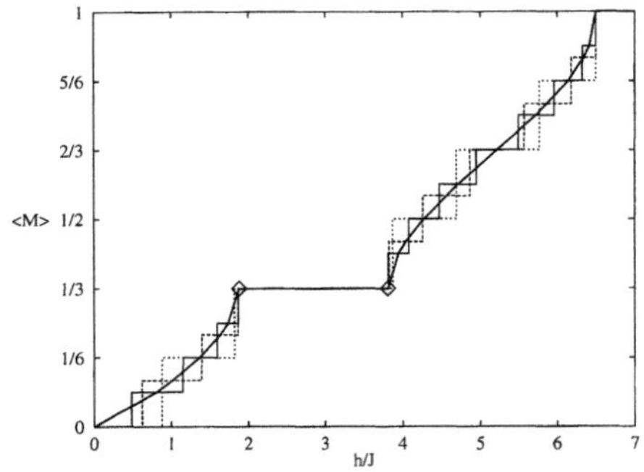

Fig. 20. The evolution of the magnetization, M, for a 3-leg ladder in an external field as determined by exact diagonalization on finite length ladders. The plateau is due to the saturation of the effective $S = 1/2$ single chain spins arising from the lowest doublet on each rung

6 Coupled Ladders and the Transition to AF Order

If an array of two leg ladders is coupled in an unfrustrated way, a transition to AF long range order from the spin liquid state can be triggered with increasing interladder coupling. This transition is an example of a quantum critical point (QCP) which separates a quantum disordered spin liquid phase from a AF long range ordered phase at $T = 0$. The nature of this QCP has been studied both by mean field treatments [15,36] and by extensive numerical simulations [36].

The simplest case is when an array of parallel 2-leg ladders are simply coupled to form a square lattice with an interladder coupling J'. This system is described by a model Hamiltonian

$$\mathcal{H}_{\mathrm{CL}} = \sum_{\alpha} \mathcal{H}_{\mathrm{L}}^{\alpha} + J' \sum_{\langle \alpha,\beta \rangle} \sum_{i}^{N} \boldsymbol{S}_{\alpha,i,1} \cdot \boldsymbol{S}_{\beta,i,2} \tag{22}$$

where $\langle \alpha, \beta \rangle$ denotes n.n. ladders and $\mathcal{H}_{\mathrm{L}}$ is given by (10).

The addition of the interladder interaction broadens the triplet magnon band further thereby reducing the spin gap which is determined by the minimum excitation energy. This effect can be straightforwardly treated by extending the mean field treatment for a single ladder outlined in Sect. (3). At a critical value of J' the triplet magnon mode goes soft and the spin gap goes to zero. Normand and Rice [15] obtained a critical value for a 2D array of isotropic ladders ($J_{\parallel} = J_{\perp} = J$) of $J'_c \cong 0.4J$. When $J' > J'_c$ the nature of the condensate in the groundstate changes from a singlet $\bar{s}$ to a superposition of a singlet and one of the triplets, e.g. $\alpha\bar{s} + \beta\bar{t}_z$ with $|\alpha|^2 + |\beta|^2 = 1$. The admixture of a triplet component gives rise to a AF staggered moment on each site, so that the QCP separates a AF ordered phase from a spin liquid phase at the point where the triplet magnon mode goes soft.

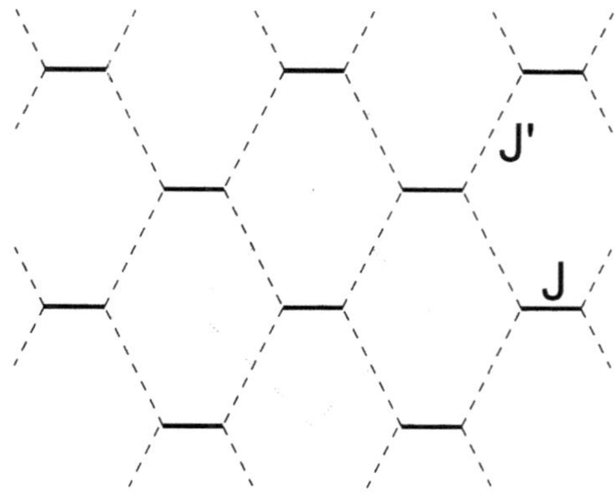

Fig. 21. The three-dimensional unfrustrated pattern of interladder couplings (J') in the plane perpendicular to the 2-leg ladders that occurs in the cuprate compound $La_2Cu_2O_5$

The QCP in the case of unfrustrated coupling in a 3D array of the form shown schematically in Fig. 21, has been studied most extensively because of its relation to the cuprate compound $La_2Cu_2O_5$ [15,36]. This material has a structure with an array of two leg ladders of this form and was initially thought to have a spin liquid groundstate [37]. But later it was found to undergo a phase transition to an AF ordered state (see below) [38,39]. Mean field treatments of this 3D structure have been supplemented with extensive QMC calculations using the loop algorithm by Troyer, Zhitomirsky and Ueda [36]. The value of the critical ratio in the QMC simulations is $J'_c/J \simeq 0.12$ in good agreement with the mean field value [15,?]. As the QCP is approached from the spin liquid phase the mean field treatment gives a triplet magnon dispersion which when expanded near its minimum takes the form $\omega_{\boldsymbol{k}}^2 = c_{\parallel}^2 k_z^2 + c_{\perp}^2 k_{\perp}^2 + \Delta^2$ where $c_{\parallel}(c_{\perp})$ are the velocities along (perpendicular to) the ladders. Both velocities remain finite at $J' = J'_c$. The spin gap vanishes as $\Delta \sim (J'_c - J')^{1/2}$ as $J' \to J'_c$. The linear dispersion of the triplet magnon mode at the QCP gives a T^3-law for the number of thermally excited magnons. Therefore calculating the spin susceptibility as we did earlier as a Curie form for a dilute gas of thermally excited magnons gives an asymptotic dependence $\chi(T) = T^2/6c_{\parallel}c_{\perp}$ as $T \to 0$. The extensive simulations by Troyer et al. [36] also show a T^2-law in agreement with this simple soft mode prediction (Fig. 22). Note that in the AF ordered phase there are only two gapless spin wave modes. This means that one of the three gapless modes of the triplet magnon at the QCP evolves to an amplitude mode of the AF order and therefore develops a gap in the AF phase. Thus the approach to the QCP from the AF side is characterized by a softening of the AF amplitude mode, while at the same time the velocity of the transverse spin waves remains finite.

As mentioned above $La_2Cu_2O_5$ was initially believed to be a coupled ladder compound with a spin liquid groundstate because of the clear drop in $\chi(T)$ found by Hiroi and Takano [37]. However subsequent NMR [38] and μSR

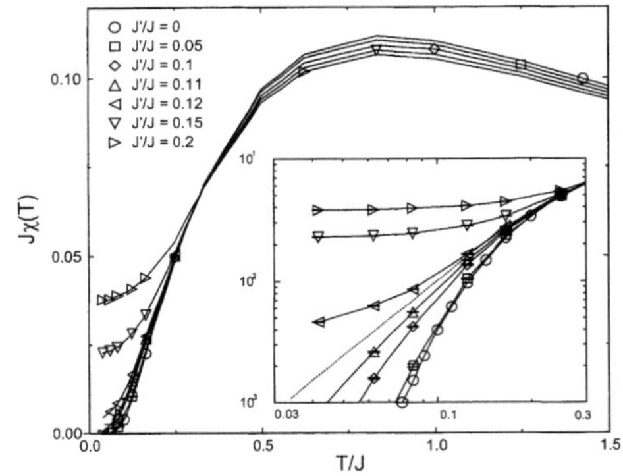

Fig. 22. The simulatd values of the spin susceptibility $\chi(T)$, obtained by Troyer et al. [36] using a loop algorithm QMC method in the vicinity of the quantum critical point, separating spin liquid behavior from long range AF order, for the 3D lattice shown in Fig. 21

[39] experiments showed a Neel transition to an AF order at a Neel Temperature $T_N = 117$ K. Normand and Rice [15] then proposed that the compound $La_2Cu_2O_5$ lies near to, but on the AF side of the QCP. The numerical calculations of Troyer et al. [36] for $\chi(T)$ could be fit to the experiments supporting this proposal. The lack of a clear anomaly in $\chi(T)$ at $T = T_N$ was attributed to the very strong quantum fluctuations in the vicinity of the QCP. Recently a more careful reexamination by Johnston et al. [28] showed that the best fit to $\chi(T)$ is obtained if the assumption of isotropy ($J_\| = J_\perp$) on the ladders is relaxed and a ratio $J_\|/J_\perp = 0.5$ was found to give the best fit (Fig. 23).

The large values of J in the cuprates with 180° Cu-O-Cu bonds makes them ill suited to studies in an applied magnetic field. So far as I am aware theoretical studies of the effect of an external field on the QCP which controls the transition to 3D AF order have not been carried out. However the closely related problem of the effect of a weak 3D coupling between spin ladders on the transition to a state with finite magnetization has been examined by Giamarchi and Tsvelik [40] and also by Nikumi et al. [41] and by Wessel, Olshanii and Haas [42]. in this case only one component of the triplet magnon goes soft and there is a close analogy to the Bose-Einstein condensation transition. The condensate gives rise to a 3D AF order with a staggered magnetization which is oriented transverse to the external field.

7 Conclusions

The aim of these lectures has been to give a simple and intuitive introduction to the properties of low dimensional quantum magnets. The focus has been on the novel spin liquid states that occur in spin ladders. The application of an external magnetic field can cause the spin gap to the triplet magnon mode to

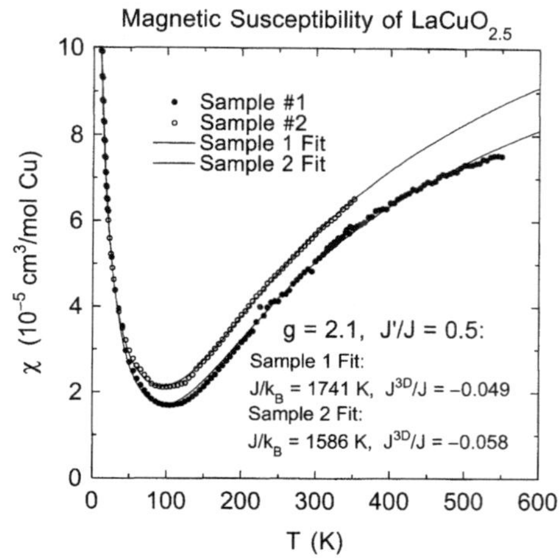

Fig. 23. The spin susceptibility, $\chi(T)$, versus temperature T for $La_2Cu_2O_5$ and fits to the theoretical form for 3D coupled 2-leg ladders including also an impurity term. The best fit is obtained with a rung coupling $J_\perp (= J')$ and a leg coupling $J_{//} (= J)$ in the ratio $J_\perp / J_{//} = 0.5$. The optimum absolute values are slightly sample dependent, from Johnston et al. [28]

close leading to a quantum critical point which marks the transition to a state with finite magnetization at zero temperature and also with long range AF order when three dimensional couplings are included.

Acknowledgements

I would especially like to express my gratitude to a number of colleagues and collaborators on the study of spin ladders over the past years: S. Gopalan, A. Honecker, A. Läuchli, F. Mila, B. Normand, M. Sigrist, M. Troyer and M. Zhitomirsky.

References

1. D.C. Mattis: *The Theory of Magnetism I* (Springer-Verlag, Berlin, 1981)
2. A. Auerbach: *Interacting Electrons and Quantum Magnetism* (Springer-Verlag, Berlin, 1994)
3. E. Manousakis: Rev. Mod. Phys. **63**, 1 (1991)
4. E. Dagotto, T.M. Rice: Science **271**, 618 (1996)
5. For recent reviews see H.J. Schulz: in *Mesoscopic Quantum Physics*, Les Houches, Session LXI, 1994, eds. E. Akkermans, G. Montambaux, J.-L. Pichard, J. Zinn-Justin (Elsevier, Amsterdam, 1995) p. 333 cond-mat/9503150 and H.J. Schulz, G. Cuniberti, P. Pieri cond-mat/9807366
6. C.K. Majumdar, D.K. Ghosh: J. Math. Phys. **10**, 1388 (1969)
7. B. Shastry, B. Sutherland: Phys. Rev. Lett. **47**, 964 (1981)

8. I. Affleck: in *Fields, Strings and Critical Phenomena*, eds. E. Brezin, J. Zinn-Justin, Les Houches 1990 p. 583
9. M. Reigrotzki, H. Tsunetsugu, T.M. Rice: J. Phys. C **6**, 9235 (1994)
10. E. Dagotto, J. Riera, D.J. Scalapino: Phys. Rev. B **45**, 5944 (1992)
11. S. Sachdev, R.N. Bhatt: Phys. Rev. B **41**, 9323 (1990)
12. S. Gopalan, T.M. Rice, M. Sigrist: Phys. Rev. B **49**, 8901 (1994)
13. J. Piekarewicz, J.R. Shepard: Phys. Rev. B **60**, 9456 (1999)
14. The earlier mean field calculation [12] has a numerical error which is corrected in [13,**?**]
15. B. Normand, T.M. Rice: Phys. Rev. B **54**, 7180 (1996)
16. M. Troyer, H. Tsunetsugu, D. Würtz: Phys. Rev. B **50**, 13515 (1994)
17. M. Greven, R.J. Birgeneau, U.T. Wiese: Phys. Rev. Lett. **77**, 1865 (1996)
18. D.G. Shelton, A.A. Nersesyan, A.M. Tsvelik: Phys. Rev. B **53**, 8521 (1996)
19. N. Hatano, Y. Nishiyama: J. Phys. A **28**, 3911 (1995)
20. T. Barnes, E. Dagotto, J. Riera, E.S. Swanson: Phys. Rev. B **47**, 3196 (1993)
21. N. Motoyama, H. Eisaki, S. Uchida: Phys. Rev. Lett. **76**, 3212 (1996)
22. T.M. Rice, S. Gopalan, M. Sigrist: Europhys. Lett. **23**, 445 (1994)
23. M. Azuma, Z. Hiroi, M. Takano, K. Ishida, Y. Kitaoka: Phys. Rev. Lett. **73**, 3663 (1994)
24. D.C. Johnston et al.: cond-mat/0001147
25. G. Chaboussant, P.A. Crowell, L.P. Levy, O. Piovesana, A. Madouri, D. Mailly: Phys. Rev. B **55**, 3046 (1997)
26. G. Chaboussant et al.: Phys. Rev. Lett. **80**, 2713 (1998)
27. G. Chaboussant et al.: Eur. Phys. J. B **6**, 167 (1998)
28. F. Mila: Eur. Phys. J. B **6**, 201 (1998)
29. A. Honecker, F. Mila, M. Troyer: Eur. Phys. J. B **15**, 227 (2000)
30. F.D.M. Haldane: Phys. Rev. Lett. **61**, 1029 (1988)
31. B. Frischmuth, B. Ammon, M. Troyer: Phys. Rev. B **54**, R3714 (1996)
32. B. Frischmuth, S. Haas, G. Sierra, T.M. Rice: Phys. Rev. B **55**, R3340 (1997)
33. K. Nomura: Phys. Rev. B **48**, 16814 (1993)
34. S. Eggert, I. Affleck, M. Takahashi: Phys. Rev. Lett. **73**, 332 (1994)
35. D.C. Cabra, A. Honecker, P. Pujal: Phys. Rev. B **55**, 6241 (1997)
36. M. Troyer, M.E. Zhitomirsky, K. Ueda: Phys. Rev. B **55**, R6117 (1997)
37. Z. Hiroi, M. Takano: Nature, **377**, 41 (1995)
38. S. Matsumoto, Y. Kitaoka, K. Ishida, K. Asayama, Z. Hiroi, N. Kobayashi, M. Takano: Phys. Rev. B **53**, 11942 (1996)
39. R. Kadoro et al.: Phys. Rev. B **54**, R9628 (1996)
40. T. Giamarchi, A.M. Tsvelik: Phys. Rev. B **59**, 11398 (1999)
41. T. Nikumi, M. Oshikawa, A. Oosawa, H. Tanaka: Phys. Rev. Lett. **84**, 5868 (2000)
42. S. Wessel, M. Olshanii, S. Haas: Phys. Rev. Lett. **87**, 206407 (2001)

Frustrated Quantum Magnets

Claire Lhuillier[1] and Grégoire Misguich[2]

[1] Laboratoire de Physique Théorique des Liquides,
Université P. et M. Curie and UMR 7600 of CNRS,
case 121, 4 Place Jussieu, 75252 Paris Cedex
[2] Service de Physique Théorique,
CEA Saclay,
F-91191 Gif-sur-Yvette Cedex

Abstract. In these lectures we present a summary of the $T = 0$ properties of the ground state and first excitations of two-dimensional frustrated quantum magnets. Up to now four phases have been observed at $T = 0$: a Néel ordered phase with spontaneous symmetry breaking and soft modes (the antiferromagnetic magnons) and three phases with no $SU(2)$ symmetry breaking and a spin gap. The first four sections are devoted to the characterization of these four phases. We first explain the two main quantum effects which can take place in the semi-classical Néel phase: reduction of the order parameter by quantum fluctuations and in the case of a competition between various ordered phases: selection of the most symmetric one by quantum fluctuations. Amongst the three spin gapped phases, the Valence Bond Crystal has long range order in $S = 0$ valence bonds and all its excitations have integer spins. The other two, the Resonating Valence Bond Spin Liquids have no local order parameter and fractionalized excitations. After a characterization of these different phases we use information gathered from exact diagonalizations, toy models and large N expansions to emphasize criteria that seem important to find one or another of these spin gapped phases. The last section of this paper is concerned with the special behavior of frustrated magnets in high magnetic fields: metamagnetic behavior and quantized plateaus. Some insights are given on Chiral Spin Liquids and on a proposed connexion between Quantum Hall Effect and magnetization plateaus.

In this chapter we sketch a rapid survey of recent theoretical advances in the study of frustrated quantum magnets with a special emphasis on two dimensional magnets. One dimensional problems are only very briefly discussed: this field has been extraordinarily flourishing during the past twenty years both experimentally and theoretically. In contrast the understanding of two dimensional quantum magnets is much more limited. The number of unexplained experimental results is probably extremely large, unfortunately the theoretical tools to deal with exotic quantum phases in 2 dimensions are still rather limited. In the absence of a significant amount of exact results, the picture drawn in the following sections is based on the comparison of various approaches to $SU(2)$ spins: series expansions, exact diagonalizations, Quantum Monte-Carlo calculations and the hints got from large-N generalizations.

From these comparisons we suggest that one could expect at least 4 kinds of different low temperature physics in two-dimensional systems:

- semi-classical Néel like phases
- and three kinds of purely quantum phases.

These 4 phases are the subject of the first 4 sections of this paper: their properties are summarized in Table 1. The quantum phases appear in situations where there are competing interactions, a high degree of frustration and a rather low coordination number. In section 5 we discuss the very rich magnetic phase diagram of frustrated quantum magnets, which can exhibit hysteretic metamagnetic

Table 1. The four 2-dimensional phases described in the four first sections.

Phases	G.-S. Symmetry Breaking	Order Parameter	First excitations
Semi-class. Néel order	SU(2) Space Group Time Reversal	Staggered Magnet.	Gapless magnons
Valence Bond Crystal	Space Group	dimer-dimer LRO or S=0 plaquettes LRO	all excitations are gapped Confined spinons thermally activated C_v and χ
R.V.B. Spin Liquid (Type I)	topological degeneracy	No local order parameter	all excitations are gapped Deconfined spinons thermally activated C_v and χ
R.V.B. Spin Liquid (Type II)	topological degeneracy	No local order parameter	Singlet excitations are gapless Triplet excitations are gapped Deconfined spinons T=0 entropy thermally activated χ C_v insensitive to magn. field at low T

transitions and magnetic plateaus at zero and at rational values of the magnetization. Chiral phases, and the possible relation between magnetization plateaus and Hall conductance plateaus specifically studied in some other lectures of this School are briefly discussed.

To complete Table 1 let us underline that Néel ordered magnets are characterized by a unique energy scale (essentially given by its Curie-Weiss temperature θ_{CW}, which is directly related to the coupling constant of the Hamiltonian). All the quantum phases studied up to now display a second energy scale which can be an order or two order of magnitude lower than θ_{CW} and mainly associated with a spin gap. It is the range of temperatures and energies under study in this lectures.

1 Semi-classical Groundstates

1.1 Heisenberg Problem

In most of the cases the ground state of the antiferromagnetic spin-1/2 Heisenberg Hamiltonian

$$\mathcal{H} = 2J \sum_{\langle ij \rangle} \boldsymbol{S}_i \cdot \boldsymbol{S}_j, \tag{1}$$

with $J > 0$ and the sum limited to first neighbors, is Néel ordered at $T = 0$. In 3 dimensions, when increasing the temperature the Néel ordered phase gives place to a paramagnet through a 2nd order phase transition ($T_N \sim \mathcal{O}(J)$). In 2 dimensions the ordered phase only exists at $T = 0$ (Mermin–Wagner theorem [1]). In one dimension, even at $T = 0$, there is only quasi-long range order with algebraically decreasing correlations (for an introduction to spin systems see [2–4]).

Néel ground states break the continuous $SU(2)$ symmetry and support low lying excitations which are the Goldstone modes associated with this broken symmetry. These excitations, called magnons, are $\Delta S^z = 1$ bosons that can be pictured as long wave-length twists of the order parameter (the staggered magnetization). They form isolated branches of excitations described by their dispersion relation $\omega(\mathbf{k})$. $\omega(\mathbf{k})$ vanishes in reciprocal space at the set of $\mathbf{k}$ vectors $\{\tilde{\mathbf{k}}\}$ characteristic of the long range order: i.e. the wave vector $\mathbf{k} = \mathbf{0}$ and the peaks in the structure function

$$\mathcal{S}(\mathbf{q}) = \int d^D\mathbf{r}\, \mathrm{e}^{\mathrm{i}\mathbf{q}\cdot\mathbf{r}} \mathbf{S}_0 \cdot \mathbf{S}_{\mathbf{r}}, \tag{2}$$

where D is the lattice dimensionality.

The geometry of the Néel order parameter can in general be determined through a classical minimization of (1). It is a two-sublattice up-down order (noted *ud* in the following) in the case of the square lattice, a three-sublattice with magnetization at 120 degrees from each other in the triangular case.

This classical approach neglects the quantum fluctuations: the classical solution *ud* is an eigenstate of the Ising Hamiltonian but not of the Heisenberg

Hamiltonian, the extra $X-Y$ terms of the Heisenberg Hamiltonian induce pairs of spin-flips and reduce the value of the staggered magnetization. For a moderate reduction, the spin-wave calculation is usually a good approximation to take this quantum effect into account [5]. Introduction of quantum fluctuations

- renormalizes the ground state energy through the zero point fluctuations
- decreases the sublattice magnetization by a contribution

$$\delta \propto \int \frac{g(k)}{\omega(k)} d^D k \tag{3}$$

where $\omega(k)$ is the dispersion law and $g(k)$ is a smooth function of k, depending on the lattice, which is non zero in the whole Brillouin zone. It is a straightforward exercise to show that $\omega(k) \propto k$ for small k. Equation (3) indicates the absence of long range order (LRO) in $D=1$ dimension through the breakdown of the spin-wave approximation (divergence of the integral).

The reduction of the order parameter with respect to the classical saturation value M/M_{cl} increases when the coordination number decreases; it is larger on triangular based lattices than on bipartite ones (see Table 2). This last feature is sometimes called "geometrical frustration". In the framework of the Heisenberg model this should be understood in the following sense: the ground state of the Heisenberg problem on the triangular lattice is less stable than the ground state of the same problem on the square lattice (and bipartite lattices), both in the classical regime (where $\langle \boldsymbol{S}_i \cdot \boldsymbol{S}_j \rangle^{\rm cl}_{\rm sq} = -0.25$ and $\langle \boldsymbol{S}_i \cdot \boldsymbol{S}_j \rangle^{\rm cl}_{\rm tri} = -0.125$), as well as in the $SU(2)$ quantum regime ($\langle \boldsymbol{S}_i \cdot \boldsymbol{S}_j \rangle^{\rm qu}_{\rm sq} = -0.3346$, whereas $\langle \boldsymbol{S}_i \cdot \boldsymbol{S}_j \rangle^{\rm qu}_{\rm tri} = -0.1796$).

The ground state of the spin 1/2 Heisenberg model (1) does not have Néel order on the following lattices:

- the Bethe Chain (exact result)
- the Kagomé lattice (at the classical level there is probably some kind of nematic order [6,7])
- and the pyrochlore lattice [8,9][1] .

Table 2 summarizes quantitative properties of the spin-1/2 Heisenberg model on different simple lattices.

1.2 Semi-classical Groundstates with Competing Interactions

Another way to frustrate Néel order and try to destabilize it, consists in adding competing interactions. The archetype of such a problem is the $J_1 - J_2$ model:

$$\mathcal{H} = 2J_1 \sum_{\langle ij \rangle} \boldsymbol{S}_i . \boldsymbol{S}_j + 2J_2 \sum_{\langle\langle ij \rangle\rangle} \boldsymbol{S}_i . \boldsymbol{S}_j , \tag{4}$$

[1] It should be remarked that in the last two cases, the divergence of (3) comes from the existence of a zero mode on the whole Brillouin zone. This zero mode is due to a local degeneracy of the classical ground state [7].

Table 2. Energy per bond and sublattice magnetization in the ground state of the spin-1/2 Heisenberg Hamiltonian on various lattices. The sq-hex-dod. is a bipartite lattice formed with squares, hexagons and dodecagons.

Lattices	Coordination number	$2\langle \boldsymbol{S}_i.\boldsymbol{S}_j \rangle$ per bond	M/M_{cl}	
dimer	1	–1.5		
1 D Chain	2	–0.886	0	
honeycomb [10]	3	–0.726	0.44	bipartite
sq-hex-dod. [11]	3	–0.721	0.63	lattices
square [12]	4	–0.669	0.60	
one triangle	2	–0.5		
Kagomé [13]	4	–0.437	0	frustrating
triangular [14]	6	–0.363	.50	lattices

where the first sum runs on first neighbors and the second on second neighbors. As an example let us consider the square lattice case. The nearest neighbor antiferromagnetic coupling favors a (π,π) order, which can be destabilized by a 2nd neighbor J_2 coupling $> J_{2c} \sim 0.35$. For much larger J_2, the system recovers some Néel long range ordering. With a simple classical reasoning one would then expect an order parameter with a 4 sublattice geometry and an internal degeneracy [15–17][2]. As a consequence of quantum or thermal fluctuations a collinear configuration ($(\pi, 0)$ or $(0,\pi)$) is selected via the mechanism of "order through disorder" [18,19]: the collinear order is stabilized by fluctuations because it is a state of larger symmetry with a larger reservoir of soft fluctuations nearby. In the classical problem the stabilization is entropy driven (whence the name of the mechanism), in the quantum problem the softer fluctuations induce a smaller zero point energy and thus a stabilization of the more symmetric state. In all these cases, the first excitations are gapless $\Delta S^z = 1$ magnons.

1.3 One-Dimensional Problems

As already mentioned, the Heisenberg problem in that case (Bethe Chain) is critical. Its low-lying excitations are free spin-1/2 excitations called spinons (first discovered by Fadeev). Due to quantum selection rules they appear in pairs, the spectrum of excitation is thus a continuum with two soft points (0 and π). A frustrating J_2 coupling, larger than the critical value $J_2^c \sim 0.2411$, opens a gap in the spectrum and the ground state becomes dimerized (and two fold degenerate)

[2] Notice that the extra degeneracy is a global one and not a local one as in the Kagomé problem.

with long range order in singlet-singlet correlations. Majumdar and Gosh [20] have shown that the ground state of this problem for $J_2 = 0.5$ is an exact product of singlet wave-functions [3]. The low-lying excitations are pairs of solitons [21], forming a continuum. For specific values of the coupling there appears an isolated branch of singlets around the point $\pi/2$ [22].

2 Valence Bond Crystals

2.1 Effect of a Frustrating Interaction on a Néel State

Coming back to the $J_1 - J_2$ model of (4), we examine the possibility of destroying Néel ground states by a frustrating interaction. At the point of maximum frustration ($J_2 \sim 0.5 J_1$) between the two Néel phases ((π,π) and $(\pi,0)$) the system can lower its energy by forming $S = 0$ valence bonds between nearest neighbors. This maximizes the binding energy of the bonds involved in singlets and decreases the frustrating couplings between different singlets. But contrary to Néel states, nearest neighbors not involved in singlets do not contribute to the stabilization of this phase: so the presence of such a phase between two different Néel phases is not mandatory and should be studied in each case. The nature of the arrangements of singlets is also an open question: the coverings could have long range order or not. We call the phases with long range order in the dimer arrangements: Valence Bond Crystals (noted in the following VBC). The simplest picture of a Valence Bond Crystal ground state is given in Fig. 1. The other kinds of dimer covering ground states will be studied in the next sections.

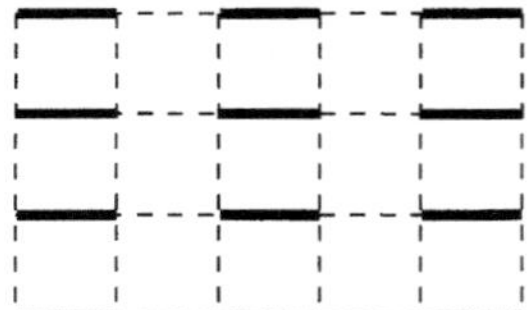

Fig. 1. A simple "columnar" valence-bond crystal on the square lattice: fat links indicate the pair of sites where the spins are combined in a singlet $|\uparrow\downarrow\rangle - |\downarrow\uparrow\rangle$. A realistic VBC will also include fluctuations of these singlet positions so that the magnetic correlation $\langle \boldsymbol{S}_i \cdot \boldsymbol{S}_j \rangle$ will be larger than $-\frac{3}{4}$ along fat links and will not be zero along the dashed ones

2.2 Groundstate Properties of the Valence Bond Crystals

- The ground state of a VBC is a pure $S = 0$ state: it does not break $SU(2)$ rotational invariance and has only short range spin-spin correlations,
- It has long range dimer-dimer or $S = 0$ plaquettes-plaquettes correlations[3],

[3] A $S = 0$ plaquette is an ensemble of an even number of nearby spins arranged in a singlet state: a 4-spin S=0 plaquette long range order has been suggested for the $J_1 - J_2$ model on the square lattice [23,24], and a 6-spin S=0 plaquette long range order for the same model on the hexagonal lattice [25,10,26].

- In the thermodynamic limit, it breaks the symmetries of the lattice: translational symmetry and possibly rotational symmetry.

Remarks:

Such a ground state is highly reminiscent of the dimerized ground state appearing in the $J_1 - J_2$ chain for $J_2 > J_2^c$. But, as explained below, the spectrum of excitations of the two-dimensional VBC is expected to be different from the one-dimensional case.

On a finite sample the exact ground state of a VBC cannot be reduced to the symmetry breaking configuration A of Fig. 1: it is the symmetric superposition of configuration A and of its three transforms (B, C and D) in the symmetry operations of the lattice group. In the thermodynamic limit, the 4 different superpositions of the A, B, C, and D configurations are degenerate, allowing the consideration of a symmetry breaking configuration like A as "a thermodynamic ground state" of the system. In the general case of a VBC, the degeneracy of the thermodynamic ground state is finite and directly related to the dimension of the lattice symmetry group. On a finite square sample, the 4 different superpositions of the A, B, C, and D configurations are non degenerate[4], but their energies are expected to approach zero exponentially with the system size.

2.3 Elementary Excitations

Spin–spin correlations are exponentially decreasing and we thus expect that the system will have a spin-gap. The nature of the first excitations is not totally settled: it is believed that they should be $S = 1$ bosons (some kinds of optical magnons). The possibility of $S = 1/2$ excitations (spinons) is dismissed on the following basis: the creation of a pair of spins 1/2 by excitation of a singlet is certainly the basic mechanism for producing an excitation. The question is then: can these spins 1/2 be separated beyond a finite distance? In doing so they create a string of misaligned valence bonds, the energy of which increases with its length. It is the origin of an elastic restoring force which binds the two spins 1/2. This simple picture shows the influence of dimensionality; creating a default in the $J_1 - J_2$ chain only affects the neighboring spins of the chain, the perturbation does not depend on the distance between the two spins 1/2, in this case the spinons are de-confined giving rise to a continuum of two-particle excitations [21,22]. Because of the confinement of spinons in the VBCs, the excitation spectrum of this kind of states will exhibit isolated modes ("optical magnons") below the continuum of multi-particle excitations.

An open question is whether the effective coupling between the two spinons is preferentially ferromagnetic or antiferromagnetic? This might be connected to the preferred position of the spinons on the same (or different) sublattice(s)? Numerical results on the $J_1 - J_2$ model on the hexagonal lattice plead in favor

---

[4] These 4 superpositions will have the following momenta: $(0,0)$, $(0,0)$, $(\pi,0)$ and $(0,\pi)$. The two later will be degenerate (because of the $\pi/2$-rotation symmetry) so that 3 nearly-degenerate energies will appear in the finite-size spectrum.

of an antiferromagnetic coupling between spinons, manifesting itself by a gap in the $S = 0$ sector smaller than in the $S = 1$ sector.

As a consequence of the gap(s), both the spin-susceptibility χ and the specific heat C_v of the VBCs are thermally activated.

2.4 A Toy Model

The simplest 2-dimensional quantum model exhibiting VBC phases is the quantum hard-core dimer (QHCD) model introduced by Rokhsar and Kivelson in the context of high T_c super-conductivity [27]. It is not strictly speaking a spin model since the Hamiltonian directly operates on nearest-neighbor dimer coverings. The Hamiltonian is defined by its leading non-zero matrix elements within this subspace. On the square lattice it reads:

$$\mathcal{H} = \sum_{\text{Plaquette}} \left[-J \left(|\,\|\,\rangle\langle\, = \,| + \text{h.c.} \right) + V \left(|\,\|\,\rangle\langle\,\|\,| + |\, = \,\rangle\langle\, = \,| \right) \right] \quad (5)$$

The connection between this model and a spin model such as the Heisenberg model is in the general case a difficult issue, because it involves the overlap matrix[5] of dimer configurations (we will come back to this point in section 4). However, this model can be usefully seen as an effective description of the low-energy physics of some non Néel phases. The kinetic term ($J > 0$) favors resonances between dimer configurations which differ by two dimers flips around a plaquette. The potential term ($V > 0$) is a repulsion between parallel dimers. A strong negative V favors a Valence Bond columnar phase (Fig. 1), whereas a strong positive V forbids crystallization in the columnar phase and favors a staggered one (Fig. 2). From our present point of view, both of these phases are gapped VBC. They are separated by a quantum critical point at $J = V$. At that point the exact ground state is the equal-amplitude superposition of all nearest-neighbor dimer configurations [27], the correlations are algebraic [28], the gap closes and the system supports $S = 0$ (quasi-)Goldstone modes (called "resonons" in ref. [27]).

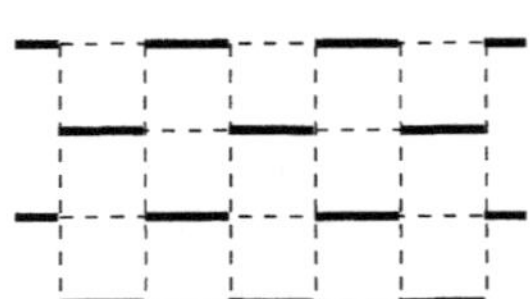

Fig. 2. Staggered dimer configuration stabilized by a strong dimer-dimer repulsion $V > 0$ in the QHCD on the square lattice

[5] Consider the restriction of the Heisenberg Hamiltonian to the subspace of nearest-neighbor dimer coverings. Due to the non-orthogonality between dimer coverings, this effective Hamiltonian is complicated and non-local. However, it can be formally expanded in powers of $x = 1/\sqrt{N}$ where N counts the number of states a spin can have at each site (of course in our case $N = 2$). When only the lowest order ($\mathcal{O}(x^4)$) terms are kept, the effective Hamiltonian is given by (5).

Moessner and Sondhi have recently studied the same kind of model on the hexagonal lattice: it has a richer phase diagram of VBC phases, and essentially the same generic physics [26].

2.5 Possible Realizations in Spin-1/2 $SU(2)$ Models

The $J_1 - J_2$ model on the square lattice near the point of maximum frustration has been studied by different approaches: exact diagonalizations, series expansions, Quantum Monte-Carlo and Stochastic Reconfiguration. There is a general agreement that the $J_1 - J_2$ model on the square [17,23,29,24], hexagonal [10] and checker board lattices [30] has at least one VBC phase between the two semi-classical Néel phases. Long range order might involve plaquettes of 4 spins (on the square lattice or on the checker board lattice) and even 6 spins on the hexagonal lattice.

Two elements should be noticed:

- All these lattices are bipartite and their coordination number is not extremely large,
- These phases appear when both J_1 and J_2 are antiferromagnetic (see next section for a ferromagnetic phase destabilized by an antiferromagnetic coupling).

2.6 Large-N Limits

Introduced by Affleck and Marston [31] and Arovas and Auerbach [32], large-N limits are powerful analytical methods. The $SU(2)$ algebra of a spin S at one site can be represented by $N = 2$ species of particles $a^\dagger_\sigma$ (with $\sigma = \uparrow, \downarrow$), provided that the total number of particles on one site is constrained to be $a^\dagger_\uparrow a_\uparrow + a^\dagger_\downarrow a_\downarrow = 2S$. The raising operator S^+ (or S^-) is simply represented by $a^\dagger_\uparrow a_\downarrow$ (or $a^\dagger_\downarrow a_\uparrow$, respectively). These particles can be chosen to be fermions (Abrikosov fermions) or bosons (Schwinger bosons). The Heisenberg interaction is a four-body interaction for these particles.

The idea of large-N limits is to generalize the $SU(2)$ symmetry of the spin$-S$ algebra to an $SU(N)$ (or $Sp(N)$) symmetry by letting the flavors index σ go from 1 to N. The $SU(N)$ (or $Sp(N)$) generalization of the Heisenberg model is solved by a saddle point calculation of the action, which decouples the different flavors. It is equivalent to a mean-field decoupling of the four-body interaction of the physical $N = 2$ model.

Whether the large-N limit of some $N = 2$ model is an accurate description of the physics of "real" $N = 2$ spins is a difficult question but the phase diagrams obtained using these approaches (where the value of the "spin" S can be varied) are usually coherent pictures of the competing phases in the problem. In particular, these methods can describe both Néel ordered states and Valence-Bond Crystals [25], as well as short-range Resonating Valence Bonds phases (section 3) or Valence-Bond Solids [33]. Read and Sachdev [34] studied the $J_1 - J_2 - J_3$ model on the square lattice by a Sp(N) bosonic representation and predicted a columnar VBC phase (as in Fig. 1) near $J_2 \simeq 0.5 J_1$.

2.7 Experimental Realizations of Valence Bond Crystals

In the recent years many experimental realizations of VBCs have been studied. The most studied prototype in 1d is $CuGeO_3$: it is not a pure realization of the dimerized phase of a pure $J_1 - J_2$ model insofar as a Spin-Peierls instability induces a small alternation in the Hamiltonian. This compound has received a lot of attention both from the experimental and theoretical points of view (see [22] and references therein). In two dimensions two compounds are 2D VBCs CaV_4O_9 [35–47] and $SrCu_2(BO_3)_2$. This last compound is a good realization of the Shastry Sutherland model [48,49]. However, in both case the ground state is non-degenerate because the Hamiltonian has an integer spin in the unit cell (4 spins 1/2) and the dimerization does not break any lattice symmetry.

3 Short-Range Resonating Valence-Bond Phases: Type I SRRVB Spin Liquid

3.1 Anderson's Idea

Inspired by Pauling's idea of "resonating valence bonds" (RVB) in metals, P. W. Anderson [50] introduced in 1973 the idea that antiferromagnetically coupled spins 1/2 could have a ground state completely different from the two previous cases.

An RVB state can be viewed as a linear superposition of an exponential number of disordered valence bond configurations where spins are coupled by pairs in singlets (contrary to the Valence Bond Crystal where a finite number of ordered configurations dominate the ground state wave-function and the physics of the phase). Because of these singlet pairings, many configurations are expected to have a reasonably low energy for an antiferromagnetic Hamiltonian. However, these energies are not necessarily lower that the variational energy of a competing Néel state. What lower the energy of a RVB state with respect to the energy of one particular lattice dimerization, are the resonances between the exponential number of dimer configurations which are energetically very close or degenerate. These resonances are possible because the Hamiltonian (the Heisenberg one for instance) has non-diagonal matrix elements between almost any pair of dimer coverings.

One central question is to characterize the kinds of dimer configurations which have the most important weights in the wave-function. The low-energy physics will crucially depend on the separation between the spins which are paired in singlets[6]. Both ideas of short - and long-ranged RVB states have been developed in the literature. Liang et al. have shown that long range order on the

[6] The set of all dimer coverings, including singlet dimers of *all* lengths, is in fact an over-complete basis of the whole $S = 0$ subspace. Without specifying which dimer configurations do enter in the wave-function, a linear superposition of dimer configurations can in fact be *any* $S = 0$ state, including a state with long-range spin-spin correlations!

square lattice can be recovered with dimer wave functions as soon as the weight of configurations with bonds of length l decreases more slowly than l^{-5} [51]. In this section we concentrate on the case where only *short-ranged* (i.e. a finite number of lattice spacings but not necessarily first-neighbors) dimer singlets participate significantly in the ground state wave-function. In such a situation, taken apart the singular case of the Valence Bond Crystals, the short range RVB state has no long range order, it is fully invariant under $SU(2)$ rotations: it is a genuine *Spin Liquid*[7].

3.2 Groundstate Properties of Type I SRRVB Spin Liquids

Consider a spin model with a spin-$\frac{1}{2}$ in the unit cell. Our definition of a short-range RVB state is a wave-function which has the following properties:

(A) it can be written as a superposition of short-ranged dimer configurations,
(B) it has exponentially decaying spin-spin correlations, dimer-dimer correlations and any higher order correlations,
(C) its ground state displays a subtle topological degeneracy [52,27,53–57].

Let us make some remarks on this definition. First, property (A) does not imply (B). As an example, the equal-amplitude superposition of all nearest-neighbor dimer configurations on the *square* lattice has algebraic spin-spin correlations [28]. However such equal-amplitude superposition on the *triangular* lattice was recently shown to satisfy (B) [58]. Property C is specific to systems with half odd integer spins in the crystallographic unit cell.

3.3 Elementary Excitations of Type I SRRVB Spin Liquids

If all kinds of correlations decay exponentially with distance over a finite correlation length ξ, all symmetry sectors are expected to be gapful[8]. The simplest heuristic argument is the following: low-energy excitations are usually obtained by long wavelength deformations of the ground state order parameter. If the correlation length is finite, elementary excitations will have a size of order ξ (or smaller), which is the largest distance in the problem. Such excitations will therefore have a finite energy (uncertainty principle).

Magnons and Spinons

A conventional Néel antiferromagnet has elementary excitations called magnons (or spin-waves) which carry an integer spin $\Delta S^z = \pm 1$. In one dimension, there are free spin-$\frac{1}{2}$ elementary excitations (spinons). In two dimensions, spinons

[7] Some authors use the word spin liquid with a less restrictive meaning for all spin systems exhibiting a spin gap inasmuch as they do not break SU(2). In view of the importance of the LRO in the ground state and low lying excitations of the Valence Bond Crystals described in the previous section, we prefer the present definitions.

[8] and χ and C_v are thermally activated

are confined in VBC phases but the possibility of unconfined spinons exists in a short-range RVB phase (naively speaking there are no more elastic forces to bind spinons in disordered dimer coverings). A field-theoretic description (Large$-N$ limit and gauge-theory) of spinon (de-)confinement in spin liquids was carried out by Read and Sachdev [34]. Unconfined spinons are fractional excitations in the sense that they carry a quantum number (total spin) which is a fraction of the local degrees of freedom, namely the $\Delta S = \pm 1$ spin flips. A first example of a *2D system* with unconfined spin-$\frac{1}{2}$ excitations might have been observed experimentally [59]. From the theoretical point of view, only toy models have been rigorously demonstrated to exhibit spinons [58,60], although they are a generic feature in large-N ($Sp(N)$) approaches to frustrated spin-liquids (see for instance Refs. [34,61,62]) at least at mean-field level.

3.4 The Hard Core Quantum Dimer Model on the Triangular Lattice

The QHCD model was originally introduced to look for an RVB phase. As explained in the previous section, it turns out that on the square lattice it displays only (gapped or critical) VBC phases. It has been recently generalized to the triangular lattice by Moessner and Sondhi [58]. Contrary to the original square lattice case, the QHCD model on the triangular lattice provides a short-range RVB spin-liquid with a finite correlation length at zero temperature. This phase survives in a finite interval of parameter $V_c \leq V \leq J$. The triangular version of the QHCD model is probably the simplest microscopic model which exhibits a short-range RVB phase, and quasi particle deconfinement.

3.5 Realizations of a Type I Spin Liquid in $SU(2)$ Spin Models

Since the pioneering work of Anderson, there has been an intense theoretical activity on the RVB physics, specially after Anderson [63] made the the proposition that such an insulating phase could be closely related to the mechanism of high-T_c superconductivity. Short-range RVB states are certainly a stimulating theoretical concept, their realization in microscopic spin models is a more complicated issue. Up to now, two-dimensional models which could exhibit a short-range RVB phase are still few: Ising-like models in a transverse magnetic field [64] which are closely related to QHCD models by duality [60,65], a quasi 1d model [66], a spin-orbital model [67], and the two short-range RVB phases in $SU(2)$ models, that we will now present.

The Multiple-Spin Exchange Model

The multiple-spin exchange model (called MSE in the following) was first introduced by Thouless [68] for the nuclear magnetism of three-dimensional solid He^3 [69] and by Herring [70] for the Wigner crystal. It is an effective Hamiltonian which governs the spin degrees of freedom in a crystal of fermions. The Hamiltonian is a sum of permutations which exchange the spin variables along rings of neighboring sites. It is now largely believed that MSE interactions on the triangular lattice also describe the magnetism of solid He^3 mono-layers adsorbed on graphite [71–73] and that it could be a good description of the two dimensional Wigner crystal of electrons [74]. In the He^3 system, exchange terms including up to 6 spins are present [71]. Here we will only focus on 2- and 4-spin interactions which constitute the minimal MSE model where a short-range RVB ground state is predicted from exact diagonalizations [55]. The Hamiltonian reads:

$$\mathcal{H} = J_2 \sum_{\bullet\!-\!\bullet} P_{ij} + J_4 \sum_{\lozenge} (P_{ijkl} + P_{lkji}) \tag{6}$$

The first sum runs over all pairs of nearest neighbors on the triangular lattice and P_{ij} exchanges the spins between the two sites i and j. The second sum runs over all the 4-sites plaquettes and P_{ijkl} is a cyclic permutation around the plaquette. The 2-spin exchange is equivalent to the Heisenberg interaction since $P_{ij} = 2\boldsymbol{S}_i \cdot \boldsymbol{S}_j + 1/2$, but the four-spin term contains terms involving 2 and 4 spins and makes the model a highly frustrated one.

As $J_2 < 0$ and $J_2/J_4 \simeq -2$ in low-density solid He^3 films, the point $J_2 = -2$ $J_4 = 1$ has been studied by means of exact diagonalizations up to $N = 36$ sites [55]. These data point to a spin-liquid with a short correlation length and a spin gap. No sign of a VBC could be found. In addition, a topological degeneracy which characterizes short-range RVB states was observed. From the experimental point of view, early specific heat and spin susceptibility measurements are not inconsistent with a spin-liquid phase in solid He^3 films. Indeed very low temperature magnetization measurements [73] suggest a small but non-zero spin gap in low-density mono-layers.

An explanation of the origin of this short-range RVB phase in the MSE model can be guessed from the analogy between multiple-spin interactions and QHCD models. From the analysis of QHCD models we understand that RVB phases are possible when VBC are energetically unstable. Columnar VBC are stabilized by strong parallel dimer attraction and staggered [58] VBC appear when the repulsion between these parallel dimers is strong. In between, an RVB phase can arise[9]. From this point of view, tuning the dimer-dimer interactions is of great importance and the four-spin interaction of model (6) plays this role [55].

A Type I SRRVB Phase on the Hexagonal Lattice

A second RVB phase has recently been found on the hexagonal lattice in a highly frustrating regime where the three first neighbors are coupled ferro-magnetically whereas the six second neighbor couplings are antiferromagnetic [10]. For weak second neighbor coupling, the ground state is a ferromagnet. Increasing the second neighbor coupling leads to an instability toward a short range RVB phase

[9] In Ref. [60] an RVB state is selected by introducing defects in the lattice in order to destabilize the competing VBC states.

which has all the above-mentioned properties (except the topological degeneracy, because there are two spins 1/2 in the unit cell). This kind of phase has possibly been observed years ago by L.P. Regnault and J. Rossat-Mignod in $BaCo_2(AsO_4)_2$[75].

These two RVB phases appear in the vicinity of a ferromagnetic phase: is it or is it not an essential ingredient to form an RVB phase? One might argue that this feature helps in favoring RVB phases against VBC ones, because the plausible VBCs would have large elementary plaquettes and would be very sensitive to resonances between the different forms of plaquettes, thus disrupting long range order. It should also be noticed that the first neighbor coupling being ferromagnetic, the short range RVB phase will predominantly form on second neighbors, and thus on a triangular lattice. So the properties of the dimer coverings on the triangular lattice might at the end be the essential ingredient to have a RVB phase!

3.6 Chiral Spin-Liquid

The definition of a short-range RVB spin-liquid proposed in section 3.2 excludes any long range order. However, a state with broken *time-reversal* symmetry and chiral long range order could accommodate all the other properties of a spin-liquid (the chiral observable is the triple product of three spins: see ref. [76] for various equivalent definitions). Such a chiral phase would have a doubly degenerate ground state in the thermodynamic limit. Inspired from Laughlin's fractional quantum Hall wave functions, Kalmeyer and Laughlin [77,78] have build a spin-$\frac{1}{2}$ state on the triangular lattice which exhibits some chiral long-range order (see also [79]). This *complex* wave function is directly obtained from the bosonic $m = 2$ ($\nu = \frac{1}{m}$) Laughlin wave-function. Such a state is a spin singlet with unconfined spinon excitations which have anyonic statistics (right in between Bose and Fermi). This chiral liquid was initially proposed for the triangular-lattice Heisenberg antiferromagnet but the later turned out to be Néel long-range ordered [14]. Wen et al. [76] discussed the properties expected for a chiral spin-liquid, its excitations and some possible mean-field descriptions. To our knowledge, there is no example of a microscopic $P-$ and $T-$symmetric spin model which exhibits a chiral spin-liquid phase. The possibility of realizing a chiral phase in the presence of an external magnetic field (which explicitly breaks the time-reversal invariance) is discussed in Sect. 5.4.

4 Type II SRRVB Phases: The "Kagomé-Like" Magnets

On triangular based lattices, destabilization of the coplanar 3-sublattice Néel order *either by an increase of the quantum fluctuations* (through a decrease of the coordination number when going from the triangular lattice to the Kagomé one) *or by adding competing interactions* (4-spin exchange processes) leads to an unexpected situation where the degeneracy of the exponential number of short range dimer coverings is only marginally lifted by quantum resonances, giving rise to a quantum system with a continuum of singlet excitations adjacent to the ground state. This property has first been shown by exact diagonalizations of the Heisenberg Hamiltonian on the Kagomé lattice [13], then for the MSE Hamiltonian on the triangular lattice [80].

4.1 Description of the Groundstate and the First Excitations in the $S = 0$ Sector

The ground state is a trivial superposition of an exponential number of singlets, like in any RVB ground state described in the previous section. But contrary to the situations described in the previous sections there is no gap above the ground state in the singlet sector.

Mambrini and Mila [81] have shown that the qualitative properties of the ground state and the first excitations are well described in the restricted basis of nearest neighbor couplings: to this extent, this second spin-liquid is a real short-range RVB state (indeed dressing these states with longer dimer coverings improves quantitatively the energy, but does not change the picture).

To understand the mechanism of (non)formation of the gap in the Kagomé spin-liquid, it is interesting to compare Mambrini's results to the earlier work of Zeng and Elser [82]. This comparison shows that the non orthogonality of the dimer basis is an essential ingredient to produce the continuum of singlets adjacent to the ground state. The above-mentioned QHCD model which implicitly truncates the expansion in the overlaps of dimers is by the fact unable to describe such a phase. On the other hand, taking into account the non orthogonality of dimer configurations would generate a QHCD model involving an infinite expansion of n-dimers kinetic and potential terms. In this basis the effective Hamiltonian describing the original Heisenberg problem has an infinite range of exponentially decreasing matrix elements.

This system has a $T = 0$ residual entropy in the singlet sector ($\sim \ln(1.15)$ per spin) [13,83]. The $S = 0$ excitations cannot be described as Goldstone modes of a quasi long range order in dimers (similar to the critical point of the R.K. QHCD model of subsection 2.4). In such a system, the density of states would increase as $\exp(N^{\alpha/(\alpha+D)})$, where D is the dimensionality of the lattice, and α the power index of the dispersion law of the excitations ($\epsilon(k) = |k|^{\alpha}$), whereas it increases as $\exp(N)$ in the present case (this represents a large numerical difference [84]).

4.2 Excitations in the $S \neq 0$ Sectors

The magnetic excitations are probably gapped: this assumption is a weak one. The spin gap if it exists is small (of the order of $J/20$).

In each $S \neq 0$ sector the density of low lying excitations increase exponentially with the system size as the $S = 0$ density of states, but with extra prefactors (as for example a N prefactor in the case of the $S = 1/2$ sector [85,13,83,81]).

The elementary excitations are de-confined spinons [13]. An excitation in the $S = 1/2$ sector could be seen as a dressed spin-1/2 in the sea of dimers [83]. The picture of Uemura et al. [86] drawn from the analysis of muon data on SrCrGaO is perfectly supported by exact diagonalizations results. The analytic description of such a phase remains a challenge. The fermionic $SU(N)$ description [87,88] might give a good point of departure: here the $S = 0$ ground state is indeed degenerate in the saddle point approximation. But the difficulty of the analysis of this degenerate ground state in a $1/N$ expansion remains to be solved! A recent attempt to deal with such problems in a dynamical mean field approach looks promising [89].

4.3 Experimental Realizations

No perfect $S = 1/2$ Kagomé antiferromagnet has been up to now synthesized.

An organic composite $S = 1$ system has been studied experimentally [90]: it displays a large spin gap (of the order of the supposed-to-be coupling constant and thus much larger than what is expected on the basis of the $S = 1/2$ calculations). It is difficult to claim that it is an experimental manifestation of an even-odd integer effect, because the ferromagnetic binding of the spins 1/2 in a spin 1 is not so large that the identities of the underlying compounds could not play a role. (From a theoretical point of view, it would be extremely interesting to have exact spectra of a spin-1 Kagomé antiferromagnet: if topological effects are essential to the physics of the spin-1/2 Kagomé one might expect completely different spectra for the spin-1 system.)

A spin-3/2 bilayer of Kagomé planes, the SrCrGa oxide has been extensively studied [91]. It displays some features that could readily be explained in the present framework of the spin-1/2 theoretical model:

- Dynamics of the low lying magnetic excitations [86]
- Vanishing elastic scattering at low temperature [92]
- Very low sensitivity of the low-T specific heat to very large magnetic fields [93, 94]

but some features (essentially the anomalous spin glass behavior [95,96]) remain to be explained in a consistent way.

There are also a large number of magnetic compounds with a pyrochlore lattice (corner sharing tetrahedra). At the classical level such Heisenberg magnets have ground states with a larger degeneracy than the Kagomé problem [9]. They are expected to give spin-liquids [8], and indeed some of them display no frozen magnetization [97]. Whether the Heisenberg nearest neighbor problem for spin-1/2 has the same generic properties on the pyrochlore lattice and on the Kagomé lattice is still an open question. Contrary to some expectations [98], the Heisenberg problem on the checker-board lattice (a 2-dimensional pyrochlore) has a VBC ground state [30,99]. Nevertheless it is up to now totally unclear if the checker-board problem is a correct description of the 3d pyrochlores, there are even small indications that this could be untrue [30].

5 Magnetization Processes

In this section we discuss some aspects of the behavior of Heisenberg spin systems in the presence of an external magnetic field. Frustrated magnets either with a semi-classical ground state or in purely quantum phases exhibit a large number of specific magnetic behaviors: metamagnetism, magnetization plateaus. In the first subsection we describe the free energy patterns associated with these various behaviors. In Sect. 5. 2, we discuss a classical and a quantum criterion for the appearance of a magnetization plateau. In the following Sect. we describe two mechanisms recently proposed to explain the formation of a plateau.

5.1 Magnetization Curves and Free-Energy Patterns

The simplest quantum antiferromagnet in 2D is the spin$-\frac{1}{2}$ Heisenberg model on the square lattice:

$$\mathcal{H} = J \sum_{\langle i,j \rangle} \boldsymbol{S}_i \cdot \boldsymbol{S}_j - B \sum_i S_i^z \tag{7}$$

The system has two sublattices with opposite magnetizations in zero field and at zero temperature. The full magnetization curve of this model has been obtained by numerical and analytical approaches [100]. The sublattice magnetizations gradually rotates toward the applied field direction as the magnetic field is increased (Fig. 3). At some finite critical field B_{sat} the total magnetization reaches saturation[10].

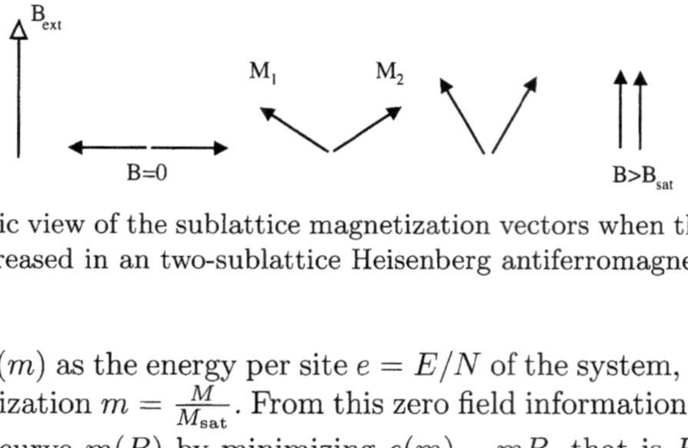

Fig. 3. Schematic view of the sublattice magnetization vectors when the external magnetic field is increased in an two-sublattice Heisenberg antiferromagnet

We define $e(m)$ as the energy per site $e = E/N$ of the system, as a function of the net magnetization $m = \frac{M}{M_{\mathrm{sat}}}$. From this zero field information, we get the full magnetization curve $m(B)$ by minimizing $e(m) - mB$, that is $B = \frac{\partial e}{\partial m}$. In the square-lattice antiferromagnet case discussed above, $e(m)$ is almost quadratic $e(m) = \frac{m^2}{2\chi_0}$ and the corresponding magnetization is almost linear (Fig. 4).

[10] As in most models, the saturation field can be computed exactly by comparing the energy of the ferromagnetic state ($E_F/N = J/2 - B/2$) with the (exact) energy of a ferromagnetic magnon ($E(\boldsymbol{k} \neq \boldsymbol{0}) = E_F + B + J(\cos(k_x) + \cos(k_y) - 2)$). The magnon energy is minimum in $\boldsymbol{k}_{\mathrm{min}} = (\pi, \pi)$. The saturation field is obtained when $E_F = E(\boldsymbol{k}_{\mathrm{min}})$, that is $B_{\mathrm{sat}} = 4J$. The calculation is unchanged for an arbitrary value of the spin S and one finds $B_{\mathrm{sat}} = 8JS$.

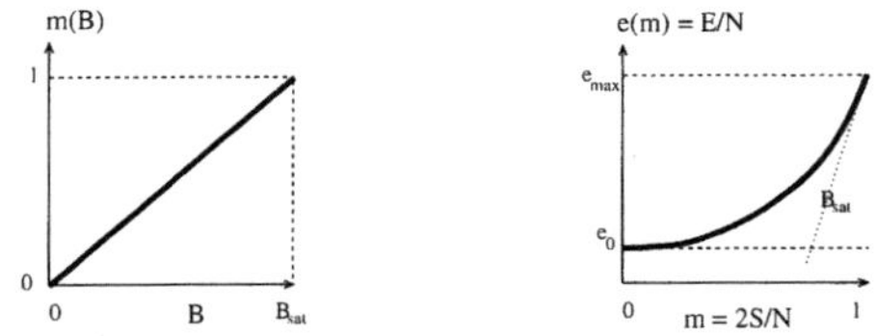

Fig. 4. Linear response obtained with $e \simeq \frac{m^2}{2\chi}$ as in an AF system with collinear LRO

When the system is more complicated, because of frustration for instance, the magnetization process can be more complex. In particular, magnetization plateaus or metamagnetic transitions can occur. For instance, the addition of a second-neighbor coupling on the square-lattice antiferromagnet opens a plateau at one half of the saturated magnetization [101]. It is useful to translate these anomalies of the magnetization curve into properties of $e(m)$.

A plateau at $m_0 = 0$ (also called spin gap) is equivalent to $\frac{\partial e}{\partial m}_{|m_0=0} > 0$ (Fig. 5). At the field where the magnetization starts growing the transition can be critical or first order. In one dimension, exact results on the Bose condensation of a dilute gas of interacting magnons lead to $m \sim \sqrt{\delta B}$ for integer spin chains [102]. In that case $e(m) = \Delta m + am^3 + o(m^3)$ and the system is critical at $B = \Delta$.

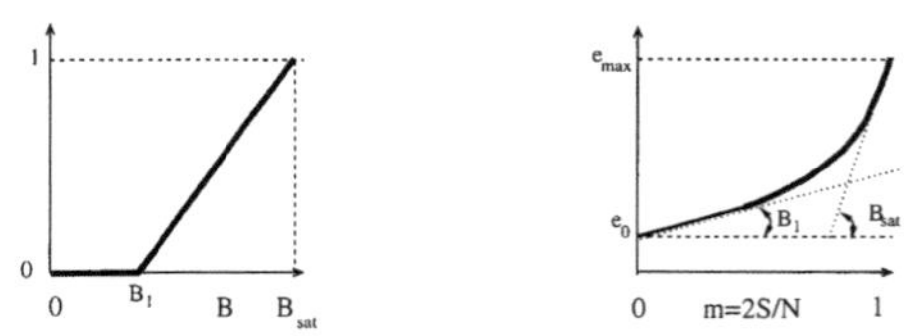

Fig. 5. $m = 0$ magnetization plateau due to a linear $e(m) = \Delta m + o(m)$

A plateau at finite magnetization $m_0 > 0$ for $B \in [B_1, B_2]$ is a discontinuity of $\frac{\partial e}{\partial m}$ (Fig. 6). Such a behavior can arise in a frustrated one (see next Sect. 5. 2). The vanishing susceptibility when B is inside the plateau comes from the fact that magnetic excitations (which increase or decrease the total magnetization) are gapped when the magnetic field lies in the interval $[B_1, B_2]$. As before, the system can be critical at the "edges" of the plateaus. In the following subsections we examine the origins of such plateaus in classical and quantum spin systems.

A metamagnetic transition is a discontinuity of the magnetization as a function of the applied field, it is a first order phase transition and is equivalent to a concavity in $e(m)$ (Fig. 7). As in any first order transition, this can give rise to an hysteretic behavior in experiments. Such a behavior is highly probable in frustrated magnets [72,75] for essentially two reasons:

- Due to the frustration, different configurations of spins corresponding to phases with different space symmetry breakings are very near in energy,

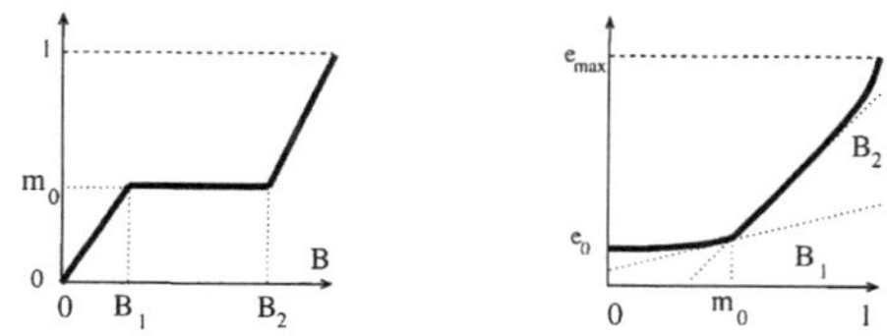

Fig. 6. A magnetization plateau originates from a discontinuity in the slope of $e(m)$. The simplest example of quantum Heisenberg model with such a plateau is the triangular-lattice antiferromagnet (see Fig. 10 and text), a more sophisticated example is the behavior of the same model on a Kagomé lattice, which equally displays a magnetization plateau at $M/M_{\mathrm{sat}} = 1/3$

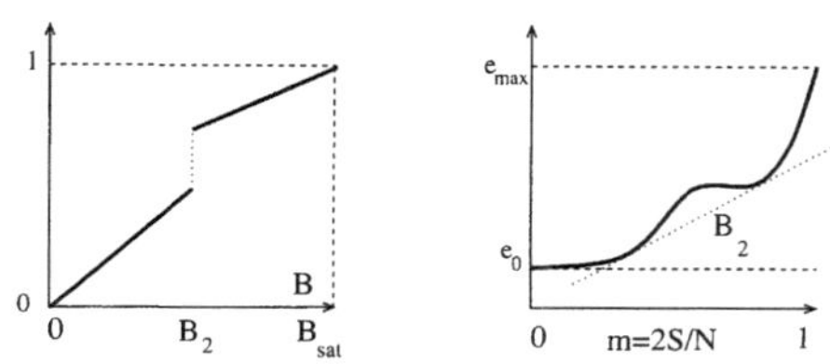

Fig. 7. Metamagnetic transition associated to a concavity in the $e(m)$ curve

- Impurities in magnetic compounds pin the existing structures and hinder the first order phase transitions, at variance with their role in the standard liquid-gas transition.

As a last general, but disconnected, remark let us underline that the magnetic studies of quantum frustrated magnets could also help in solving elusive questions relative to the existence of an exotic H=0 ground state. We have already seen that the extraordinary thermo-magnetic behavior of SrCrGaO could be a signature of a type II Spin Liquid [93,94]. Years ago L.P. Regnault and J. Rossat-Mignod studied $BaCo_2(AsO_4)_2$: the knowledge of its magnetic phase diagram convinced them that something queer was going on in this material at $H = 0$ and in fact exact diagonalizations now point to a type I Spin Liquid! In fact in very large magnetic fields the magnets become increasingly classical and a semi-classical approach is justified: any deviation from a semi-classical behavior when H is decreasing is thus an important indication of a possible quantum exotic ground state!

5.2 Magnetization Plateaus

Classical Spins: Collinearity Criterion

Magnetization plateaus are often believed to be a purely quantum-mechanical phenomenon which is sometimes compared in the literature to Haldane phases of integer spin chains. This is certainly not always true since some *classical* spin models have magnetization plateaus at zero temperature. For instance, as shown by Kubo and Momoi [103], the MSE model on the triangular lattice has a large range of parameters where magnetization plateaus at $M/M_{\text{sat}} = \frac{1}{3}$ and $M/M_{\text{sat}} = \frac{1}{2}$ appear at zero temperature[11]. The ground state of the system at $M/M_{\text{sat}} = \frac{1}{3}$ is the so-called *uud* structure where two sublattices have spins pointing "up" along the field axis and the third one has down spins. At $M/M_{\text{sat}} = \frac{1}{2}$ the ground state is of *uuud* type (Fig. 8).

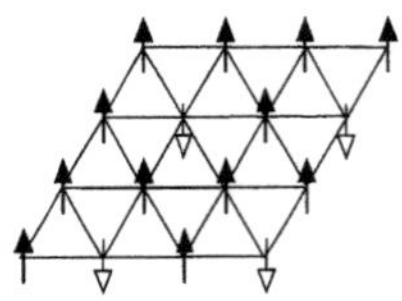

Fig. 8. Collinear spin structure of type *uuud* on the triangular lattice. This state with $M/M_{\text{sat}} = \frac{1}{2}$ is realized in the classical and quantum MSE model under magnetic field (see text and Fig. 9)

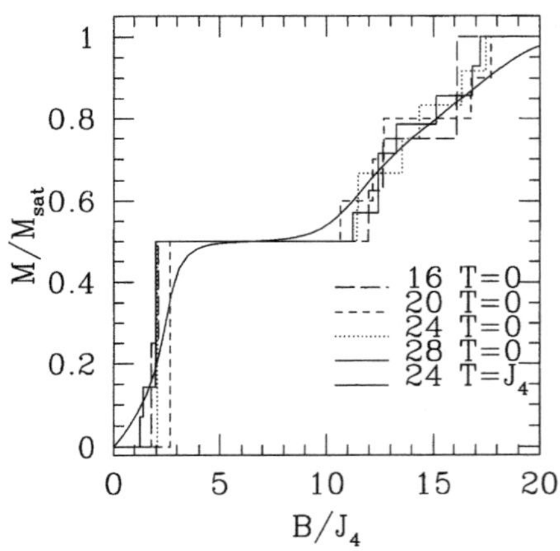

Fig. 9. Magnetization curve of the MSE model at $J_2 = -2$ and $J_4 = 1$ computed numerically [104,105] for finite-size samples with N up to 28 sites. The magnetization plateau at $M/M_{\text{sat}} = \frac{1}{2}$ (due to the *uuud* state of Fig. 8) clearly appears even at finite temperature (full line [105])

In fact, we show below that if a classical spin system exhibits a magnetization plateau then all the spins are necessarily collinear to the magnetic field. This restricts possible spin configurations to states of the $u^{n-p}d^p$ kind with $n - p$ spins "up" and p spins "down" in the unit cell. As a consequence, the total magnetization per site must be of the form: $M/M_{\text{sat}} = 1 - 2p/n$ where n is the size of the unit cell, and p an integer. This commensuration between the

[11] As discussed below, these plateaus are also present in the quantum $S = \frac{1}{2}$ model [72,104,80].

magnetization value at a plateau and the total spin in the unit cell will turn out to hold equally for quantum spins.

The proof of this classical collinearity condition is sketched below. We consider the function $e(m)$ (which is defined without external field). Let us suppose that the ground state Ψ_0 at m_0 is *not* a collinear configuration. We will show that $e(m)$ has a continuous derivative in m_0, *i.e.* no plateau occurs at m_0. From a non-collinear state, one can chose an angle θ to deform Ψ_0 into a new configuration $\Psi(\theta)$ with a different magnetization

$$m(\theta) = m_0 + a\theta + \mathcal{O}(\theta^2) \tag{8}$$

and $a \neq 0$. This can be done, for instance, by rotating the spins of a sublattice whose magnetization is not collinear to the field[12]. Here we just require that the spins can be rotated of an infinitesimal angle, as it is the case for three-component classical spins. $\Psi(\theta)$ is a priori no longer the ground state and its energy is:

$$e(\Psi(\theta)) = e(m_0) + \alpha\theta + \mathcal{O}(\theta^2) \tag{9}$$

No assumption on α is needed. As a variational state of magnetization $m(\theta)$, $\Psi(\theta)$ has an energy $e(\Psi(\theta))$ larger or equal to the the ground state energy at magnetization $m(\theta)$:

$$e(m_0) + \alpha\theta + \mathcal{O}(\theta^2) \geq e(m(\theta)) \tag{10}$$

With (8) we see that:

$$e(m_0) + \frac{\alpha}{a}\left(m(\theta) - m_0\right) + \mathcal{O}\left((m(\theta) - m_0)^2\right) \geq e(m(\theta)) \tag{11}$$

If we assume that no metamagnetic transition occurs in the neighborhood of m_0, then $e(m)$ is convex about m_0 and (11) insures the differentiability of $e(m)$ in $m = m_0$. We have $\frac{\partial e}{\partial m}_{|m=m_0^-} = \frac{\partial e}{\partial m}_{|m=m_0^+}$ and no plateau occurs at m_0.

This proof is not restricted to rotation invariant interactions. We have only made the (extremely weak) assumption that the energy is a continuous and differentiable function of the spins directions (9). If the couplings are written $J^x \boldsymbol{S}_i^x \boldsymbol{S}_j^x + J^y \boldsymbol{S}_i^y \boldsymbol{S}_j^y + J^z \boldsymbol{S}_i^z \boldsymbol{S}_j^z$, a plateau must correspond to a collinear state with respect to the magnetic field direction, whatever the easy-plane or easy-axis might be. This property remains also true if multiple-spin interactions are present.

Fluctuations

Zero-temperature magnetization plateaus will of course be smeared out at very high temperature. However, in some cases, thermal fluctuations can enhance magnetizations plateaus. In the case of the triangular-lattice classical Heisenberg antiferromagnet the magnetization is perfectly linear at zero temperature

[12] If Ψ_0 was a collinear state, the deviation of magnetization would be smaller: $m(\theta) = m_0 + b\theta^2 + \mathcal{O}(\theta^3)$ and the argument would fail.

for isotropic Heisenberg interactions. It is thanks to thermal fluctuations that a *uud* structure is stabilized and that the susceptibility is reduced (but not zero) about $M/M_{\rm sat} \simeq \frac{1}{3}$ [106] (quasi plateau). This phenomenon is due to the higher symmetry of collinear states compared with other nearby spin configurations. This symmetry enlarges the available phase space volume for low-energy fluctuations and selects the *uud* phase.

Quantum fluctuations can play a very similar role: the quantum $S = \frac{1}{2}$ Heisenberg model on the triangular lattice has an exact magnetization plateau at $M/M_{\rm sat} = \frac{1}{3}$ at zero temperature [107–109]. Again, the ground state at $M/M_{\rm sat} = \frac{1}{3}$ is a *uud*-like state. This *uud* plateau has a simple origin in the Ising limit [107,109] ($\mathcal{H} = J^z \sum_{\langle i,j\rangle} S_i^z S_j^z)$) but survives up to the isotropic point $J^z = J^{xy}$. What should be stressed is that the introduction of quantum fluctuations through couplings in the xy plane renormalizes the magnetic field interval where the susceptibility vanishes but the total magnetization at the plateau remains *exactly* $\frac{1}{3}$. The sublattice magnetizations of the two u sublattices and the d sublattice are reduced from their classical values to $M_u = 1-\eta$ and $M_d = -(1-2\eta)$ but the total magnetization is unaffected $M = (2M_u + M_d) = \frac{1}{3} M_{\rm sat}$. The same phenomenon was observed numerically for the *uuud* plateau of the the MSE model on the triangular lattice which also exists in the classical limit [103] and survives quantum fluctuations in the spin$-\frac{1}{2}$ model [72,104,80] without change in the magnetization value. In the following paragraphs we discuss the origin of this robustness of the magnetization value at a plateau.

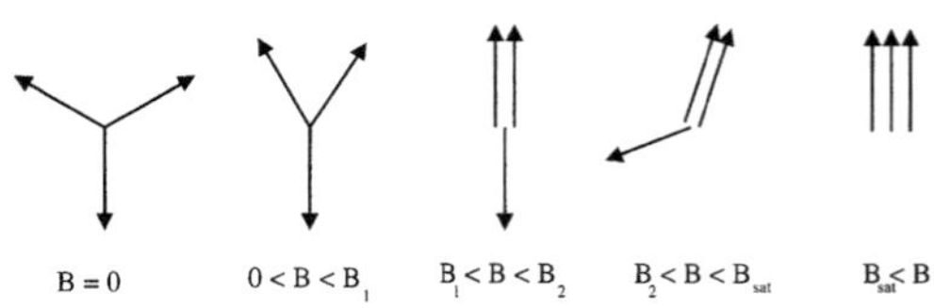

Fig. 10. Magnetization process of the spin$-\frac{1}{2}$ Heisenberg antiferromagnet on the triangular lattice. The magnetic field is along the vertical and the three vectors represent the three sublattice magnetizations. A plateau at $M/M_{\rm sat} = \frac{1}{3}$ is present for magnetic field between B_1 and B_2 [107–109]

Quantum Spins and Oshikawa's Criterion

The Lieb-Schultz-Mattis (LSM) theorem [110] proves that in one dimension and in the absence of magnetic field the ground state is either degenerate or gapless if the spin S at each site is half-integer. The proof relies on the construction of a low energy variational state which is orthogonal to the ground state. Oshikawa, Yamanaka and Affleck [111] realized that this proof can be readily extended to magnetized states as long as $n(S - s^z)$ is *not* an integer, where s^z is the magnetization per site and n the period of the ground state[13]. As a magnetization

[13] n can be larger than what is prescribed by the Hamiltonian if the groundstate *spontaneously* breaks the translation symmetry.

plateau requires a spin gap, this suggests that plateaus can only occur when $n(S - s^z)$ is an integer. However the low-energy state produced by LSM has the same magnetization s^z as the ground state, - it is a non-magnetic excitation-, and does not exclude a gap to *magnetic* excitations when $n(S - s^z)$ is not a integer. In fact arguments based on the bosonization technique were provided by Oshikawa et al. [111] (see also Ref. [112]) to support the hypothesis that $n(S - s^z) \in \mathbb{Z}$ is indeed a necessary condition to have a magnetization plateau in one-dimensional systems. Since then numerous studies of magnetization plateaus in one-dimensional systems [113–115] confirmed that all the plateaus satisfy $n(S - s^z) \in \mathbb{Z}$. We will see below that this criterion appears also to apply in higher dimension.

The LSM construction does not give a low energy state in dimension higher than one but Oshikawa [56] developed a different approach to relate the magnetization value to the number of spins in the unit cell. His result applies to itinerant particles on lattice models when the number of particles is a conserved quantum number. It states that a gap is only possible when the number of particle in the unit cell is an integer. Since spins S are exactly represented by interacting bosons, his results applies to quantum magnets and reduces to $n(S - s^z) \in \mathbb{Z}$. The hypothesis that one has to make to establish the result is the following: the system has periodic boundary conditions (it is a $d-$dimensional torus) and the gap, if any, does not close when adiabatically inserting one fictitious flux quantum inside the torus[14]. In the spin language, this amounts to say that the spin gap does not close when twisting the boundary conditions from 0 to 2π [57]. This hypothesis can only be checked numerically [57] but it is completely consistent with the idea that the gapped system is a liquid with a finite correlation length and which physical properties do not depend on boundary conditions in the thermodynamic limit.

All known examples of magnetization plateaus indeed do satisfy Oshikawa's criterion. It is interesting to notice the close resemblance between the requirement that $n(S - s^z)$ is an integer and the collinearity condition discussed in Sect. 5.2. Let us assume that we can represent one spin S with $2S$ classical spins of length $\frac{1}{2}$ and require that these spins are in a collinear configuration. If the unit cell has n sites it contains $n' = 2Sn$ small spins, p of which are down $(u^{n'-p}d^p)$. For one unit cell the magnetization is $M = n'/2 - p$ so that the magnetization per site is $s^z = M/n = S - p/n$. The classical criterion reads $n(s^z - S) = -p \in \mathbb{Z}$ which is the same condition as Oshikawa's result. In spite of this striking analogy, the classical collinearity condition is an exact result valid in any dimension and it does not involve a topological property (as periodic boundary conditions on the sample) unlike the quantum version. Ultimately indeed it rests on the property that the discretization of a spin S only involves spin-1/2 units.

[14] The number of sites in the $(d-1)$-dimensional section of the torus is supposed to be odd.

5.3 Magnetization Plateaus as Crystal of Magnetic Particles

Magnetization plateaus can often be understood simply if the system has some unit cell where spins are strongly coupled (exchange J_1) and weak bonds $J_2 \ll J_1$ between the cells. In this case, magnetization plateaus are governed by the quantized magnetization of a single cell, and the plateaus can be continuously connected to the un-coupled cells limit $J_2 = 0$. A pertubative calculation in powers of J_2/J_1 will in general predict the plateau to be stable over a finite interval of J_2. This situation appears, for instance, in one-dimensional ladder systems [114]. In 2D, an example is the $M/M_{\rm sat} = \frac{1}{2}$ plateau predicted [116] in the 1/5-depleted square lattice realized in CaV_4O_9. This lattice is made of coupled square plaquettes and the $M/M_{\rm sat} = \frac{1}{2}$ plateau comes from the magnetization curve of a single plaquette (which magnetization can of course take only three values: $M/M_{\rm sat} = 0, \frac{1}{2}$ and 1).

However, in a translation invariant 2D system, with a single spin in the unit cell, the mechanism for the appearance of plateaus is not so simple. The $M/M_{\rm sat} = \frac{1}{3}$ plateau of the triangular-lattice Heisenberg antiferromagnet cannot be understood in such a strong-coupling picture. The magnetization plateaus predicted at magnetizations smaller than $\frac{1}{2}M_{\rm sat}$ in the 1/5-depleted square lattice mentioned above cannot either be understood within such a picture.

Totsuka [112] and Momoi and Totsuka [117] have associated magnetization plateaus with the crystallization of "magnetic particles". In their picture the zero-field ground state is a vacuum which is populated by bosonic particles when the external magnetic field is turned on. Depending on the model and its zero-field ground state these bosons can represent different microscopic degrees of freedom: a spin flip on a single site, an $S^z = 1$ state on a link or a magnetic state of a larger number of spins. These bosons obey an hard-core constraint and carry one quantum of magnetization so that the magnetization of the system is proportional to the particle density. The magnetic field couples to the bosons as a chemical potential and the boson compressibility is the magnetic susceptibility of the spins. As for the antiferromagnetic interactions (such as $J\boldsymbol{S}_i \cdot \boldsymbol{S}_j$), they generate kinetic as well as repulsive interactions terms for these bosons. On general grounds, Momoi and Totsuka expect this gas of interacting bosons to be either a Bose condensate (superfluid) or a crystal (or charge-density wave). On one hand, the finite compressibility of the superfluid leads to a continuously varying magnetization as a function of magnetic field[15]. On the other hand, the underlying lattice will make the crystal incompressible (density fluctuations are gapped) which precisely corresponds to a magnetization plateau. Such a crystalline arrangement of the magnetic moments is also consistent with the quasi-classical picture provided by $u^{n-p}d^p$-like states.

In this approach, the densities (*i.e* magnetizations) at which the bosons crystallize (*i.e* form plateaus) is mainly determined by the range and strength of the boson-boson repulsion and the geometry of the lattice. In some toy models where the kinetic terms for the bosons vanishes (hopping is forbidden by the lattice

[15] In some cases, the superfluid component is believed [117] to coexist with a crystal phase between magnetization plateaus (supersolid).

geometry), this allows to demonstrate rigorously the existence of magnetization plateaus [118]. Looking at the spatial structures which minimizes the repulsion (and thus neglecting kinetic terms in the boson Hamiltonian) gives useful hints of the possible plateaus. Since these structures are stabilized by the repulsive interactions, they can be compared to a Wigner crystal (except that particles are bosons, not fermions).

The magnetization curve of the quasi 2D oxide $SrCu_2(BO_3)_2$ has been measured at very low temperatures up to 57 Tesla [49,119], which corresponds to $M/M_{\rm sat} \simeq \frac{1}{3}$. The magnetization curve displays a large spin gap and plateaus at $\frac{1}{4}$ and $\frac{1}{3}$. A small one at $\frac{1}{8}$ is also reported. This compound is the first experimental realization of the Shastry-Sutherland model [48] (see Fig. 11). In this system magnetic excitations have a very small dispersion. This can be understood from the geometry of this particular lattice [120,121] and has been confirmed experimentally by inelastic neutron scattering experiments [122]. This low kinetic energy makes the bosons quasi-localized objects and explains their ability to crystallize [121,123] and give magnetization plateaus.

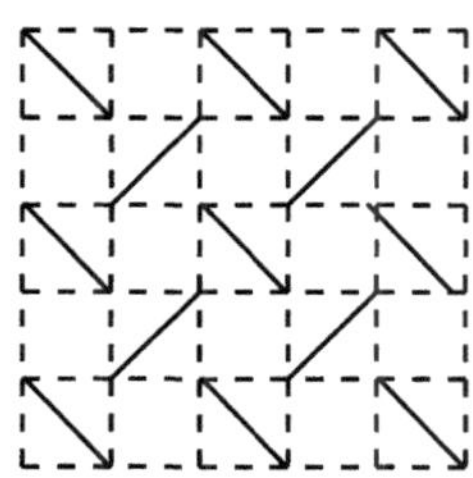

Fig. 11. Shastry-Sutherland lattice. Full lines correspond to Heisenberg exchange J and dotted ones to J'. When $0 < J' < 0.7J$ the ground state is an exact product of singlet states along all the J links. This ground state is very likely to be realized in $SrCu_2(BO_3)_2$ (in zero field) where the couplings are evaluated to be $J'/J \simeq 0.68$ [120]

5.4 More Exotic States – Analogy with Quantum Hall Effect

A recent piece of work on magnetization plateaus in two-dimensional spin-$\frac{1}{2}$ magnets by Misguich, Jolicœur and Girvin [124] establishes a connection between this phenomenon and quantized plateau of the (integer) quantum Hall effect. Although both phenomena show up as quantized plateaus in a 2D interacting system, they are apparently not directly related : magnetization plateaus involve spins on a lattice whereas the quantum Hall effect appears with fermions in the continuum with plateaus in their transverse Hall conductance σ_{xy}. The link between the two problems appears when the spins are represented as fermions attached to one quantum of fictitious magnetic flux, as explained below.

A spin-$\frac{1}{2}$ model is equivalent to hard-core bosons[16]. It is possible to map exactly these bosons to a model of fermions in the presence of a fictitious gauge

[16] Raising and lowering operators S_i^+ and S_i^- are equivalent to bosonic creation and annihilation operators b_i^+ and b_i for bosons satisfying an hard-core constraint $b_i^+ b_i = S_i^z + \frac{1}{2} \leq 1$.

field[17]. The idea is that the hard-core constraint will be automatically satisfied by the Pauli principle and the fictitious flux quantum attached to each fermion will transmute the Fermi statistics into a bosonic one [125–127].

In this framework a down spin is an empty site and an up spin is a composite object made of one (spinless) fermion and one vortex in the fictitious gauge field centered on a neighboring plaquette. This vortex can be simply pictured as an infinitely thin solenoïd piercing the plane through a plaquette adjacent to the site of the fermion. As two of these fermion+flux objects are exchanged adiabatically, the -1 factor due to the Pauli principle is exactly compensated by the factor -1 of the Aharonov-Bohm effect of one charge making half a turn around a flux quantum. Consequently, these objects are bosons and represent faithfully the spin-$\frac{1}{2}$ algebra. Technically, the flux attachment is performed by adding a Chern-Simons term in the Lagrangian of the model, the role of which is to enforce the constraint that each fictitious flux is tied to one fermion.

At this stage the spin problem is formulated as fermions interacting with a Chern-Simons gauge theory. A mean field approximation which is not possible in the original spin formulation is now transparent: the gauge field can be replaced by its static mean value [125,79]. Since this static flux Φ per plaquette comes from the flux tubes initially attached to each fermion, it is proportional to the fermion density $\Phi = 2\pi \langle c^+c \rangle$. Because of this flux, each energy band splits into sub-bands with a complicated structure. When this magnetic field is spatially uniform the band structure has a fractal structure as a function of the flux which is called a Hofstadter "butterfly" [128]. The mean-field ground state is obtained by filling these lowest energy sub-bands with fermions until their density satisfies $\Phi = 2\pi \langle c^+c \rangle$. For a given flux, one can then integrate over the density of states to get the ground state energy. This energy as a function of the flux (or the fermion density) is equivalent to $e(m)$ since $2m + \frac{1}{2} = \langle c^+c \rangle$ and one can compute the magnetization curve. Magnetization plateaus open when some particular band-crossings appear in the Hofstadter spectrum [124].

This approximation scheme was applied to the Shastry-Sutherland model (Fig. 11) and a good quantitative agreement [124] was found with the experimental results of Onizuka et al. [119] on $SrCu_2(BO_3)_2$. In particular, the magnetization plateaus at $M/M_{\rm sat} = 0, \frac{1}{4}$ and $\frac{1}{3}$ were reproduced. The $M/M_{\rm sat} = \frac{1}{3}$ due to the *uud* state on the triangular-lattice antiferromagnet has also been described with this technique [124].

This method is indeed very similar to the Chern-Simons approach to the fractional quantum Hall effect [129] in which real fermions (electrons) are represented as hard-core bosons carrying m (odd integer) flux quanta[18]. In particular, as in

---

[17] In one dimension, this mapping from spins to fermions is the famous Jordan-Wigner transformation and no extra degree of freedom is required. It is a bit more involved in 2D where introducing a fictitious magnetic field is necessary.

[18] These bosons interact with a fictitious magnetic field as well as with the real magnetic field. In the mean-field approximation mentioned above the fictitious field is replaced by its static average. The real and averaged fictitious magnetic field exactly cancel when the Landau level filling factor ν is $\nu = \frac{1}{m}$. When this cancelation occurs the

the quantum Hall effect, the system is characterized by a quantized response coefficient. In the quantum Hall effect this quantity is the transverse conductance σ_{xy} which relates the electric field to the charge current in the perpendicular direction and in the spin system it relates the spin current with a Zeeman field gradient in the perpendicular direction [130]. In the mean field approximation this transverse spin conductance is obtained from the TKNN integers [131] of the associated Hofstadter spectrum.

As in the quantum Hall effect, the topological nature of the quantized conductance protects the plateaus from Gaussian fluctuations [79,124] of the (fictitious) gauge field around its mean-field value. If this mean-field theory captures the physics of magnetization plateaus, this topological picture of the quantized magnetization establishes a deep connection with the conductance plateaus in the Quantum Hall Effect.

There could not be now any definitive conclusion, neither on a set of conditions both necessary and sufficient to have a plateau in two-dimensional frustrated systems, nor on the existence of other mechanisms than those presented in sections 5.3 and 5.4. But magnetization plateaus as well as metamagnetic transitions are probably rather ubiquitous properties of frustrated magnets which still deserve both experimental and theoretical studies.

Acknowledgments

We thank Chitra and R. Moessner for their careful reading of the manuscript and their valuable suggestions.

References

1. N.D. Mermin, H. Wagner: Phys. Rev. Lett. **17**, 1133 (1966)
2. D.C. Mattis: *The Theory of Magnetism I*, Vol. 17 of Springer Series in Solid-State Sciences (Springer-Verlag, Berlin, Heidelberg, New York, Tokyo, 1981)
3. W.J. Caspers: *Spin Systems* (World Scientific, Singapore, 1989)
4. A. Auerbach: *Interacting Electrons and Quantum Magnetism* (Springer-Verlag, Berlin Heidelberg New York, 1994)
5. P. Anderson: Phys. Rev. **86**, 694 (1952)
6. J. Chalker, P.C.W. Holdsworth, E.F. Shender: Phys. Rev. Lett. **68**, 855 (1992)
7. I. Richtey, P. Chandra, P. Coleman: Phys. Rev. B **47**, 15342 (1993)
8. B. Canals, C. Lacroix: Phys. Rev. Lett. **80**, 2933 (1998)
9. R. Moessner, J.T. Chalker: Phys. Rev. B **58**, 12049 (1998)
10. J.-B. Fouet, P. Sindzingre, C. Lhuillier: Eur. Phys. J. B **20**, 241 (2001)
11. P. Tomczak, J. Richter: Phys. Rev. B **59**, 107 (1999)
12. N. Trivedi, D. Ceperley: Phys. Rev. B **41**, 4552 (1990)
13. C. Waldtmann et al.: Eur. Phys. J. B **2**, 501 (1998)
14. B. Bernu, P. Lecheminant, C. Lhuillier, L. Pierre: Phys. Rev. B **50**, 10048 (1994)
15. P. Chandra, B. Doucot: Phys. Rev. B **38**, 9335 (1988)

---

bosons no longer feel any magnetic field and they can Bose-condense. A gap opens and gives rise to the fractional quantum Hall effect.

16. P. Chandra, P. Coleman, A. Larkin: J. Phys. Cond. Matt. **2**, 7933 (1990)
17. H. Schultz, T. Ziman: Europhys. Lett. **8**, 355 (1992)
18. J. Villain, R. Bidaux, J. Carton, R. Conte: J. Phys. Fr. **41**, 1263 (1980)
19. E.F.Shender: Sov. Phys. J.E.T.P. **56**, 178 (1982)
20. C.K. Majumdar, D.K. Ghosh: J. of Math. Phys. **10**, 1388 (1969)
21. B. Shastry, B. Sutherland: Phys. Rev. Lett. **47**, 964 (1981)
22. H. Yokoyama, Y. Saiga: J. Phys. Soc. Jpn. **66**, 3617 (1997)
23. M.E. Zhitomirsky, K. Ueda: Phys. Rev. B **54**, 9007 (1996)
24. L. Capriotti, S. Sorella: Phys. Rev. Lett. **84**, 3173 (2000)
25. N. Read, S. Sachdev: Phys. Rev. B **42**, 4568 (1990)
26. R. Moessner, S.L. Sondhi, P. Chandra: Phys. Rev. B **64**, 144416 (2001)
27. D. Rokhsar, S. Kivelson: Phys. Rev. Lett. **61**, 2376 (1988)
28. M.E. Fisher, J. Stephenson: Phys. Rev. **132**, 1411 (1963)
29. V.N. Kotov, J. Oitmaa, O. Sushkov, Z. Weihong: Phys. Rev. **60**,14613 (1999)
30. J.-B. Fouet, M. Mambrini, P. Sindzingre, C. Lhuillier: cond-mat/0108070
31. I. Affleck, J. Marston: Phys. Rev. B **37**, 3774 (1988)
32. D. Arovas, A. Auerbach: Phys. Rev. B **38**, 316 (1988)
33. I. Affleck, T. Kennedy, E. Lieb, H. Tasaki: Phys. Rev. Lett. **59**, 799 (1987)
34. N. Read, S. Sachdev: Phys. Rev. Lett. **66**, 1773 (1991)
35. N. Katoh, M. Imada: J. Phys. Soc. Jpn. **63**, 4529 (1994)
36. M. Troyer, H. Tsnunetsgu, D. Wurte: Phys. Rev. B **50**, 13515 (1994)
37. S. Taniguchi et al.: J. Phys. Soc. Jpn. **64**, 2758 (1995)
38. Y. Fukumoto, A. Oguchi: J. Phys. Soc. Jpn. **65**, 1440 (1996)
39. M. Albrecht, F. Mila: Europhys. Lett **34**, 145 (1996)
40. M. Albrecht, F. Mila: Phys. Rev. B **53**, 2945 (1996)
41. T. Miyasaki, D. Yoshioka: J. Phys. Soc. Jpn. **65**, 2370 (1996)
42. S. Sachdev, N. Read: Phys. Rev. Lett. **77**, 4800 (1996)
43. K. Ueda, H. Kontani, M. Sigrist, P.A. Lee: Phys. Rev. Lett. **76**, 1932 (1996)
44. K. Kodama et al.: J. Phys. Soc. Jpn. **65**, 1941 (1996)
45. M. Troyer, H. Kontani, K. Ueda: Phys. Rev. Lett. **76**, 3822 (1996)
46. K. Kodama et al.: J. Phys. Soc. Jpn. **66**, 28 (1997)
47. T. Ohama, H. Yasuoka, M. Isobe, Y. Ueda: J. Phys. Soc. Jpn. **66**, 23 (1997)
48. B. Shastry, B. Sutherland: Physica B (Amsterdam) **108**, 1069 (1981)
49. H. Kageyama et al.: Phys. Rev. Lett. **82**, 3168 (1999)
50. P. Anderson: Mater. Res. Bull. **8**, 153 (1973)
51. S. Liang, B. Doucot, P. Anderson: Phys. Rev. Lett. **61**, 365 (1988)
52. B. Sutherland: Phys. Rev. B **37**, 3786 (1988)
53. N. Read, B. Chakraborty: Phys. Rev. B **40**, 7133 (1989)
54. X. Wen: Phys. Rev. B **44**, 2664 (1991)
55. G. Misguich, C. Lhuillier, B. Bernu, C. Waldtmann: Phys. Rev. B **60**, 1064 (1999)
56. M. Oshikawa: Phys. Rev. Lett. **84**, 1535 (2000)
57. G. Misguich, C. Lhuillier, M. Mambrini, P. Sindzingre: cond-mat/0112360 (to appear in Eur. Phys. J. B)
58. R. Moessner, S.L. Sondhi: Phys. Rev. Lett. **86**, 1881 (2001)
59. R. Coldea, D.A. Tennant, A.M. Tsvelick, Z. Tylczynski: Phys. Rev. Lett. **86**, 1335 (2001)
60. C. Nayak, K. Shtengel: Phys. Rev. B **64**, 064422 (2001)
61. C.H. Chung, J.B. Marston, S. Sachdev: Phys. Rev. B **64**, 134407 (2001)
62. C.H. Chung, J.B. Marston, R.H. McKenzie: J. Phys.: Condens. Matter **13**, 5159 (2001)

63. P. Anderson: Science **235**, 1196 (1987)
64. R. Moessner, S.L. Sondhi, P. Chandra: Phys. Rev. Lett. **84**, 4457 (2000)
65. R. Moessner, S.L. Sondhi, E. Fradkin: Phys.Rev. B **65**, 024504 (2002)
66. A. Parola, S. Sorella, Q. Zhong: Phys. Rev. Lett. **71**, 4393 (1993)
67. G. Santoro et al.: Phys. Rev. Lett. **83**, 3065 (1999)
68. D. Thouless: Proc. Phys. Soc. **86**, 893 (1965)
69. M. Roger, J. Hetherington, J. Delrieu: Rev. Mod. Phys. **55**, 1 (1983)
70. C. Herring: in *Magnetism* vol. IV, G.T. Rado, H. Suhl eds. (Academic press, New York and London, 1966)
71. M. Roger et al.: Phys. Rev. Lett. **80**, 1308 (1998)
72. G. Misguich, B. Bernu, C. Lhuillier, C. Waldtmann: Phys. Rev. Lett. **81**, 1098 (1998)
73. E. Collin et al.: Phys. Rev. Lett. **86**, 2447 (2001)
74. B. Bernu, L. Candido, D. Ceperley: Phys. Rev. Lett. **86**, 870 (2001)
75. L. Regnault, J. Rossat-Mignod: in *Phase transitions in quasi two-dimensional planar magnets*, ed. L.J.D. Jongh (Kluwer Academic Publishers, The Netherlands, 1990), pp. 271–320
76. X. Wen, F. Wilczek, A. Zee: Phys. Rev. B **39**, 11413 (1989)
77. V. Kalmeyer, R. Laughlin: Phys. Rev. Lett. **59**, 2095 (1987)
78. V. Kalmeyer, R. Laughlin: Phys. Rev. B **39**, 11879 (1989)
79. K. Yang, L. Warman, S. Girvin: Phys. Rev. Lett. **70**, 2641 (1993)
80. W. LiMing, G. Misguich, P. Sindzingre, C. Lhuillier: Phys. Rev. B **62**, 6372 (2000)
81. M. Mambrini, F. Mila: Eur. Phys. J. B **17**, 651,659 (2001)
82. C. Zeng, V. Elser: Phys. Rev. B **42**, 8436 (1990)
83. F. Mila: Phys. Rev. Lett. **81**, 2356 (1998)
84. C. Lhuillier, P. Sindzingre, J.-B. Fouet: cond-mat/0009336
85. P. Lecheminant et al.: Phys. Rev. B **56**, 2521 (1997)
86. Y. Uemura et al.: Phys. Rev. Lett. **73**, 3306 (1994)
87. J. Marston, C. Zeng: J. Appl. Phys. **69**, 5962 (1991)
88. R. Siddharthan, A. Georges: Phys. Rev. B **65**, 014417 (2002)
89. A. Georges, R. Siddharthan, S. Florens: Phys. Rev. Lett. **87**, 277203 (2001)
90. N. Wada et al.: J. Phys. Soc. Jpn. **66**, 961 (1997)
91. A.P. Ramirez: Annu. Rev. Matter. Sci. **24**, 453-480, (1994) and references therein.
92. S.-H. Lee et al.: Europhys. Lett **35**, 127 (1996)
93. A.P. Ramirez, B. Hessen, M. Winkelmann: Phys. Rev. Lett. **84**, 2957 (2000)
94. P. Sindzingre et al.: Phys. Rev. Lett. **84**, 2953 (2000)
95. A. Keren et al.: Phys. Rev. B **53**, 6451 (1996)
96. A. Wills et al.: Phys. Rev. B **62**, R9264 (2000)
97. M.J. Harris et al.: Phys. Rev. Lett. **73**, 189 (1994)
98. S. Palmer, J. Chalker: condmat/0102447
99. R. Moessner, O. Tchernyshyov, S.L. Sondhi: cond-mat/0106286
100. M.E. Zhitomirsky, T. Nikuni: Phys. Rev. B **57**, 5013 (1998)
101. A. Honecker: cond-mat/0009006
102. I. Affleck: Phys. Rev. B **43**, 3215 (1991)
103. K. Kubo, T. Momoi: Z. Phys. B. Condens. Matter **103**, 485 (1997)
104. T. Momoi, H. Sakamoto, K. Kubo: Phys. Rev. B **59**, 9491 (1999)
105. G. Misguich: Ph.D. thesis, Université Pierre et Marie Curie. Paris. France, 1999
106. H. Kawamura, S. Miyashita: J. Phys. Soc. Jpn. **54**, 4530 (1985)
107. H. Nishimori, S. Miyashita: J. Phys. Soc. Jpn. **55**, 4448 (1986)
108. A.V. Chubukov, D.I. Golosov: J. Phys.: Condens. Matter **3**, 69 (1991)

109. A. Honecker: J. Phys. Condens. Matter **11**, 4697 (1999)
110. E.H. Lieb, T.D. Schultz, D.C. Mattis.: Ann. Phys. (N.Y) **16**, 407 (1961)
111. M. Oshikawa, M. Yamanaka, I. Affleck: Phys. Rev. Lett. **78**, 1984 (1997)
112. K. Totsuka: Phys. Rev. B **57**, 3454 (1998)
113. D.C. Cabra, A. Honecker, P. Pujol: Phys. Rev. Lett. **79**, 5126 (1997)
114. A. Honecker: Phys. Rev. B **59**, 6790 (1999)
115. A. Honecker, F. Mila, M. Troyer: Eur. Phys. J. B **15**, 227 (2000)
116. Y. Fukumoto, A. Oguchi: J. Phys. Soc. Jpn. **68**, 3655 (1999)
117. T. Momoi, K. Totsuka: Phys. Rev. B **61**, 3231 (2000)
118. E. Müller-Hartmann, R.R.P. Singh, C. Knetter, G.S. Uhrig: Phys. Rev. Lett. **84**, 1808 (2000)
119. K. Onizuka et al.: J. Phys. Soc. Jpn. **69**, 1016 (2000)
120. S. Miyahara, K. Ueda: Phys. Rev. Lett. **82**, 3701 (1999)
121. S. Miyahara, K. Ueda: Phys. Rev. B **61**, 3417 (2000)
122. H. Kageyama et al.: Phys. Rev. Lett. **84**, 5876 (2000)
123. T. Momoi, K. Totsuka: Phys. Rev. B **62**, 15067 (2000)
124. G. Misguich, T. Jolicœur, S.M. Girvin: Phys. Rev. Lett. **87**, 097203 (2001)
125. E. Fradkin: Phys. Rev. Lett. **63**, 322 (1989)
126. D. Eliezer, G.W. Semenoff: Ann. Phys. (New York) **217**, 66 (1992)
127. A. Lopez, A.G. Rojo, E. Fradkin: Phys. Rev. B **49**, 15139 (1994)
128. D.R. Hofstadter: Phys. Rev. B **14**, 2239 (1976)
129. S.C. Zhang: Int. J. Mod. Phys B **6**, 25 (1992)
130. F. Haldane, D.P. Arovas: Phys. Rev. B **52**, 4223 (1995)
131. D. Thouless, M. Kohmoto, M.P. Nightingale, M. den Nijs: Phys. Rev. Lett. **49**, 405 (1982)

NMR Studies of Low-Dimensional Quantum Antiferromagnets

Mladen Horvatić[1] and Claude Berthier[1,2]

[1] Grenoble High Magnetic Field Laboratory, CNRS and MPI-FKF, B.P. 166, F-38042 Grenoble Cedex 9, France
[2] Laboratoire de Spectrométrie Physique, Université Joseph Fourier, B.P. 87, F-38402 St. Martin d'Hères, France

Abstract. Selected (high magnetic field) NMR studies of low-dimensional spin systems are presented as textbook examples of how NMR observables (magnetic hyperfine shift, electric field gradient, nuclear spin–lattice relaxation and nuclear spin–spin relaxation) are used to reveal local electronic configuration and static and dynamic spin susceptibility. We discuss NMR data for the doped Haldane chain $Y_2BaNi_{1-x}Mg_xO_5$, the Kagomé based $SrCr_8Ga_4O_{19}$ compound, the spin-Peierls chain $CuGeO_3$ and the related α-NaV_2O_5 system, the organo-metallic spin ladder $Cu_2(C_5H_{12}N_2)_2Cl_4$, and the spin 1/2 Heisenberg chain Sr_2CuO_3.

1 Low-Dimensional Spin Systems, Magnetic Fields and NMR

Strictly speaking, quantum spin systems are compounds in which relevant degrees of freedom can be reduced to a spin variable $\boldsymbol{S}_i$, such as in the Heisenberg Hamiltonian, $\mathcal{H}_{\mathrm{H}} = \sum_{i,\delta} J_{i,\delta}\, \boldsymbol{S}_i \cdot \boldsymbol{S}_{i+\delta}$. However, a strong motivation for studying such systems in the last decade came from much more complicated systems, copper-oxide high T_C superconductors (SC). In these compounds superconductivity takes place within CuO_2 planes in which well localised, antiferromagnetically (AF) coupled, $S = 1/2$ spins on copper sites coexist with mobile holes dominantly carried by oxygen orbitals. While this composite, strongly interacting, 2D system is so-far clearly too complicated to be understood, it has stimulated numerous theoretical and experimental studies of related simpler AF quantum spin systems, starting from 1D, "undoped", i.e. true spin "chains" and "ladders". Studying ladders with an increasing number of legs has been one way to approach a 2D system using better known descriptions in 1D [1,2].

An important property specific to spin 1/2 chains is that, using the Jordan-Wigner canonical transformation, their magnetic Hamiltonians can be transformed into that of strongly interacting 1D spinless fermions, in which the magnetic field plays the role of the chemical potential and the band width is of the order of the coupling J. This makes quantum spin chains remarkable models for the investigation of quasi-1D physics, in which arbitrary band filling is simply controled in situ by varying the magnetic field. Finally, low-dimensional spin compounds in general can be regarded as the simplest model systems allowing us to study phenomena of general interest: (magnetic field induced) quantum phase transitions and critical points, the frustration (i.e., the effect of competing interactions), and the effects of doping impurities.

For a spin system the magnetic field trivially appears as a relevant variable, the field range of interest being of the order of the spin coupling J. Indeed, a common feature of a number of low-dimensional AF quantum spin systems is the existence of a collective singlet ($S = 0$) ground state, separated from the triplet ($S = 1$) excitations by a gap $\Delta \lesssim J$ the value of which can be strongly reduced by frustration [3,1]. The application of a magnetic field H lifts the Zeeman triplet degeneracy and lowers the energy of the $M = -1$ branch down to the ground state. It thus induces at $T = 0$ and $H_C = \Delta/g\mu_B$ a quantum phase transition from a gapped spin-liquid ground state to a gapless magnetic phase. Novel effects are expected since the magnetic correlation length becomes large as the spin-gap is closed, and in quasi-1D systems an additional magnetic soft mode appears at an incommensurate wave vector. Note that in the chemical synthesis of new compounds the J values cannot be controlled, so, very often, this interesting physics appears only in very high magnetic field.

The most pertinent experimental techniques are, of course, those sensitive to magnetism: magnetisation measurements, nuclear magnetic resonance (NMR) and neutron diffraction, given here in order of complexity. Typically, the first information about the magnetism of a new system is obtained from magnetisation data, which gives a *macroscopic* information, can be obtained on very small samples well below 1 mg, but very often suffers from the presence of impurities in the sample. Neutron diffraction [4] provides the most detailed microscopic information on the magnetic structure and on the $(\boldsymbol{q}, \omega)$-dependence of excitations in the system (with $\hbar\omega \gtrsim 1\,\text{meV}$), the data are obtained in the reciprocal space, on relatively large samples (down to about 100 mg) that preferably should not contain protons, and measurements in magnetic field are performed in big split-coils limited presently to a maximal field of only 14.5 T. The NMR, presented in detail in this lecture, provides microscopic *local* information taken precisely at the position of chosen atomic nucleus. It is a bulk technique that is not affected by impurities or surface effects, and can be performed on small single crystals (~ 1 mg). In spin systems one can obtain local spin value, as well as $\boldsymbol{q}$-integrated information on spin excitations at essentially zero energy (μeV range), with only a partial access to its $\boldsymbol{q}$-dependence. As the sample space needs to be connected to a spectrometer only by one semi-rigid coaxial cable, it is relatively easy to realise probe-heads providing access to "extreme conditions" such as very low (dilution fridge) temperature, high pressure or high magnetic field, or even to all of these simultaneously. In particular, NMR has already been performed in fields up to 45 T [5], which is presently the strongest continuous magnetic field available. Note that NMR covers an extremely wide range of applications, employing very different techniques and conditions of measurement. While "high resolution NMR" is addressed in [6], this chapter is devoted to "solid state NMR" and, more specifically, to systems with "localised" spins. The purpose is not to cover the whole literature concerning NMR studies of low-D quantum antiferromagnets, but rather, through specific examples, to demonstrate the possibilities of

NMR technique in that field, putting the emphasis on magnetic field dependent phenomena.

2 NMR Observables

Without entering into details of how the NMR is actually performed [7–10], we will limit the presentation to its basic principles in order to explain what physical quantities can be observed. In a typical configuration, NMR relies on the Zeeman interaction $\mathcal{H}_Z = -\boldsymbol{\mu}_n \cdot \boldsymbol{H}_n$ of the magnetic moment of nuclei (of selected atomic species) $\boldsymbol{\mu}_n = \hbar\gamma_n \boldsymbol{I}_n$, where γ_n and $\boldsymbol{I}_n$ are the gyromagnetic ratio and the spin of the nucleus, to obtain an information on the local magnetic field value $\boldsymbol{H}_n$ at this position. The experiment is performed in magnetic field $H_0 \sim 10\,\mathrm{T}$ whose value is precisely known (calibrated by NMR), and which is perfectly constant in time and homogeneous over the sample volume. A resonance signal is observed at the frequency corresponding to transitions between adjacent Zeeman energy levels $\omega_{\mathrm{NMR}} = \gamma_n H_n$, allowing very precise determination of H_n, and therefore of the local, induced, so-called "hyperfine field" $\boldsymbol{H}_{\mathrm{hf}} = \boldsymbol{H}_n - \boldsymbol{H}_0$ (as γ_n is known from calibration on a convenient reference sample). This hyperfine field, produced by the electrons surrounding the chosen nuclear site, is used as a signature of local electronic environment.

High Magnetic Field and NMR

The spectral resolution of NMR is directly limited by the temporal and spatial homogeneity of external magnetic field H_0. In the experiments where NMR is used for the determination of complex molecular structures [6], the H_0 field variations over the nominal sample dimension of 1 cm should be $\sim 10^{-9}$ for studies in liquid solutions or 10^{-6}–10^{-8} in solid state compounds. In both cases the magnetic field is produced by commercially available, "high-resolution" superconducting (SC) magnets, providing *fixed* field, limited by the present SC technology to a maximum field of 21 T. (Standard field value is 9 T, while fields above 12 T are considered to be "high" which is related to the cost of installations that is rapidly increasing as field is approaching 20 T.) The interest of high fields for structural investigations mostly consists in an important improvement of resolution and sensitivity.

When NMR is used as a probe of magnetic properties in solid state physics, the required field homogeneity is typically much lower, 10^{-3}–10^{-5}, but the field should preferably be variable (sweepable). Up to 20 T such a field is available from commercial SC "solid state NMR magnets", with homogeneity of 10^{-5}–10^{-6}. Higher fields (up to a maximum of 45 T) are available from big resistive or hybrid (SC+resistive) magnets, but their homogeneity is not optimised for NMR. Still, a typical value of 40×10^{-6} over a 1–2 mm sample positioned precisely in the field centre is satisfactory for a great majority of solid state NMR studies. However, because of small sample size requirement and very high running cost, it is reasonable to use these big magnets only for NMR studies of magnetic field dependent phenomena. In order to enable also some very high field "high resolution NMR" studies, several current projects are aiming at improving the temporal and spatial homogeneity of big resistive magnets to or below $10^{-6}/1\,\mathrm{cm}$.

2.1 Local Static Magnetic Field or Spin Value

In the absence of quadrupole coupling (see Sect. 2.2), the frequency in an NMR spectrum directly measures the local magnetic field at the position of the chosen nucleus. More precisely, we get an average of local field on the time scale of the measuring process, ~ 10–$100\,\mu\mathrm{s}$ for solid state NMR. For systems with localised electronic spins in particular, it is easy to see that the induced field is linearly dependent on the spin polarisation of the nearest electronic spin(s) [10]

$$\omega_{\mathrm{NMR}}/\gamma_{\mathrm{n}} = |\boldsymbol{H}_0 + \langle \boldsymbol{H}_{\mathrm{hf}} \rangle| = \left| \boldsymbol{H}_0 + \sum\nolimits_k - A_{\mathrm{n},k} \langle \boldsymbol{S}_k \rangle + \boldsymbol{C}_{\mathrm{n}} \right| . \tag{1}$$

This linear dependence defines the hyperfine coupling constant (tensor) $A_{\mathrm{n},k}$ of nucleus "n" to electronic spin $\langle \boldsymbol{S}_k \rangle$ at position k, while $\boldsymbol{C}_{\mathrm{n}}$ accounts for quadratic (second order orbital or van Vleck) contributions which are not sensitive to the spin direction, as well as (generally much smaller) contribution of other (unpolarised) closed-shell electrons. Hyperfine coupling will be very different according to the distance between nuclear and electronic spins:

- On-site ($\mathrm{n} = k$) hyperfine coupling is strong, $A \sim 1$–$100\,\mathrm{T}$, and is approximately known for a given spin configuration (standard reference is [11]).
- When the coupling is "transferred" or "supertransferred" by an exchange process (i.e., due to overlap of wave functions) from the first or second neighbour site, its value is generally impossible to predict.
- For any distant spins ($\mathrm{n} \neq k$), there is also direct magnetic dipole coupling, which is thus precisely known for given geometry ($\propto |\boldsymbol{r}_{\mathrm{n}} - \boldsymbol{r}_k|^{-3}$), and is generally smaller than (super)transferred hyperfine coupling.

The interaction Hamiltonian corresponding to hyperfine coupling is $\mathcal{H}_{\mathrm{n},k} = \hbar\gamma_{\mathrm{n}} \boldsymbol{I}_{\mathrm{n}} \cdot A_{\mathrm{n},k} \cdot \boldsymbol{S}_k$, and the experimentally measured "magnetic hyperfine shift" K is defined as the frequency shift with respect to the reference:

$$K(T) \equiv \omega_{\mathrm{NMR}}/\left(\gamma_{\mathrm{n}} H_0\right) - 1 = \sum\nolimits_k A_{\mathrm{n},k} \left(g_k \mu_{\mathrm{B}}\right)^{-1} \chi_k(T) + K_0 \, , \tag{2}$$

where g_k and χ_k are the g-tensor and the magnetic susceptibility (per site!) tensor of the k spins, μ_{B} the Bohr magneton and K_0 the shift corresponding to $\boldsymbol{C}_{\mathrm{n}}$ term in (1). K is a tensor whose different components are obtained for different orientations of $\boldsymbol{H}_0$. Regarding the left-hand side of (2), we remark that NMR spectra can be *equivalently* obtained either in a fixed external field H_0 as a function of frequency, or at a fixed frequency ω_{NMR} as a function of magnetic field. This latter configuration is more convenient for very wide spectra, except when the physical properties of the sample strongly vary with H_0. Equation (2), which is equivalent to $\mathcal{H}_{n,k}$ or (1), indicates how the A tensor can be measured by NMR; when the temperature dependence of "bulk" magnetic susceptibility χ_{macro} is dominated by the spatially homogeneous contribution of a single spin species, A is calculated from the slope $\Delta K(T)/\Delta\chi_{macro}(T)$. If $\langle \boldsymbol{S} \rangle$ is taken to be a number, then A is given in units of magnetic field; for historical reasons, the number that is usually declared is the "hyperfine field" $= A/g$ (in gauss/μ_B).

Finally, we remark that (1) and (2) are quite general and can take into account itinerant electrons as well; the summation over k is then rather written in the reciprocal space, using $\sum_{\boldsymbol{r}} f(\boldsymbol{r})g(\boldsymbol{r})^* = \sum_{\boldsymbol{q}} f(\boldsymbol{q})g(\boldsymbol{q})^*$.

The Direct Determination of the Correlation Length in the Doped Haldane Chain $Y_2BaNi_{1-x}Mg_xO_5$

Simple AF Heisenberg spin chains have very different low temperature behaviours according to their spin value; while there is no gap for low-energy excitations of half-integer chains (see Sect. 2.4), the *integer* spin chains are characterised by a gap between the lowest energy levels, predicted by Haldane. A very good example of such a "Haldane" compound is Y_2BaNiO_5, where Ni^{++} ions carrying $S = 1$ spins form a nearly perfect 1D Heisenberg AF chain, with $J = 285\,K$. Transversal (2D) coupling is negligible, $J'/J = (1\text{–}5)\times10^{-4}$, and the anisotropy terms of the spin Hamiltonian are small, inducing only a small anisotropy of the Haldane gap, $\Delta_H^{a,b,c} = 89, 102, 112\,K$. When this system is doped by Mg, these non-magnetic ($S = 0$) impurities replace the Ni atoms, and thus "cut" an ideal infinite chain to an ensemble of open segments of finite length. The ^{89}Y NMR [12] spectrum of such a doped system (see Fig. 1) clearly shows that the spin-polarisation is spatially inhomogeneous. The main (central) peak corre-

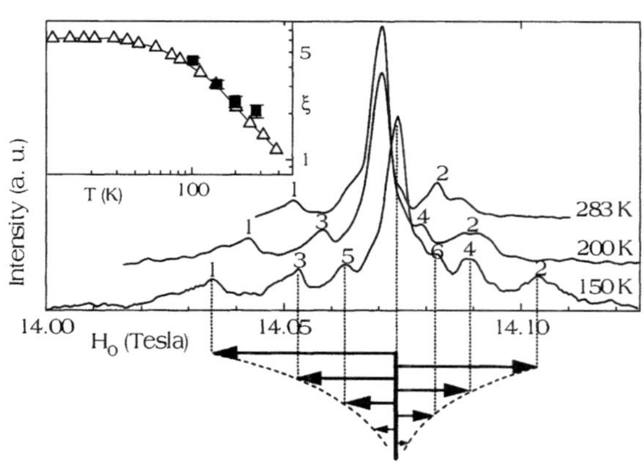

Fig. 1. ^{89}Y NMR spectra in the doped Haldane chain $Y_2BaNi_{0.95}Mg_{0.05}O_5$ recorded at fixed frequency of 29.4 MHz by sweeping magnetic field [12]. Resolved satellite peaks allow the reconstitution of the local magnetisation near the chain end, i.e., Mg impurity, site by site. Thus defined experimental correlation length ξ (*full squares*) is shown in the inset together with the theoretical prediction (*triangles*) for the infinite chain correlation length ξ_∞

sponds to the majority spin sites which are far from impurities; indeed, its shift $K(T)$ is found to be independent of doping and can therefore be used to follow the temperature dependence of the spin polarisation intrinsic to a *pure* system even in a very "dirty" sample! Note that bulk magnetic susceptibility data show an impurity induced Curie upturn at low temperature even for undoped samples; this contribution has to be subtracted in order to get the intrinsic Haldane $\chi_{\mathrm{macro}}(T)$. In this way it is confirmed that for the central NMR line $K(\chi_{\mathrm{macro}})$ is indeed a linear function, defining the hyperfine coupling constant. NMR shift thus becomes a "calibrated" probe of the local spin polarisation.

The structure around the main peak in the NMR spectra clearly corresponds to spins near the impurities (chain ends), as its intensity is proportional to the doping. Down to 100 K within this structure we can distinguish peaks (whose shifts are again nearly independent of doping) that correspond to *individual* spin sites, experiencing different spin polarisations. Labelling these peaks according to their decreasing distance to the central peak, we can reconstitute the spin polarisation site by site; we thus experimentally determine that the spin polarisation is indeed staggered and exponentially decaying as expected from theory, defining the characteristic decay length ξ [12]. Furthermore, the values of ξ in the experimentally accessible temperature range are found to be equal to the theoretical prediction for the correlation length ξ_∞ of an *infinite* chain. This result confirms that the perturbation developed around spinless impurity (Mg ion) indeed "reflects" very well, in a broad temperature range, the bulk properties of the ideal unperturbed Haldane system. This has been subsequently confirmed by both classical [13] and quantum [14] Monte Carlo techniques at finite field and temperature, with remarkable agreement between calculated and observed NMR spectra. However, what happens below 100 K ($\simeq \Delta_{\mathrm{H}}/k_{\mathrm{B}}$) is less clear; the peaks in the "end-of-chains structure" of ^{89}Y NMR spectra disappear, and there are indications of some unexpected dynamics [15].

(Pseudo-) Gap Opening in the Kagomé-Based $SrCr_8Ga_4O_{19}$ Compound

A Kagomé lattice is a 2D arrangement of corner sharing triangles, which can be obtained from a simple triangular 2D lattice if in every second "line" one removes every second site. AF Heisenberg spins placed on Kagomé lattice are *frustrated*, and classical description predicts highly degenerate ground state. On the other hand, for $S = 1/2$ spins quantum description [3] seems to predict singlet ground state, separated from triplet excitations by a gap – which has never been observed experimentally. The $SrCr_8Ga_4O_{19}$ compound is considered as an archetype of Kagomé-based systems, its structure containing $S = 3/2$ (Cr^{3+}) spins on a "pyrochlore block" consisting of two Kagomé planes and one triangular plane in-between. Unfortunately, this compound always presents a quite important percentage of Cr→Ga substitution defects; as a consequence, low temperature χ_{macro} always shows a strong increase, and it is impossible to separate the "impurity contribution" from the intrinsic behaviour of a hypothetically pure system. However, NMR spectra (see also Sect. 2.2) show that the ^{69,71}Ga hyperfine shift (corresponding to the gallium site which is coupled to Kagomé planes [16]) starts to decrease below 50 K, giving an indication for the possible opening of a spin-gap [17]. At the same time, the *width* of NMR line continues to increase at low-T in the same way as χ_{macro}, confirming that this increase is related to some "defects". This is similar to the behaviour of doped Haldane system shown in Fig. 1, where the total, impurity-induced width of NMR spectrum increases on lowering the temperature. However, in $SrCr_8Ga_4O_{19}$ NMR has not revealed the real low-T behaviour so far, as the signal is rapidly lost below 20 K. Moreover, it is not clear whether the observed behaviour is representative of a Kagomé plane or of the pyrochlore block.

The Magnetic Field Dependence of the Soliton Lattice in the IC Phase of the Spin–Peierls System $CuGeO_3$

A spin–Peierls chain is a Heisenberg, AF, $S = 1/2$ chain on an elastic lattice, in which the J coupling depends on the spin position $J_{i,i+1} = J(x_i - x_{i+1})$ [18]. At low temperature, this spin chain can gain energy by spontaneous dimerisation (deformation) of the lattice, which allows the opening of a gap in the low-energy magnetic excitations (absent in simple Heisenberg half-integer spin chains). This dimerised phase has a non-magnetic collective singlet ground state and an energy gap towards triplet excitations. Application of magnetic field reduces the gap and, above a critical field H_C, drives the system into a magnetic phase with spatially inhomogeneous magnetisation (Fig. 2). In this field-induced phase, magnetisation appears as an incommensurate (IC) lattice of magnetisation peaks (solitons), where each soliton is bearing a total spin 1/2. The most studied spin–Peierls system is the inorganic compound $CuGeO_3$ [19], presenting a spin–Peierls transition at 14–10 K (depending on H), and a critical field $H_C \simeq 13$ T. The NMR in $CuGeO_3$ has been performed on the "on-site" copper nuclei which are directly and very strongly coupled to the electronic spin. In the

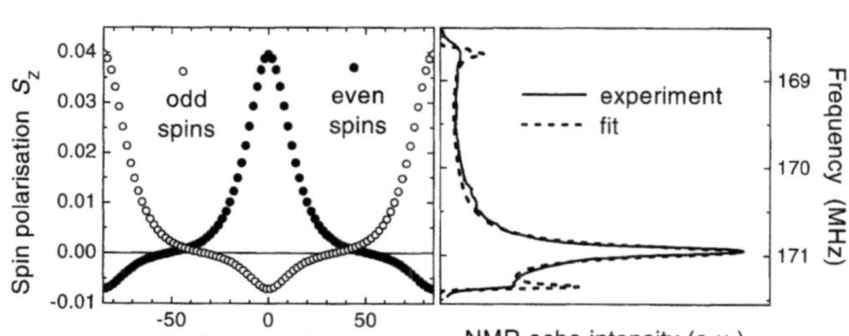

Fig. 2. Reconstitution of the real space spin-polarisation profile (*left*) and the corresponding fit to Jacobi elliptic functions [22] (*right, dashed line*) superposed on experimental NMR lineshape (*right, full line*), taken in the incommensurate high magnetic field phase of the spin–Peierls compound $CuGeO_3$ (here just above the critical field). In this way, NMR lineshape has provided the precise magnetic field dependence of average and staggered spin-polarisation and the magnetic correlation length up to $2H_C$

high-temperature and in the dimerised phase, symmetric NMR lines are observed (see also Sect. 2.2), reflecting spatially uniform magnetisation. Comparing $K(T)$ vs. $\chi_{\rm macro}(T)$ the complete hyperfine coupling tensor as well as the orbital shift tensor $K_0 = K(T=0)$ could be determined, and were found in good agreement with the values expected for a $d_{X^2-Y^2}$ orbital of Cu^{++} ion [20,11].

Above H_C each line is converted to a very wide asymmetric spectrum (Fig. 2) corresponding to a spatially non-uniform distribution of magnetisation. We remark here that the NMR lineshape is in fact the density distribution of the local magnetisation, i.e., of the spin polarisation. Therefore, for a periodic function in 1D, it can be directly converted to the corresponding real-space spin-polarisation profile (soliton lattice, one period of which is shown in Fig. 2) [21,22]. It was thus possible to obtaine a full *quantitative* description of the H dependence of the spin-polarisation profile in the range from H_C to $2H_C = 26\,\mathrm{T}$ [22], clearly showing how the modulation of magnetisation evolves from the limit of nearly independent solitons just above H_C, up to the high magnetic field limit, where it becomes nearly sinusoidal. Analysis of these data proved that the staggered component of magnetisation is reduced in the NMR image by phason-type motion of the soliton lattice [23,24]. The magnetic correlation length is found to be smaller than that associated to the lattice deformation (measured by X-rays [25]), which is a direct consequence of the frustration due to the second neighbour interaction in the system [23]. The observed field dependence of correlation length remains to be understood.

2.2 Local Static Electric Field Gradient

For nuclei with $I > 1/2$, the charge distribution is no longer spherical, and has therefore a quadrupole moment Q_{ab} which is coupled to the second derivative of the local potential, i.e., to the electric field gradient (EFG) $\partial^2\Phi/\partial a \partial b$, by the quadrupole Hamiltonian $\mathcal{H}_Q = \frac{1}{6}\sum_{a,b} Q_{ab}\partial^2\Phi/\partial a \partial b$. As the EFG tensor is symmetric and traceless (according to the Poisson's equation), we customary use the coordinate system of its principal axes (X, Y, Z) to reduce the whole tensor to only *two* parameters: the maximum EFG value and the tensor asymmetry parameter [7]:

$$\begin{aligned} \mathcal{H}_Q &= \tfrac{1}{6} h\nu_Q \left[3I_Z^2 - I(I+1) + \tfrac{1}{2}\eta(I_+^2 + I_-^2)\right] \ , \text{ where} \qquad (3) \\ \nu_Q &= \frac{3eQ/h}{2I/(2I-1)} \partial^2\Phi/\partial Z^2 \ , \ \eta = \frac{\partial^2\Phi/\partial X^2 - \partial^2\Phi/\partial Y^2}{\partial^2\Phi/\partial Z^2} \ , \end{aligned}$$

and Q is the nuclear quadrupole moment $Q = \langle I, I_z{=}I \,|\sum_i 3z_i^2 - r_i^2|\, I, I_z{=}I\rangle$. For crystal sites with non-zero EFG and $I > 1/2$ nuclei, $\mathcal{H}_Q$ will modify the nuclear levels and therefore split otherwise single NMR line (corresponding to equidistant Zeeman levels) to $2I$ lines, which are nearly equidistant for $\mathcal{H}_Q \ll \mathcal{H}_Z$. Furthermore, $\mathcal{H}_Q$ enables the detection of "nuclear quadrupole resonance" (NQR) even in *zero* magnetic field; e.g., for $I = 3/2$ nuclei (such as $^{69,71}Ga$ and $^{63,65}Cu$ nuclei in the two previous examples) a single NQR resonance frequency is observed at $\nu_{\rm NQR} = \nu_Q\sqrt{1+\eta^2/3}$. For $\mathcal{H}_Q \sim \mathcal{H}_Z$ the NMR spectra are quite "irregular", and a numerical solution of the complete $\mathcal{H}_Q + \mathcal{H}_Z$ Hamiltonian is needed to deduce hyperfine shift and EFG tensors.

The EFG tensor (i.e., ν_Q, η and orientation of principal axes) provides valuable information about the local symmetry and electronic configuration. While the absolute value of ν_Q is often quite difficult to predict (due to "anti-shielding" effects), simple models can be used to quantitatively describe modifications of electronic structure corresponding to, e.g., the doping dependence of ν_Q. In many cases the ν_Q frequency can be determined more precisely than the hyperfine shift, and is used to detect even small changes in the local electronic configuration. Finally, note that the static EFG, being a traceless tensor, is always zero for sites of cubic symmetry.

α-NaV_2O_5 Is *Not* a Simple Spin–Peierls Compound

Following the discovery of the first inorganic spin–Peierls system, namely famous $CuGeO_3$, the α-NaV_2O_5 compound was proposed to be the second member of the same family [26]. However, it finally turned out that opening of the spin gap in α-NaV_2O_5 (as observed by χ_{macro}) is at the same time accompanied (or rather induced) by an important change in the electronic configuration at the spin sites. From NMR data this is clearly seen from the strong discontinuous change of both the EFG tensor at the ^{23}Na site and of the hyperfine coupling constant of the ^{51}V (spin carrying) site, observed as the slope of $K(\chi_{\text{macro}})$ which is very different above and below the transition [27]. This is in sharp contrast to what is observed in $CuGeO_3$ where, for the "on-site" ^{63,65}Cu, the EFG tensor is perfectly constant across the spin–Peierls transition (within error bars of only $\simeq 0.5\,\%$) [28], and the $K(\chi_{\text{macro}})$ data above and below the transition lay on the same line [20]. In this case the dimerisation corresponds to very small lattice distortions ($\sim 0.2\,\%$), and the local spin configuration is unchanged.

2.3 Nuclear Spin–Lattice Relaxation Rate ($1/T_1$) or Dynamic Spin Susceptibility

When the nuclear "longitudinal" magnetisation M_z ($z \parallel H_0$), i.e. the spin energy, is changed, the nuclear system will return to its equilibrium by transferring this extra energy to electrons or phonons (called the "lattice"). The rate of energy transfer will be proportional to the fluctuations of the Hamiltonian coupling nuclear spins to the lattice, $\mathcal{H}_Z + \mathcal{H}_Q$. By measuring the characteristic time T_1 of this process, we therefore obtain an information on the *fluctuations* of the local magnetic field (i.e., of nearby electronic spins) and/or of EFG.

Although in principle any perturbation of the equilibrium $M_z(\infty)$ value would be suitable, a typical pulsed NMR T_1 measurement starts either by a pure inversion of the nuclear magnetisation, $M_z(0) = -M_z(\infty)$, generated by a "π" pulse, or by its cancellation, $M_z(0) = 0$, obtained by a "comb" of many "$\pi/2$" pulses. After a variable recovery time t, M_z is measured (by any convenient technique, FID or spin-echo) in order to define the experimental $M_z(t)$ dependence [7].

The relaxation towards equilibrium is described by a *linear rate equation* for the nuclear levels populations, which leads to (multi-)exponential relaxation. For $I = 1/2$, there are only two levels (one transition), and the relaxation is simply $\propto \exp(-t/T_1)$, where $1/T_1 = 2W$, W being the transition probability. (For $I > 1/2$, the number of transitions, and therefore the number of exponentials in the relaxation may be bigger than one, but for relaxation of pure magnetic origin there is still only one characteristic probability W [29,30].) To calculate W, we use the general expression (obtained from the first order perturbation for the equation of motion of the density matrix) for the probability of transitions between discrete levels, induced by a time dependent perturbation that is spread in frequency [7]

$$W_{m\leftarrow k} = \hbar^{-2} \int_{-\infty}^{\infty} \mathrm{d}t\, \mathrm{e}^{\mathrm{i}\omega_{mk}t} \, \langle <m|\mathcal{H}_{\mathrm{int}}(t)|k><k|\mathcal{H}_{\mathrm{int}}(0)|m> \rangle \ , \tag{4}$$

where $\langle ... \rangle$ denotes statistical average. As in spin systems the EFG contribution is mostly negligible, we substitute pure *magnetic* interaction Hamiltonian $\mathcal{H}_{\mathrm{int}} = -\hbar\gamma_{\mathrm{n}}\boldsymbol{I}_{\mathrm{n}} \cdot \delta\boldsymbol{h}(t)$, where $\delta\boldsymbol{h}(t) = \boldsymbol{h}(t) - \langle \boldsymbol{h}(t) \rangle = -A \cdot \boldsymbol{S}(t)$, in order to get:

$$T_{1z}^{-1} = \tfrac{1}{2}\gamma_{\mathrm{n}}^2 \int_{-\infty}^{\infty} \mathrm{d}t\, \mathrm{e}^{\mathrm{i}\omega_{\mathrm{NMR}}t} \, \langle \delta h_{+}(t)\delta h_{-}(0) \rangle \ . \tag{5}$$

Expanding (5) we explicitly get spin-spin correlation functions

$$T_{1z}^{-1} = \tfrac{1}{2}\gamma_{\mathrm{n}}^2 \int_{-\infty}^{\infty} \mathrm{d}t\, \mathrm{e}^{\mathrm{i}\omega_{\mathrm{NMR}}t} \, \{ (A_{xz}^2 + A_{yz}^2) \, \langle S_z(t)S_z(0) \rangle + \tag{6}$$
$$+ (A_{xx}^2 + A_{yx}^2) \, \langle S_x(t)S_x(0) \rangle + (A_{yy}^2 + A_{xy}^2) \, \langle S_y(t)S_y(0) \rangle \} \ .$$

Note that for a hyperfine coupling tensor A that is *non*-diagonal (in the laboratory frame) both parallel and transverse (to external field $H_0 \parallel z$) spin–spin correlation functions contribute to relaxation, while only the latter contribution is active if A is diagonal. In general, $A_{\alpha\neq\beta} \neq 0$ as soon as H_0 is *not* parallel to a symmetry axis of the A tensor, leading to a complicated angular dependence of T_1.

To take into account a coupling to several electronic spins we replace $A \cdot \boldsymbol{S}$ by $\sum_{\boldsymbol{r}} A(\boldsymbol{r}) \cdot \boldsymbol{S}(\boldsymbol{r}) = \sum_{\boldsymbol{q}} A(\boldsymbol{q}) \cdot \boldsymbol{S}(-\boldsymbol{q})$ to get (using inversion and translation symmetry):

$$T_{1z}^{-1} = \tfrac{1}{2}\gamma_{\mathrm{n}}^2 \sum_{\boldsymbol{q}} \sum_{\alpha=x,y,z} \left(A_{x\alpha}^2(\boldsymbol{q}) + A_{y\alpha}^2(\boldsymbol{q}) \right) \int_{-\infty}^{\infty} \mathrm{d}t\, \mathrm{e}^{\mathrm{i}\omega_{\mathrm{NMR}}t} \, \langle S_\alpha(\boldsymbol{q},t)S_\alpha(-\boldsymbol{q},0) \rangle \ . \tag{7}$$

While (7) is in fact more appropriate for localised spin systems, we can also use the fluctuation–dissipation theorem to replace the spin–spin correlation functions by the imaginary part of magnetic susceptibility χ'', and obtain a standard expression used for both localised and itinerant (metallic) electrons:

$$T_{1z}^{-1} = k_{\mathrm{B}}T \left(\gamma_{\mathrm{n}}^2/\mu_{\mathrm{B}}^2 \right) \sum_{\boldsymbol{q}} \sum_{\alpha=x,y,z} \left(A_{x\alpha}^2(\boldsymbol{q}) + A_{y\alpha}^2(\boldsymbol{q}) \right) \chi_{\alpha\alpha}''(\boldsymbol{q},\omega_{\mathrm{NMR}}) / \left(\omega_{\mathrm{NMR}} g_{\alpha\alpha}^2 \right) \ . \tag{8}$$

In (7) and (8) we see that the $\boldsymbol{q}$-dependence of hyperfine coupling provides a so-called "form factor" which can modify the "sensitivity" of T_1 to different $\boldsymbol{q}$-components of spin dynamics. For example, in the CuO_2 plane of high T_C superconductors, oxygen (^{17}O) is situated in-between two copper spins, and is coupled to each of them by hyperfine coupling constant C. In the reciprocal space the total coupling is therefore $^{17}A(\boldsymbol{q}) = C\exp(-\mathrm{i}q_x a/2) + C\exp(+\mathrm{i}q_x a/2) = 2C\cos(q_x a/2)$, and the oxygen form factor $^{17}A(\boldsymbol{q})^2$ is filtering-out (removing) the AF component of copper spin fluctuations in $^{17}T_1$ measurements. On the other hand, AF fluctuations will certainly be seen in the copper $^{63}T_1$, because the on-site coupling is $\boldsymbol{q}$-independent. This way, measuring copper spin fluctuations at different positions we can obtain a *partial* information on their $\boldsymbol{q}$-dependence.

While the notion of form factors in NMR relaxation is much older [31], it has been "reintroduced" and extensively used in NMR studies of copper-oxide superconductors to separate the AF and the $\boldsymbol{q} = 0$ contributions to fluctuations of copper spins [32–36]. In these systems the analysis of ^{63,65}Cu nuclear relaxation is somewhat complicated due to the fact that both the on-site coupling and the (supertransfered) nearest-neighbour hyperfine coupling (usually denoted by B) have the *same* size [37].

In spin systems, T_1 measurements are often used for the determination of the singlet–triplet gap value Δ, from the activation energy E_A of the Arhenius type temperature dependence $T_1^{-1} \propto \exp(-E_A/k_B T)$. From (6) we recall that, according to the hyperfine coupling tensor, T_1 may be sensitive to both transverse "+−" *and* longitudinal "$z\,z$" (to H_0) spin fluctuations, and thus to *different* relaxation processes. Figure 3 summarises the situation in 1D; depending on the process, the observed E_A values are very different [38,39,20]. Only for a 2-magnon *intra*branch process (Fig. 3a) one observes directly the field-dependent gap of

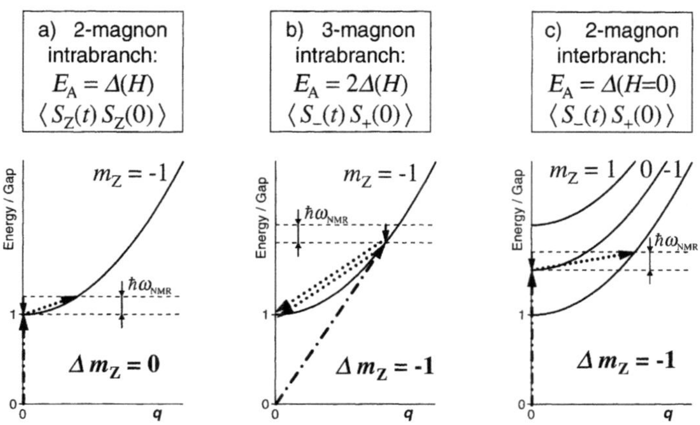

Fig. 3. Different magnon-scattering processes involved in the nuclear spin–lattice relaxation. Dash–dotted arrow points to the incoming magnon, solid line arrow denotes the absorbed nuclear energy (here strongly oversized for visibility), and dotted arrow points to the outgoing magnon. Depicted situations correspond to the minimal final energy, i.e., the activation energy, which is different in each of the three processes

the lowest laying magnon branch, $T_1^{-1} \propto \mathrm{K}_0(\hbar\omega_{\mathrm{NMR}}/2k_{\mathrm{B}}T)\exp(-\Delta(H)/k_{\mathrm{B}}T)$, where K_0 is the modified Bessel function. Note that this longitudinal contribution is expected to be *absent* from T_1 measured for H_0 applied along the crystal symmetry axes! Only the transverse contributions are expected for this orientation, leading to $E_{\mathrm{A}} = 2\Delta_m(H)$ for the 3-magnon *intra*branch process (Fig. 3b), or to nearly magnetic field independent $E_{\mathrm{A}} \cong \Delta_0 = \Delta(H=0)$ in the 2-magnon *inter*branch process (Fig. 3c). More precisely, in the latter case we get $T_1^{-1} \propto (1+\exp(-g\mu_{\mathrm{B}}H/k_{\mathrm{B}}T))\,\mathrm{K}_0(g\mu_{\mathrm{B}}H/2k_{\mathrm{B}}T)\exp(-(2\Delta_0 - g\mu_{\mathrm{B}}H)/2k_{\mathrm{B}}T)$. Expressions given here [20] correspond to the integral over the phase-space for the scattering from purely parabolic magnon dispersion in 1D, and neglect any eventual q- and H-dependence of the matrix elements.

H–T Phase Diagram of the Organo-Metallic Spin Ladder $Cu_2(C_5H_{12}N_2)_2Cl_4$

The organo–metallic $Cu_2(C_5H_{12}N_2)_2Cl_4$ compound [40,41] (if one considers only the shortest bonds connecting copper spins,) can be viewed as a 1D spin-ladder structure (inset to Fig. 4). This system has been thought to be the first physical realisation of a strong coupling (along rungs, $J_\perp \gg J_\parallel$), $S = 1/2$, spin ladder in which the exchange integrals $J_\perp$ and $J_\parallel$ are low enough so that the full H–T phase diagram can be covered by experimentally accessible magnetic fields (Fig. 4). It is predicted that for $H = 0$ the ground state is a gapped spin liquid, in which the singlet ground state is separated by a gap $\Delta = J_\perp - J_\parallel$ from the lowest laying triplet excitations. As long as $H < H_{\mathrm{C1}} = \Delta/g\mu_{\mathrm{B}}$, the system remains gapped with $\Delta(H) = \Delta - g\mu_{\mathrm{B}}H$. The point $(H_{\mathrm{C1}}, T = 0)$ is a quantum critical point which separates the singlet spin liquid state from a magnetic state, which in a purely 1D approach is equivalent to a Luttinger liquid (LL), or an "*XXZ*" spin-1/2 chain. Within this phase, one expects a divergence

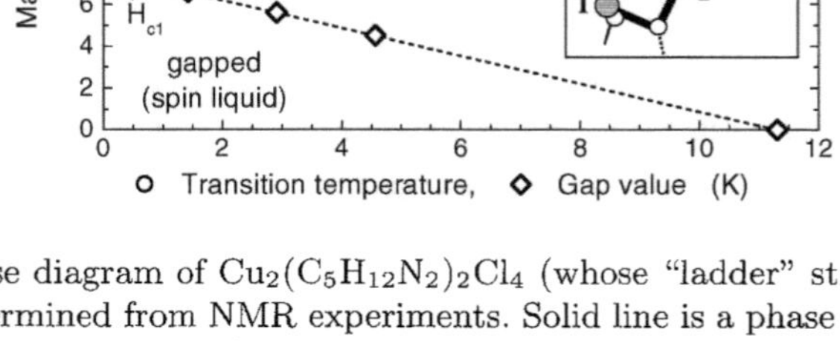

Fig. 4. The phase diagram of $Cu_2(C_5H_{12}N_2)_2Cl_4$ (whose "ladder" structure is shown in the inset) determined from NMR experiments. Solid line is a phase boundary to the 3D ordered phase, determined (*open circles*) from the changes in the NMR lineshape [43]. Open diamonds give the gap value determined from the T_1 measurements [42]. The dotted lines separate different regimes

of the nuclear spin-lattice relaxation rate $1/T_1 \propto T^{-\alpha(H)}$; in practice, there is always a transverse coupling between the chains, which induces a 3D transition at a finite temperature between H_{C1} and $H_{C2} = J_\perp + 2J_\parallel$. The point $(H_{C2}, T = 0)$ separates this 3D phase issued from the LL from a fully polarized phase which again is gapped with $\Delta = g\mu_B(H - H_{C2})$.

All the expected features have indeed been observed through the behaviour of the spin-lattice relaxation of protons in single crystals of $Cu_2(C_5H_{12}N_2)_2Cl_4$. In the gaped phase, measurements of $T_1(T)$ at two different proton sites revealed two different activation energies. At one of the sites the value was nearly the same as the E_A obtained from the shift (i.e., static susceptibility) measurements (on both sites), while two times larger E_A was observed on the other site. This was associated to two different relaxation processes, namelly 2-magnon (zz) and 3-magnon $(+-)$ intrabranch scattering, "selected" by different hyperfine coupling constants (see (6) and Fig 3) [39,41].

As shown in Fig. 4, the linear dependence of the gap on magnetic field (diamonds) is confirmed both below $H_{C1} = 7.5\,\mathrm{T}$ in a spin liquid state, and above $H_{C2} = 13.5\,\mathrm{T}$ in a fully polarised state [42,41]. Between H_{C1} and H_{C2}, and above 1.5 K, $1/T_1$ indeed shows divergence associated to the LL regime, while at lower temperature appears a 3D ordered state. Transition line to this state has been established by the modifications in the proton lineshape [43]. Near the H_{C1} critical point, the field dependence of transition temperature is well described by a power law with an exponent of 2/3, in agreement with the predictions of Giamarchi and Tsvelik [44]. Modifications of the proton spectra, as well as the temperature dependence of $1/T_1$, strongly suggest that below $\simeq 2\,\mathrm{K}$ the longitudinal couplings $J_\parallel$ are modified, and finally modulated in the 3D ordered phase, in a scenario which is equivalent to the spin-Peierls transition [43].

Finally, from recent neutron data quite *different* magnetic exchange paths have been proposed, leading to a *frustrated*, *3D* quantum spin liquid [45]. However, the neutron data cannot provide complete magnetic Hamiltonian and, for the moment, it is not clear how a 3D Hamiltonian could be related to the observed phase diagram.

2.4 Nuclear Spin–Spin Relaxation Rate ($1/T_2$) or Non-local (AF) Static Spin Susceptibility

The nuclear spin–spin relaxation time T_2 is the characteristic time for the decay of the *transverse* ($\perp$ to $H_0 \parallel z$) component of the nuclear magnetisation $M_\perp$. In this lecture we will refer only to T_2 values obtained from a *standard* spin-echo measurement excited by the "$\pi/2 - \tau - \pi$" pulse sequence [7]. This sequence removes from the decay of $M_\perp$ any contribution due to static local field inhomogeneities. In contrast to the relaxation of longitudinal magnetisation (T_1), where the process implies the exchange of energy with the lattice (see Sect. 2.3), the decrease of $M_\perp$ is an energy conserving process of the loss of the phase coherence, due to the local "field" fluctuations. (The local "field" means here *any* coupling of each nuclear spin to its "environment" of other nuclear spins and electrons.) The dominant process is usually the spin exchange between two nuclear spins,

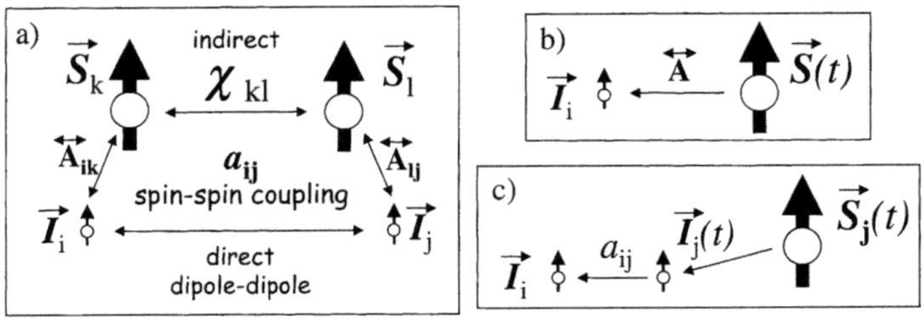

Fig. 5. The processes involved in the nuclear spin–spin relaxation: (**a**) the indirect coupling via the non-local electronic spin susceptibility and the direct dipole coupling between "like" nuclear spins, (**b**) the Redfield or "T_1" process, and (**c**) "corrections" due to fluctuations of other nuclear spins

which is therefore a measure of their mutual coupling a_{ij}. The trivial component of this coupling, which is always present, is the direct nuclear dipole–dipole coupling ($\propto r_{ij}^{-3}$), and its contribution to T_2 relaxation can be precisely predicted from the crystal structure. More interesting is the so-called indirect nuclear coupling (see Fig. 5a), through the coupling of nuclei to electronic spins and the mutual coupling of electronic spins, i.e., the static (non-local) spin susceptibility $\chi'(\boldsymbol{q})$. When this contribution is dominant, in AF spin systems where χ' is strongly enhanced at the AF wave vector $\boldsymbol{Q}_{\mathrm{AF}}$, it can be used as a direct measure of $\chi'(\boldsymbol{Q}_{\mathrm{AF}})$. While the $M_\perp(t)$ relaxation is in general a complex function of a_{ij} couplings and difficult to calculate, the expression for the *second moment* of this function is relatively simple. This is not only of academic interest, as $M_\perp(t)$ is quite often close to a Gaussian, $\propto \exp(-(t/T_{2\mathrm{G}})^2/2)$, for which the second moment $T_{2\mathrm{G}}^{-2}$ is experimentally easy to determine.

We suppose that the nuclear states are (nearly) pure Zeeman spin-states of spin I, where only the $(m,\ m-1)$ transition is excited (other transitions being split by quadrupole Hamiltonian for $I > 1/2$). Only pairs for which *both* nuclei are excited by NMR pulses contribute to the T_2 spin-echo relaxation; if their pair interaction is given by

$$\mathcal{H}_{ij}/\hbar = a_{ij}^{zz} I_i^z I_j^z + a_{ij}^{\mp}(I_i^+ I_j^- + I_i^- I_j^+) + \text{non-secular terms}\ , \tag{9}$$

the corresponding second moment of the $M_\perp(t)$ dependence for the NMR spin-echo is a simple sum over all the nuclear couplings (squared) [46],

$$T_{2\mathrm{G}}^{-2} = C_{\mathrm{active}}/[2(2I+1)] \sum\nolimits_j' [a_{ij}^{zz} - 2(I+m)(I-m+1)a_{ij}^{\mp}]^2\ . \tag{10}$$

Here C_{active} is the abundance of the observed nucleus, and $a_{ij}^{\mp}$ should be put to zero when the spins i and j are "unlike", which means that their local energy is different (e.g. $h_i^z I_i^z$, where $h_i^z \neq h_j^z$), so that the spin-exchange process is forbidden by energy conservation.

The comprehensive expressions for the a_{ij} tensor due to indirect interactions (see Fig. 5a) were first derived intuitively in [47], and from the linear response theory in [48]:

$$a_{ij}^{\alpha\beta} = \sum\nolimits_{k,l,\gamma} - A_{i,k}^{\alpha\gamma} \chi_{k,l}^{\prime\gamma\gamma} A_{j,l}^{\beta\gamma} (\gamma_{\mathrm{n}}/g\mu_{\mathrm{B}})^2 \hbar\ . \tag{11}$$

where $\alpha, \beta, \gamma = x, y, z$. However, the final *practical* formula for T_{2G}^{-2} was only obtained when it was realised that the sum in (10) is more easily carried out in the reciprocal space [49]; the coupling (11) is there simplified to

$$a^{zz}(\boldsymbol{q}) = -\chi'(\boldsymbol{q})|A_{zz}(\boldsymbol{q})|^2(\gamma_{\mathrm{n}}/g\mu_{\mathrm{B}})^2\hbar\,, \tag{12}$$

and (10), for the case of ^{63}Cu spin ($C_{\mathrm{active}} = 0.69$, $I = 3/2$) and $a_{ij}^{\mp} = 0$, becomes [49]

$$T_{2G}^{-2} = \frac{0.69}{8\hbar^2}\left(\frac{\hbar\gamma_{\mathrm{n}}}{g\mu_{\mathrm{B}}}\right)^4 \left\{\sum\nolimits_{\boldsymbol{q}} \left[\chi'(\boldsymbol{q})|A_{zz}(\boldsymbol{q})|^2\right]^2 - \left[\sum\nolimits_{\boldsymbol{q}} \chi'(\boldsymbol{q})|A_{zz}(\boldsymbol{q})|^2\right]^2\right\}. \tag{13}$$

The second term in the braces removes from the sum (10) the $i = j$ term $\left(\sum_{\boldsymbol{r}}' f(\boldsymbol{r}) = \sum_{\boldsymbol{r}} f(\boldsymbol{r}) - f(r=0)\right)$, and ensures that T_{2G}^{-2} is non-zero only for *non-local* coupling, i.e., when either χ or A (or both) are $\boldsymbol{q}$-dependent. When in AF spin systems the $\chi'(\boldsymbol{q})$ is strongly peaked at $\boldsymbol{Q}_{\mathrm{AF}}$, the leading term is the first one (called the "scaling part" in the following example [50]), and the second term is therefore often left out, which is sometimes unjustified.

Note that the above described relaxation process is *not* the only one leading to phase decoherence; any other fluctuation of the local "field" may contribute as well. In particular, the fluctuations of nearby electronic spin (see Fig. 5b) that are responsible for the T_1 relaxation, also provide a decay of $M_\perp(t)$. This "Redfield" or "T_1" process [7] always leads to a *single*-exponential relaxation $M_\perp(t) \propto \exp(-t/T_{2\mathrm{R}})$. For purely magnetic relaxation [51]

$$T_{2\mathrm{R}z}^{-1} = \tfrac{1}{2}(T_{1x}^{-1}+T_{1y}^{-1}-T_{1z}^{-1})|_{\omega_{\mathrm{NMR}}=0}+[I(I+1)-m(m-1)-\tfrac{1}{2}]T_{1z}^{-1}(\omega_{\mathrm{NMR}})\,. \tag{14}$$

So, when T_1 does not depend on frequency, the Redfield contribution can be predicted from the T_1 data. Typically it is "removed" from raw experimental $M_\perp$ data by considering the product $\exp(+t/T_{2\mathrm{R}}) \times M_\perp(t)$

Finally, as shown in Fig. 5c, *any* neighbouring nuclear spin (coupled by a_{ij}), which is driven to fluctuate by, e.g., nearby electronic spins (or by any other mechanism) also contributes to the relaxation. The resulting relaxation depends on the coupling and the strength of these fluctuations, and is difficult to calculate. As a rule of the thumb, if the T_2 data are used as a measure of χ', one would chose the nucleus for which this contribution is minimal, that is the on-site nucleus which is the most strongly coupled to the electronic spin. For example, in copper-oxide spin systems the measurements of T_{2G} are performed on one of the two ^{63,65}Cu isotopes, and the contribution of the other isotope is taken into account as a relatively small "correction to T_{2G}". For more details on these corrections within the Gaussian formalism we refer to a review [52] and Refs. therein, as well as to the Appendix in [53]. More general stochastic approach is given in [54].

We also mention that for a proper determination of T_{2G} the *whole* NMR line has to be excited by the radio frequency pulses (to satisfy the condition mentioned above Eq. (9)), which is incompatible with very wide lines. To solve this problem, a T_{2G} determination from the "stimulated echo" pulse sequence has been proposed [55].

The applications of T_{2G} measurements to quantify $\chi'(\boldsymbol{Q}_{AF})$ in AF quantum spin systems are (almost) exclusively limited to ^{63,65}Cu NMR in copper oxides: spin chains (see the next subsection), ladders (Sect. 2.5) and high T_C cuprates (to prove the opening of the pseudo-gap [36] and for the determination of the dynamic exponent "z" [56,36,55]).

Spin 1/2 Heisenberg Chain Sr_2CuO_3 and the Logarithmic Corrections to the Field-Theory Description

Sr_2CuO_3 is considered to be the most ideal representative of a simple, $S = 1/2$, AF Heisenberg chain. Relatively high coupling, $J \simeq 2200$ or 2850 K (from susceptibility and optical absorption, respectively), makes it ideal for experimental studies of the low temperature $T \ll J$ behaviour. In particular, T_1 and T_{2G} data measured on the on-site ^{63}Cu spin have been compared to the theoretical predictions (see [50] and Refs. therein) in order to reveal the "exact" temperature dependence, beyond the leading behaviour given by the field theory, $T_1^{-1} = \text{const.}$ and $T_{2G}^{-1} \propto 1/\sqrt{T}$. Indeed, the first predictions claimed the same logarithmic correction factor for both T_1^{-1} and $\sqrt{T}/T_{2G}$, i.e., $\ln^{1/2}(T_0/T)$, where $T_0 \sim J$ is the high energy cut-off. However, this temperature dependence was clearly stronger than the experimental one. Therefore, these "logarithmic corrections" were revised and checked against Quantum Monte Carlo (QMC) numerical calculations [57]. Concerning the interpretation of the T_{2G} data, there are several difficulties both on theoretical and experimental side. The analytical expression for logarithmic corrections could only be obtained for the first, so-called "scaling" term of (13), while the second term in (13) had to be estimated from QMC calculation. On the experimental side, raw T_{2G} data had to be revised by taking into account "corrections to T_{2G}" mentioned in the previous Subsect. [52,53]. The remaining uncertainty comes from the determination of the hyperfine coupling tensor A (here determined in an unconventional way) and of the J coupling. This is partly removed by considering the $T_{2G}/T_1\sqrt{T}$ ratio, depending only on the (better known) anisotropy of the A tensor. Altogether, very good quantitative agreement between the theory and experiment has been established; small remaining discrepancies may be due to spin-phonon interactions. This example shows how NMR results helped the theoretical description of "logarithmic" corrections and, at the same time, clearly demonstrates how the quantitative interpretation of the T_{2G} data is quite involved.

2.5 Other Important Examples

A great number of Cu–O based spin chains and ladders [1] have been studied by hyperfine shift, T_1 and T_{2G} NMR measurements. As these studies were aimed at understanding of the role of the spin gap in the copper-oxide high T_C superconductors, particular attention was paid to the existence of the spin gap and the effect of doping. A detailed account of these works is beyond the scope of these lectures, and we just mention few selected references [58,52,60,61,69], all fine examples of how NMR can be used to reveal the low-energy spin dynamics. While the first one [58] compares several types of chains and ladders, all others [52,60,61,69] are devoted to the most studied compound $Sr_{14-x}Ca_xCu_{24}O_{41}$, the only 1-D spin system in which the superconductivity has been discovered (under pressure). In its structure this compound contains doped two-leg ladders, for which the possibility of superconductivity has indeed been predicted. The superconducting phase appears at very high pressure, which makes the NMR studies [69] technically quite difficult.

Another subject which has recently raised considerable attention is the study of "magnetisation plateaus" (see Sect. 5 in [3]). When they are associated to a commensurate spin-ordered state, this latter can be easily detected by the NMR lineshape. An example is the "1/8 plateau" of the 2D spin compound $SrCu_2(BO_3)_2$ [63], representative of the Shastry–Sutherland Hamiltonian [3]. From ^{63,65}Cu NMR spectra it has been confirmed that this phase is indeed ordered, all the local spin values have been determined and compared to the theoretically predicted spin-structures [64].

Finally, we also mention the class of "zero-D" spin systems which are the molecular rings [65]. An interesting example is the magnetic field induced crossing of molecular levels detected by a peak in the T_1 relaxation, which is ascribed to the cross-relaxation between nuclear and molecular magnetic levels [66].

3 Conclusions

Being a local probe for magnetism, NMR is particularly appropriate for studying systems with localised electronic spins. In this chapter we have presented the physical quantities that can be accessed by NMR in such systems, and have given few selected examples of NMR studies of low-D quantum AF spin systems.

First, the resonance frequency can be used to distinguish local electronic configurations and it provides a direct measure of the *local* average *spin-polarisation value*. From the NMR line-shape one is also able to investigate and resolve spatially inhomogeneous spin polarisation. Homogeneous intrinsic magnetisation can thus be distinguished from a perturbation created by impurities and studied "separately" in the same sample. In favourable cases, one can obtain precise, site-by-site image of local spin values, either in the case of a perturbation around impurity, as well as in the spin-ordered phases. In the same way one can get a precise image of the real-space profile of an incommensurate modulation of the spin polarisation and, e.g., follow it as a function of temperature or/and magnetic field. In general, any kind of the phase transition to a 3D ordered magnetic state will usually induce a sizeable change in the NMR lineshape. However, the level of the obtained useful information strongly depends on each particular system.

By measuring T_1 relaxation time one obtains information on nearly zero-energy spin-excitations: the electronic spin–spin correlation functions or, equivalently, the imaginary part of spin-susceptibility. In some cases a partial information on their $\boldsymbol{q}$-dependence can be provided from the comparison of T_1's recorded at different crystallographic sites. When a spin-gap is present in the system, its value can be determined from the activated temperature dependence of T_1^{-1}. However, one must take care that different processes can be involved in nuclear relaxation and lead to different activation energies. A sharp divergence of T_1^{-1} is a standard signature of strong (critical) spin fluctuations appearing in the vicinity of second order phase transitions. But, low temperature divergence of T_1^{-1} can also be due to a special gapless phase, such as the Luttinger liquid regime in a 1D system.

Under certain conditions, the T_{2G} relaxation time in AF spin systems can also give access to the staggered (AF) static spin susceptibility, but a quantitative analysis may be quite involved. Via the Kramers–Kronig relations this quantity can be regarded as the energy-integrated imaginary part of spin-susceptibility (whose zero frequency limit is given by T_1 data), and comparing temperature dependence of the T_1 and T_{2G} data one can detect the opening of the pseudo-gap in the AF spin excitations.

Altogether, a full set of NMR data may provide a complete picture of static and (low-energy) dynamic spin susceptibility. These data are usually very complementary to the neutron scattering data, when available. However, the possibility to perform measurements up to the highest available static magnetic field (presently 45T) makes the NMR a unique tool for the investigation of spin systems.

Acknowledgement

We would like to express our gratitude to a number of colleagues and collaborators on NMR studies of low-D AF spin systems, and in particular to: G. Chaboussant, Y. Fagot-Revurat, M. Hanson, M.-H. Julien, H. Mayaffre, P. Ségransan, and F. Tedoldi.

References

1. T.M. Rice: 'Low Dimensional Magnets'. In this volume
2. E. Dagotto, T.M. Rice: Science **271**, 618 (1996) E. Dagotto: Rep. Prog. Phys. **62**, 1525 (1999)
3. C. Lhuillier, G. Misguich: 'Frustrated Magnetism'. In this volume
4. C.L. Broholm: 'Neutron Studies of 1 and 2D Magnets'. In this volume
5. P. Vonlanthen et al.: preprint cond-mat/0112203 (2001) W.G. Clark et al.: unpublished
6. J.H. Prestegard: 'High Resolution NMR of Biomolecules' and D. Massiot: 'High Resolution Solid State NMR'. In this volume
7. C.P. Slichter: *Principles of Magnetic Resonance*, 3rd edn (Springer, Berlin 1990) A. Abragam: *Principles of Nuclear Magnetism*, (Oxford Univ. Press, New York 1961)
8. M. Mehring, V.A. Weberruß: *Object-Oriented Magnetic Resonance : Classes and Objects, Calculations and Computations*, (Academic Press, San Diego 2001)

9. E. Fukushima, S.B.W. Roeder: *Experimental Pulse NMR. A nuts and bolts approach*, (Addison–Wesley, London Amsterdam 1981)
10. A. Narath: 'Nuclear Magnetic Resonance in Magnetic and Metallic Solids'. In *Hyperfine interactions*, ed. A. Freeman, R.B. Frankel (Academic Press, New York London 1967) pp. 287–363
11. A. Abragam, B. Bleaney: *Electronic paramagnetic resonance of transition ions*, (Clarendon Press, Oxford 1970)
12. F. Tedoldi, R. Santachiara, M. Horvatić: Phys. Rev. Lett. **83**, 412 (1999) F. Tedoldi et al.: App. Magn. Res. **19**, 381 (2000)
13. S. Botti et al.: Phys. Rev. B **63**, 012409 (2000)
14. F. Alet, E.S. Sørensen: Phys. Rev. B **62**, 14116 (2000)
15. F. Tedoldi et al.: unpublished
16. A. Keren et al.: Phys. Rev. B **57**, 10745 (1998).
17. P. Mendels et al.: Phys. Rev. Lett. **85**, 3496 (2000)
18. W. Bray et al.: 'The Spin–Peierls Transition'. In *Extended Linear Compounds*, ed. by J.C. Miller (Plenum, New York 1982) pp. 353–415
19. M. Hase et al.: Phys. Rev. Lett. **70**, 3651 (1993)
20. Y. Fagot-Revurat et al.: Phys. Rev. B **55**, 2964 (1997)
21. Y. Fagot-Revurat et al.: Phys. Rev. Lett. **77**, 1816 (1996)
22. M. Horvatić et al.: Phys. Rev. Lett. **83**, 420 (1999)
23. G. Uhrig et al.: Phys. Rev. B **60**, 9468 (1999)
24. H.M. Rønnow et al.: Phys. Rev. Lett. **84**, 4469 (2000)
25. K. Kiryhukin et al.: Phys. Rev. Lett. **76**, 4608 (1996)
26. M. Isobe, Y. Ueda: J. Phys. Soc. Jpn. **65**, 1178 (1996) Y. Fujii et al.: J. Phys. Soc. Jpn. **66**, 326 (1997)
27. Y. Fagot-Revurat et al.: Phys. Rev. Lett. **84**, 4176 (2000)
28. J. Kikuchi et al.: J. Phys. Soc. Jpn. **63**, 872 (1994)
29. A. Narath: Phys. Rev. **162**, 320 (1967)
30. M. Horvatić: J. Phys.: Condens. Matter **4**, 5811 (1992)
31. E.P. Maarschall: 'Critical behaviour of the fluorine NMR linewidth'. In *Magnetic Resonance and Related Phenomena, Proceedings of the XVI th Congress AMPERE* ed. I. Ursu, (Publishing House of the Academy of the Socialist Republic of Romania 1971) pp. 485–492
32. A.J. Millis, H. Monien, D. Pines: Phys. Rev. B **42**, 167 (1990)
33. P.C. Hammel et al.: Phys. Rev. Lett. **63**, 1992 (1989)
34. M. Horvatić et al.: Phys. Rev. B **48**, 13848 (1993).
35. C.P. Slichter: 'Experimental Evidence for Spin Fluctuations in High Temperature Superconductors'. In *Strongly Correlated Electronic Materials, The Los Alamos Symposium 1993*, ed. K.S. Bedell, Z. Wang, D.E. Meltzer (Addison–Wesley, New York 1994) pp. 427
36. C. Berthier et al.: J. Phys. I France **6**, 2205 (1996)
37. F. Mila, T.M. Rice: Physica C **157**, 561 (1989)
38. J. Sagi, I. Affleck: Phys. Rev. B **53**, 9188 (1996)
39. G. Chaboussant et al.: Phys. Rev. Lett. **79**, 925 (1997)
40. B. Chiari et al.: Inorg. Chem. **29**, 1172 (1990)
41. C. Chaboussant et al.: Eur. Phys. J. B **6**, 167 (1998)
42. G. Chaboussant et al.: Phys. Rev. Lett. **80**, 2713 (1998)
43. H. Mayaffre et al.: Phys. Rev. Lett. **85**, 4795 (2000)
44. T. Giamarchi, A. Tsvelik: Phys. Rev. B **59**, 11398 (1999)
45. M.B. Stone et al.: Phys. Rev. B **65**, 064423 (2002)

46. J.A. Gillet et al.: Physica C **235–240**, 1667 (1994)
47. C.H. Pennington, C.P. Slichter: Phys. Rev. Lett. **66**, 381 (1991)
48. Y. Itoh et al.: J. Phys. Soc. Jpn. **61**, 1287 (1992)
49. D. Thelen, D. Pines: Phys. Rev. B **49**, 3528 (1994) M. Takigawa: Phys. Rev. B **49**, 4158 (1994)
50. M. Takigawa et al.: Phys. Rev. B **56**, 13681 (1997)
51. M. Horvatić: unpublished
52. C.H. Recchia, K. Gorny, C.H. Pennington: Phys. Rev. B **54**, 4207 (1996)
53. N.J. Curro et al.: Phys. Rev. B **56**, 877 (1997)
54. A. Keren et al.: Phys. Rev. Lett. **78**, 3547 (1997) and Phys. Rev. B **60**, 9279 (1999)
55. S. Fujiyama, et al.: Phys. Rev. B **60**, 9801 (1999)
56. A. Sokol, D. Pines: Phys. Rev. Lett. **71**, 2813 (1993)
57. O.A. Starykh, R.R.P. Singh, A.W. Sandvik: Phys. Rev. Lett. **78**, 539 (1997)
58. K. Ishida et al.: Phys. Rev. B **53**, 2827 (1996)
59. M. Takigawa et al.: Phys. Rev. B **57**, 1124 (1998)
60. K. Magishi et al.: Phys. Rev. B **57**, 11533 (1998)
61. T. Imai et al.: Phys. Rev. Lett. **81**, 220 (1998)
62. H. Mayaffre et al.: Science **279**, 345 (1998) Y. Piskunov et al.: Eur. Phys. J. B **24**, 443 (2001)
63. K. Onizuka et al.: J. Phys. Soc. Jpn. **69**, 1016 (2000)
64. K. Kodama et al.: unpublished
65. D. Gatteschi et al.: Science **265**, 1054 (1994) B. Barbara, L. Gunther: Phys. World **12**, No. 3, 35 (1999)
66. M.-H. Julien et al.: Phys. Rev. Lett. **83**, 227 (1999) M. Affronte et al.: Phys. Rev. Lett., (2002) in press

Magnetized States of Quantum Spin Chains

C. Broholm[1,3], G. Aeppli[2], Y. Chen[1], D.C. Dender[3], M. Enderle[4], P.R. Hammar[1], Z. Honda[5], K. Katsumata[6], C.P. Landee[7], M. Oshikawa[8], L.P. Regnault[9], D.H. Reich[1], S.M. Shapiro[10], M. Sieling[11], M.B. Stone[1], M.M. Turnbull[7], I. Zaliznyak[10], and A. Zheludev[12]

[1] Department of Physics and Astronomy, Johns Hopkins University, Baltimore, MD 21218, USA
[2] NEC Research Institute, 4 Independence Way, Princeton, NJ 08540, USA
[3] National Institute of Standards and Technology, Center for Neutron Research, Gaithersburg, MD 20899, USA
[4] Institute Max Von Laue Paul Langevin, F-38042 Grenoble, France
[5] Faculty of Engineering, Saitama University, Urawa, Saitama 338-8570, Japan
[6] RIKEN Harima Institute, Mikazuki, Sayo, Hyogo 679-5148, Japan
[7] Carlson School of Chemistry and Department of Physics, Clark University, Worcester, MA 01610, USA
[8] Department of Physics, Tokyo Institute of Technology, Oh-oka-yama, Meguro-ku, Tokyo, 152-8551 Japan
[9] CEA-Grenoble, DRFMC-SPSMS-MDN, 17 rue des Martyrs, 38054 Grenoble cedex 9, France
[10] Physics Department, Brookhaven National Laboratory, Upton, NY 11973, USA
[11] Physikalisches Institut, Universität Frankfurt, Robert-Mayer-Str. 2-4, D-60054 Frankfurt, Germany
[12] Solid State Division, Oak Ridge National Laboratory, Oak Ridge, TN 37831, USA

Abstract. Quantum spin chains display complex cooperative phenomena that can be explored in considerable detail through theory, numerical simulations, and experiments. Here we review neutron scattering experiments that probe quantum spin chains in high magnetic fields. Experiments on copper-containing organometallic systems show that the uniform antiferromagnetic spin-1/2 chain has a gapless continuum of magnetic excitations and is critical in zero field. Application of a magnetic field creates incommensurate soft modes with a characteristic wave-vector that grows in proportion to the magnetization. These experimental results are evidence that the spins-1/2 chain maps to a one dimensional Luttinger liquid. Experiments on antiferromagnetic spin chains built from spin-1 nickel atoms show a Haldane gap to bound triplet excitations and a finite critical field that must be exceeded to induce magnetization at low temperatures. These results indicate that the integer spin chain has an isolated singlet ground state with hidden topological order. For both spin-1/2 and spin-1 systems, site alternation leads to a field induced gap in the excitation spectrum.

1 Introduction

Cooperative phenomena in magnetism generally involve mesoscopic spin clusters with classical dynamics. Consequently, quantum effects are seldom apparent in macroscopic properties. Quasi-one-dimensional magnets are an exception because their long-range correlations are controlled by microscopic domain walls with quantum dynamics [1]. In these systems, quantum effects can therefore have a profound impact on physical properties on all length scales. Consider for example, a uniform spin chain with antiferromagnetic Heisenberg interactions at $T = 0$. While half odd integer spin chains have quasi-long range order and a gapless spectrum, integer spin chains have a finite correlation length and an isolated singlet ground state [2].

These one-dimensional quantum effects have consequences for macroscopic properties of three-dimensional solids. In particular, there are transition metal oxides and organometallic materials that contain arrays of weakly interacting spin chains. For Cu-spins that represent spin-1/2 degrees of freedom, such an array generally acquires long-range order at sufficiently low temperatures as slowly fluctuating spin chains with quasi-long-range one-dimensional order develop long-range inter-chain phase coherence. However, for quasi-one-dimensional spin systems built from spin-1 Ni atoms, there is a critical value for inter-chain coupling below which low temperature magnetic order cannot be achieved. Systems with sufficiently weak inter-chain coupling adopt a quasi-one-dimensional cooperative singlet ground state at low temperatures without undergoing any phase transitions.

Quantum effects also strongly affect the response of quasi-one-dimensional magnetic systems to an applied magnetic field. For spin chains with a gap in their excitation spectrum, there is a critical field that must be exceeded to induce magnetization at the absolute zero temperature. In addition, the magnetized states of many isotropic quantum spin chains are expected to have incommensurate low energy fluctuations, an effect that has no analogy in magnets with classical dynamics [3].

In this paper, we shall review experiments that probe the magnetized states of uniform antiferromagnetic spin-1/2 and spin-1 chains. The emphasis is on neutron scattering experiments that can provide detailed information about atomic scale dynamic correlations. After a brief summary of the neutron scattering technique, separate sections for spin-1/2 and spin-1 chains describe model systems, zero field data, and finite field data. The experimental results are compared to relevant theories. We end with a summary of results and a discussion of challenges that remain.

1.1 Magnetic Neutron Scattering

Neutrons interact with nuclei through the strong interaction and with electrons through electromagnetic dipole-dipole interactions [4]. The corresponding scattering cross sections for an atom with a partially filled electronic shell are of order r_0^2 for both interactions, $r_0 = (e^2/m_e c^2) = 2.82$ fm being the classical electron radius. The cross sections are sufficiently small that the Born approximation is valid and the scattering processes can be described in terms of a differential scattering cross section. The magnetic part of the cross section, which is of interest here, can be written as follows [5]

$$\frac{d^2\sigma}{d\Omega dE'} = (\gamma r_0)^2 |\frac{g}{2}F(Q)|^2 \sum_{\alpha\beta}(\delta_{\alpha\beta} - \hat{Q}_\alpha\hat{Q}_\beta)\mathcal{S}^{\alpha\beta}(\boldsymbol{Q},\omega). \tag{1}$$

The scattering process is specified by the wave-vector transfer, $\mathbf{Q} = \mathbf{k_i} - \mathbf{k_f}$ and the energy transfer $\hbar\omega = E_{\mathrm{I}} - E_{\mathrm{f}}$ to the sample. Furthermore, $\gamma = 1.913$ and $g \approx 2$ are the spectroscopic g-factors of the neutron and the magnetic atom respectively, and $F(Q)$ is the magnetic form factor [6]. The interesting part of Eq. (1) is the Fourier transformed two point dynamic spin correlation function:

$$\mathcal{S}^{\alpha\beta}(\boldsymbol{Q},\omega) = \frac{1}{2\pi\hbar}\int dt e^{i\omega t}\frac{1}{N}\sum_{\boldsymbol{RR'}} < S^{\alpha}_{\boldsymbol{R}}(t) S^{\beta}_{\boldsymbol{R'}}(0) > e^{-i\boldsymbol{Q}\cdot(\boldsymbol{R}-\boldsymbol{R'})} \tag{2}$$

$\mathcal{S}^{\alpha\beta}(\boldsymbol{Q},\omega)$ can be be related to the generalized spin susceptibility through the fluctuation dissipation theorem[5].

$$\mathcal{S}(\boldsymbol{Q},\omega) = \frac{1}{1-e^{-\beta\hbar\omega}}\frac{\chi''(\boldsymbol{Q},\omega)}{\pi(g\mu_{\mathrm{B}})^2} \tag{3}$$

where $\beta = 1/k_{\mathrm{B}}T$ and χ'' denotes the imaginary part of the generalized spin susceptibility. $\mathcal{S}(\boldsymbol{Q},\omega)$ is the natural juncture for theoretical and experimental work on cooperative magnetic phenomena and magnetic neutron scattering is the experimental technique that provides the most complete access to it.

Many of the neutron scattering experiments to be described in this article were performed on the SPINS cold neutron spectrometer at the NIST Center for Neutron Research, which is shown in Fig. 1. Neutrons from the cold source of the 20 MW NBSR reactor are transported to the instrument through an evacuated rectangular glass structure with inner dimensions 6 cm by 12.5

Fig. 1. The SPINS cold neutron spectrometer at the NIST Center for Neutron Research in Gaithersburg, MD, USA where many of the experiments described in this paper were performed. The tall cylindrical item marked Oxford, is a high field dilution refrigerator cm. The inner walls of this neutron guide are coated with ^{58}Ni, which completely reflects neutrons with an angle of incidence less than a critical angle $\theta_c = 2.04 \cdot 10^{-3}\text{Rad/Å} \times \lambda$, where λ is the neutron wave-length. A horizontally segmented pyrolytic graphite monochromator Bragg reflects a monochromatic and vertically focused beam to the sample position. The incident beam energy range is 2.3 meV to 14 meV and the relative energy resolution 2-8% depending on the configuration. The neutron beam easily penetrates the aluminum vacuum cans of cryogenic systems that provide appropriate thermodynamic conditions for the sample.

The detection system is contained within a shielded volume to reduce background contributions from epithermal neutrons. The neutron detector is a ^{3}He proportional counter with greater than 90% detection efficiency for neutrons with energies less than 15 meV. To reach the detector, neutrons scattered from the sample must reflect from a pyrolytic graphite crystal. This ensures that, the energy of detected neutrons is known by virtue of the fact that they must have satisfied the Bragg conditions at this "analyzer" crystal. The SPINS instrument offers much flexibility in configuring the detection system. In particular, the analyzer consists of 11 vertical blades with a height of 15 cm and widths of 2.1 cm that can rotate independently about vertical axes. The analyzer is placed 91 cm from the sample where it can rotate about a vertical axis. The entire detection system in turn can rotate about the sample to access neutrons scattered in different directions by the sample.

We used three basic detector configurations for the experiments reported here. In the conventional "triple axis" mode, Soller collimators surrounding the analyzer determine the scattering angle at the analyzer. Only three vertical analyzer blades are used in this mode and the detector is then sensitive only to a specific fixed neutron energy. To vary energy transfer, the incident beam energy is changed by adjusting the scattering angle at the monochromator.

In the monochromatic focusing mode the final energy is also fixed, however, the eleven blades of the segmented analyzer are set up to reflect neutrons with a range of scattering angles from the sample onto a single channel detector. The loss of wave-vector resolution perpendicular to the direction of the scattered beam is not a problem for one-dimensional systems as the chain axis can be oriented along $\mathbf{k}_f$ where the wave-vector resolution is excellent. The monochromatic focusing mode enhances sensitivity by approximately a factor five over the conventional triple axis configuration without increasing the fast neutron background. It requires a weakly dispersive direction for the system of interest and a single crystalline sample with low mosaic distribution.

The third configuration that we use is denoted the energy dispersive analyzer mode. All analyzer blades are oriented parallel to the analyzer assembly so they form a flat reflecting surface that subtends a solid angle of approximately 0.04 Sr to the sample. Neutrons with varying sample scattering angles also have different angles of incidence on the analyzer and hence the Bragg condition for reflection varies across the surface of the analyzer. The beam reflected from the analyzer in this mode has a horizontal divergence angle of approximately 10 degrees. After passage through a radial collimator for background suppression, neutrons are detected by a position sensitive detector. Splayed horizontally over the detector are thus neutrons corresponding to a range of different values of $\mathbf{Q}$ and $\hbar\omega$, with $\mathbf{Q}$ varying along the reciprocal lattice direction defined by the analyzer surface. The dispersive analyzer mode enhances efficiency by approximately a factor five over the conventional triple axis mode. Background tends to be higher than for the monochromatic focusing mode but the configuration can be used for samples that are dispersive throughout the horizontal plane or where the sample mosaic distribution covers several degrees.

2 Antiferromagnetic Spin-1/2 Chain

The uniform antiferromagnetic spin-1/2 chain plays a special role in physics because it is one of very few interacting quantum many body system, where exact analytical results are available. The spin Hamiltonian reads

$$\mathcal{H} = J\sum_n \mathbf{S}_n \cdot \mathbf{S}_{n+1} - g\mu_{\mathrm{B}}H\sum_n S_n^z, \tag{4}$$

where $J > 0$ is the antiferromagnetic exchange constant and $\mathbf{S}_n$ are vector spin operators for spin-1/2 degrees of freedom.

2.1 Theoretical Results for the Spin-1/2 Chain

There is a steadily growing body of exact results available about the uniform spin-1/2 chain, most of them based on the Bethe ansatz [7,8]. This correct "guess"for the ground state of the spin-1/2 chain enabled calculation of the ground state energy per spin in the absence of a magnetic field [9] $\langle\mathcal{H}\rangle/N = |J|2(\frac{1}{4} - \ln 2)$. Bethe's ansatz was later used to derive thermodynamic properties including the equal time spin correlation function

$$\langle \mathbf{S}_i \cdot \mathbf{S}_{i+n}\rangle = (-1)^n n^{-1}, \tag{5}$$

which indicates quasi-long-range antiferromagnetic order. It is also possible to derive the wave functions and energies for the lowest energy excited states as a function of wave-vector transfer [10]:

$$\epsilon_{\mathrm{L}}(q) = \frac{\pi}{2}J|\sin q| \tag{6}$$

The functional form of this expression is similar to the dispersion relation for spin waves in a putative spin-1/2 one-dimensional antiferromagnet with exchange constant $\pi J/2$. However, Eq. (6) is not the dispersion relation for a long lived quasiparticle, rather it represents the lower bound of a continuum of excited states. It has been shown that the fundamental quasi-particles of the spin-1/2 chain are fermionic "spinons" that occupy a cosine band [11,12]

$$\epsilon(q) = \frac{\pi}{2}J\cos q. \tag{7}$$

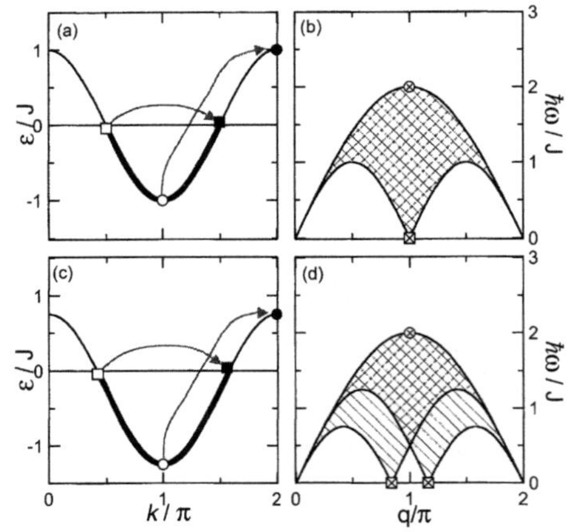

Fig. 2. Exact zero and finite field results for the XY spin-1/2 chain obtained through the Jordan-Wigner transformation. The Heisenberg spin chain has qualitatively similar properties as described in Ref. [14]. (a) Shows the band structure for spinons in the unmagnetized state. (b) shows the corresponding exact kinematical limits of the two spinon continuum. Arrows in (a) indicate scattering processes that map to the points on frame (b). (c) shows the band structure for spinons in a magnetized state. The exact boundaries [14] of the corresponding continuum in the longitudinal part of the dynamic spin correlation function, $\mathcal{S}^{\ddagger\ddagger}(q,\omega)$, are shown in (d). The transverse part of the dynamic spin correlation function is gapless for $q = \pi$

For the easy plane spin-1/2 chain, spinons do not interact and thus they form an ideal one-dimensional degenerate fermi gas [13]. For isotropic interactions (the Heisenberg case), there are strong interactions and spinons form a one-dimensional Luttinger liquid. For the un-magnetized spinchain, the spinon band is half filled as indicated in Fig. 2(a). Low energy excitations with wave-vector transfer, q, correspond to annihilating a spinon with wave-vector $q_1 \approx n\pi$ and creating a spinon with wave-vector $q_2 = n\pi + q$. Arrows in Fig. 2(a) indicate such spinon scattering processes while points in Fig. 2(b) indicate the corresponding values of wave-vector and energy transfer. The solid lines in Fig. 2(b) indicate the kinematical limits for spinon excitations.

Neutron scattering can probe the kinematical limits of the spinon excitation spectrum. The experiments are also sensitive to the matrix elements for spin excitations. Specifically, the dynamic spin correlation function measured by neutron scattering at zero temperature can be written as

$$\mathcal{S}^{\mu\mu}(q,\omega) = 2\pi \sum_{\lambda} |\langle 0|S_q^{\mu}|\lambda\rangle|^2 \delta(\hbar\omega - (\epsilon_\lambda - \epsilon_0)) \tag{8}$$

Recent mathematical breakthroughs have enabled analytical calculation of this important quantity for the dominant two spinon excitations in zero magnetic field [8].

Because an (un)occupied spinon state represents a spin up (down) the magnetized spin-1/2 chain maps to a spinon band away from half filling as shown in Fig. 2(c). Fig. 2(d) shows the corresponding kinematical limits for low energy excitations, which are significantly more complicated than for the unmagnetized state. The principal feature of interest, are the low energy incommensurate modes of excitation with wave-vectors that are related to the magnetization as follows

$$q_i = n\pi \pm 2\pi\langle S_z\rangle \tag{9}$$

The field induced incommensurate modes have an alternate phenomenological interpretation. As the incommensurate modes are gapless, they represent a spin density modulation that persists on an arbitrarily long time scale. The magnitude of the incommensurate wave-vector further indicates that the magnetization is carried by magnetized defects in quanta of spin-1/2 and with density $\rho = 2\langle S_z\rangle$. The band versus "defect lattice" description of the magnetized spin-1/2 chain can be compared to the band versus striped model for the incommensurate spin correlations in the copper oxide two-dimensional lattice of high temperature superconductors [15,16].

2.2 Experimental Model Systems

To examine spin correlations in the one-dimensional spin-1/2 chain through experiments requires suitable model systems that faithfully realize the spin-hamiltonian in Eq. (4). Table 1 lists some of the materials that have served as model experimental systems. It is important that the ordering temperature is small compared to the exchange constant as that indicates a highly

Table 1. Key characteristics of $S = 1/2$ linear chain Heisenberg antiferromagnets ordered according to the magnitude of the exchange constants. T_N refers to the temperature below which long range magnetic order develops. $T_{\rm SP}$ indicates the critical temperature for a spin-Peierls transition. $g\mu_{\rm B}H/J$ is calculated with $g = 2$ and $H = 10$ Tesla. The incoherent cross section, $\sigma_{\rm inc}$, is in units of 10^{-24}cm^2 per spin-1/2. Further tabulations of $S = 1/2$ HAFM chains can be found in references [49] and [50].

Chemical Formula	Common Name	J (meV)	$T_{\rm N}$ (K)	$g\mu_{\rm B}H/J$	$\sigma_{\rm inc}$ barn	Refs.
$SrCuO_2$		280(20)	<1.5	0.007	0.61	[17–19]
Sr_2CuO_3		260(10)	5	0.006	0.67	[18,20–23]
Ca_2CuO_3		≈86	8	0.013	0.65	[24,25]
$BaCu_2Ge_2O_7$		46.5	8.5	0.024	0.808	[26,27]
$BaCu_2Si_2O_7$		24.1	9.2	0.048	0.632	[26,28–30]
$KCuF_3$		17.5	39	0.067	0.82	[31–35]
$CuGeO_3$		10.4	$T_{\rm SP} = 14$	-	0.73	[36–38]
$CuCl_2 \cdot 2N(C_5H_5)$	CPC	2.31	1.14	0.50	110.9	[39,40]
$Cu(C_6H_5COO)_2 \cdot 3(H_2O)$	Cu-Benzoate	1.57	0.8	0.73	158.5	[41,42]
$[(CH_3)_2SO_2]CuCl_2$	CDC	1.43	0.91	0.81	70.4	[43,44]
$Cu(C_4H_4N_2)(NO_3)_2$	CuPzN	0.90(1)	<0.05	1.29	42.0	[45]
$CsCuCl_4$		0.34(2)	0.62	3.40	22.0	[46]
$CuSO_4 \cdot 5(H_2O)$		0.25	0.100	4.63	99.3	[47,48]
$CuSeO_4 \cdot 5(H_2O)$		0.15	0.125	7.72	99.6	[47,48]

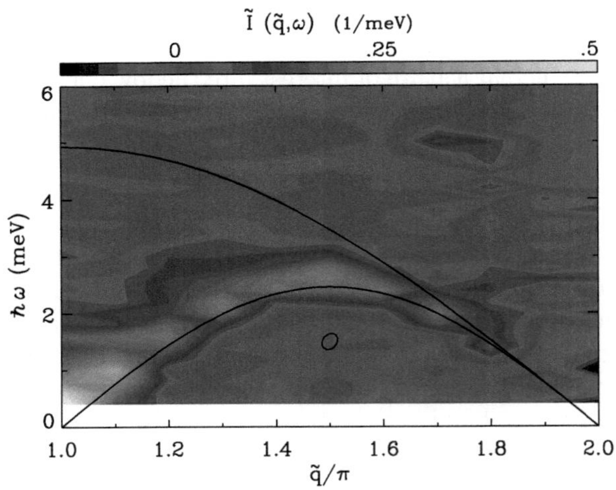

Fig. 3. Image of the dynamic spin correlation function for the spin-1/2 chain copper benzoate in the non-magnetized state. The ellipsoid shows the half width at half maximum instrumental resolution. Reproduced from Ref. [42]

one-dimensional materials where inter-chain interactions can be neglected for $\hbar\omega > k_B T_N$. The absolute value of the exchange constant determines the experimental conditions that are necessary to access a certain range of normalized temperature $k_B T/J$, normalized field $g\mu_B H/J$, and normalized energy transfer, $\hbar\omega/J$. For the present work, we are interested in large values of normalized magnetic field in the ≈ 10 T superconducting magnets that are available for neutron scattering experiments. At the same time, we want to resolve the magnetic excitation spectrum using a cold neutron triple axis spectrometer with energy resolution of order 0.1 meV. This implies an exchange constant, $J \approx 1$ meV. To optimize the signal to noise ratio it is desirable to minimize the nuclear incoherent scattering cross section per magnetic ion, a quantity also listed in the table. The two model systems that we shall focus on in this article are copper benzoate ($Cu(C_6H_5COO)_2 \cdot 3(H_2O)$) and copper pyrazine dinitrate ($Cu(C_4H_4N_2)(NO_3)_2$). Both are highly one-dimensional and have exchange constants small enough to enable large values of the reduced magnetic field. One difference that will turn out to be important for the high field experiments is that copper benzoate has two magnetic atoms per one-dimensional unit cell, while copper pyrazine dinitrate has only one.

2.3 Zero Field Properties

The zero field dynamic spin correlation function for the spin-1/2 chain was measured using cold neutron scattering from copper benzoate and is shown in Fig. 3. Similar data are also available for $KCuF_3$ [51] and copper pyrazine dinitrate [52]. While most of the intensity is accumulated along the lower boundary of the two spinon continuum (Eq. (6)), the distribution is significantly wider than

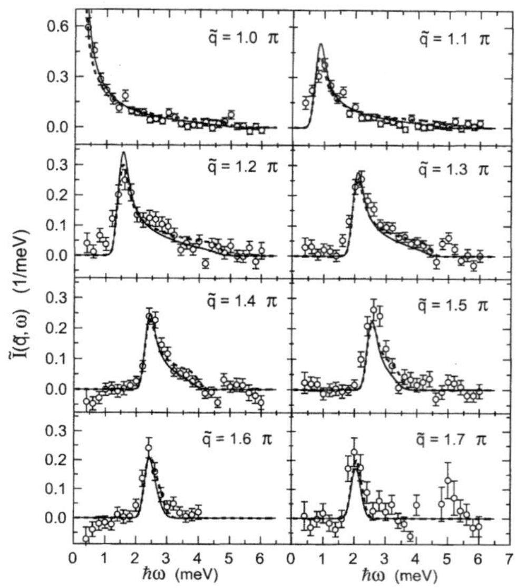

Fig. 4. Constant wave-vector cuts through the zero field neutron scattering data for copper benzoate shown in Fig. 3. The solid lines show the resolution convoluted exact two spinon contribution to the dynamic structure factor. The dashed line shows the resolution convoluted Müller approximation. Just two parameters were extracted from the combined data set. An overall pre-factor and the exchange constant $J = 1.57(1)$ meV. Adapted from Ref. [42]

can be accounted for by the instrumental resolution function (solid ellipse). The data is instead consistent with a bounded two spinon continuum.

Comparison to theories of the detailed energy dependence of the dynamic spin correlation function is best accomplished by comparing cuts through the data to resolution convoluted intensity distributions as shown in Fig. 4. The dashed lines show a phenomenological approximation that provides a good fit to the data [14,53]. The exact two spinon contribution to the dynamic spin correlation function shown with the solid line is difficult to distinguish from the approximate model [54]. The distinction lies in the higher energy part of the spectrum where the exact two spinon contribution to $\mathcal{S}(q,\omega)$ falls off more smoothly than the phenomenological form.

2.4 Spin Correlations in the Magnetized State

We now turn to the experimental evidence for incommensurate spin correlations in the magnetized state of the spin-1/2 chain. Fig. 5 shows constant energy scans at low energies ($\hbar\omega/J = 0.13$) and for values of the applied field ranging from 0 to $g\mu_B H/J = 0.51$. At zero field the data provide clear evidence for a two spinon continuum since long lived spin wave excitations would give rise to two resolution limited peaks rather than a single flat top maximum.

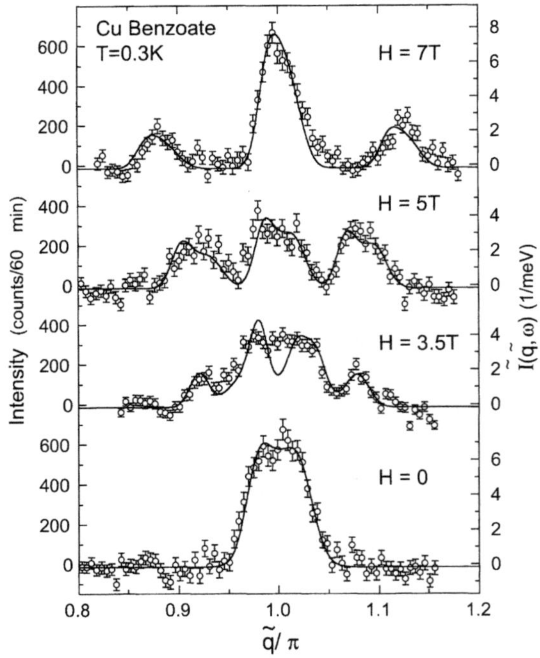

Fig. 5. Constant energy scans for $\hbar\omega = 0.21$ meV at various values of applied magnetic field at $T = 0.3$ K for copper benzoate. The data show the development of incommensurate soft modes in the magnetized spin-1/2 chain. Solid lines show a resolution convoluted model calculation as described in Ref. [41]

At finite field two small peaks split off symmetrically about $q = \pi$ and move progressively further from this location with increasing applied field. From these data, we derived the field dependence of the soft mode wave-vector that is plotted in Fig. 6. The solid line in the figure is based on Eq. (9) and a calculation of the field dependent magnetization of the spin-1/2 chain. Clearly, the wave-vector dependence of the quasi-one-dimensional soft modes in copper benzoate is consistent with the spinon theory for the uniform spin-1/2 chain. However, while the theory predicts a gapless spectrum Fig. 6(b) shows that there is a field-induced gap for copper benzoate.

Exchange anisotropy that defines an easy plane parallel to the field direction could in principle account for the gap as application of a magnetic field creates an easy axis perpendicular to the field. However, the magnitude of exchange anisotropy in copper benzoate can be estimated from the g-tensor anisotropy to be of order 1%, which is insufficient to account for the magnitude of the field induced gap [55]. Instead, it appears that staggering of the g-tensor along the spin chain as well as an alternating Dzyaloshinski-Moriya interaction causes the field-induced gap [56,57]. While neither of these terms have significant impact on zero field properties, they lead to unit cell doubling in a field and transverse long-range antiferromagnetic order. Because low energy excitations in the spin-

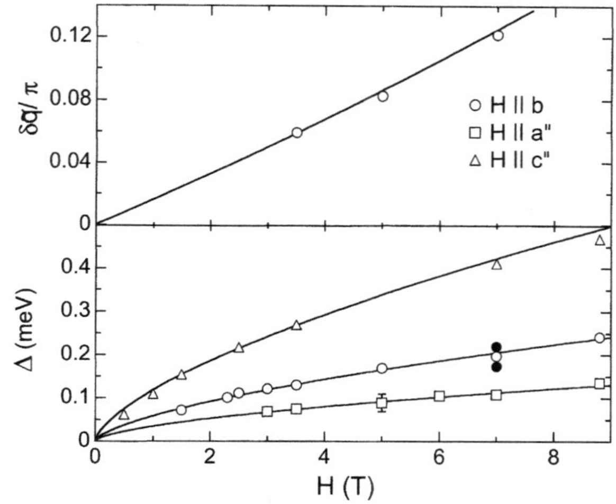

Fig. 6. (a) Field dependence of the incommensurate wave-vector for low energy spin excitations in the spin-1/2 chain copper benzoate. (b) Field dependence of the gap in the excitation spectrum of same material from a combination of neutron scattering and specific heat data. Reproduced from Ref. [41]

1/2 chain correspond to a sliding spin density wave, the formerly gapless two-spinon continuum acquires a gap when odd and even sites of the spin chain become distinguishable in a field. In field theoretical language, a staggered field is a relevant perturbation for the spinon Luttinger liquid. It can readily be shown using bosonization techniques that a staggered field induces a gap in the excitation spectrum that grows in proportion to H_s^α, where $\alpha = 2/3$ in the limit of a small staggered field, $H_s s$ and a small uniform field [56]. As the staggered g-tensor induces a transverse staggered field in proportion to the uniform field the result is a spin gap that grows in proportion to the applied field to this same power $\alpha \approx 2/3$. The solid line in Fig. 6(b) is a power-law fit that yields $\alpha_{\rm exp} = 0.65(3)$, a value that is in excellent agreement with the theory.

It has been shown that the effective low energy theory for a spin-1/2 chain with a staggered field is the Sine-Gordon model [56–58]. Originating as a mathematical model in particle physics, this model has been used to analyze condensed matter systems ranging from one-dimensional Josephson Junctions to quasi-one-dimensional easy-plane ferromagnets. However, the spin-1/2 chain with a staggered field may prove to be the best system yet in which to explore the rich excitation spectrum of this model through experiments [57].

It is interesting to note that despite the transition from a gapless spinon Luttinger liquid to a gapfull Sine-Gordon model, the field dependent incommensurability is perfectly accounted for by Eq. (9). This indicates that the appearance of incommensurate soft modes in magnetized quantum spins is a general phenomenon that should find a general explanation. One possibility is that magnetization is carried by repulsive defects in the zero field singlet ground state much as flux is carried by flux lines in the mixed phase of type II superconductors [3].

The existence of such repulsive defects (alias solitons) is well demonstrated in the spin-1/2 Heisenberg chain in presence of sizable spin-lattice couplings. In that case, the 3D lattice undergoes a structural instability induced by the magnetic fluctuations below a well-defined temperature T_{SP}, the so-called spin-Peierls transition temperature. Consequently, the chains dimerize, giving rise to a non-magnetic singlet ground state and an energy gap $\Delta_{SP} \approx 1.74 k_B T_{SP}$ opens in the excitation spectrum at $q = \pi$ [59]. Under field, the spin-Peierls system undergoes a transition to an incommensurate magnetic phase at a critical field H_C directly related to Δ_{SP} [60,61]. In the high-field phase, a soliton lattice forms, with both structural and magnetic components characterized by an incommensurate (IC) wave-vector shifted from $q = \pi$ by a quantity $\delta k_{SP} \propto \langle S_z \rangle$. Neutron diffraction measurements on $CuGeO_3$ have provided direct evidence for a long-ranged dual (magnetic and structural) soliton structure above $H_C \approx 12.5$ T [62]. Figure 7 shows the field dependences of δk_{SP} and the soliton width Γ, compared to the field-theory prediction [60] and to x-ray data [63]. Contrary to the pure Heisenberg chain, incommensurate excitations exist only above H_C for the spin-Peierls system [64].

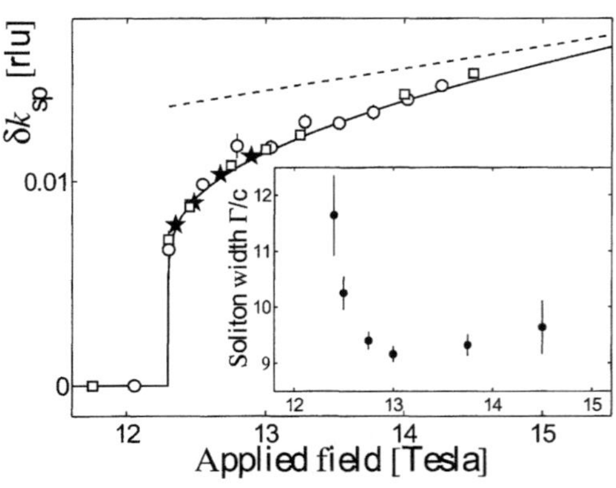

Fig. 7. Field dependence of the incommensurate component of the wave-vector (δk_{SP}) and soliton width Γ in the high-field phase of $CuGeO_3$. Open symbols: Neutron diffraction. Stars: X-ray data [63]. The lines are calculated curves according to the theories of Buzdin et al. [60] (solid line), and Cross et al. [59] (dashed line)

3 Uniform Antiferromagnetic Spin-1 Chain

Haldane discovered that integer spin chains are qualitatively different from half odd integer spin chains [2,65,66]. His approach was a large S mapping of the spin Hamiltonian to the $O(3)$ non-linear sigma model with a topological term associated with the Berry phase of the spin. For integer spins the topological term vanishes, while for half odd integer spins the topological term is important. When there is no topological term the non linear sigma model maps to a classical two-dimensional ferromagnet at finite temperature. This model has short-range order, which implies short-range order for the integer spin chain as well. For the half odd integer spin chain however, the topological term leads to a variant of the sigma model with instanton quark confinement and a gapless excitation spectrum consistent with the results described above [65].

Given the fact that the sigma model mapping relies on $S \to \infty$ one might question the applicability of results derived from it for spin chains with $S \leq 1$. While there are certainly quantitative differences, there are arguments against qualitative changes in the behavior of integer spin chains as S becomes smaller such that the main conclusions derived from the $S \to \infty$ mapping should persist in the small S limit. In addition, there are various other theoretical and numerical results that support a gap for integer spin chains. One is the Lieb Schultz Mattis theorem, which states that a gap can only be present in a quantum spin chain if the spin quantum number per unit cell is integral [67–69]. There is also much numerical work, which provides compelling evidence for spin gaps in integer spin chains [70,44].

A simple explanation for a spin gap in the spin-1 chain was provided by Affleck, Kennedy, Lieb, and Tasaki [72]. They identified a valence bond solid state in which each of the spin-1/2 degrees of freedom that in pairs make up atomic spin-1 are contracted with neighboring spin-1/2 degrees of freedom to form a singlet. Such valence bond solid states are possible when twice the spin quantum number matches the nearest neighbor coordination number. The valence bond solid for a spin-1 chain is the exact ground state of a projection spin hamiltonian of the following form

$$\mathcal{H}'_{\mathrm{VBS}} = \sum_n P_2(\mathbf{S}_n, \mathbf{S}_{n+1}), \tag{10}$$

The projection operator, P_2, is unity when the argument spin operators combined form a spin-2 degree of freedom and zero otherwise. Clearly the ground state for $\mathcal{H}'_{\mathrm{VBS}}$ has no near neighbor spin pairs in a spin 2 state and this exactly defines the valence bond solid state. Projection operators can in general be expanded as a power series of order $2S$ in spin dot products [73]. In particular, for spin-1 the spin-2 pair projection operator takes the form

$$P_2(\mathbf{S}_i, \mathbf{S}_j) = \frac{1}{2}\mathbf{S}_i \cdot \mathbf{S}_j + \frac{1}{6}(\mathbf{S}_i \cdot \mathbf{S}_j)^2 + \frac{1}{3} \tag{11}$$

Thus the hamiltonian that has the valence bond solid ground state can be written in terms of spin-1 operators as follows:

$$\mathcal{H}_{\mathrm{VBS}} = \sum_n \left[\mathbf{S}_n \cdot \mathbf{S}_{n+1} - \beta(\mathbf{S}_n \cdot \mathbf{S}_{n+1})^2\right], \tag{12}$$

where $\beta = -1/3$ and trivial terms and factors have been removed for simplicity. Studies of this hamiltonian as a function of β indicate that there are no phase transitions at $T = 0$ as β is driven to zero. The implication is that the valence bond solid is a good initial ansatz for describing the conventional Heisenberg spin-1 chain itself. Of course these arguments only hold for the spin-1 case.

In particular, for spin-1/2 the condition that $2S$ equals the number of nearest neighbors is not fulfilled so that a valence bond solid state is not possible in that case.

The valence bond solid state for an even length chain with $\beta = -1/3$ can be written as follows [74]

$$|\Psi_{\rm VBS}\rangle = \sqrt{2}|000...\rangle + \sum_{m=2,4,6,...} 2^{(m-1)/2} \sum_{perm} |0+0-0+0-00+00-...\rangle \quad (13)$$

where m counts the number of sites with projection quantum numbers different from 0 and the superposition is over all states where removal of sites with no spin projection along the quantization axis leaves behind a perfectly ordered m-site Néel state. A similar exact ground state wave function can be written for odd length chains. The valence bond solid is accordingly said to have perfect string order. String order is gauged by the expectation value of the following non-local operator

$$\mathcal{O}_{jk} = S_j^z \exp(i\pi \sum_{n=j}^{k-1} S_n^z) S_k^z \quad (14)$$

$\mathcal{H}_{\rm VBS}$ has perfect long range string order: $\lim_{|j-k|\to\infty}\langle\mathcal{O}_{jk}\rangle_{\rm VBS} = \frac{4}{9}$. It has been shown that there is also long range string order in the spin-1 chain with $\beta = 0$ though the value of the order parameter is reduced to approximately 0.37 [75].

Excited states of the spin-1 chain correspond to breaking a valence bond in the valence bond solid state [76]. Broken valence bonds can propagate coherently through the string ordered phase where they only scatter from other string order defects. Creation of a broken valence bond at rest requires an energy of $\Delta = 0.41050(2)J$ for the bi-linear spin-1 chain ($\beta = 0$) [44].

Owing to the isolated singlet ground state of the spin-1 chain and the gap in the excitation spectrum, there is also a minimum critical field $H_c = \Delta/g\mu_{\rm B}$ required to magnetize the system. Above the critical field there is a gapless critical phase [77] that could have similarities to the magnetized state of the spin-1/2 chain [3]. One of the goals of the present research has been to explore the possibility of incommensurate spin correlation in this magnetized phase of the spin-1 chain.

3.1 Experimental Model Systems

Table 2 contains key characteristics of experimental model systems for the Haldane spin-1 chain. As opposed to the spin-1/2 chain, a spin-1 degree of freedom is affected by single ion anisotropy. This turns out to be a greater obstacle to finding good model system than interchain coupling. While single ion anisotropy splits the triplet excited state, inter-chain coupling must exceed a threshold of order the Haldane gap energy to have a significant impact on the properties of a spin-1 chain. Here we shall focus on NENP and NDMAP for which high field experiments have been performed. We shall see that the materials have very different behavior in a field owing to g-factor alternation, which is present in the case of NENP but absent for NDMAP.

Table 2. Key characteristics of quasi-one-dimensional $S = 1$ Haldane antiferromagnets. $g\mu_B H/J$ is calculated with $g = 2$ and $H = 10$ Tesla.

Chemical Formula	Common Name	J meV	$\lvert J'/J\rvert$ 10^{-3}	D/J 10^{-3}	T_N K	$g\mu_B H/J$	Reference
$AgVP_2S_6$		58(4)	0.01	5.8	< 2	0.020	[78]
Y_2BaNiO_5		21	< 0.5	-39	< 0.05	0.055	[79–81]
$Ni(C_3H_10N_2)_2N_3(ClO_4)$	NINAZ	10.7	< 0.7	170	< 0.06	0.11	[82–84]
$Ni(C_3H_10N_2)_2NO_2(ClO_4)$	NINO	4.5		250	< 1.2	0.21	[85,86]
$Ni(C_2H_8N_2)_2NO_2(ClO_4)$	NENP	4.1(3)	0.8	180	< 0.0003	0.28	[87–89]
$Ni(C_5D_{14}N_2)_2N_3(PF_6)$	NDMAP	2.85	0.6	250	< 0.25	0.41	[90–94]
$CsNiCl_3$		2.275	17	-1.9	4.9	0.45	[103,95,96]

3.2 Zero Field Properties

A comprehensive map of the zero field excitation spectrum for the Haldane spin-1 chain has been obtained through neutron scattering experiments on NENP [87,88,100]. Figure 8(a) shows the zero field dispersion relation for long lived magnetic excitations. The absence of reflection symmetry about the $q = \pi/2$ line indicates that translation symmetry remains identical to the paramagnetic phase for $T \ll J/k_B$ despite antiferromagnetic interactions. Figure 8(b) shows the wave-vector dependence of the energy integrated intensity, which is a measure of the equal time spin correlation function. For a spin system with long lived excitations, the dispersion relation and equal time spin correlation function are related through an exact sum-rule[101]. Specifically it can be shown that,

$$\mathcal{S}^{\alpha\alpha}(q) = \hbar \int_{-\infty}^{\infty} \mathcal{S}^{\alpha\alpha}(q,\omega)d\omega \tag{15}$$

$$\simeq \frac{\hbar}{\omega^{\alpha\alpha}(q)} \int_{-\infty}^{\infty} \omega\mathcal{S}^{\alpha\alpha}(q,\omega)d\omega \tag{16}$$

$$\simeq -\frac{2}{3}\frac{\langle\mathcal{H}\rangle/L}{\hbar\omega^{\alpha\alpha}(q)}(1-\cos q) \tag{17}$$

where $\langle\mathcal{H}\rangle/L$ is the ground state energy per spin. The solid line in Fig. 8(b) shows that this so called single mode approximation [102] provides an excellent account of neutron scattering data through most of the zone. Subsequent higher resolution data for $CsNiCl_3$ show spin wave damping setting in for $q < \pi/2$ [103].

Figure 9 shows an energy scan at $q = \pi$ for NDMAP in zero magnetic field. The data illustrate the effects of strong easy plane anisotropy in this material. The upper mode corresponds to spin fluctuations polarized along the spin chain, while the lower peak consists of two near degenerate excitations that are polarized in the perpendicular plane. These three modes would be degenerate for an isotropic system but for NDMAP they are split by single ion and/or exchange anisotropy. The solid lines in the figure were calculated based on the SMA and the known resolution of the neutron scattering spectrometer.

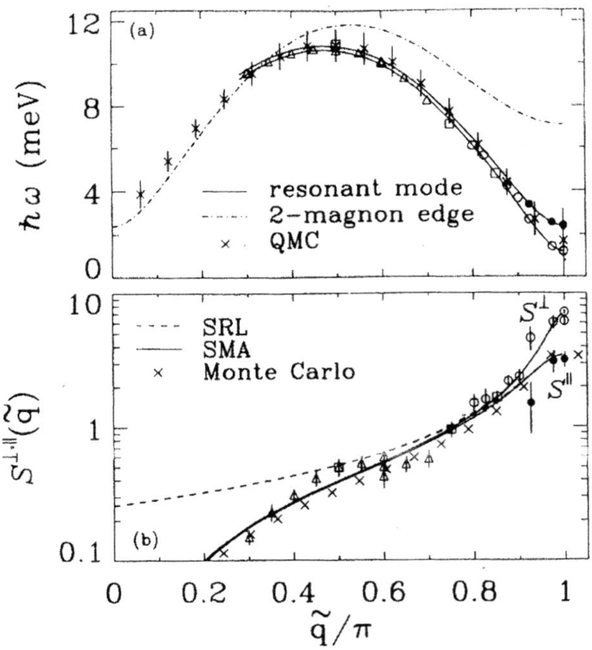

Fig. 8. (a) Circles, triangles, squares, and solid lines: dispersion relation $\omega(q)$ for long-lived modes in NENP. Dot-dashed line: Lower edge of corresponding two-magnon continuum. Crosses: QMC results [97] scaled by $J = 4.1$ meV. (b) Instantaneous spin correlation function $\mathcal{S}^{\perp,\parallel}(q)$. Crosses: Monte Carlo results [98,99]. Dashed lines: square-root Lorentzians with $\xi = 8.5$ and 4.2. Solid lines: single mode approximation based on $\omega(q)$. In both (a) and (b), open and filled symbols for $q > 0.9\pi$ correspond to polarizations perpendicular and parallel to the chain

3.3 Spin Ccorrelations in the Magnetized State

Figure 10 shows the field dependence of excitations at $q \approx \pi$ in NENP for fields applied along the c axis [104]. There are two Ni atoms per unit cell along the spin chains in this material so just like in copper benzoate, application of a uniform external field implies that the spin system effectively is also subject to a staggered field. The staggered field immediately induces transverse long range antiferromagnetic order as was clearly observed through neutron diffraction [17] and ESR [105]. In addition, a staggered field is a relevant perturbation on the high field gapless phase of the ideal Haldane spin chain. It is apparent from Fig. 10, that NENP does not have a finite field phase transition from a gap-full to a gap-less phase. Rather there is induced Néel order at any finite field and a cross over from a low field phase with a Haldane gap to a high field phase with a gap induced by the staggered field.

As a model system for studying the magnetized Haldane spin chain, NDMAP has the advantage that there is only one Ni-site per unit cell along the spin chain. Consequently it is possible to apply a pure uniform magnetic field. Figure 11

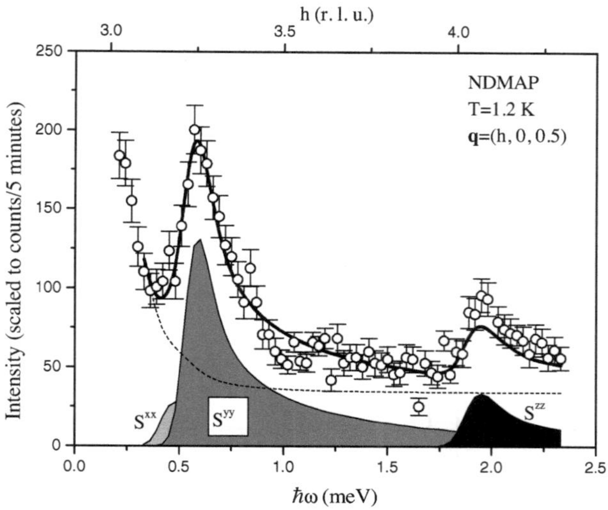

Fig. 9. Constant $q = \pi$ scan in NDMAP at zero field and T=0.3 K. The data was collected with a horizontally focusing analyzer so the spin chain was oriented along the scattered beam direction throughout the scan. x, y, and z refer to the a, b, and c directions respectively of the orthorhombic structure. The solid lines were calculated based on the SMA and known resolution effects. Reproduced from Ref. [92]

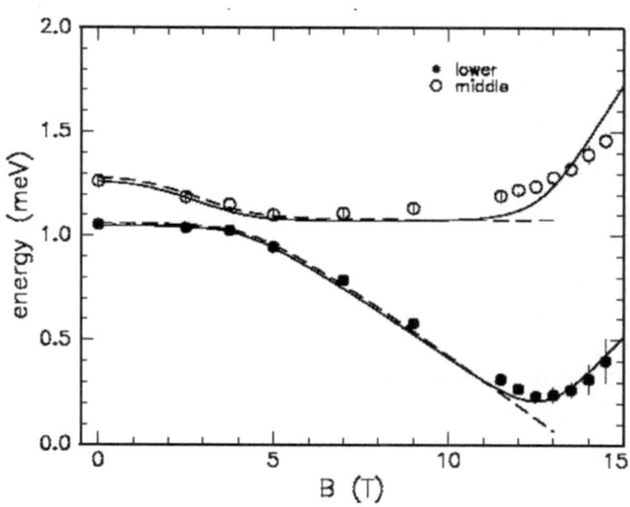

Fig. 10. Field dependence of the energies of easy a-c plane excitation at $q \approx \pi$ in NENP for fields applied along the c direction at $T = 0.3$ K. In zero field the upper mode shown with open circles is polarized along the a-direction. Reproduced from Ref. [104]

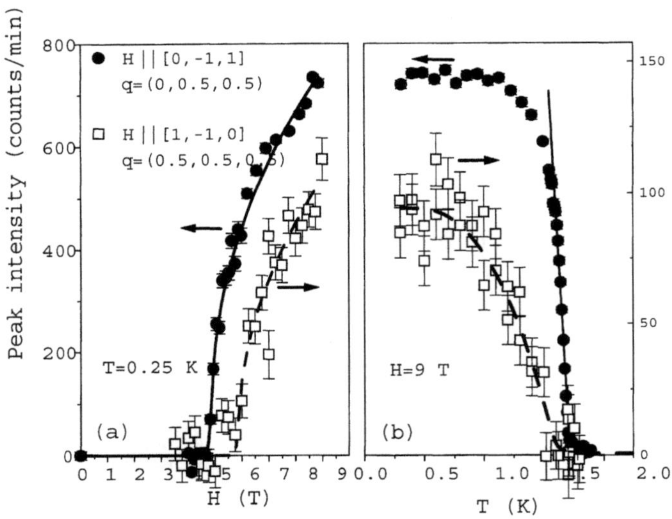

Fig. 11. Field and temperature dependence of elastic magnetic scattering in NDMAP close to the critical field. Field and wave-vectors are indicated on the figure. Only for fields with a component along the hard c axis do we find three-dimensional long-range order. The open circles recorded for fields in the easy plane measure the intensity of Bragg rods that correspond to quasi-two-dimensional magnetic order. Reproduced from Ref. [91]

shows that NDMAP has a bona-fide critical phase transition at a finite field. As a consequence of inter-chain interactions, the ordered phase persists to finite temperatures. A critical field induced phase transition has also been found in the spin-1 chain NDMAZ [106]. Interestingly, the nature of the ordered phase in NDMAP appears to be strongly dependent on the direction of the applied magnetic field. For fields applied strictly in the easy plane, the elastic magnetic scattering in the high field phase takes the form of rods in reciprocal space rather than resolution limited Bragg peaks. This indicates quasi-two-dimensional magnetic order. However, when the field has a component along the hard chain axis, there are resolution limited Bragg peaks indicating long range magnetic order. The high field phase diagram also is highly anisotropic and indicates qualitatively different cooperative properties for fields parallel and perpendicular to the chain axis [107,108].

Figrue 12 shows the excitation spectrum close to $q = \pi$ at three fields bracketing the critical field. The energy range accessed includes only the easy plane modes and the field was applied in the easy plane. As expected the mode perpendicular to the field direction in the easy plane is driven down in energy, while the mode corresponding to fluctuations along the field direction stays constant up to the critical field. This behavior is consistent with perturbation theory based on the Wigner–Eckart theorem [110,111]. Above the critical field, the energy of the $||H$ mode is driven up as the system is magnetized, while the lower mode

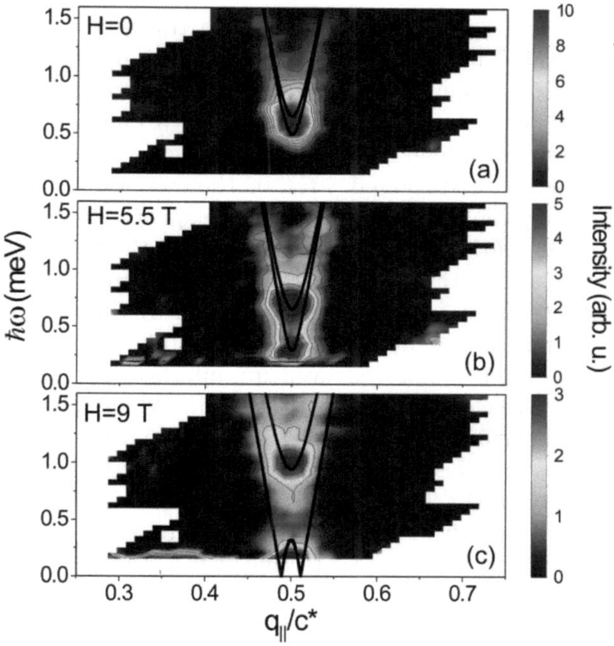

Fig. 12. Inelastic magnetic neutron scattering measured in NDMAP at $T = 2.5$ K for different values of magnetic field applied along the b axis. The extra intensity seen below $q \approx 0.4c^*$ at low energies in (**c**) may be an artifact of imperfect background subtraction. From Zheludev et al. [109]

remains quasi-elastic. On general grounds it is to be expected that at $T = 0$ there should be a gap in the excitation spectrum above the critical field because the applied field in conjunction with the easy plane anisotropy leaves no continuous degree of freedom for the spin variables [77]. In order to remain in a disordered phase, the inelastic neutron scattering experiment was carried out at finite temperatures. Consequently the quasi-elastic scattering may correspond to scattering from thermally excited solitons as has previously been seen in high field experiments for classical easy plane antiferromagnets [112].

Low energy constant energy scans through $q = \pi$ show broadening beyond resolution in the high field phase. The characteristic Half Width at Half Maximum, κ, could either be associated with the thermal soliton density or it could be a measure of the deviation from half filling of a high field spinon Luttinger liquid. Further experiments will be required to distinguish between these two scenarios.

4 Conclusions

Zero field properties of integer and half odd integer spin chains are now well understood both from a theoretical and experimental standpoint. The spin-1/2 chain has gapless two-spinon continuum, while the spin-1 chain has a bound state that is separated from the singlet ground state by the Haldane gap. The magnetized state of the spin-1/2 chain has a new characteristic length scale that can be associated with the Fermi wave-vector of a one-dimensional Luttinger liquid shifted from half filling, or with the spacing between magnetized soliton defects. When there is a staggered g-tensor the magnetized state is in general gapfull and has transverse long range antiferromagnetic order. The spin-1 chain has a finite critical field that corresponds to closing the Haldane gap. However, when there is an alternating g-tensor, the gap actually never closes. Instead there is a cross-over from a low field phase with a Haldane gap to a high field phase with a residual gap induced by the staggered field. In addition transverse antiferromagnetic order is induced even for infinitesimal applied field. When there is no staggered field the high field phase should be gapless. Unfortunately most highly one-dimensional systems also have substantial spin space anisotropy. In our experiments on NDMAP the field was applied in the easy plane and so it may induce a gap in the $T = 0$ excitation spectrum. Instead we found quasi-elastic scattering at high fields and at temperatures above the phase transition to static order. This quasi-elastic scattering may be associated with scattering from a thermally excited soliton gas.

An exciting experimental challenge that remains is to determine whether field induced incommensurate spin correlations occur in systems other than the spin-1/2 antiferromagnetic Heisenberg spin chain. The high field phase of the spin-1 chain would be an obvious place to look. Unfortunately the experiments are difficult because a gapless high field phase is only expected to exist in the high field limit when the field is applied perpendicular to an easy plane and when there is no staggered g-tensor. While NDMAP has no staggered g-tensor the proposed experiment is complicated by the fact that the magnetic field must be directed along the chain while at the same time wave-vector transfer must have a significant component along that direction. This geometry is incompatible with the conventional superconducting split coil magnet configuration where the field is perpendicular to the scattering plane. It appears that advances in high field magnet systems for neutron scattering or new model spin-1 chain systems will be needed for progress in this area.

Acknowledgements

We thank R. Paul for help with neutron activation analysis. Work at JHU was supported by the United States National Science Foundation under DMR-9801742 and DMR-0074571. DHR acknowledges the support of the David and Lucile Packard Foundation. X-ray characterization and SQUID magnetometry was carried out using facilities maintained by the JHU MRSEC under NSF Grant number DMR-0080031. ORNL is managed by UT-Battelle, LLC for the U.S. DOE under contract DE-AC05-00OR22725. Work at BNL was carried out under DOE Contract No. DE-AC02-98CH10886. This work utilized neutron research facilities supported by NIST and the NSF under Agreement No. DMR-9986442.

References

1. M. Steiner, J. Villain, C.G. Windsor: Adv. Phys. **25**, 87 (1976)
2. I. Affleck: J. Phys. Condens. Matter **1**, 3047 (1989)
3. R. Chitra, T. Giamarchi: Phys. Rev B **55**, 5816 (1997)
4. J. Byrne: *Neutrons, Nuclei, and Matter, an Exploration of the Physics of Slow Neutrons*, Bristol, Philadelphia, Institute of Physics Publishing (1994)
5. S.M. Lovesey: *Theory of Thermal Neutron Scattering from Condensed Matter*, (Clarendon Press, Oxford) 1984
6. P.J. Brown: in 'International Tables for Crystallography', Vol. C, eds. A.J.C. Wilson, E.Prince, Kluwer Academic Publishers Boston (1999)
7. H. Bethe: Z. Phys. **71**, 205 (1931)
8. M. Karbach, G. Müller: Comp. in Phys. **11**, 36-43 (1997) M. Karbach, K. Hu, G. Müller: ibid **12**, 565-573 (1998) M. Karbach, K. Hu, G. Müller: cond-mat/0008018 (2000)
9. L. Hulthén: Arkiv. Math Aston. Fysik **26A**, 11 (1938)
10. J. des Cloizeaux, J.J. Pearson: Phys. Rev. **128**, 2131 (1962)
11. J. Lowenstein: in *Recent Advances in Field Theory and Statistical Mechanics* Les Houches Summer School 1982 session XXXIX, eds. L.B. Zuber, R. Stora, North Holland (1984)
12. L. Fadeev: in *Recent Advances in Field Theory and Statistical Mechanics* Les Houches Summer School 1982 session XXXIX, eds. L.B. Zuber, R. Stora, North Holland (1984)
13. A. Luther, I Peshel: Phys. Rev. B **12**, 3908 (1975)
14. G. Müller, H. Thomas, H. Beck, J.C. Bonner: Phys. Rev. B **24**, 1429 (1981)
15. S.M. Hayden, G. Aeppli, H.A. Mook, T.G. Perring, T.E. Mason, S.-W. Cheong, Z. Fisk: Phys. Rev. Lett. **76**, 1344 (1996)
16. J.M. Tranquada, B.J. Sternlieb, J.D. Axe, Y. Nakamura, S. Uchida: Nature **375**, 561 (1995)
17. I. Zaliznyak et al.: to be published (2002)
18. N. Motoyama, N. Eisaki, S. Uchida: Phys. Rev. Lett. **76**, 3312 (1996)
19. I.A. Zaliznyak, C. Broholm, M. Kibune, M. Nohara, H. Takagi: Phys. Rev. Lett. **83**, 5370 (1999)
20. H. Suzuura, H. Yasuhara, A. Furusaki, N. Nagaosa, Y. Tokura: Phys. Rev. Lett. **76**, 2579 (1996)
21. M. Takigawa, N. Motoyma, H. Eisaki, S. Uchida: Phys. Rev. Lett. **76**, 4612 (1996)
22. A. Keren, L.P. Le, G.M. Luke, B.J. Sternlieb, W.D. Wu, Y.J. Uemura, S. Tajima, S. Uchida: Phys. Rev. B **48**, 12926 (1993)
23. T. Ami, M.K. Crawford, R.L. Harlow, Z.R. Wang, D.C. Johnston, Q. Huang: Phys Rev. B **51**, 5994 (1995)
24. K. Yamada, J. Wada, S. Hosoya, Y. Endoh, S. Noguchi, S. Kawamata, K. Okuda: Physica C **253**, 135 (1995)
25. A. Keren, K. Kojima, L.P. Le, G.M. Luke, W.D. Wu, Y. J. Uemura, S. Tajima, S. Uchida: J. Magn. Mag. Mat. **140-144**, 1641 (1995)
26. I. Tsukada, Y. Sasao, K. Uchinokura, A. Zheludev, S. Maslov, G. Shirane, K. Kakurai, E. Ressouche: Phys. Rev. B **60**, 6601 (1999)
27. I. Tsukada, J. Takeya, T. Masuda, K. Uchinokura: Phys. Rev. B **62**, R6061 (2000)
28. M. Kenzelmann, A. Zheludev, S. Raymond, E. Ressouche, T. Masuda, P. Böni, K. Kakurai, I. Tsukada, K. Uchinokura, R. Coldea: Phys. Rev. B **64**, 54422 (2001)

29. A. Zheludev, M. Kenzelmann, S. Raymond, T. Masuda, K. Uchinokura, S.-H. Lee: Phys. Rev. B **65**, 014402 (2002)
30. A. Zheludev, M. Kenzelmann, S. Raymond, E. Ressouche, T. Masuda, K. Kakurai, S. Maslov, I. Tsukada, K. Uchinokura, A. Wildes: Phys. Rev. Lett. **85**, 4799 (2000)
31. S.K. Satija, J.D. Axe, G. Shirane, H. Yoshizawa, K. Hirakawa: Phys. Rev. B **21**, 2001 (1980)
32. S.E. Nagler, D.A. Tennant, R.A. Cowley, T.G. Perring, S.K. Satija: Phys. Rev. B **44**, 12361 (1991)
33. M.T. Hutchings, E.J. Samuelsen, G. Shirane, K. Hirakawa: Phys. Rev. **188**, 919 (1969)
34. B. Lake, D.A. Tennant, S.E. Nagler: Phys. Rev. Lett. **85**, 832 (2000)
35. D.A. Tennant, R.A. Cowley, S.E. Nagler, A.M. Tsvelik: Phys. Rev. B **52**, 13368 (1995)
36. B. Grenier, L.P. Regnault, J.E. Lorenzo, J.P. Boucher, A. Hiess, G. Dhalenne, A. Revcolevschi: Phys. Rev. B **62**, 12206 (2000)
37. L.P. Regnault, M. Aïn, B. Hennion, G. Dhalenne, A. Revcolevschi: Phys. Rev. B **53**, 5579 (1996)
38. M. Arai, M. Fujita, M. Motokawa, J. Akimitsu, S.M. Bennington: Phys Rev. Lett. **77**, 3649 (1996)
39. Y. Endoh, G. Shirane, R.J. Birgeneau, P.M. Richards, S.L. Holt: Phys. Rev. Lett. **32**, 170 (1974)
40. I.U. Heilmann, G. Shirane, Y. Endoh, R.J. Birgeneau, S.L. Holt: Phys. Rev. B **18**, 3530 (1978)
41. D.C. Dender, P.R. Hammar, D.H. Reich, C. Broholm, G. Aeppli: Phys. Rev. Lett. **79**, 1750 (1997)
42. D.C. Dender, D. Davidović, D.H. Reich, C. Broholm: Phys. Rev. B **53**, 2583 (1996)
43. Y. Chen: Johns Hopkins University, PhD. Thesis
44. C.P. Landee, A.C. Lamas, R.E. Greeney, K.G. Bücher: Phys. Rev. B **35**, 228 (1987)
45. P.R. Hammar, M.B. Stone, D.H. Reich, C. Broholm, P.J. Givson, M.M. Turnbull, C.P. Landee, M. Oshikawa: Phys. Rev. B **59**, 1008 (1999)
46. R. Coldea, D.A. Tennant, R.A. Cowley, D.F. McMorrow, B. Dorner, Z. Tylczynski: Phys. Rev. Lett. **79**, 151 (1997)
47. M.W. VanTol, N.J. Poulis, Physica **69**, 341 (1973)
48. J.P. Groen, T.O. Klaassen, N.J. Poulis, G. Müller, H. Thomas, H. Beck: Phys. Rev. B **22**, 5369 (1980)
49. W.E. Hatfield, W.E. Estes, W.E. Marsh, M.W. Pickens, L.W. ter Haar, R.R. Weller: in *Extended Linear Chain Compounds*, ed. J.S. Miller (Plenum, New York, NY 1983), p. 43
50. L.J. de Jongh, A.R. Miedema: Adv. Phys. **23**, 1 (1974)
51. D.A. Tennant, R.A. Cowley, S.E. Nagler, A.M. Tsvelik: Phys. Rev. B **52**, 13368 (1995)
52. M.B. Stone et al.: to be published (2002)
53. G. Müller, H. Beck, J.C. Bonner: Phys. Rev. Lett. **43**, 75 (1979)
54. M. Karbach, G. Muller, A.H. Bougourzi, A. Fledderjohann, K.H. Mutter: Phys. Rev. B **55**, 12510 (1997)
55. A.J. Millis: Private Communications (1997)
56. I. Affleck, M. Oshikawa: Phys. Rev. B **60**, 1038 (1999)

57. F.H.M. Essler, A.M. Tsvelik: Phys. Rev. B **57**, 10592 (1998)
58. M. Oshikawa, I Affleck: Phys. Rev. Lett. **79**, 2883 (1997)
59. E. Pytte: Phys. Rev. B **10**, 4637 (1974) M.C. Cross, D.S. Fischer: Phys. Rev. B **19**, 402 (1979) J.W. Bray, L.V. Interrante, I.S. Jacobs, J.C. Bonner: in *Extended Linear Chain Compounds*, ed. J.S. Miller (Plenum, New York, 1982), Vol. 3, pp. 353-415
60. A.I. Buzdin, M.L. Kulic, V.V. Tugushev: Solid State Comm. **48**, 483 (1983)
61. M. Fujita, K. Machida: J. Phys. C: Solid State Phys. **21**, 5813 (1988)
62. H.M. Rønnow, M. Enderle, D.F. McMorrow, L.P. Regnault, G. Dhalenne, A. Revcolevschi, A. Hoser, K. Prokes, P. Vorderwisch, H. Schneider: Phys. Rev. lett. **84**, 4469 (2000)
63. V. Kiryukhin, B. Keimer, J.P. Hill, A. Vigliante: Phys. Rev. Lett. **76**, 4608 (1996)
64. M. Enderle, H.M. Rønnow, L.P. Regnault, G. Dhalenne, A. Revcolevschi, P. Vorderwisch, H. Schneider, P. Smeibidl, M. Meissner: Phys. Rev. Lett. **87**, 177203 (2001)
65. F.D.M. Haldane: Phys. Lett. **93A**, 464 (1983)
66. F.D.M. Haldane: Phys. Rev. Lett. **50**, 1153 (1983)
67. E. Lieb, T.D. Schultz, D.C. Mattis: Ann. Phys. **16**, 407 (1962)
68. M. Yamanaka, M. Oshikawa, I. Affleck: Phys. Rev. Lett. **79**, 1110 (1997)
69. M. Oshikawa: Phys. Rev. Lett. **84**, 3370 (2000)
70. M.P. Nightingale, H.W.J. Blöte: Phys. Rev. B **33** 659 (1986)
71. S.R. White, D.A. Huse: Phys. Rev. B **48** 3844 (1993)
72. I. Affleck, T. Kennedy, E.H. Lieb, H. Takaki: Phys. Rev. Lett. **59**, 799 (1987)
73. D.P. Arovas, A. Auerbach, F.D.M. Haldane: Phys. Rev. Lett. **60**, 531 (1988)
74. M. den Nijs, K. Rommelse: Phys. Rev. B **40**, 4709 (1989)
75. F.C. Alcaraz, Y. Hataugai: Phys. Rev. B **46**, 13914 (1992)
76. R. Scharf, H.-J. Mikeska: J. Phys. Cond. Matt. **7**, 5083 (1995)
77. I. Affleck: Phys. Rev. B **41**, 6697 (1990) Phys. Rev. B **43**, 3215 (1991)
78. H. Mutka, J.L. Soubeyroux, G. Bourleaux, P. Colombet: Physica B **39**, 4820 (1989)
79. J. Darriet, L.P. Regnault: Solid State Commun. **86**, 409 (1993)
80. J.F. DiTusa, S.-W. Cheong, J.-H. Park, G. Aeppli, C. Broholm, C.T. Chen: Phys. Rev. Lett. **73**, 1857 (1994)
81. G. Xu, J.F. DiTusa, T. Ito, K. Oka, T. Ito, H. Takagi, C. Broholm, G. Aeppli: Phys. Rev. B **54**, 6827 (1996)
82. A. Zheludev, S.E. Nagler, S.M. Shapiro, L.K. Chou, D.R. Talham, M.W. Meisel: Phys. Rev. B **53**, 15004 (1996)
83. J.P. Renard, V. Gadet, L.P. Regnault, M. Verdaguer: J. Magn. Magn. Mater. **90**, 213 (1990)
84. L.K. Chou, D.R. Talham, W.W. Kim, P.J.C. Signore, M.W. Meisel: Physica B **194-196**, 313 (1994)
85. J.P. Renard, M. Verdaguer, L.P. Regnault, W.A.C. Erkelens, J. Rossatmignod, J. Ribas, W.G. Stirling, C. Vettier: J. Appl. Phys. **63**, 3538 (1988)
86. K. Hirota, S.M. Shapiro, K. Katsumata, M. Hagiwara: Physica B **213**, 173 (1995)
87. S. Ma, C. Broholm, D.H. Reich, B.J. Sternlieb, R.W. Erwin: Phys. Rev. Lett. **69**, 3571 (1992)
88. L.P. Regnault, I. Zaliznyak, J.P. Renard, C. Vettier: Phys. Rev. B **50**, 9174 (1994)
89. O. Avenel et al.: Phys. Rev. B **46**, 8655 (1992)
90. Z. Honda, H. Asakawa, K. Katsumata: Phys. Rev. Lett. **81**, 2566 (1998)
91. Y. Chen, Z. Honda, A. Zheludev, C. Broholm, K. Katsumata, S.M. Shapiro: Phys. Rev. Lett. **86**, 1618 (2001)

92. A. Zheludev, Y. Chen, C.L. Broholm, Z. Honda, K. Katsumata: Phys. Rev. B **63**, 104410 (2001)
93. Y. Koike, M. Metoki, Y. Morii, T. Kobayashi, T. Ishii, M. Yamashita: J. Phys. Soc. Jpn. **69**, 4034 (2000)
94. A. Zheludev, Y. Chen, C.L. Broholm, Z. Honda, K. Katsumata: Phys. Rev. B **63**, 104410 (2001)
95. R.M. Morra, W.J.L. Buyers, R.L. Armstrong, K. Hirakawa: Phys. Rev. B **38**, 543 (1988)
96. W.J.L. Buyers, R.M. Morra, R.L. Armstrong, M.J. Hogan, P. Gerlach, K. Hirakawa: Phys. Rev. Lett. **56**, 371 (1986)
97. M. Takahashi: Phys. Rev. Lett. **62**, 2313 (1989)
98. K. Nomura: Phys. Rev. B **40**, 2421 (1989)
99. S. Liang: Phys. Rev. Lett. **64**, 1597 (1990)
100. J.P. Renard, M. Verdaguer, L.P. Regnault, W.A.C. Erkelens, J. Rossatmignod, W.G. Stirling: Europhys. Lett. **3**, 945 (1987)
101. P.C. Hohenberg, W.F. Brinkman: Phys. Rev. B **10**, 128 (1974)
102. S.M. Girvin, A.H. MacDonald, P.M. Platzman: Phys. Rev. B **33**, 2481 (1986)
103. I. Zaliznyak, S.-H. Lee, S.V. Petrov: Phys. Rev. Lett. **87**, 017202 (2001)
104. M. Enderle, L.P. Regnault, C. Broholm, D.H. Reich, I. Zaliznyak, M. Sieling, H. Ronnow, D.F. McMorrow: Physica B **276**, 560 (2000)
105. M. Sieling, U. Low, B. Wolf, S. Schmidt, S. Zvyagin, B. Luthi: Phys. Rev. B **61**, 88 (2000)
106. A. Zheludev, Z. Honda, K. Katsumata , R. Feyerherm, K. Prokes: Europhys. Lett. **55**, 868 (2001)
107. Z. Honda, H. Asakawa, K. Katsumata: Phys. Rev. Lett. **81**, 2566 (1998)
108. Z. Honda, K. Katsumata, Y. Nishiyama, I. Harada: Phys. Rev. B **63**, 064420 (2001)
109. A. Zheludev, Z. Honda, Y. Chen, C.L. Broholm, K. Katsumata, S.M. Shapiro: Phys. Rev. Lett. **88**, 077206 (2002)
110. O. Golinelli, T. Jolicoeur, R. Lacaze: J. Phys. Condens. Matter **5**, 7847 (1993)
111. L.-P. Regnault, I.A. Zaliznyak, S.V. Meshkov: J. Phys. Condens. Matter **5**, L677 (1993)
112. H.J. Mikeska, M. Steiner: Adv. in Phys. **40**, 191 (1991)

Electronic Phases of Low-Dimensional Conductors

C. Bourbonnais

Centre de Recherche sur les Propriétés Électroniques de Matériaux Avancées,
Département de Physique Université de Sherbrooke,
Sherbrooke, Québec, Canada J1K 2R1

Abstract. We briefly review the physics of electronic phases in low dimensional conductors. We begin by introducing the properties of the one-dimensional electron gas model using bosonization and renormalization group methods. We then tackle the influence of interchain coupling and go through the different instabilities of the electron system to the formation of higher dimensional states. The connection with observations made in quasi-one-dimensional organic and inorganic conductors is discussed.

1 Introduction

There is a *consensus generalis* about the impact of reducing spatial dimension in systems of interacting electrons: correlation effects are magnified and range of electronic behaviors expanded. This is well exemplified in one spatial dimension where low energy electronic excitations turn out to be entirely collective in character so that a description in terms of Fermi liquid quasi-particles, which proved to be so successful in isotropic systems, becomes simply inapplicable.

Organic conductors belong to a class of crystals for which the planar conformation of the molecular constituents combined with their packing as weakly coupled chains in the solid state yield close realizations of one-dimensional solids. Quasi-one dimensional electronic structures are also found in inorganic materials such as the molybdenum bronzes, the chalcogenides and ladder cuprates,[1] as a result of their peculiar molecular or atomic arrangements. In all these quasi-one-dimensional crystals, we are thus faced with a twofold difficulty which combines the objective of determining the temperature range where one-dimensional physics applies with the one of finding the origin of low temperature higher dimensional states. In this review, we will be mainly concerned with these two closely bound issues that are at the heart of the description of the rich phase diagram of compounds like the Bechgaard salt series ($(TMTSF)_2X$) and their sulfur analogs, the Fabre salt series ($(TMTTF)_2X$, where X = PF_6, AsF_6 ... is a monovalent anion).

In the first part of this review, we shall briefly outline the non Fermi liquid properties of the one-dimensional electron gas using bosonization and renormal-

[1] See the reviews of D. Jérome and T. M. Rice on ladder systems in this volume. Luttinger liquid behavior from edge states of two-dimensional electron gas in quantum well structures is discussed by C. Glattli in this volume.

ization group methods. We then consider the problem of instabilities of one-dimensional electronic states by introducing the coupling between chains. We discuss subsequently how the concepts of low-dimensional physics prove relevant when one tries to construct a coherent picture of electronic states that are actually found in quasi-one-dimensional organic compounds. We close this review with a brief discussion on the importance of one-dimensional physics in inorganic metals like the molybdenum bronzes.

2 Interacting Electrons in One Dimension

2.1 The Tomanaga-Luttinger Model

When one tries to understand the origin of non-Fermi-liquid behavior in one dimension,[2] it is instructive to look first at the possible elementary excitations such a low dimensional system can sustain. Consider the transfer of an electron in the final state of wave vector $k+q$ above a filled Fermi sea. Together with the hole left at k, both particles form an electron-hole elementary excitation of energy $\omega(q) = \epsilon(k+q) - \epsilon(k)$ ($\hbar = 1$). In one dimension, the available phase space below $\omega(q)$ for the decay of such excitations shrinks to zero at low energy, namely where the spectrum

$$\epsilon(k) \to \epsilon_p(k) = v_{\rm F}(pk - k_{\rm F}) \tag{1}$$

can be considered as *linear.* A linear spectrum with an unbounded interval of k values for each branch $p = \pm$ (refering to right $(+)$ and left $(-)$ moving electrons) defines the spectrum of the Luttinger model [4]. In these conditions, electron-hole excitations acquire a high degree of degeneracy $\sim qL/2\pi$. It follows that stable charge and spin-density excitations can be formed from the superpositions of excitations

$$\rho_\pm(q) = \frac{1}{\sqrt{2}} \sum_{\alpha,k} a^\dagger_{\pm,k+q,\alpha} a_{\pm,k,\alpha} \tag{2}$$

for the charge and

$$\sigma_\pm(q) = \frac{1}{\sqrt{2}} \sum_{\alpha,k} \alpha\, a^\dagger_{\pm,k+q,\alpha} a_{\pm,k,\alpha} \tag{3}$$

for the spin. This connotes that a free – bosonic – Hamiltonian may exist for the description of collective charge and spin modes. Actually for a Luttinger spectrum with all negative energy states filled [5], these composite objects obey the commutation rules

$$[\rho_p(q), \rho_{p'}(-q)] = \; p\delta_{pp'}\, q\frac{L}{2\pi}$$

[2] We refer to the excellent reviews of J. Voit [28], H. J. Schulz [2] and V. J. Emery [3] for a more exhaustive discussion of the one-dimensional gas model using bosonization method.

$$[\sigma_p(q), \sigma_{p'}(-q)] = p\delta_{pp'}\, q\frac{L}{2\pi}, \tag{4}$$

which are compatible with bosons. Another key feature is the commutation relation between the one-electron Hamiltonian

$$\mathcal{H}_0 = \sum_{k,p,\alpha} \epsilon_p(k) a^\dagger_{p,k,\alpha} a_{p,k,\alpha} \tag{5}$$

and the spin and charge-density operators:

$$\begin{aligned} [\mathcal{H}_0, \rho_p(q)] &= p v_{\mathrm{F}} q\, \rho_p(q) \\ [\mathcal{H}_0, \sigma_p(q)] &= p v_{\mathrm{F}} q\, \sigma_p(q), \end{aligned} \tag{6}$$

which reminds the algebra of operators for the harmonic oscillator. A fermion to boson correspondance for the excitations can thus be established in which $\mathcal{H}_0$ can be written as

$$\mathcal{H}_0 = \sum_{p,q} p v_{\mathrm{F}} q \big(\bar{\rho}_p(q)\bar{\rho}_p(-q) + \bar{\sigma}_p(q)\bar{\sigma}_p(-q)\big), \tag{7}$$

which is quadratic in the charge $\bar{\rho}_p \equiv (2\pi/Lq)^{\frac{1}{2}}\rho_p$ and spin $\bar{\sigma}_p \equiv (2\pi/Lq)^{\frac{1}{2}}\sigma_p$ operators. All the excited states of the fermion system can then be described in terms of bosonic variables.

In the Tomanaga-Luttinger model [4,6], electrons interact through the exchange of small momentum transfer. This allows us to define two scattering processes usually denoted g_2 and g_4 couplings with respect to the Fermi points $\pm k_{\mathrm{F}}$ (here taken for simplicity as $q-$independent interactions [5,7]). The total Hamiltonian becomes $\mathcal{H}_{\mathrm{TL}} = \mathcal{H}_0 + \mathcal{H}_{\mathrm{I}}$, in which the interacting part $\mathcal{H}_{\mathrm{I}}$ can also be expressed in terms of density operators:

$$\begin{aligned} \mathcal{H}_{\mathrm{I}} &= \frac{1}{L}\sum_{\alpha_{1,2}}\sum_{k_1,k_2,q} g_2\, a^\dagger_{+,k_1+q,\alpha_1} a^\dagger_{-,k_2-q,\alpha_2} a_{-k_2,\alpha_2} a_{+,k_1,\alpha_1} \\ &+ \frac{1}{L}\sum_{\alpha,p}\sum_{k_1,k_2,q} g_4\, a^\dagger_{p,k_1+q,\alpha} a^\dagger_{p,k_2-q,-\alpha} a_{pk_2,-\alpha} a_{p,k_1,\alpha} \\ &= 2g_2 \sum_{p,q} v_{\mathrm{F}} q \bar{\rho}_p(q)\bar{\rho}_{-p}(-q) + g_4 \sum_{p,q} v_{\mathrm{F}} q \big(\bar{\rho}_p(q)\bar{\rho}_p(-q) - \bar{\sigma}_p(q)\bar{\sigma}_p(-q)\big). \end{aligned} \tag{8}$$

The TL Hamiltonoian is therefore still quadratic in terms of the bosonic operators. Owing to the presence of the g_2 term which couples the density on different branches, a transformation is needed to diagonalize the Hamiltonian. We thus have

$$\begin{aligned} \mathcal{H}_{\mathrm{TL}} &= \mathcal{H}_\sigma + \mathcal{H}_\rho \\ &= \sum_{p,q} \omega_\sigma(q)\, b^\dagger_{\sigma,q} b_{\sigma,q} + \omega_\rho(q) b^\dagger_{\rho,q} b_{\rho,q}, \end{aligned} \tag{9}$$

in which the new operators $b_\sigma^{(\dagger)}$ for spin and $b_\rho^{(\dagger)}$ for charge obey boson commutation rules. The spin and charge spectra of collective excitations have a Debye form $\omega_\nu(q) = u_\nu \mid q \mid$ $(\nu = \sigma, \rho)$, where the velocities are

$$u_\sigma = v_{\rm F}\left(1 - \frac{g_4}{2\pi v_{\rm F}}\right) \tag{10}$$

for the spin and

$$u_\rho = v_\rho \left(1 - \left(\frac{g_2}{\pi v_\rho}\right)^2\right)^{1/2} \tag{11}$$

for the charge, with $v_\rho = v_{\rm F}(1 + g_4/2\pi v_{\rm F})$. A key property of the above Hamiltonian is the commutation relation $[\mathcal{H}_\sigma, \mathcal{H}_\rho] = 0$, which yields the decoupling between $\mathcal{H}_\sigma$ and $\mathcal{H}_\rho$ that is termed *separation between spin and charge degrees of freedom*. The corresponding split-off of acoustic excitations will have a profound influence on the properties of the system which becomes a Luttinger liquid.[8] As regards thermodynamics, the free energy will consist of two separate contributions, which yields the property of additivity for the specific heat (per unit of length) $C = C_\sigma + C_\rho$. Here $C_{\nu=\sigma,\rho} = \pi T/(3u_\nu)$ is the linear temperature dependent specific heat of each branch of acoustic excitations ($k_{\rm B} = 1$).

2.2 Phase Variables Description

Collective excitations in a Luttinger liquid are reminiscent of those of a vibrating string. This relation can be further sharpened if one introduces the pair of conjugate phase variables

$$\begin{aligned} \phi_\nu(x) &= -i\frac{\pi}{L}\sum_q \frac{e^{-iqx}}{q} e^{-\alpha_0|q|/2}[\nu_+(q) + \nu_-(q)] \\ \Pi_\nu(x) &= i\frac{1}{L}\sum_q e^{-iqx} e^{-\alpha_0|q|/2}[\nu_+(q) - \nu_-(q)], \end{aligned} \tag{12}$$

which satisfy commutation relation $[\phi_\nu(x_1), \Pi_{\nu'}(x_2)] = i\delta_{\nu,\nu'}\delta(x_1 - x_2)$ in the limit $\alpha_0 \to 0$ for the short distance cut-off.

The phase variable representation allows us to rewrite the Tomonaga-Luttinger Hamiltonian in the harmonic form

$$\mathcal{H}_{\rm TL} = \sum_{\nu=\rho,\sigma} \frac{1}{2} \int \left[\pi u_\nu K_\nu \Pi_\nu^2 + u_\nu (\pi K_\nu)^{-1} \left(\frac{\partial \phi_\nu}{\partial x}\right)^2\right] dx, \tag{13}$$

where K_ν is the stiffness constants of acoustic excitations. The properties of the model are then entirely governed by the set of parameters $\{u_\nu, K_\nu\}$. These are functions of the microscopic coupling constants. Thus for a rotationally invariant system (spin independent interactions), one has $K_\sigma = 1$ and

$$K_\rho = \left(\frac{2\pi v_{\rm F} + g_4 - 2g_2}{2\pi v_{\rm F} + g_4 + 2g_2}\right)^{1/2}. \tag{14}$$

2.3 Properties of the Luttinger Liquid State

One-Particle. Consider the Matsubara time-ordered single-particle Green's function $G_p(x,\tau) = -\langle T_\tau\, \psi_{p,\alpha}(x,\tau)\psi^\dagger_{p,\alpha}(0,0)\rangle$, which is expressed as a statistical average over fermion fields. It can be evaluated explicitly by using the harmonic phase Hamiltonian (13), with the aid of the relation between the Fermi and bosonic fields [9–11]:

$$\begin{aligned}\psi_{p,\alpha}(x) &= L^{-\frac{1}{2}} \sum_k a_{p,k,\alpha}\, e^{ikx} \\ &\sim \lim_{\alpha_0\to 0} \frac{e^{ipk_{\mathrm{F}}x}}{\sqrt{2\pi\alpha_0}} \exp\Big(-\frac{i}{\sqrt{2}}[p(\phi_\rho+\alpha\phi_\sigma)+(\theta_\rho+\alpha\theta_\sigma)]\Big),\end{aligned} \tag{15}$$

where $\theta_\nu(x) = \pi\int \Pi_\nu(x')dx'$. One finds

$$G_p(x,\tau) = \frac{e^{ipk_{\mathrm{F}}x}}{\alpha_0^{-\theta}} \prod_\nu [\xi_\nu \sinh(x+iu_\nu\tau)/\xi_\nu]^{-\frac{1}{2}-\theta_\nu}[\xi_\nu \sinh(x-iu_\nu\tau)/\xi_\nu]^{-\theta_\nu} \tag{16}$$

where $\theta_\nu = \frac{1}{4}(K_\nu + 1/K_\nu - 2)$. As a correlation function of the electron with itself, the Green's function gives useful information about the spatial and time decay of single-particle quantum coherence in the presence of collective oscillations of the Luttinger liquid. At equal Matsubara time, which amounts to put $\tau = 0$ in the above expression, we observe that G_p depends on two characteristic length scales $\xi_\sigma = u_\sigma/\pi T$ and $\xi_\rho = u_\rho/\pi T$, corresponding to the de Broglie quantum lengths for spin and charge acoustic excitations. Thus for $\alpha_0 \ll x \ll \xi_\nu$, the fermion coherence decays according to the power law

$$G_p(x) \approx \frac{e^{ipk_{\mathrm{F}}x}}{\alpha_0^{-\theta}} \frac{1}{x^{1+\theta}}, \tag{17}$$

where the exponent $\theta = \theta_\sigma + \theta_\rho$ is called the anomalous dimension of the Green's function. It is non-zero in the presence of interaction (the canonical dimension of the Green's function is unity in a free electron gas). For non-zero interaction, the spatial decay of quasi-particle coherence is therefore faster. The existence of a anomalous power law is also the mark of scaling (a situation analogous to correlations of the order parameter at the critical point), namely the absence of particular length scale of the fermion coherence between α_0 and ξ_ν.

For large distances $x \gg \xi_\nu$, we have

$$G_p(x) \propto e^{-x/\xi}, \tag{18}$$

indicating that thermal fluctuations lead to an exponential decay of coherence and the absence of scaling. The effective coherence length $\xi^{-1} = 1/\xi_\sigma + 1/\xi_\rho$ combines the spin and the charge quantum lengths.

The absence of ordinary quasi-particle states in a Luttinger liquid will also show up in the one-electron spectral properties. These can be extracted from the Fourier transform of the retarded Green's function. The quantity of interest is the spectral weight defined as the imaginary part of the Green's function:

$$A_p(q,\omega) = -\frac{1}{\pi}\mathrm{Im}G_p(pk_{\mathrm{F}}+q,\omega), \tag{19}$$

which gives the probability of having a single-particle state of wave vector pk_F+q with a energy ω measured from the Fermi level. The spectral function takes on particular importance since it can be probed at $q<0\,(q>0)$ by photoemission (inverse photoemission) experiments. We will focus here on spin independent interactions for which $K_\sigma = 1$ and $\theta_\sigma = 0$. The presence of collective modes with two different velocities in a Luttinger liquid has a pronounced influence on the spectral function in comparison to that of a Fermi liquid. In the latter case, the spectral weight $A_p(\omega) = z\delta(\omega)$ is simply a delta function at the Fermi edge indicating the presence of well defined quasi-particle excitations of weight z at zero temperature. At finite temperature or finite q, there is the usual broadening of the quasi-particle peak ($\sim T^2$ or $v_F^2q^2$).

The progress made to achieve the Fourier transform of $G_p(x,\tau\to it)$ at zero temperature [12,13], indicates instead the absence of quasi-particle peak in the spectral weight. In effect, collective modes suppresses the delta function, which is replaced at not too large θ by power law singularities

$$\begin{aligned} A_+(k_F+q,\omega) &\sim_{\omega\to u_\rho q} \mid \omega - u_\rho q \mid^{\frac{1}{2}\theta-\frac{1}{2}} \\ &\sim_{\omega\to u_\sigma q} \Theta(\omega-u_\sigma q)(\omega-u_\sigma q)^{\theta-\frac{1}{2}} \\ &\sim_{\omega\to -u_\sigma q} \Theta(-\omega-u_\rho q)(-\omega-u_\rho q)^{\frac{1}{2}\theta}, \end{aligned} \tag{20}$$

for $q>0$ [12]. At finite temperature [14,15], thermal broadening will round singularities and cusps.

Another physical quantity of interest is the one-electron density of states (per spin)

$$\begin{aligned} N(\omega) &= \sum_p \int \frac{dq}{2\pi} A_p(q,\omega) \\ &\propto \mid \omega \mid^\theta . \end{aligned} \tag{21}$$

In a Luttinger liquid, the density of states is not constant but presents a dip close to the Fermi level [16]. Strickly at the Fermi level, the density of states is zero showing once again the absence of quasi-particles at $T=0$. At finite temperature, $N(T)\propto T^\theta$, the dip partly fills at the Fermi level due to thermal fluctuations.

Two-Particle Response. The two-particle response function in Matsubara-Fourier space is defined by

$$\chi(q,\omega_m) = \iint dx d\tau\, \chi(x,\tau)\, e^{-iqx+i\omega_m\tau}, \tag{22}$$

where $\chi(x,\tau) = -\langle T_\tau O(x,\tau)O^\dagger\rangle$ is the two-particle correlation function. At small q and ω, the dynamic magnetic susceptibility (or compressibility) can be calculated using the spin (charge) operator $O=(\sigma_+ + \sigma_-)/\sqrt{2}$ ($O=(\rho_+ + \rho_-)/\sqrt{2}$) and (15), with the result after analytic continuation

$$\chi_\nu(q, i\omega_m\to\omega+i0^+) = -\frac{1}{\pi u_\nu}\sum_p \frac{p u_\nu q}{\omega - p u_\nu q} + i\pi\sum_p q\delta(\omega - p u_\nu q) \tag{23}$$

for both branches at zero temperature. The simple pole structure of the real part of this expression is analogous to the one found for acoustic phonons. Correspondingly, the absence of damping shown by the imaginary part emphasizes once more that spin and charge acoustic excitations are eigenstates of the system. In the static ($\omega = 0$) and uniform ($q \to 0$) limits, $\chi_\nu \to 2(\pi u_\nu)^{-1}$. A non-zero susceptibility at zero temperature then occurs despite the absence of density of states at the Fermi level. The proportionality between the uniform response and the density of states that holds for a Fermi liquid is meaningless for a Luttinger liquid due to the absence of quasi-particles. The finite uniform response rather probes the density of states of acoustic boson modes in the spin or charge channel. At non zero temperature for the TL model, $\chi_\nu(T)$ is only very weakly temperature dependent on the scale of the Debye energy $\omega_\nu = u_\nu \alpha_0^{-1}$ of acoustic modes, which is of the order of the Fermi energy for not too large couplings [17,18].

Other quantities like staggered density-wave (close to wave vector $2k_{\rm F}$) responses are also of practical importance in the analysis of X-ray and NMR experiments [19–22]. In the following we will focus on spin-spin correlation function for $q \sim 2k_{\rm F}$; the latter can be evaluated using the spin-density operator $\boldsymbol{O} = \psi_- \boldsymbol{\sigma} \psi_+$ in the definition of the two-particle correlation function given above. At equal-time for example, one gets the power law decay

$$\begin{aligned} \chi(x) &= \langle \boldsymbol{O}(x) \cdot \boldsymbol{O}(0) \rangle \\ &\sim \frac{\cos(2k_{\rm F}x)}{x^{1+K_\rho}}, \end{aligned} \tag{24}$$

which is governed by the LL parameter K_ρ. The temperature dependence of the antiferromagnetic response is given by the Fourier transform of $\chi(x,\tau)$ evaluated at $2k_{\rm F}$ and in the static limit

$$\chi(2k_{\rm F}, T) \sim T^{-\gamma}. \tag{25}$$

The power law exponent $\gamma = 1 - K_\rho > 0$ is non universal and increases with the strength of interactions up to its highest value $\gamma = 1$ corresponding to the Heisenberg universality class. A similar expression is found for the $2k_{\rm F}$ charge-density-wave response in the TL model. For atttractive couplings, a power law singularity is to be found in the superconducting channel, where K_ρ in (24-25) is simply replaced by $1/K_\rho$.

The imaginary part $\mathrm{Im}\chi(q + 2k_{\rm F}, \omega)$ of the dynamic response is another related quantity that plays an important part in experimental situations giving an experimental access to the Luttinger liquid parameter K_ρ [20,22]. In the spin-density-wave channel of a Luttinger liquid at non zero temperature, one has the power law enhanced form

$$\mathrm{Im}\chi(q + 2k_{\rm F}, \omega) \sim (\pi u_\sigma)^{-1} \frac{\omega}{T} \left(\frac{T}{E_{\rm F}} \right)^{-\gamma} \tag{26}$$

for small (real) frequency and q close to 0. We shall revert to this below in the context of NMR.

2.4 The One-Dimensional Electron Gas Model

When large momentum transfer ($\sim 2k_{\rm F}$) is allowed for scattering events, we have an additional coupling parameter which is the backscattering process denoted g_1. Moreover, when the band is half-filled (one electron per lattice site), $4k_{\rm F}$ coincides with the reciprocal lattice vector $G = 2\pi/a$ ($a \sim \alpha_0$ is the lattice constant) and Umklapp scattering becomes possible. Another coupling g_3 is then added to the Hamiltonian for which two electrons are transferred from one side of the Fermi surface to the other. The total Hamiltonian, known as the 1D fermion gas problem now becomes $\mathcal{H} = \mathcal{H}_{\rm TL} + \mathcal{H}'$, where

$$\begin{aligned}\mathcal{H}' &= \frac{1}{L}\sum_{\{k,q,\alpha\}} g_1\, a^\dagger_{+,k_1+2k_{\rm F}+q,\alpha} a^\dagger_{-,k_2-2k_{\rm F}-q,\alpha'} a_{+,k_2,\alpha'} a_{-,k_1,\alpha} \\ &+ \frac{1}{2L}\sum_{\{p,k,q,\alpha\}} g_3\, a^\dagger_{p,k_1+p(2k_{\rm F}+q),\alpha} a^\dagger_{p,k_2-p(2k_{\rm F}+q)+pG,-\alpha} a_{-p,k_2,-\alpha} a_{-p,k_1,\alpha} \end{aligned} \quad (27)$$

corresponds to the additional terms expressed in the fermion representation. In terms of phase variables, the part for antiparallel spins reads

$$\mathcal{H}' = \int dx \Big\{ \frac{2g_1}{(2\pi\alpha_0)^2}\cos(\sqrt{8}\phi_\sigma) + \frac{2g_3}{(2\pi\alpha_0)^2}\cos(\sqrt{8}\phi_\rho)\Big\}, \quad (28)$$

whereas the parallel part of g_1 goes into an additional renormalization of u_σ and K_ν in $\mathcal{H}_{\rm TL}$. From this expression, one first observes that g_1 (g_3) solely depends on the spin (charge) phase variable. Therefore the spin and charge parts of the total Hamiltonian $\mathcal{H} = \mathcal{H}_\sigma + \mathcal{H}_\rho$ still commute and thus preserve spin-charge separation.

An exact solution of $\mathcal{H}$ cannot be found in the general case, except at a particular value of the coupling constants for each sector, corresponding to the Luther-Emery solutions [3,23]. However, one can seek an approximate solution using scaling theory. In the framework of the Tomanaga-Luttinger model, we have already emphasized that anomalous dimensions found in single and pair correlation functions are hallmarks of scaling. In effect, a Luttinger liquid is a self-similar system when it is looked at different length x (and $u_\nu\tau$) scales. We can profit by this property for the more general Hamiltonian by looking at the evolution or the flow of the couplings $g_{1,3}$ and Luttinger liquid parameters as a function of successive change of space and time scales. In practice, the corresponding space-time variations of the phase variables ϕ_ν are integrated out using $g_{1,3}$ as perturbations. Then by rescaling both the initial length ($x \to xe^{-l}$) and time ($u_\nu\tau \to u_\nu\tau e^{-l}$) scales yields the renormalization group flow equations

$$\begin{aligned}\frac{dK_\nu}{dl} &= -\frac{1}{2}K_\nu^2 g_\nu^2 \\ \frac{dg_\nu}{dl} &= g_\nu(2-2K_\nu)\end{aligned} \quad (29)$$

for spin and charge parameters, where $g_\sigma \equiv g_1$ and $g_\rho \equiv g_3$. In the repulsive case, where $g_{i=1\ldots4} > 0$, g_1 is marginally irrelevant in the spin sector, that is to say

$g_1^* \to 0$ when $l \to \infty$. For a rotationally invariant system, we have $K_\sigma^* \to 1$ and $u_\sigma^* \to v_\sigma = v_{\rm F}(1 - g_4/2\pi v_{\rm F})$. If the band filling is incommensurate, $g_3 = 0$ and we recover in this repulsive case the physics of a Luttinger liquid for both spin and charge at large distance. At half-filling, however, g_3 is non zero at $l = 0$ and becomes a marginally relevant coupling that scales to large values as l grows; in turn, the charge stiffness $K_\rho \to 0$ and velocity $u_\rho^* \to 0$ at large l. Strong coupling in g_3 and vanishing K_ρ signals the presence of a charge gap [2], which is given by

$$\Delta_\rho \sim E_{\rm F}\left(\frac{g_3}{E_{\rm F}}\right)^{1/[2(1-n^2K_\rho)]}, \tag{30}$$

where $n = 1$ at half-filling. The physics here corresponds to the one of a 1D Mott insulator [24]. The presence of a gap is also confirmed by the fact that when l increases K_ρ decreases and the combination of $2g_2(l) - g_1(l)$ will invariably crosses the so-called Luther-Emery line at $2g_2(l_{\rm LE}) - g_1(l_{\rm LE}) = 6/5$, where an exact diagonalization of the charge Hamiltonian $\mathcal{H}_\rho$ can be carried out [3,23].

A charge gap is not limited to half-filling but may be present for other commensurabilities too [2]. At quarter-filling for example, the transfer of four particles from one side of the Fermi surface to the other leads (instead of the above g_3 term) to the Umklapp coupling

$$\mathcal{H}_{1/4} \simeq \frac{2g_{1/4}}{(2\pi\alpha_0)^2}\int dx\cos(2\sqrt{8}\phi_\rho). \tag{31}$$

The scaling dimension of the operator $e^{i2\sqrt{8}\phi_\rho}$ is now $8K_\rho$, while each term of $\mathcal{H}_{\rm TL}$ in (13) has a scaling dimension of 2, so that the flow equation becomes

$$\frac{dg_{1/4}}{dl} = (2 - 8K_\rho)g_{1/4}, \tag{32}$$

which goes to strong coupling if $K_\rho < 1/4$, which corresponds to sizeable couplings with longer spatial range [25]. The value of the insulating gap is given by (30) by taking $n = 2$ at quarter-filling [2,25]. It worth noting that in the special situation where the quarter-filled chains are weakly dimerized, both half-filling and quarter-filling Umklapp are present with different bare amplitudes – a situation met in some charge transfer salts [26,33].

Let us look at the consequences of a charge gap on correlation functions. In the single-particle case, we see that taking $K_\rho \to 0$ at $T < \Delta_\rho$ yields large θ_ρ. Therefore the single electron coherence will become vanishingly small at large distance. Each electron is confined within the characteristic length scale $\xi_\rho \sim v_\rho/\Delta_\rho$, which can be seen as the size of bound electron-hole pairs of the Mott insulator. A large value of θ alters the spectral properties by producing a gap in the spectral weight and in turn the density of states [28,29].

The impact is different on spin-spin correlation functions. We first note that the uniform magnetic susceptibility, which uniquely depends on the spin velocity, remains *unaffected* by the charge gap, as a consequence of the spin-charge separation (see § 2.5). As regards $2k_{\rm F}$ antiferromagnetic spin correlations, their amplitude increases and shows a slower spatial decay $\chi(x) \sim 1/x$; this corresponds to a stronger power law singularity in temperature $\chi(2k_F, T) \sim T^{-1}$.

2.5 A Many-Body Renormalization Group Approach

Having described the basic properties of the electron gas from the bosonic standpoint, we can now proceed to its renormalization group description from the many-body point of view. Although the latter works well in weak coupling, it gives a different depth of perspective in the one-dimensional case and it proves particularly useful in the complex description of instabilities of one-dimensional electronic states when interchain coupling is included.

Renormalization Group. When we try to analyze the properties of the 1D electron gas using perturbation theory, we are faced with infrared singularities. These correspond to the logarithmically singular responses $\chi^0 \sim \ln E_F/T$ of a free electron gas to Cooper *and* $2k_F$ electron-hole (Peierls) pair formations. One dimension is special in that both share the same phase space [30]. Their presence with different phase relations in the electron-electron scattering amplitudes indicates that both pairings counterbalance one another by interference to ultimately yield a Luttinger liquid in leading order.

Another property of the Cooper and Peierls logarithmic divergences is the lack of particular energy scale in the interval between E_F and T, a feature that signals scaling. Renormalization group ideas can be applied to the many-body formulation in order to obtain the low-energy properties of the electron gas model [31,32]. In the following, we briefly outline the momentum shell Kadanoff-Wilson approach developed in Refs. [32–34]. The partition function Z is first expressed as a functional integral over the fermion (Grassman) fields ψ

$$Z = \iint D\psi^* D\psi \; e^{S^*[\psi^*,\psi]}. \tag{33}$$

In the Fourier-Matsubara space, the action $S = S_0 + S_I$ consists of a free part

$$S_0[\psi^*, \psi] = \sum_{p,\alpha,\widetilde{k}} [G_p^0(\widetilde{k})]^{-1} \psi^*_{p,\alpha}(\widetilde{k}) \psi_{p,\alpha}(\widetilde{k}) \tag{34}$$

where

$$G_p^0(\widetilde{k}) = [i\omega_n - \epsilon_p(k)]^{-1} \tag{35}$$

is the bare electron propagator with $\widetilde{k} = (k, \omega_n)$; and an interacting part

$$S_I[\psi^*, \psi] = -\frac{T}{2L} \sum_{\{\alpha,p,\widetilde{k}\}} g^{\alpha_1\alpha_2;\alpha_3\alpha_4}_{p_1p_2;p_3p_4} \psi^*_{p_1,\alpha_1}(\widetilde{k}_1)\psi^*_{p_2,\alpha_2}(\widetilde{k}_2)\psi_{p_3,\alpha_3}(\widetilde{k}_3)\psi_{p_4,\alpha_4}(\widetilde{k}_4), \tag{36}$$

in which the couplings constants of the electron gas model are $g^{\alpha\alpha';\alpha'\alpha}_{+-;+-} = g_1$, $g^{\alpha\alpha';\alpha'\alpha}_{+-;-+} = g_2$, $g^{\alpha\alpha';\alpha'\alpha}_{\pm\pm;\mp\mp} = g_3$, and $g^{\alpha\alpha';\alpha'\alpha}_{\pm\pm;\pm\pm} = 2g_4$ [35]. The relevant parameter space of the action for the electron gas will be denoted

$$\mu_S = (G_p^0, g_1, g_2, g_3, g_4). \tag{37}$$

The RG tool is used to look at the influence of high-energy states on the electron-electron scattering near $\pm k_F$ at low energy. We will focus on the RG results at the one-loop level, which will be sufficient for our purposes. The method consists of successive partial integrations of fermion degrees of freedom ($\bar{\psi}^{(*)}$) in the outer energy shell (o.s) $\pm E_0(\ell)d\ell/2$ above and below the Fermi points as a function of ℓ [32]. Here $E_0(\ell) = E_0 e^{-\ell}$ with $\ell > 0$, is the scaled bandwidth cutoff $E_0(\equiv 2E_F)$ imposed to the spectrum (1). We can write

$$\begin{aligned} Z &\sim \iint_< D\psi^* D\psi \, e^{S[\psi^*,\psi]_\ell} \iint_{\mathrm{o.s}} D\bar{\psi}^* D\bar{\psi} \, e^{S_0[\bar{\psi}^*,\bar{\psi}]} \, e^{S_{\mathrm{I},2}+\dots} \\ &\propto \iint_< D\psi^* D\psi \, e^{S[\psi^*,\psi]_\ell \, + \, \frac{1}{2}\langle S_{I,2}^2\rangle_{\mathrm{o.s}} \, + \, \cdots} \quad , \end{aligned} \tag{38}$$

where $S_{\mathrm{I},2}$ is given by the interaction term with two $\bar{\psi}^{(*)}$ in the outer momentum shell in the Cooper and Peierls channels ($2k_F$ electron-hole and zero momentum Cooper pairs), while the other two remain fixed in the inner ($<$) shell.

At the one-loop level, the averages $\langle (S_{\mathrm{I},2})^2\rangle_{\mathrm{o.s}}$ in the outer momentum shell are calculated with respect to $S_0[\bar{\psi}^*,\bar{\psi}]$, which ultimately leads to the scaling transformation of μ_S as a function of ℓ

$$\mathcal{R}_{d\ell}[\mu_S(\ell)] = \mu_S(\ell + d\ell). \tag{39}$$

The outer momentum shell contributions to the Peierls and Cooper channels have different signs and lead to the aforementioned interference in the renormalization flow, which is governed by the following set of equations

$$\begin{aligned} \frac{d\tilde{g}_1}{d\ell} &= -\tilde{g}_1^2 + \dots \\ \frac{d}{d\ell}(2\tilde{g}_2 - \tilde{g}_1) &= \tilde{g}_3^2 + \dots \\ \frac{d\tilde{g}_3}{d\ell} &= \tilde{g}_3(2\tilde{g}_2 - \tilde{g}_1) + \dots , \end{aligned} \tag{40}$$

where the influence of g_4 has been included through the normalization $\tilde{g}_1 = g_1/\pi v_\sigma$, $2\tilde{g}_2 - \tilde{g}_1 = (2g_2 - g_1)/\pi v_\rho$, and $\tilde{g}_3 = g_3/\pi v_\rho$. These scaling equations [35], are consistent with those obtained in (29) from the bosonization technique by expanding the stiffness constants $K_{\nu=\sigma,\rho}$ to leading order in the coupling constants. Therefore from the many-body standpoint, the interference between the Cooper and Peierls channels appears as an indispensable building block of Luttinger and Luther-Emery liquids.

Magnetic Susceptibility. The description sketched above can also be used for the calculation of uniform responses at small q and ω when the couplings g_1 and g_3 are present [17,21,35]. We will be mainly concerned here with the spin susceptibility (a similar approach also applies for compressibility). We will see that the flow of $g_1(\ell)$ is responsible for a temperature dependence of susceptibility.

The first thing that needs to be said is that thermally excited spin excitations involved in the uniform magnetic response do not contribute to the logarithmic singularities Cooper and Peierls channels. In effect, these last singularities refer to electron and hole states located outside the thermal width $\sim 2T$ around the Fermi points, while it is the other way around for the uniform spin response of the Landau channel. The fact that the Landau channel does not interfere directly with the other two constitutes an advantage in the calculation. One can indeed use the renormalization group method to first integrate quantum degrees in the interfering Cooper and Peierls channels, namely down to $\ell_T = \ln E_F/T$, after which the resulting low-energy action can be used to calculate uniform spin susceptibility [17,21]. If we try to outline this way of doing, we first note that the interacting part of the action at ℓ can be written

$$S_{\rm I} = \sum_{p,\widetilde{q}} (2g_2 - g_1)(\ell)\rho_p(\widetilde{q})\rho_{-p}(-\widetilde{q}) - g_1(\ell)\sum_{p,\widetilde{q}} \boldsymbol{S}_p(\widetilde{q})\cdot \boldsymbol{S}_{-p}(-\widetilde{q}) + S_{\rm I}[g_4], \quad (41)$$

in which we have defined the composite fields $\rho_\pm = \frac{1}{2}(L)^{-\frac{1}{2}}\sum_{\widetilde{k}^*,\alpha}\psi^*_{\pm,\alpha}\psi_{\pm,\alpha}$ for charge and $\boldsymbol{S}_\pm = \frac{1}{2}(L)^{-\frac{1}{2}}\sum_{\widetilde{k}^*}\psi^*_{\pm,\alpha}\boldsymbol{\sigma}^{\alpha\beta}\psi_{\pm,\beta}$ for spin. Here the sum on k satisfies $|\,\epsilon_p(k)\,| \leq E_0(\ell)/2$; we have omit the g_3 term since it gives no direct contribution to the Landau channel. The above expression for the action being quadratic in spin and charge fields, it can be linearized using an Hubbard-Stratonovich transformation which allows us to express the partition function as a functional integral over auxilliary charge ϕ and spin $\boldsymbol{M}$ fields

$$\begin{aligned} Z = Z(g_4)\iint \mathcal{D}\phi\mathcal{D}\boldsymbol{M}\,\exp\Big\{-\sum_{\widetilde{q},p,p'} &[\phi_p(\widetilde{q})A_{p,p'}(\widetilde{q})\phi_{p'}(-\widetilde{q}) \\ &+ \boldsymbol{M}_p(\widetilde{q})B_{p,p'}(\widetilde{q})\boldsymbol{M}_{p'}(-\widetilde{q})]\Big\}, \end{aligned} \quad (42)$$

where $Z(g_4)$ is the partition function of the system with g_4 interaction only, which can be treated in RPA in the spin and charge sectors [18]. The effective low-energy free energy density is thus essentially quadratic in both ϕ and $\boldsymbol{M}$ – mode-mode coupling terms vanish for a linear spectrum – and can be seen as an approximate harmonic representation of the electron gas model at low energy. From the expressions of matrix elements $A_{p,p}(\widetilde{q}) = \frac{1}{2}(2\tilde{g}_2 - \tilde{g}_1)(\ell)$, $A_{\pm,\mp} = 1$ and $B_{\pm,\pm}(\widetilde{q}) = -\frac{1}{2}\tilde{g}_1(\ell)$, $B_{\pm,\mp} = 1$, the uniform magnetic response, when expressed as statistical averages over auxiliary fields, is given by

$$\begin{aligned} \chi_\sigma(\widetilde{q}) &= \frac{1}{g_1(\ell)}[\langle\frac{1}{6}\sum_{p,p'}\boldsymbol{M}_p(\widetilde{q})\cdot\boldsymbol{M}_{p'}(-\widetilde{q})\rangle - 1] \\ &= -\frac{2}{\pi}\frac{1}{\bar{u}_\sigma(\ell)}\frac{u_\sigma^2(\ell)q^2}{[\omega - u_\sigma(\ell)q][\omega + u_\sigma(\ell)q]} \\ &\quad + i\,\frac{1}{\bar{u}_\sigma(\ell)}\sum_p u_\sigma(\ell)q\,\delta(\omega - pu_\sigma(\ell)q), \end{aligned} \quad (43)$$

where $\bar{u}_\sigma(\ell) = v_\sigma(1 + \frac{1}{2}\tilde{g}_1(\ell))$. The spectrum of low energy acoustic spin excitations now becomes

$$\omega_\sigma = u_\sigma(\ell)\,|\,q\,|$$

$$= v_\sigma \big(1 - \tilde{g}_1^2(\ell)/4\big)^{\frac{1}{2}} \mid q \mid, \tag{44}$$

which, owing to the presence of g_1, shows ℓ dependent corrections with respect to the Tomanaga-Luttinger limit (Eqn.(10)) (a similar expression holds for the charge spectrum following the substitution $v_\sigma \to v_\rho$ and $-\tilde{g}_1(\ell) \to (2\tilde{g}_2 - \tilde{g}_1)(\ell)$). The temperature dependence of the static and uniform spin susceptibility is obtained by putting $\widetilde{q} = 0$ and $\ell \to \ln \omega_\sigma / T$ in (44) with the result

$$\chi_\sigma(T) = \frac{2}{\pi v_\sigma} \frac{1}{1 - \frac{1}{2}\tilde{g}_1(T)}, \tag{45}$$

where

$$\tilde{g}_1(T) = \frac{\tilde{g}_1}{1 + \tilde{g}_1 \ln \omega_\sigma / T} \tag{46}$$

is the solution of the first equation of (40). For repulsive couplings, the reduction of $g_1(T)$ imparts a temperature dependence to the susceptibility which is shown in Fig. 1. As one can see, the logarithmic corrections make the χ approaching its $T = 0$ value with an infinite slope [17,21,35,36]. This singularity in the derivative occurs only at very low temperature and can be hard to detect in practice. Finite magnetic field or interchain hopping between stacks tends to suppress the singularity.

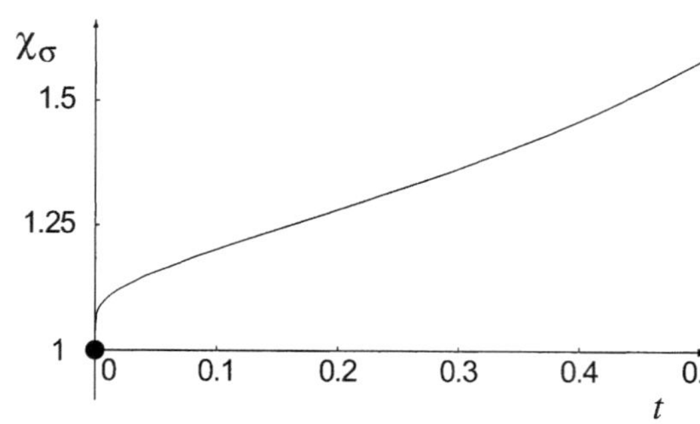

Fig. 1. The temperature variation of the magnetic susceptibility expressed in $\pi v_\sigma/2$ units as a function of the reduced temperature $t = T/\omega_\sigma$ for the electron gas model ($\tilde{g}_1 \sim 1$)

Nuclear Relaxation Rate. The nuclear spin-lattice relaxation rate measured in NMR experiments is another quantity of practical importance if one tries to gain information about spin correlations. It is given by the Moriya expression

$$T_1^{-1} = \mid A \mid^2 T \int \frac{\chi''(\boldsymbol{q}, \omega)}{\omega} \, d^D q, \tag{47}$$

which is taken in the zero Larmor frequency limit ($\omega \to 0$) and where A is proportional to the hyperfine matrix element. Thus the relaxation of nuclear spins gives in principle relevant information about the static, dynamics and dimensionality D of electronic spin correlations. This gives in turn a relatively easy access to the parameters K_ρ, $\bar{u}_\sigma(T)$ and $u_\sigma(T)$ of the electron gas [37]. According to Eqns. (43) and (26), the enhancements of χ'' occur at $q \sim 0$ and in the interval $q \sim 2k_{\rm F} \pm T/v_{\rm F}$ close to $2k_{\rm F}$. The integration is then readily done to give

$$T_1^{-1} \simeq C_0(T) T \chi_\sigma^2(T) + C_1(T) T^{K_\rho}. \tag{48}$$

Owing to the presence of g_1, we have $C_0(T) = C_0\big(u_\sigma(T)/\bar{u}_\sigma(T)\big)^{\frac{1}{2}}$ and $C_1(T) = C_1(1 + \tilde{g}_1 \ln \omega_\sigma/T)^{\frac{1}{2}}$. As a function of temperature, two different behaviors can be singled out. At high temperature, where uniform spin correlations dominate and the $2k_{\rm F}$ ones are small, the relaxation rate is then governed by the $C_0(T) T \chi_\sigma^2(T)$ term. In the low temperature domain, however, $2k_{\rm F}$ spin correlations are singularly enhanced while uniform correlations remain finite so that $T_1^{-1} \sim C_1(T) T^{K_\rho}$. The temperature dependence of T_1^{-1} over the whole temperature range thus contrasts with that of a Fermi liquid where $(T_1 T)^{-1} \sim$ cst., as found for the Korringa law in ordinary metals.

It is interesting to consider the case where a gap is present in the charge part and for which $K_\rho = 0$. We thus have

$$T_1^{-1} \simeq C_0(T) T \chi_\sigma^2(T) + C_1(T). \tag{49}$$

The relaxation will tend to show a finite intercept as $T \to 0$ (here logarithmic corrections in $C_1(T)$ will give rise to an upturn in the low temperature limit). The temperature profile is summarized in Figure 2.

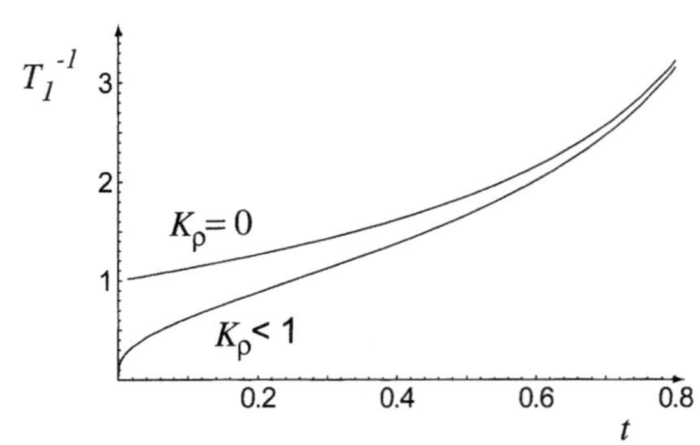

Fig. 2. Nuclear spin-lattice relaxation rate (expressed in arbitrary units) as a function of the reduced temperature $t = T/\omega_\sigma$, when the system scales to the Luttinger liquid ($1 > K_\rho > 0$) and strong coupling ($K_\rho = 0$) sectors

3 Instabilities of 1D Quantum Liquids: Interchain Coupling

Electronic materials like organic conductors can be only considered as close realizations of 1D interacting fermion systems. In the solid state, molecular stacks are not completely independent, that is to say *interchain coupling*, though small, must be taken into account in their description. Two different kinds of interchain coupling are generally considered. First, we have potential coupling like Coulomb interaction which introduces scattering of particles on different stacks. In certain conditions, potential coupling may give rise to long-range (density-wave) order at finite temperature. Second, there is the kinetic coupling, commonly denoted $t_\perp$, which allows an electron to hop from one stack to another. In the following, we shall confine ourselves to the latter coupling, which is the most studied and by far the most complex. In effect, it is from $t_\perp$ that most instabilities of 1D quantum liquids discussed thus far can occur. These are also called *crossovers* which form the links between the physics in one and higher dimensions as either the restoration of a Fermi liquid component or the onset of long-range order. In the following, we will review the physics of both types of crossover induced by $t_\perp$ using the renormalization group method that was described earlier.

3.1 One-Particle Dimensionality Crossover and Beyond

The Route Towards the Restoration of a Fermi Liquid. The overlap of molecular orbitals allowing electrons to hop from one chain to the next modifies the electron spectrum. In the tight-binding picture of a linear array of $N_\perp$ chains we have

$$E_p(\mathbf{k}) = \epsilon_p(k) - 2t_\perp \cos k_\perp , \tag{50}$$

where $\mathbf{k} = (k, k_\perp)$. Generalizing the functional-integral representation of the partition function in the presence of a non-zero but small $t_\perp \ll E_\mathrm{F}$, the propagator of the free part of the action is now $G_p^0(\mathbf{k}, \omega_n) = [i\omega_n - E_p(\mathbf{k})]^{-1}$. The non-interacting situation can trivially serve to illustrate how a dimensionality crossover of the quantum coherence in the single-particle motion is achieved as a function of temperature. We first note that since $t_\perp$ enters on the same footing as ω_n (or T) and ϵ_p (or k) in the bare propagator, it is a relevant perturbation; that is to say its importance grows according to $t'_\perp = st_\perp$ following the rescaling of energy $\epsilon'_p = s\epsilon_p$ and temperature $T' = sT$ by a factor $s > 1$. The temperature scale at which the crossover occurs can be readily obtained by equating s with the ratio of length scales ξ/a and by setting $t'_\perp \sim E_\mathrm{F}$ at the crossover. This condition of isotropy yields the crossover temperature $T_{\mathrm{x}^1} \sim t_\perp$, which is not a big surprise since $t_\perp$ acts as the only characteristic energy scale introduced in the interval between T and E_F.

In the presence of interactions, however, the flow of the enlarged parameter space $\mu_\mathrm{S} = (G_p^0(k, \omega_n), t_\perp, g_1, g_2, g_3)$ of the action, under the transformation (39) will modify this result. As shown in great detail elsewhere [32,34], the partial trace operation (38), when carried out beyond the one-loop level, not only alters the scattering amplitudes but also the single-particle propagator. At sufficiently high energy, the corresponding one-particle self-energy corrections keep in first approximation their 1D character and then modify the purely one-dimensional part of the propagator through the renormalization factor $z(\ell)$. Thus the effective bare propagator at step ℓ reads

$$G_p^0(\mathbf{k}, \omega_n, \mu_{\rm S}(\ell)) = \frac{z(\ell)}{i\omega_n - \epsilon_p(k) + 2z(\ell)t_\perp \cos k_\perp}. \tag{51}$$

Detailed calculations show that $z(\ell)$ obeys a distinct flow equation at the two-loop level which depends on the couplings constants (the generalization of Eqn. (40) at the two-loop level) [32]. Its integration up to ℓ_T leads to

$$z(T) \sim \left(\frac{T}{E_{\rm F}}\right)^\theta, \tag{52}$$

where the exponent $\theta > 0$ for non-zero interaction and is consistent with the one given by the bosonization method (cf. Eq. (21)) in lowest order. Being the residue at the single-particle pole of the 1D propagator, $z(T)$ coincides with the reduction factor of the density of states at the Fermi level (Eqn. (21)). The reduction of the density of states along the chains also modifies the amplitude of interchain hopping. The crossover criteria mentioned above now becomes $(\xi/a)z(T)t_\perp \sim E_{\rm F}$, which leads to the usual downward renormalization of the one-particle crossover temperature [38]:

$$T_{\rm x^1} \sim t_\perp \left(\frac{t_\perp}{E_{\rm F}}\right)^{(1-\theta)/\theta}. \tag{53}$$

According to this expression, $T_{\rm x^1}$ decreases when the interaction – which can be parametrized by θ – increases (Figure 3); it is non-zero as long as $\theta < 1$ for which $t_\perp$ remains a relevant variable. The system then undegoes a crossover to the formation of a Fermi liquid with quasi-particle weight $z(T_{\rm x^1})$. For strong coupling, $T_{\rm x^1}$ vanishes at the critical value $\theta_{\rm c} = 1$ and becomes undefined for $\theta > 1$, $t_\perp$ being then marginal in the former case and irrelevant in the latter (Figure 3). Consequently, in the latter case, no transverse band motion is possible and the single-particle coherence is spatially confined along the stacks. These large values of θ cannot be attained from the above perturbative renormalization group. They are found on the Luther-Emery line at half-filling or at quarter-filling in the presence of a charge gap [3,25,28]; in the gapless Tomanaga-Luttinger model for sufficiently strong coupling constants or when the range of interaction increases [7].

The above scaling approach to the deconfinement temperature is obviously not exact and corresponds to a random phase approximation with respect to the transverse one-electron motion [38–40]. This can be seen easily by just rewriting (51) in the RPA form

$$G_p^0(\mathbf{k}, \omega_n, \mu_{\rm S}(\ell)) = \frac{z(\ell)G_p^0(k, \omega_n)}{1 + z(\ell)G_p^0(k, \omega_n)\, 2t_\perp \cos k_\perp}, \tag{54}$$

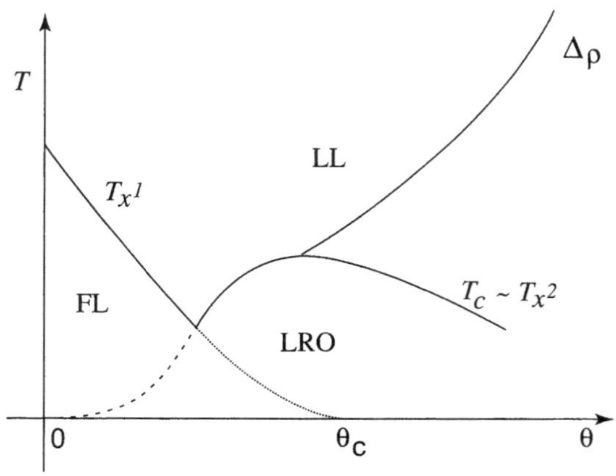

Fig. 3. Characteristic temperature scales of the quasi-one-dimensional electron gas model as a function of interaction, parametrized by the exponent θ. In the Fermi liquid sector, perfect nesting conditions prevail

where $z(\ell)G_p^0(k,\omega_n)$ is the 1D propagator at the step ℓ of the RG. Transverse RPA becomes essentially exact, however, in the limit of infinite range $t_\perp$ [39]. In regard to this approximation, it should be stressed that the above renormalization group treatment of deconfinement does not take into account the dynamics of spin-charge separation, namely the fact that the spin and charge excitations travel at different velocities (§ 2.1). It was inferred that this mismatch in the kinematics may suppress T_{x^1} and in turn the formation of a Fermi liquid [41]. As shown by Boies et al. [39] using a functional-integral method, the use of the Matsubara-Fourier transform of the complete expression (16) in the RPA allows one to overcome this flaw of the RG. However, the calculation shows that in the general case it still yields a finite crossover temperature

$$T_{x^1} \approx t_\perp \left(\frac{t_\perp}{E_F}\right)^{\theta/(1-\theta)} \left(\frac{v_F}{u_\rho}\right)^{\theta_\rho/(1-\theta)} \left(\frac{v_F}{u_\sigma}\right)^{\theta_\sigma/(1-\theta)} F[(u_\sigma/u_\rho)^{1/2}], \tag{55}$$

where $F(x)$ is a scaling function. Distinct velocities for spin and charge do give rise to additional corrections but the difference takes place at the quantitative level. Electronic deconfinement is therefore robust at this level of approximation and even beyond [40,42–46]. It is worth stressing that the renormalization of $t_\perp$ by intrachain interactions has been recently confirmed on numerical grounds for two-chain (ladder) fermion systems [47,48].

As we will see later, the above picture is modified significantly when the influence of $t_\perp$ on pair correlations is taken into account, especially when the amplitude of interactions increases and T_{x^1} becomes small. As regards the temperature interval over which one-particle crossover is achieved, it is not expected to be very narrow. In comparison with crossovers in ordinary critical phenomena, which are confined to the close vicinity of a phase transition [49], deconfinement of single-particle coherence in quasi-1D system is likely to be spread out over a sizeable temperature domain. This is so because the temperature interval

$\sim [T, E_{\mathrm{F}}]$ of the 1D quantum critical domain, which is linked to the primary Luttinger liquid fixed point, is extremely large. Recent calculations using dynamical mean-field theory seem to corroborate the existence of a sizeable temperature interval for the crossover [45].

Instability of the Fermi Liquid. Let us now turn our attention to the question of whether or not a Fermi liquid component in a quasi-one-dimensional metal remains stable well below T_{x^1}. Here we will neglect all the aforementionned transients to deconfinement and consider T_{x^1} as a sharp boundary between the Luttinger and Fermi liquids. By looking at the effective spectrum of the above model in which $t_\perp \to t_\perp^* = zt_\perp$ in (50), we observe that the whole spectrum obeys the relation $E_-^*(\mathbf{k}) = -E_+^*(\mathbf{k}+\mathbf{Q}_0)$, showing electron-hole symmetry or perfect nesting at $\mathbf{Q}_0 = (2k_{\mathrm{F}}, \pi)$. The response of the free quasi-1D electron gas is still logarithmicaly singular $\chi^0(\mathbf{Q}_0, T) \sim \ln T_{\mathrm{x}^1}/T$ in the Peierls channel for $\mathbf{Q}_0$ electron-hole excitations within the energy shell $\sim T_{\mathrm{x}^1}$ above and below the coherent *warped* Fermi surface. This singularity is also to be found in the perturbation theory of the scattering amplitudes, and can therefore lead to an instability of the Fermi liquid. For repulsive interactions, the most favorable instability is the one that yields a spin-density-wave state. The temperature at which the SDW instability occurs can be readily obtained by extending the renormalization group method of § 2.5 below T_{x^1} (or $\ell > \ell_{\mathrm{x}^1} = \ln E_{\mathrm{F}}/T_{\mathrm{x}^1}$). When perfect nesting prevails, a not too bad approximation consists of neglecting the interference between the Cooper and Peierls channels [50] (we shall revert to the problem of interference below T_{x^1} later in § 4.1). Thus by retaining the outer shell decomposition of $S_{I,2}$ in the latter channel only, one can write down a ladder flow equation

$$\frac{d\tilde{J}}{d\ell} = \frac{1}{2}\tilde{J}^2 + ... \tag{56}$$

for an effective coupling constant $\tilde{J} = \tilde{g}_2 + \tilde{g}_3 - \tilde{V}_\perp \ldots$ that defines the net attraction between an electron and a hole separated by $\mathbf{Q}_0$ (the origin of the exchange term $V_\perp$ will be discussed in § 3.2). This equation is integrated at once

$$\tilde{J}(T) = \frac{\tilde{J}^*}{1 - \frac{1}{2}\tilde{J}^* \ln T_{\mathrm{x}^1}/T}, \tag{57}$$

where $\tilde{J}^*$ is the effective SDW coupling at T_{x^1}, resulting from the integration of 1D many-body effects at $\ell < \ell_{\mathrm{x}^1}$. The above expression leads to a simple pole singularity at the temperature scale

$$T_{\mathrm{c}} = T_{\mathrm{x}^1} e^{-2/J^*}, \tag{58}$$

which corresponds to a BCS type of instability of the Fermi liquid towards a SDW state. As long as the nesting conditions are fulfilled, it invariably occurs for any non-zero interaction (dashed line of Fig. 3).

Nesting frustration is therefore required to suppress the transition. When nesting deviations are sufficiently strong, we will see, however, that the Fermi liquid is not stabilized after all. Actually, when interference between the Peierls and the Cooper channels is restored, the system turns out to become unstable to superconducting pairing, a mechanism akin to the Kohn-Luttinger mechanism for superconductivity in isotropic systems [51–54,66]. We shall return to this in § 4.1.

3.2 Two-Particle Dimensionality Crossover and Pair Deconfinement

When one examines the properties of the one-dimensional electron gas, one observes that the exponent γ of the pair response is not simply equal to twice the anomalous dimension θ of the single-particle Green function. Although both exponents depend on the Luttinger liquid parameter K_ρ, one-electron and pair correlations are governed by distinct power law decays. Thus the effect of an increase in the strength of interaction (K_ρ is decreasing for repulsive interactions) leads to a faster spatial decay of single-particle coherence (Eqn. (17)), whereas the opposite is true for triplet electron-hole pair (antiferromagnetic) correlations (Eq. (24)). The question now arises whether $t_\perp$ can promote interchain pair propagation besides single-particle coherence. Actually, this possibility exists and results from *interchain pair-hopping* processes [56–58], a mechanism that is not present in the Hamiltionian at the start but which emerges when interactions along the stacks combine with $t_\perp$ in the one-dimensional region. The renormalization group approach proved to be particularly useful in this respect making possible a unified description of both modes of propagation [32–34,54,56].

For repulsive interactions, the most important pair hopping contribution is the interchain exchange which favors antiferromagnetic ordering of neighboring chains. Roughly speaking, from each partial trace operation in (38), there is a "seed" $f(\ell)d\ell$ of interchain exchange that builds up as a result of combining perturbatively the effective hopping ($zt_\perp$) and the couplings ($g's$) in the shell of degrees of freedom to be integrated out. This can be seen as a new relevant interaction for the system, which in its turn is magnified by antiferromagnetic correlations. The net interchain exchange term generated by the flow of renormalization can be written as

$$S_\perp = -\frac{1}{4}\sum_{\langle i,j\rangle}\sum_{\widetilde{q}} V_\perp(\ell)\,\mathbf{O}_i(\widetilde{q})\cdot\mathbf{O}_j(\widetilde{q}), \tag{59}$$

which favors antiferromagnetic of spins on neighboring chains i and j. Going to transverse Fourier space, $V_\perp$ corresponds to the exchange amplitude at the ordering wave vector $\mathbf{Q}_0 = (2k_\mathrm{F}, \pi)$. In the one-dimensional regime, it is governed at the one-loop level by the distinct flow equation

$$\frac{d}{d\ell}\tilde{V}_\perp = f(\ell) + \tilde{V}_\perp\gamma(\ell) - \frac{1}{2}(\tilde{V}_\perp)^2, \tag{60}$$

where $\tilde{f}(\ell) \simeq -2[(\tilde{g}_2(\ell) + \tilde{g}_3(\ell))t_\perp/E_0]^2 e^{(2-2\theta(\ell))\ell}$. Here $\theta(\ell)$ and $\gamma(\ell)$ are the power law exponents of the one-particle propagator (Eqns.(21) and (52)) and antiferromagnetic response (Eqn. (25)) respectively (these are scale dependent due to the presence of Umklapp scattering). One observes from the right-hand-side of the above equation that the seed term resulting from the perpendicular delocalization of the electron and hole within the pair competes with the second term due to antiferromagnetic correlations along the chains. The outcome of this competition will be determined by the sign of $2 - 2\theta(\ell) - \gamma(\ell)$. Regarding the last term, it is responsible for a simple pole singularity of $J_\perp$ at a non-zero $\ell_c = \ln(E_F/T_c)$, signaling the onset of long-range order at T_c [59]. The temperature at which the change from the one-dimensional regime to the onset of transverse order occurs can be equated with a distinct dimensionality crossover denoted by $T_{x^2} \simeq 2T_c$ for pair correlations. The latter makes sense as long as $zt_\perp$ is still a perturbation, that is to say for $T_{x^2} > T_{x^1}$, which defines the region of validity of (60).

For repulsive interactions and in the presence of relevant Umklapp scattering, one can therefore distinguish two different situations. The first one corresponds to the presence of a charge gap well above the transition. As we have seen earlier, it defines a domain of ℓ where $\theta(\ell)$ is large and $\gamma(\ell) = 1$, that is $2 - 2\theta(\ell) - \gamma(\ell) < 0$. The physics of this strong coupling regime bears some resemblance to the problem of weakly coupled Heisenberg spin chains. However, in the Luther-Emery liquid case or at quarter-filling with a gap, each electron is not confined to a single site as in the Heisenberg limit but is delocalized over a finite distance $\xi_\rho \sim v_F/\Delta_\rho$, corresponding to the size of bound electron-hole pairs. A simple analysis of (60) shows that these pair effectively hop through an effective coupling $\tilde{J}_\perp \approx (\xi_\rho/\alpha_0)(t_\perp^{*2}/\Delta_\rho)$. When coupled to singular correlations along the chains, this leads to the antiferromagnetic transition temperature

$$T_c \approx \frac{t_\perp^{*2}}{\Delta_\rho} \sim T_{x^2}, \tag{61}$$

where $t_\perp^* = z(\Delta_\rho)t_\perp$ is the one-particle hopping at the energy scale of the charge gap.

A characteristic feature of strong coupling is the increase of T_c when the gap Δ_ρ decreases (Fig. 3). The above behavior of T_c continues up to the point where $T_{x^2} \sim \Delta_\rho$, namely when the insulating behavior resulting from the charge gap merges into the critical domain of the transition. θ and γ take smaller values in the normal metallic domain so that $2 - 2\theta(\ell) - \gamma(\ell)$ will first reach zero after which it will become positive corresponding to interchain pair-hopping in weak coupling. The growth of the seed term then surpasses the one due to pair vertex corrections in (60). An approximate expression of the transition temperature in this case is found to be

$$T_c \approx g^{*2} t_\perp^*, \tag{62}$$

where $g^* = g_2^* + g_3^*$ and $t_\perp^* = t_\perp z(T_c)$. Again this expression makes sense as long as $T_{x^2} > T_{x^1}$, which on the scale of interaction should not correspond to a wide interval. Still, it is finite and shows a decrease of T_c for decreasing interactions. This leads to a maximum of T_c at the boundary between strong and weak coupling domains (Fig. 3).

As soon as $T_{x^2} < T_{x^1}$, the single particle deconfinement occurs first at $zt_\perp \approx E_0(\ell)$ and interchain hopping can no longer be treated as a perturbation. This invalidates (60), and we have seen earlier that a Fermi liquid component forms

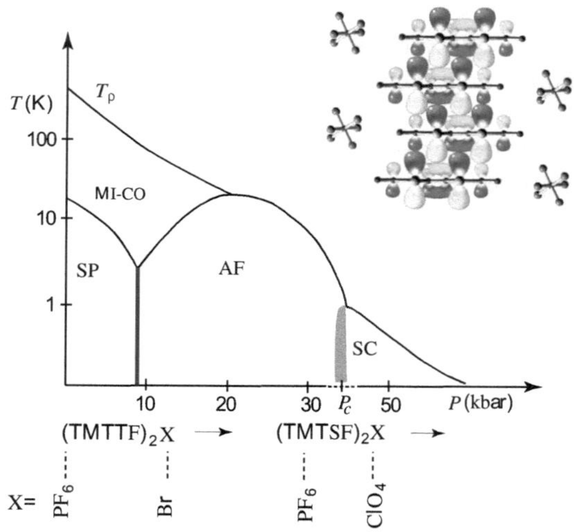

Fig. 4. Temperature-Pressure phase diagram of the Fabre ($(TMTTF)_2X$) and Bechgaard ($(TMTSF)_2X$) salts series. Inset: a side view of the of the crystal structure with the electronic orbitals of the stacks

under these conditions. An instability towards SDW is still possible under good nesting conditions for the Fermi surface. The exchange mechanism then smoothly evolves towards the condensation of electron-hole pairs from a Fermi liquid. In this regime, the residual pair-hopping amplitude $V_\perp(\ell_{x^1})$ contributes to the effective coupling J^* in (58). Figure 4 summarizes the various temperature scales characterizing the quasi-one-dimensional electron gas problem in the presence of Umklapp scattering.

4 Applications

4.1 The Fabre and Bechgaard Transfer Salt Series

The series of Fabre ($(TMTTF)_2X$) and Bechgaard ($(TMTSF)_2X$) transfer salts show striking unity when either hydrostatic or chemical pressure (S/Se atom or anion X=PF_6, AsF_6, Br, ..., substitutions) is applied. Electronic and structural properties naturally merge into the universal phase diagram depicted in Fig. 4 [33,60–63]. Its structure reveals a characteristic sequence of ground states enabling compounds of both series to be linked one to another [22,64–67,21,69–71,35,73]. In this way, Mott insulating sulfur compounds like $(TMTTF)_2PF_6$, AsF_6... were found to develop a charge-ordered (CO) state and a lattice distorted spin-Peierls (SP) state. The SP state is suppressed under moderate pressure and replaced by an antiferromagnteic (AF) Néel state similar to the one found in the $(TMTTF)_2Br$ salt in normal conditions; the Mott state is in turn suppressed under pressure and antiferromagnetism of sulfur compounds then acquires an itinerant character analogous to the spin-density wave (SDW) state of the $(TMTSF)_2X$ series at low pressure. Around some critical pressure P_c, the SDW state is then removed as the dominant ordering and forms a common boundary with organic superconductivity which closes the sequence of ordered states.

Within the bounds of this review, we shall not attempt a detailed discussion of the whole structure of the phase diagram but rather place a selected emphasis on the description of antiferromagnetic and superconducting orderings together with their respective normal phases. A detailed discussion of the spin-Peierls instability and charge ordering can be found elsewhere [19,33,73–77].

Electrical Transport and Susceptibility. A convenient way to broach the description of the phase diagram is to first examine the normal phase of the Fabre salts $(TMTTF)_2X$ for the inorganic monovalent anions X= PF_6 and Br. [78]. As shown in Fig. 5, there is a clear upturn in electrical resistivity at temperatures $T_\rho \approx 220$ K (PF_6) and $T_\rho \approx 100$ K (Br), which depicts a change from metallic to insulating behavior.[64] In both cases, T_ρ is a much higher temperature scale than the one connected to long-range order whose maximum is around 20 K. Below T_ρ, charge carriers become thermally activated. In a band picture of insulators, a thermally activated behavior should be present for spins too. For the compounds shown in Fig. 5, spin excitations are instead unaffected and remain gapless. This is shown by the regular temperature dependence of the spin susceptibility χ_σ at T_ρ (inset of Fig. 5). Resistivity data tell us that the gap in the charge is about $\Delta_\rho \approx 2 \ldots 2.5 T_\rho$, which exceeds the values of $t_{\perp b}$ given by band calculations. According to the discussion given in § 3.2, this would correspond to a situation of strong electronic confinement along the chains. Confinement is confirmed by the absence of a plasma edge in the reflectivity of both compounds when the electric field is oriented along the transverse b direction [70,38].

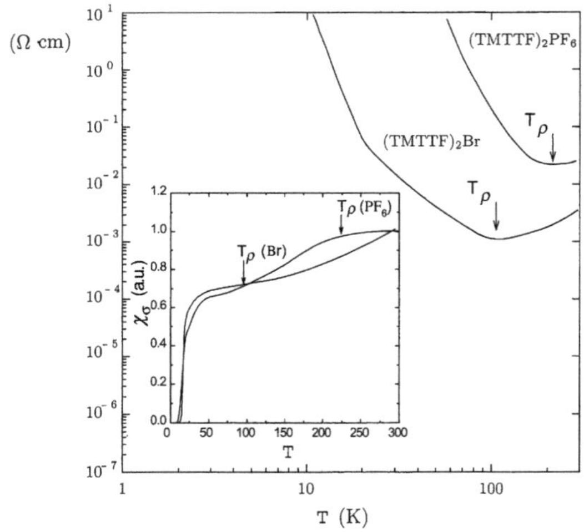

Fig. 5. Temperature dependence of resistivity and static spin susceptibility (inset) for the Fabre salts $(TMTTF)_2Br$ and $(TMTTF)_2PF_6$

The magnetic susceptibility, which decreases with the temperature, is also compatible with the one-dimensional prediction of Fig. 1. Wzietek et al. [22], analyzed in detail the NMR suceptibility data at constant volume using the expression (45). Very reasonable fits were obtained above 50 K provided that interactions $g_{1,2,4} \simeq 1$ are sizeable. In the charge sector, when the origin of the gap is attributed to half-filling umklapp alone, one has for a compound like $(TMTTF)_2PF_6$ the small bare value $\tilde{g}_3 \sim 0.1$; which is obtained from reasonable band parameters by matching the experimental T_ρ and the value ℓ_ρ at which $\tilde{g}_3(\ell_\rho) \sim 1$ in (40). [33,61] This is consistent with the fact that the stacks are weakly dimerized [26,60,80]. In the quarter-filling scenario, $g_{1/4}$ would be much larger using small K_ρ in (30).

Nuclear Magnetic Resonance. Among other measurable quantities that are sensitive to one-dimensional physics, we have the temperature dependence of the nuclear spin-lattice relaxation rate T_1^{-1} [22]. Consider for example the insulating compounds $(TMTTF)_2X$. According to the scaling pictures of § 2.4 and § 2.5, the charge stiffness $K_\rho = 0$ vanishes in the presence of a gap below T_ρ, or for $\ell > \ell_\rho$. The resulting behavior for the relaxation rate is then

$$T_1^{-1} \sim C_1 + C_0 T \chi_\sigma^2. \tag{63}$$

As shown in Fig. 6, this behavior is indeed found for $(TMTTF)_2PF_6$ salt when the relaxation rate data are combined with those of the spin susceptibility $(T\chi_\sigma^2)$ in the normal phase above 40 K, namely above the onset of spin-Peierls fluctuations [37,81]. A similar behavior is invariably found in all insulating materials down to low temperature where three-dimensional magnetic or lattice long-range order is stabilized [22,26,83]. Long-range order also prevents the observation of logarithmic corrections in C_1 which are expected to show up in the low temperature limit.

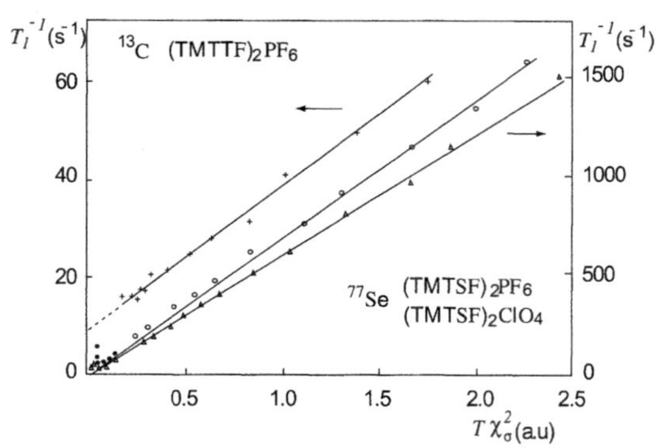

Fig. 6. Temperature dependence of the nuclear relaxation rate plotted as T_1^{-1} *vs* $T\chi_\sigma^2(T)$, where $\chi_\sigma(T)$ is the measured spin susceptibility. $(TMTTF)_2PF_6$ (crosses, left scale), $(TMTSF)_2PF_6$ (open circles, right scale) and $(TMTSF)_2ClO_4$ (open triangles, right scale). After Ref.[37]

If we now turn our attention to the effect of pressure, the phase diagram of Fig. 4 shows that hydrostatic pressure reduces T_ρ. At sufficiently high pressure, the insulating behavior merges with the critical behavior associated with the formation of a spin-density-wave state [21,23,85]. The normal phase is then entirely metallic. This change of behavior can also be achieved *via* chemical means. We have already seen the effect of chemical pressure within the Fabre salts series when for example the monovalent anion Br was put in place of PF_6 leading to a sizeable decrease of T_ρ (Fig. 5). When we substitute TMTSF for TMTTF, however, it leads to a larger shift of the pressure scale as exemplified by the normal phase of the Bechgaard salts $(TMTSF)_2X$ ($X=PF_6$, AsF_6, ClO_4, ...), which is metallic. Assuming that there is a temperature domain where a one-dimensional picture applies to mobile carriers, one must have $K_\rho > 0$. Therefore the contribution of uniform spin excitations to the relaxation rate becomes more important. For a compound like $(TMTSF)_2PF_6$, which develops a SDW state at $T_c \approx 12$ K, deviations to the $T\chi_\sigma^2$ law due to antiferromagetic correlations become visible below $T\chi_\sigma^2 \approx 1$ (≈ 200 K); whereas for a ambient pressure superconductor like $(TMTSF)_2ClO_4$ ($T_c \approx 1.2$ K), which is on the right of the PF_6 salt on the pressure scale, deviations show up at much lower temperature, $T \approx 30$ K or $T\chi_\sigma^2 \approx 0.1$ in Fig. 6. Attempts to square these non critical antiferromagnetic enhancements with the Luttinger liquid picture, however, show that $K_\rho \simeq 0.1$ [22,38]. Figure 6 serves to illustrate how the strength of antiferromagnetic correlations decrease as one moves from the left to the right side of the phase diagram.

DC Transport and Optical Conductivity. The metallic resistivity of the Fabre salts ($T > T_\rho$ in Fig. 5) and of the Bechgaard salts at high temperature has also been analyzed in the one-dimensional framework [33]. It was shown that Umklapp scattering at quarter-filling is the only mechanism of electronic relaxation that can yield metallic resistivity above T_ρ – half-filling Umklapp alone would lead at small K_ρ to an insulating behavior at all temperatures [86]. Following Giamarchi [33,87], the prediction at quarter-filling is $\rho(T) \sim T^{16K_\rho - 3}$, which can reasonably account for the constant volume metallic resistivity observed at high temperature. In the Bechgaard salts for example, an essentially linear temperature dependence $\rho(T) \sim T^\nu$ with $\nu \simeq 1$ is found down to 100 K, which would correspond to $K_\rho \simeq 0.25$ [78,38], a value not too far from NMR estimates for $(TMTSF)_2PF_6$. Below, a stronger power law sets in approaching a Fermi liquid behavior with $\nu \simeq 2$, which would indicate the onset of electronic deconfinement. Roughly similar conclusions, as to the value of the charge stiffness K_ρ and the onset of deconfinement in the Bechgaard salts, have been reached from the analysis of DC transverse resistivity measurements in the high temperature region [35,88]. These are also characterized by a marked change of behavior taking place between 50...100 K [89].

Optical conductivity measurements on members of both series have recently prompted a lot of interest in the extent to which a one-dimensional description applies to Fabre and Bechgaard salts [63,70,38,90]. As shown in Fig. 7, sulfur compounds show the absence of a Drude weight in the low frequency limit and

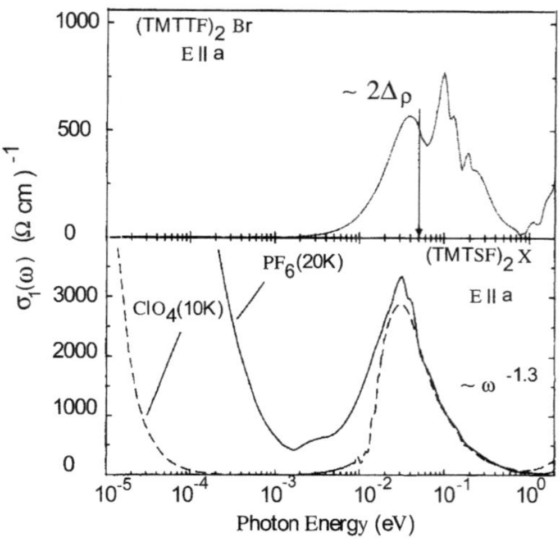

Fig. 7. Optical conductivity of the Fabre salt $(TMTTF)_2Br$ (top) and the Bechgaard salts $(TMTSF)_2PF_6$ and $(TMTSF)_2ClO_4$ After Ref. [70]

the infrared conductivity is entirely dominated by in optical gap of the charge sector as expected. It is noteworthy that the optical gap is closer to $2\Delta_\rho$ than Δ_ρ. Following recent work of Essler and Tsvelik [91], this is consistent with double solitonic excitations in the charge sector and thus a gap produced by quarter-filling Umklapp scattering.

The results for conductivity in the Bechgaard salts came as a surprise, however, since despite the pronounced metallic character of these systems [92], and the existence of a very narrow zero frequency mode, the charge gap still captures most of the spectral weight at high frequencies [70,38,93]. This behavior turns out to mimic that of a doped Mott insulator [33]. According to this picture, the high frequency tail of the conductivity above the gap behave as $\sigma_1(\omega) \sim \omega^{16K_\rho - 5}$ for quarter-filling Umklapp in 1D [87,94]. A power law $\omega^{-1.3}$ is observed for the Bechgaard salts over more than a decade in frequency (Fig. 7), which yields the value $K_\rho \simeq 0.23$ for the charge stiffness. This is consistent with the estimate made from DC resistivity. The purely one-dimensional prediction works well in the high frequency range presumably because the effect of interchain hopping is small there ($\omega > t_{\perp b}$). However, deviations from what is expected in a 1D doped Mott insulator are seen at lower frequencies and have been attributed to the influence of $t_{\perp b}$.

Photoemission Results. As mentionned earlier in Sect. 2.3, Angular Resolved Photoemission Spectroscopy (ARPES) experiments give in principle access to momentum and energy dependence of the one-particle spectral density $A(\boldsymbol{q}, \omega)$ for $\boldsymbol{q} < 0$ [95,96]. In practice, this is submitted to the experimental constraints of energy ($\Delta\omega$) and momentum ($\Delta\boldsymbol{q}$) resolutions, and to thermal broadening. The photoemission signal will then go like [96]

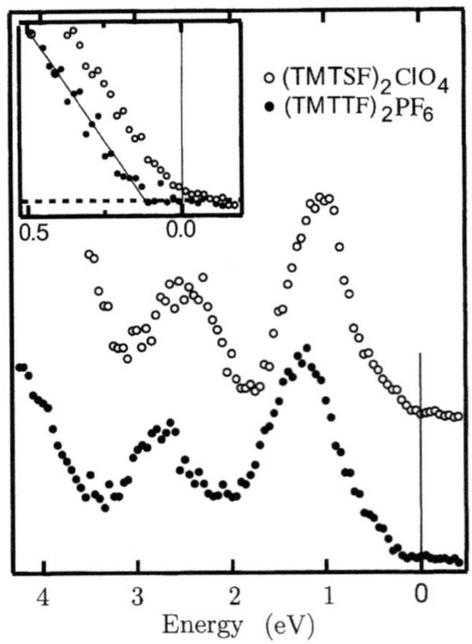

Fig. 8. ARPES spectra of $(TMTTF)_2PF_6$ and $(TMTSF)_2ClO_4$. The inset identifies an energy shift corresponding to $2\Delta_\rho$ for $(TMTTF)_2PF_6$. After Ref. [97]

$$I(\boldsymbol{q}, \omega, T) \sim \sum_{\Delta\omega'} f(\omega') \sum_{\Delta \boldsymbol{q}'} A(\boldsymbol{q}', \omega', T),$$

where f is the Fermi-Dirac distribution. ARPES measurements of Zwick et al. [97], for the Bechgaard salt $(TMTSF)_2ClO_4$ and the Fabre salt $(TMTTF)_2PF_6$ are shown in Fig. 8. The data are amazing in many respects. In $(TMTSF)_2ClO_4$ for example, which is a 1K superconductor (Fig. 4), the data reveal weak spectral intensity at the Fermi level and the absence of dispersing low-energy peaks associated to either spin or charge degrees of freedom. The exponent θ needed to describe the ($\sim$ linear) energy profile of I down to the non dispersing peak at $\sim$ 1 eV is rather large. Although the origin of these peculiar features is as yet not fully understood, it has been proposed that cleaving and radiation alteration of the surface may introduce imperfections and defects that may change the properties of the Luttinger liquid near the surface [95,98]. Defects can be seen as introducing finite segments of chains with open boundary conditions which correspond to *bounded Luttinger liquids* [98]. Their spectral weight is given by

$$A(\omega) \sim \mid \omega \mid^{(2K_\rho)^{-1} - 1/2},$$

which is $k-$independent, that is non dispersing but with an exponent that is still governed by the K_ρ of the bulk. Thus by taking $K_\rho \simeq 1/3$, which is actually not too far from the values of other experiments discused above, would lead to power law compatible with experimental findings.

The ARPES data for the Mott insulating compound $(TMTTF)_2PF_6$ reveal similar features apart a clear shift $\sim$ 100 mev of the onset towards negative energy. The shift seems to be close to $\sim 2\Delta_\rho$, as expected for quarter-filling Umklapp [46].

Dimensionality Crossovers and Long-Range Order. Our next task will be to give a brief description of the onset of long-range order in the light of ideas developed earlier in Sect. 3 for an array of weakly coupled 1D electron gas. It was pointed out that the nature of the 1D electronic state strongly influences the way the system undergoes a crossover to a higher dimensional behavior and how this may yield to long-range order. We have seen for example that in the presence of a 1D Mott insulating state, the charge gap gives rise to strong coupling conditions that prevent electronic deconfinement. Spins can still order antiferromagnetically in the transverse direction *via* the interchain exchange $V_\perp$. The case of $(TMTTF)_2Br$ compound is particularly interesting in this respect since this mechanism can be studied in a relatively narrow pressure interval where the change from strong to weak coupling actually takes place (Fig. 4).

The data of Fig. 9 obtained by Klemme et al. [69], gives the detailed pressure dependence of both T_c and T_ρ for the bromine salt up to 13 kbar. When the insulating behavior at T_ρ meets the critical domain under pressure a maximum of T_c is clearly seen. This accords well with the description of the transition given in Sect. 3.2 in terms of weakly coupled antiferromagnetic chains (right side of Fig. 3). Here pressure mainly contracts the lattice, modifying upward longitudinal bandwidth while decreasing the stack dimerization. These are consistent with the decrease of correlations along the chains whose strength has been parametrized by θ in Fig. 3. Pressure increases $t_\perp$ too – roughly at the same rate as $t_\parallel$ [99]– which contributes to the variation of θ under pressure [45].

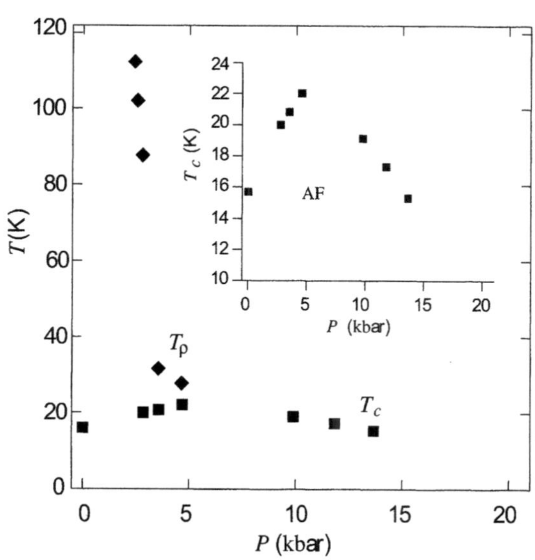

Fig. 9. The pressure profile of T_ρ (full diamonds) and the antiferromagnetic critical temperature (full squares) for $(TMTTF)_2Br$. After Ref. [69]

Similar profiles of (T_c, T_ρ) under pressure have been confirmed in other members of the Fabre salt series [23,22,101,102].

Passed the maximum, the interchain exchange is gradually replaced by an instability of the Fermi surface to form a SDW state. We have emphasized in Sect. 3.2, however, that this is closely related to the issue of where the onset of electronic deconfinement takes place in the metallic state. In the general phase diagram of Fig. 4, this fundamental issue applies to the region where the Fabre salt series overlaps with the Bechgaard salt series. In the simple picture given in Sect. 3.1, which is portrayed in Fig. 3, the scale T_{x^1} for deconfinement and restoration of a Fermi liquid component rises up in the small coupling region. On an empirical basis, however, a clear indication of such a scale is still missing so far and the figures proposed have led to conflicting views. If one agrees for example on the small value of $K_\rho \simeq 0.2$ given by optics and transport for the Bechgaard salts, the expected $T_{x^1} \sim 10$ K would be rather small. Although this would be consistent with earlier interpretation of NMR results [22,38], it contradicts others which favour a Fermi liquid description: the emergence of a T^2 law for parallel resistivity below 100 K, the gradual onset of transverse plasma edge in the same temperature range [70,103], the $t_\perp$ values extracted from angular dependence of magnetoresitance a very low temperature [104], the observation of Wiedeman-Franz Law at low temperature [105], to mention only a few (a more detailed discussion can be found in Refs. [63,78,38]). At present, it is not clear to what extent a synthesis of these conflicting figures will require a radical change of approach in setting out the deconfinement problem or if it simply reflects the fact that deconfinement takes place over a large temperature interval [45].

Before closing this discussion, we will briefly examine the mechanism commonly held responsible for the suppression of the SDW state as the critical pressure P_c is approached from below in the phase diagram (Fig. 4). By looking more closely at the effect of pressure on electronic band structure, we realize that corrections to the spectrum such as the longitudinal curvature of the band or transverse hopping $t_{\perp 2}$ to second nearest-neighbor chains magnify under pressure; their influence can no longer be neglected in the description of the SDW instability of the normal state. In effect, in the presence of an effective $t^*_{\perp 2}$, the spectrum below T_{x^1} becomes

$$E_p(\mathbf{k}) = \epsilon_p(k) - 2t^*_\perp \cos k_\perp - 2t^*_{\perp 2} \cos 2k_\perp .$$

This leads to $E_+(\mathbf{k}+\mathbf{Q}_0) = -E_-(\mathbf{k}) + 4t^*_{\perp 2} \cos 2k_\perp$, and thus to the alteration of nesting conditions of the whole Fermi surface. These deviations will cut off the infrared singularity of the Peierls channel, which becomes rhoughly $\chi^0(\mathbf{Q}_0, T) \sim N(0) \ln(\sqrt{T^2+\delta^2}/T_{x^1})$, where $\delta \sim t^*_{\perp 2}$. Its substitution in the ladder expression (57) leads to a T_c that rapidly goes down when $t^*_{\perp 2} \sim T_c$ (see Fig. 10 for the results of a detailed calculation), in qualitative agreement with experimental findings (Fig. 4).

The strongest support for the relevance of nesting frustration in this part of the phase diagram is provided by the analysis of the cascade of SDW phases

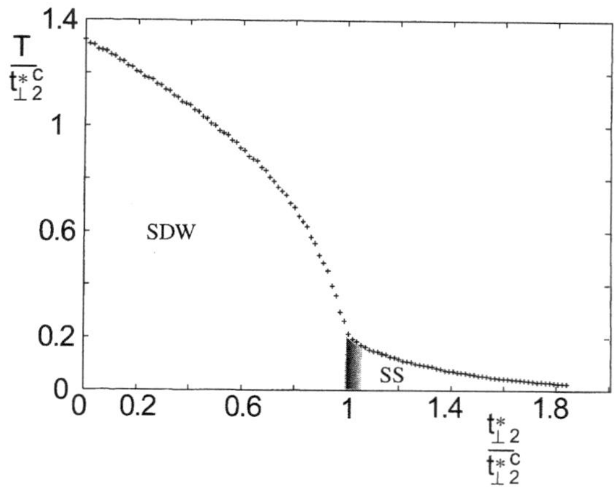

Fig. 10. Variation of the critical temperature as a function of nesting deviations $t^*_{\perp 2}$ ($\sim$ pressure). The shaded area corresponds to the crossover region between SDW and superconductivity. After Ref. [66]

at $P > P_c$, which are observed when a magnetic field oriented along the less conducting direction is cranked up beyond some threshold [26,106]. In effect, the magnetic field confines the electronic motion in the transverse direction and thus restores the infrared singularity of the Peierls channel at some discrete (quantized) values of the nesting vector, each of which characterizing a SDW phase of the cascade [107]. Besides the indisputable success of a weak coupling (ladder) description of field-induced SDW – which constitutes a whole chapter of the physics of these materials [108]– some features of the normal phase under field such as the magnetoresistance and NMR refuse to bow to a simple Fermi liquid description [109].

On the Nature of Superconductivity. Let us now turn our attention to the superconductivity that is found near P_c for both series, namely at the right hand end of the phase diagram in Fig. 4. The symmetry – singlet or triplet – of the superconducting order parameter in these charge-transfer salts is an open question that is currently much debated [110–112]. Here we will tackle this problem from a theoretical standpoint that is in line with what has been previously discussed. We shall give a cursory glance at recent progress made on the origin of organic superconductivity, namely as to whether it could be driven by electronic correlations. In this matter, it is noteworthy that superconductivity shares a common boundary and even overlaps with the SDW state [65,23,85,101]. This close proximity between the two ground states, which is an universal feature of both series of compounds, is peculiar in that it is the electrons of a single band that partake in both types of long-range order. Moreover, SDW correlations are well known to permeate deeply the normal phase above the superconducting phase [22]. All this goes to show that pairings between electrons and holes responsible for antiferromagnetism and superconductivity are not entirely exclusive and that both phenomena may have a common – electronic – origin [52–54,66,61].

We have become familiar with the mixing of Cooper and Peierls pairings in the context of a Luttinger liquid (see Sect. 2.5). The interference between the two is maximum in strictly one dimension where the Fermi surface reduces to two points. Below the scale T_{x^1}, interference was neglected in the ladder description of the SDW instability (see Sect. 3.1). However, although interference is weakened by the presence of a coherent wrapping of the Fermi surface, it still exerts an influence below T_{x^1} by becoming non uniform, that is **k**-dependent along the Fermi surface. As shown recently by Duprat et al. [66], non uniform interference can be taken into account using the renormalization group method of Sect. 2.5. This technique allows us to write down a two-variable flow equation for the SDW coupling constant

$$\begin{aligned}\frac{d\tilde{J}(k_\perp,k'_\perp)}{d\ell} =& -\frac{1}{N_\perp}\sum_{k''_\perp}\tilde{J}(k_\perp,k''_\perp)I_{\mathrm{C}}(k''_\perp,k'_\perp,\ell)\tilde{J}(k''_\perp,k'_\perp)\\ &+\frac{1}{N_\perp}\sum_{k''_\perp}\tilde{J}(k_\perp,k''_\perp)I_{\mathrm{P}}(k''_\perp,k'_\perp,t^*_{\perp 2},\ell),\tilde{J}(k''_\perp,k'_\perp)\end{aligned} \tag{64}$$

where $I_{\mathrm{C,P}}$ are the Cooper loop and the Peierls one in the presence of nesting deviations. The pair of variables $(k_\perp,k'_\perp)$ refers to transverse momenta of ingoing and outgoing particles that participate in electron-hole and electron-electron pairings close to the Fermi surface.

At small $t^*_{\perp 2}$, there is a simple pole singularity in J at $k'_\perp = k_\perp - \pi$, which signals the expected SDW instability at $\ell_{\mathrm{c}} = \ln T_{\mathrm{c}}/T_{x^1}$ and wavevector $\mathbf{Q}_0$ (Fig. 10). As far as T_{c} is concerned, this result is qualitatively similar to the ladder approximation of Sect. 3.1. When nesting deviations increase, however, T_{c} decreases rapidly and shows an inflection point at a critical $t^*_{\perp 2,\mathrm{c}}$ instead of reaching zero as for the single channel approximation (Fig. 10). The singular structure of J in $k_\perp, k'_\perp$ space then qualitatively changes, becoming modulated by a product of simple harmonics $\cos k_\perp \cos k'_\perp$. This indicates that singular pairing is now present in the singlet Cooper channel. The attraction takes place between electrons on neighboring stacks as a result of their coupling to spin fluctuations. In the framework of the present model, the singlet superconducting gap $\Delta(k_\perp) = \Delta_0 \cos k_\perp$ presents nodes at $k_\perp = \pm\pi/2$. When typical figures for deviations to perfect nesting are used, that is $t^*_{\perp 2,\mathrm{c}} \sim 10$ K $\sim T_{\mathrm{c}}(t^*_{\perp 2} \approx 0)$, the one-loop calculations is able to reproduce an important feature of the phase diagram which is the profiles of SDW and superconductivity in both series of compounds near P_{c} (Fig. 4).

4.2 The Special Case of TTF[Ni(dmit)$_2$]$_2$

Among the very few quasi-one-dimensional organic materials that do not show long-range ordering, the two-chain compound TTF[Ni(dmit)$_2$]$_2$ is interesting [113,114]. The TTF stacks remain metallic down to the lowest temperature reached for this system. Although the reason for this lack of long-range order is not well understood, band calculations revealed a very pronounced quasi-1D anisotropy of the electronic structure [115], actually stronger than the one of the Fabre and the Bechgaard salts. The band filling is not known precisely but it is incommensurate with the underlying lattice. We have here favorable conditions for the emergence of Luttinger liquid physics.

In this respect, the results of Wzietek et al. [114], given in Fig. 11 for the temperature variation of the proton (^{1}H) NMR T_1^{-1} of the TTF chains, are particularly revealing. When the data of Fig. 11 are compared with the characteristic LL shape of Fig. 2 at $K_\rho > 0$, the connection with the one-dimenional theory is striking. This is confirmed at the quantitative level by the fit (continuous line in Fig. 11) of data using an expression of the form (48), where the interaction parameter $K_\rho \simeq 0.3$ have been used. The plot of $(T_1T)^{-1}$ *vs* T (left scale and inset) allows one to isolate the enhancement at low temperature due to the 1D antiferromagnetic response.

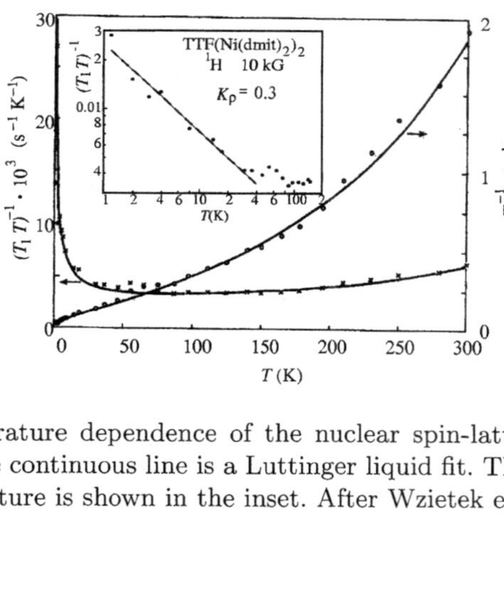

Fig. 11. The temperature dependence of the nuclear spin-lattice relaxation rate of TTF[Ni(dmit)$_2$]. The continuous line is a Luttinger liquid fit. The power law enhancement at low temperature is shown in the inset. After Wzietek et al. [114]

4.3 An Incursion into Inorganics: The Purple Bronze $Li_{0.9}Mo_6O_{17}$

The objective of finding Luttinger liquid behavior in crystals does not focus uniquely on organic conductors but constitutes an important line of research in other materials too. This is the case of molybendum bronzes which form a class of low-dimensional inorganic systems well known for their strongly anisotropic electronic structure and the rich phenomenology associated with the formation of charge-density-wave order [116]. Here we will briefly consider the $Li_{0.9}Mo_6O_{17}$ compound, which stands out as a special case of the so-called "purple bronze" series. This compound consists of molecular MoO_6 octahedra and MoO_4 tetrahedra arranged in a 3D network for which strong Mo-O-Mo interactions form zig-zag chains along a preferential direction. This strong one-dimensional character is confirmed by band calculations [117] and by experimental Fermi surface

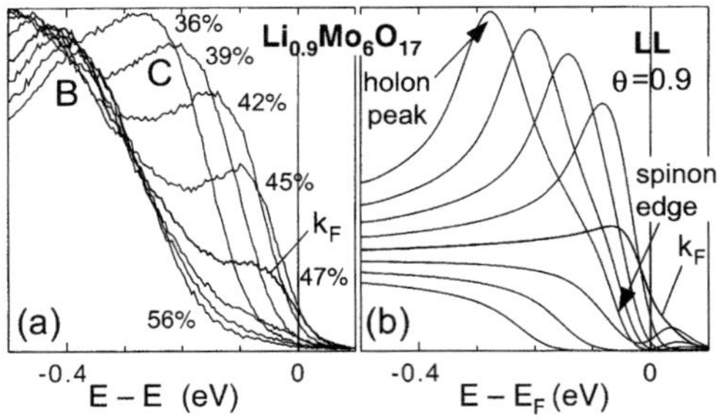

Fig. 12. (**a**) ARPES data at 250 K for $Li_{0.9}Mo_6O_{17}$ at different % of the Brillouin zone in the Γ–Y direction; (**b**) Luttinger liquid prediction at $\theta = 0.9$, and $u_\sigma = u_\rho/5$, including experimental momentum and energy resolutions, and thermal (Fermi-Dirac) broadening. After Ref. [119]

mapping [118], which both yield a rather flat Fermi surface arising from the two degenerate bands that cross the Fermi level. The metallic temperature domain is rather wide, extending down to $T_c \approx 24$ K where a phase transition occurs. Although the origin of the latter is as yet not well understood, it is not a CDW state and the normal phase does not show any sign of CDW precursors. Therefore electronic interactions dominate and this renders this material particularly appealing for ARPES studies [118].

Figure 12 shows high resolution ARPES data obtained by Gweon et al. [119], on $Li_{0.9}Mo_6O_{17}$ at $T = 250$ K. As one moves along the $\Gamma - Y$ direction in the Brillouin zone, the lineshape of the C band shifts towards the Fermi edge for increasing $\mathbf{k} < \mathbf{k}_F$ with a peak that decreases sharply in intensity and broadens significantly before reaching $\mathbf{k}_F$ to finally pull back for $\mathbf{k} > \mathbf{k}_F$ and merges in the tail of a weakly dispersing band (B) at high energy. The absence of true crossing of the band, its lack of sharpening as $\mathbf{k} \to \mathbf{k}_F$ and the low spectral weight left at $\mathbf{k}_F$ contrast with what is found in the pototypical Fermi liquid like compound $TiTe_2$ [96,120,121]. The analysis of the dispersing lineshape C in the framework of the LL theory proved to be much more satisfactory [96,118,119,122]. A large anomalous exponent $\theta = 0.9$ and differing values for the velocities u_σ and u_ρ are required to reasonably account for the data. According to the LL theory of the spectral weight [13], a large θ leads to edges of holon and spinon excitations that crosses the Fermi edge with different amplitudes as shown in Fig. 12-b. The LL theory used in Fig. 13-b is for temperature $T = 0$ and has $u_\sigma/u_\rho = 1/5$ as in the analysis of Ref. [118]. Subsequent analysis [122] of the data of Fig. 13-a using a finite temperature theory [15,123] leads to a value of $u_\sigma/u_\rho = 1/2$. A large value of θ for a system like $Li_{0.9}Mo_6O_{17}$ with incommensurate band-filling would indicate that long-range Coulomb interactions play an important role in such a system. According to the results of Sect. 3.1, a large θ will also yield strong electronic confinement along the chains. The impact on other physical properties should also be observable as for example in the poor conductivity and the absence of a plasma edge in the transverse directions [124].

5 Conclusion

A large part of the phenomenology shown by quasi-one-dimensional conductors cannot be understood in the traditional framework of solid state physics. Probably for no other crystals do we have to reckon with concepts provided by the now well understood physics of interacting electrons in one dimension. The existence of long-range order at finite temperature in many of these systems indicates that the link between one and higher-dimensional physics is also essential to their understanding.

As we have seen in this review some progress has been achieved in that direction but there are also several basic questions left to answer. Among them, let us mention how Fermi liquid quasi-particles are appearing in the normal phase for systems like the Bechgaard salts (or their sulfur analogs at high pressure) ? A clarification of this issue would certainly represent a significant advance in the comprehension of these fascinating low-dimensional solids.

Acknowledgements

The author thanks D. Jérome, L.G. Caron, R. Duprat, N. Dupuis, J. P. Pouget, A.-M. Tremblay and R. Wortis for numerous discussions on several aspects of this work; J. W. Allen and G.-H. Gweon for several remarks and correspondance about ARPES results on molybdenum bronzes. This work is supported by the Natural Sciences and Engineering Research Council of Canada (NSERC), le Fonds pour la Formation de Chercheurs et l'Aide à la Recherche du Gouvernement du Québec (FCAR), the "superconductivity program" of the Institut Canadien de Recherches Avancées (CIAR).

References

1. J. Voit: Rep. Prog. Phys. **58**, 977 (1995)
2. H.J. Schulz: in *Strongly Correlated Electronic Materials: The Los Alamos Symposium 1993*, ed. K.S. Bedell et al. (Addison-Wesley, Reading, MA, 1994), p.187
3. V.J. Emery: in *Highly Conducting One-Dimensional Solids*, eds. J.T. Devreese, R.E. Evrard, V.E. van Doren (Plenum Press, New York, 1979), p.247
4. J.M. Luttinger: J. Math. Phys. **4**, 1154 (1963)
5. D.C. Mattis, E.H. Lieb: J. Math. Phys. **6**, 304 (1965)
6. S. Tomonaga: Prog. Theor. Phys. **5**, 544 (1950)
7. H.J. Schulz: J. Phys. C: Solid State Phys. **16**, 6769 (1983)
8. F.D.M. Haldane: J. Phys. C **14**, 2585 (1981)
9. D.C. Mattis: J. Math. Phys. **15**, 609 (1974)
10. A. Luther, I. Peschel: Phys. Rev. B **9**, 2911 (1974)
11. F.D.M. Haldane: Phys. Rev. Lett. **45**, 1358 (1980)
12. J. Voit: Phys. Rev. B **47**, 6740 (1993)
13. V. Meden, K. Schonhammer: Phys. Rev. B **46**, 15 753 (1992)
14. N. Nakamura, Y. Suzumura: Prog. Theor. Phys. **98**, 29 (1997)
15. D. Orgad et al.: Phys. Rev. Lett. **86**, 4362 (2001)
16. Y. Suzumura: Prog. Theor. Phys. **63**, 51 (1980)

17. H. Nélisse et al.: Eur. Phys. J. B. **12**, 351 (1999)
18. W. Metzner, C. Di Castro: Phys. Rev. B **47**, 16107 (1993)
19. J.P. Pouget, S. Ravy: J. Phys. I (France) **6**, 1501 (1996)
20. J.P. Pouget: in *Physics and Chemistry of Low-Dimensional Inorganic Compounds*, eds. M.G.C. Schlenker, J. Dumas, S.V. Smaalen (Plenum Press, New York, 1996), p.185
21. C. Bourbonnais: J. Phys. I (France) **3**, 143 (1993)
22. P. Wzietek et al.: J. Phys. I (France) **3**, 171 (1993)
23. A. Luther, V.J. Emery: Phys. Rev. Lett. **33**, 589 (1974)
24. E. Lieb, F.Y. Wu: Phys. Rev. Lett. **20**, 1445 (1968)
25. F. Mila, X. Zotos: Europhys. Lett. **24**, 133 (1993)
26. S. Barisic, S. Brazovskii: in *Recent Developments in Condensed Matter Physics*, ed. J.T. Devreese (Plenum, New York, 1981), Vol. 1, p.327
27. T. Giamarchi: Physica B **230-232**, 975 (1997)
28. J. Voit: Eur. Phys. J. B **5**, 505 (1998)
29. E.W. Carlson, D. Orgad, S.A. Kivelson, V.J. Emery: Phys. Rev. B **62**, 3422 (2000)
30. Y.A. Bychkov, L.P. Gorkov, I. Dzyaloshinskii: Sov. Phys. JETP **23**, 489 (1966)
31. J. Solyom: Adv. Phys. **28**, 201 (1979)
32. C. Bourbonnais, L.G. Caron: Int. J. Mod. Phys. B **5**, 1033 (1991)
33. C. Bourbonnais: in *Les Houches, Session LVI (1991), Strongly interacting fermions and high-T_c superconductivity*, eds. B. Doucot, J. Zinn-Justin (Elsevier Science, Amsterdam, 1995), p.307
34. C. Bourbonnais, B. Guay, R. Wortis: in *Theoretical methods for strongly correlated electrons*, eds. A.M. Tremblay, D. Sénéchal, A. Ruckenstein, C. Bourbonnais (Springer, Heidelberg, 2002)
35. I.E. Dzyaloshinskii, A.I. Larkin: Sov. Phys. JETP **34**, 422 (1972)
36. S. Eggert, I. Affleck, M. Takahashi: Phys. Rev. Lett. **73**, 332 (1994)
37. C. Bourbonnais et al.: Phys. Rev. Lett. **62**, 1532 (1989)
38. C. Bourbonnais et al.: J. Phys. (Paris) Lett. **45**, L755 (1984)
39. D. Boies, C. Bourbonnais, A.-M. Tremblay: Phys. Rev. Lett. **74**, 968 (1995)
40. E. Arrigoni: Phys. Rev. Lett. **80**, 790 (1998)
41. P.W. Anderson: Phys. Rev. Lett. **67**, 3844 (1991)
42. L.G. Caron, C. Bourbonnais: Synt. Met. **27A**, 67 (1988)
43. H.J. Schulz: Int. J. Mod. Phys. B **5**, 57 (1991)
44. C. Castellani, C. Di Castro, W. Metzner: Phys. Rev. Lett. **72**, 316 (1994)
45. S. Biermann, A. Georges, A. Lichtenstein, T. Giamarchi: Phys. Rev. Lett. **87**, 276405 (2001) arXiv:cond-mat/0107633.
46. F.H.L. Essler, A.M. Tsvelik: arXiv:cond-mat/0108382 (unpublished)
47. S. Capponi, D. Poilblanc, E. Arrigoni: Phys. Rev. B **57**, 6360 (1998)
48. L. Caron, C. Bourbonnais: arXiv:cond-mat/0112071 (unpublished)
49. D. Nelson: Phys. Rev. B **11**, 3504 (1975)
50. V.N. Prigodin, Y.A. Firsov: Sov. Phys. JETP **49**, 369 (1979)
51. W. Kohn, J.M. Luttinger: Phys. Rev. Lett. **15**, 524 (1965)
52. V.J. Emery: Synt. Met. **13**, 21 (1986)
53. M.T. Béal-Monod, C. Bourbonnais, V.J. Emery: Phys. Rev. B **34**, 7716 (1986)
54. C. Bourbonnais, L.G. Caron: Europhys. Lett. **5**, 209 (1988)
55. R. Duprat, C. Bourbonnais: Eur. Phys. J. B **21**, 219 (2001)
56. C. Bourbonnais, L.G. Caron: Physica B **143**, 450 (1986)
57. S. Brazovskii, Y. Yakovenko: J. Phys. (Paris) Lett. **46**, L111 (1985)
58. Y.A. Firsov, Y.N. Prigodin, C. Seidel: Phys. Rep. **126**, 245 (1985)

59. Strickly speaking, long-range order at finite temperature is impossible in two dimensions for an order parameter with more than one component. However, a finite coupling in the third direction is present which enables to obtain a finite T_c
60. V.J. Emery, R. Bruisma, S. Barisic: Phys. Rev. Lett. **48**, 1039 (1982)
61. L.G. Caron, C. Bourbonnais: Physica B **143**, 453 (1986)
62. S. Brazovskii, Y. Yakovenko: Sov. Phys. JETP **62**, 1340 (1985)
63. C. Bourbonnais, D. Jérome: Science **281**, 1156 (1998)
64. C. Coulon et al.: J. Phys. (Paris) **43**, 1059 (1982)
65. R. Brusetti, M. Ribault, D. Jerome, K.Bechgaard: J. Phys. (Paris) **43**, 801 (1982)
66. J. Pouget et al.: Mol. Cryst. Liq. Cryst. **79**, 129 (1982)
67. F. Creuzet, S.P. Parkin, D. Jérome, J. Fabre: J. Phys. (Paris) Coll. **44**, 1099 (1983)
68. L. Balicas et al.: J. Phys. I (France) **4**, 1539 (1994)
69. B.J. Klemme et al.: Phys. Rev. Lett. **75**, 2408 (1995)
70. V. Vescoli et al.: Science **281**, 1181 (1998)
71. D. Chow et al.: Phys. Rev. Lett. **81**, 3984 (1998)
72. J. Moser et al.: Eur. Phys. J. B **1**, 39 (1998)
73. D.S. Chow et al.: Phys. rev. Lett. **85**, 1698 (2000)
74. L.G. Caron, C. Bourbonnais, F. Creuzet, D. Jerome: Synt. Met. **19**, 69 (1987)
75. C. Bourbonnais, B. Dumoulin: J. Phys. I (France) **6**, 1727 (1996)
76. J. Riera, D. Poilblanc: Phys. Rev. B **62**, R16243 (2000) arXiv:cond-mat/0006460
77. P. Monceau, F. Nad, S. Brazovskii: Phys. Rev. Lett. **86**, 4080 (2001)
78. C. Bourbonnais, D. Jérome: in *Advances in Synthetic Metals, Twenty Years of Progress in Science and Technology*, eds. P. Bernier, S. Lefrant, G. Bidan (Elsevier, New York, 1999), pp.206–261, arXiv:cond-mat/9903101
79. A. Schwartz et al.: Phys. Rev. B **58**, 1261 (1998)
80. K. Penc, F. Mila: Phys. Rev. B **50**, 11 429 (1994)
81. F. Creuzet et al.: Synt. Met. **19**, 289 (1987)
82. D. Jérome: in *Organic Conductors: fundamentals and applications*, ed. J.-P. Farges (Dekker, New York, 1994), pp.405–494
83. B. Gotschy et al.: J. Phys. I (France) **2**, 677 (1992)
84. H. Wilhelm et al.: Eur. Phys. J. B **21**, 175 (2001)
85. D. Jérome, A. Mazaud, M. Ribault, K. Bechgaard: J. Phys. (Paris) Lett. **41**, L95 (1980)
86. L.P. Gor'kov, I.E. Dzyaloshinskii: JETP Lett. **18**, 401 (1973)
87. T. Giamarchi, A.J. Millis: Phys. Rev. B **46**, 9325 (1992)
88. P. Fertey, M. Poirier, P. Batail: Eur. Phys. J. B **10**, 305 (1999)
89. J. Moser: (unpublished)
90. M. Dressel et al.: Phys. Rev. Lett. **77**, 398 (1996)
91. F.H.L. Essler, A.M. Tsvelik: arXiv:cond-mat/0105582 (unpublished)
92. K. Bechgaard et al.: Solid State Comm. **33**, 1119 (1980)
93. N. Cao, T. Timusk, K. Bechgaard: J. Phys. I (France) **6**, 1719 (1996)
94. D. Controzzi, F.H.L. Essler, A.M. Tsvelik: Phys. Rev. Lett. **86**, 680 (2001)
95. M. Grioni, J. Voit: in *Electron spectroscopies applied to low-dimensional materials*, ed. H. Stanberg, H. Hughes (Kluwer Academic Publ., Netherlands, 2000) p.501
96. G.-H. Gweon et al.: J. Elec. Spectro. Rel. Phenom. **117-118**, 481 (2001) arXiv:cond-mat/0103470
97. F. Zwick et al.: Phys. Rev. Lett. **79**, 3982 (1997)
98. J. Voit, Y. Wang, M. Grioni: Phys. Rev. B **61**, 7930 (2000)
99. D. Jérome, H. Schulz: Adv. Phys. **31**, 299 (1982)

100. D. Jaccard et al.: J.Phys.: Condens. Matter **13**, 89 (2001)
101. T. Adachi et al.: J. Am. Chem. Soc. **122**, 3238 (2000)
102. P. Aubaun-Senzier et al.: J. Phys. (Paris) **50**, 2727 (1989)
103. C.S. Jacobsen, D. Tanner, K. Bechgaard: Phys. Rev. Lett. **46**, 1142 (1981)
104. G.M. Danner, W. Kang, P.M. Chaikin: Phys. Rev. Lett. **72**, 3714 (1994)
105. S. Belin, K. Behnia: Phys. Rev. Lett. **79**, 2125 (1997)
106. P. Chaikin: J. Phys. I (France) **6**, 1875 (1996)
107. L. P. Gorkov, A. G. Lebed: J. Phys. (Paris) Lett. **45**, L433 (1984) M. Héritier, G. Montambaux, P. Lederer: J. Phys. (Paris) Lett. **45**, L943 (1984) K. Yamaji: Synt. Met. **13**, 19 (1986) L. Chen, K. Maki, V. Virosztek: Physica B **143**, 444 (1986)
108. T. Ishiguro, K. Yamaji: *Organic Superconductors*, Vol. 88 of *Springer-Verlag Series in Solid-State Science* (Springer-Verlag, Berlin, Heidelberg, 1990)
109. K. Behnia et al.: Phys. Rev. Lett. **74**, 5272 (1995)
110. I.J. Lee, M.J. Naughton, G.M. Danner, P.M. Chaikin: Phys. Rev. Lett. **78**, 3555 (1997)
111. I.J. Lee et al.: Phys. Rev. Lett. **88**, 17004 (2002)
112. A.G. Lebed: Phys. Rev. B **59**, R721 (1999)
113. L. Brossard et al.: C. R. Acad. Sc. Paris **205**, 302 (1986)
114. C. Bourbonnais et al.: Europhys. Lett. **6**, 177 (1988)
115. A. Kobayashi et al.: Solid State Commun. **62**, 57 (1987)
116. *Physics and Chemistry of Low-Dimensional Inorganic Compounds*, eds. C. Schlenker, J. Dumas, M. Greeblatt, S. van Smaalen (Plenum Press, New York, 1996)
117. M.H. Whangbo, E. Canadell. J. Am. Chem. Soc. **110**, 358 (1988)
118. J.D. Denlinger et al.: Phys. Rev. Lett. **82**, 2540 (1999)
119. G.-H. Gweon et al.: Proc. of Strongly Correlated Electron Systems (SCES), Univ. of Michigan (August 2001), to be published in Physica B, arXiv:cond-mat/0107211
120. T. Straub et al.: Phys. Rev. B **55**, 13473 (1997)
121. R. Claessen et al.: Phys. Rev. Lett. **69**, 808 (1992)
122. G.-H. Gweon et al.: (unpublished)
123. D. Orgad: Phlos. Mag. B **81**, 375 (2001)
124. L. Degiorgi et al.: Phys. Rev. B **38**, 5821 (1988)

Two Prototypes of One-Dimensional Conductors: $(TM)_2X$ and Cuprate Spin Ladders

Denis Jérome[1], Pascale Auban-Senzier[1], and Yuri Piskunov[1,2]

[1] Laboratoire de Physique des Solides (CNRS), Université Paris Sud, 91405, Orsay, France
[2] The Institute of Metal Physics, 620219, Ekaterienburg, Russia

Abstract. We intend to survey the physical properties of two families of materials exhibiting one dimensional metal-like conductivity, the conducting organic salts and the hole-doped cuprate spin ladders. The study of their transport and magnetic properties under pressure suggests that in both compounds a one dimensional confinement of the carriers due to strong intra-chain interactions takes place at high temperature while the application of a high pressure favours the deconfinement of these carriers followed by the onset of a superconducting instability at low temperature.

1 Introduction

Organic superconductivity was discovered in 1979 in the one dimensional organic conductor $(TMTSF)_2PF_6$ [1]. This discovery has to be considered in the framework of a long quest for new superconductors which started fifteen years earlier with the suggestion made by W. Little [2] for a new pairing mechanism in organic conductors. The extensive studies carried out over the last 20 years have revealed the existence of a vast series of isostructural compounds, $(TM)_2X$, exhibiting signatures of a non-Fermi behavior at high temperature and a variety of ground states at low temperature. Besides superconductivity, low dimensional organic conductors have brought a wealth of new concepts in physics.

The discovery of superconductivity under pressure in a hole-doped spin-ladder cuprate $Sr_{14-x}Ca_xCu_{24}O_{41}$ with $x = 13.6$ (sintered powder) in 1996 [3] and $x = 11.5$ (single crystal) [4], has renewed the interest for superconductivity in low dimensional systems and its interplay with magnetism. These systems can be viewed as a link between high T_c cuprates and quasi one dimensional organics which are two classes of exotic superconductors where the mechanism of superconductivity is still highly debated [5,6].

2 Organics

2.1 Towards TM_2X

The prerequisite for the formation of a molecular conductor is first the requirement for having charged molecules in a solid and second to allow their charge to delocalize between molecular entities. Charging the molecules can be achieved either by a charge transfer reaction or by the formation of an organic salt. The first case has been realized in TTF-TCNQ by a transfer of charge between the donor molecule TTF and the acceptor molecule TCNQ [7,8] and the second route gave rise to the synthesis of the first organic superconductor $(TMTSF)_2PF_6$ in 1979 [1]. Actually, it is the study of the two-chain charge transfer compounds (TMTSF-DMTCNQ), (TM-DM) where the donor is the tetramethyl selenide derivative of TTF which drew chemists and physicists's attention on the role of the TMTSF molecule. The outcome of this study has been decisive for the quest of organic superconductivity [9]. This 1-D conductor undergoes a Peierls transition at 42 K [10]where unlike TTF-TCNQ a distortion occurs simultaneously on both chains [11]. Several other results have triggered our attention. X-ray experiments had shown that the charge transfer is only $\rho = 0.5$ leading to a quarter-filled band situation [11]. Transport and thermopower data emphasized the dominant role played by the TMTSF chain in the mechanism driving the Peierls transition and also in its contribution to the conduction at high temperature [12]. The really new and unexpected finding has been the suppression of the Peierls transition under pressure and the conductivity remaining metal-like reaching the unprecedented value of 10^5 $(\Omega\ \mathrm{cm})^{-1}$ under 10 kbar at the temperature of liquid helium [13]. The conducting state of TM-DM was also remarkable in displaying a huge transverse magnetoresistance below 50 K [14].

Since all these phenomena, were new and unexpected the effort was put on a structure made of only one organic stack comprising the lucky TMTSF molecule. Such a structure was already known from the early work of the Montpellier chemistry group who synthesized and studied the series of isostructural $(TMTTF)_2X$ organic salts [15] where TMTTF is the sulfur analog of TMTSF and X is a monoanion such as ClO_4^-,BF_4^- or SCN^- etc., see Fig. 1. All these compounds turn into strong insulators at low temperature. This is the reason why they did not attract much interest until recently when they have been practically rediscovered after twenty years of studies which were mainly devoted to the selenide series, $(TMTSF)_2X$.

2.2 The TM_2X Era

Several exciting properties of $(TMTSF)_2PF_6$, an increase of the metal-like conductivity up to 10^5 $(\Omega\ \mathrm{cm})^{-1}$ at 12 K where an abrupt metal-insulator transition of SDW nature takes place [16], the absence of any noticeable lattice modulation in X-ray measurements [11,17], stimulated its further investigation under pressure and allowed the stabilization of a metallic state down to liquid helium temperature at a pressure of about 9 kbar. The finding of a very small and still non-saturating resistivity at 1.3 K extrapolating linearly to a practically zero value at $T = 0$ K was a sufficient motivation to trigger its study in a dilution refrigerator and a zero resistance state was stabilized below 1 K [1]. Shortly after the discovery of superconductivity in $(TMTSF)_2PF_6$ many other members of the same series with a variety of anions have also been found superconducting in the vicinity of 1 K in the 10 kbar pressure domain [18]. $(TMTSF)_2ClO_4$ is

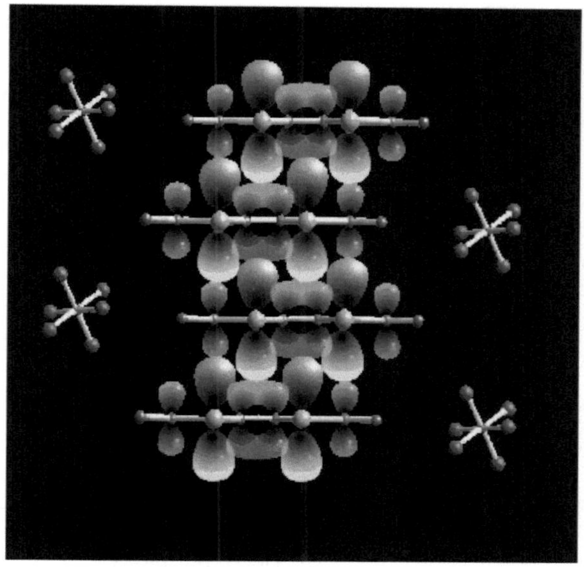

Fig. 1. View of the $(TM)_2X$ structure, courtesy of Institut des Matériaux de Nantes

the only member of the $(TMTSF)_2X$ series to show superconductivity under atmospheric pressure [19]. The discovery of superconductivity in the $(TMTSF)_2X$ family was a very exciting phenomenon since it was the first time such an instability could be stabilized in an organic compound. This happened about 15 years after the publication of Little's suggestion and 10 years after the holding of an international symposium organized by W.A. Little at Hawai on the Physical and Chemical Problems of Possible Organic Superconductors, [20]. Probably the most surprizing feature of superconductivity in $(TMTSF)_2X$ is the common border existing between antiferromagnetic (SDW) and superconducting phases with the region of maximum T_c for superconductivity located right at the border.

2.3 Organic Conductors and High Pressure Physics

The role of pressure in $(TMTTF)_2PF_6$ stabilizing an antiferromagnetic ground state instead of the spin-Peierls phase at atmospheric pressure has initiated a reinvestigation of the sulfur-analog series, $(TMTTF)_2X$, aiming at the stabilization of superconductivity in this series as well. This has been successfully achieved only recently with $(TMTTF)_2Br$ under 26 kbar [21] and $(TMTTF)_2PF_6$ at 47 kbar [22–24]. $(TMTTF)_2PF_6$ is the only member so far of the TM_2X series to exhibit all three different ground states allowing therefore to render the phase diagram a genuine property of these Q-1-D conducting salts.

Thanks to a variety of experimental studies and emergent theories we have now reached a reasonable level of understanding in the physical properties of

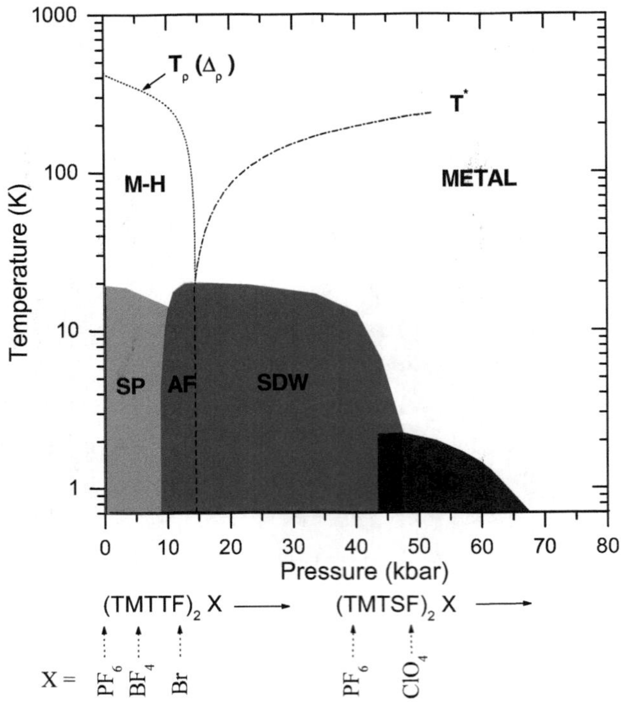

Fig. 2. The generic Pressure-Temperature phase diagram for $(TM)_2X$ salts according to the pressure data of $(TMTTF)_2PF_6$, the cross-over line is shown as $T^\star$

these 1-D conductors namely the $(TMTTF)_2X$ and $(TMTSF)_2X$ salts which both belong to the extended $(TM)_2X$ family [25], Fig. 2.

The salient feature of the $(TM)_2X$ diagram is first the existence of a wide variety of ground states occuring below 20K or so.

Moving towards the right across the $(TM)_2X$ phase diagram a succession of ground states is revealed either changing compounds or changing pressure on a given compound.

$(TMTTF)_2PF_6$ is the only system which can be driven through the entire series of ground states under high pressure: spin-Peierls, Néel antiferromagnetism, SDW phase with an incommensurate magnetic modulation and ultimately superconductivity [26].

As far as the non-ordered high temperature phase is concerned, its character in the phase diagram depends crucially on the nature of the corresponding ground state. Crudely speaking, all compounds which are located left of $(TMTTF)_2Br$ in the diagram are insulators at high temperature whereas those on the right are highly conducting and display a metal-like conductivity down to the transition into the SDW insulating phase.

2.4 Far Infrared Conduction

It is also illuminating to have a look at the conductivity in the far infrared regime. A large gap of order 2000 K is observed in the frequency dependence of the FIR conductivity of sulfur compounds.This is in line with the activation energy of the DC conductivity of those compounds. However the surprise arises for selenium compounds which behave apparently like highly conducting metals in spite of a marked gap at 200 cm^{-1} observed at low temperature in the FIR regime .

This feature suggests that all $(TM)_2X$ compounds exhibit a FIR gap, the amplitude of which depending on the actual position in the phase diagram. Given the unified phase diagram established experimentally and the recent proposal of theoretical ideas in 1-D physics applying specifically to 1-D conductors with a commensurate band filling we have reached a plausible theoretical description which we intend to summarize very briefly in the rest of this article, (for an almost up to date review see, [6]).

2.5 1-D Physics in TM_2X Compounds

What 1-D physics means is that instead of the usual description of low lying excitations in terms of quasi particle states in the Landau-Fermi liquid model, a collective mode description with decoupled spin and charge modes is a more appropriate approach [27,28]. Such a model for 1D conductors has been proposed starting from a linearized energy spectrum for excitations close to the Fermi level and adding the relevant Coulomb repulsions which are responsible for electron scattering with momentum transfer $2k_F$ and 0. This is the popular Luttinger model for a 1-D conductor in which the spatial variation of all correlation functions exhibits a power law decay at large distance, characterized by a non-universal exponent K_ρ which is a function of the microscopic coupling constants [29]. The Luttinger model requires a band filling of the conduction band which is not commensurate with the underlying lattice. Very few organic conductors fullfill this requirement. TTF[Ni(dmit)]$_2$ is one of them as the HOMO-LUMO energy difference which is large compared to the bandwith makes the system semi-metallic [30,31]. Consequently, NMR is able to measure the exponent K_ρ of the Luttinger liquid down to low temperature [32].

However, $(TM)_2X$ conductors are rather peculiar systems which deviate from the Luttiger model since the stoechiometry imposes half a carrier (hole) per TM molecule independent of the applied pressure. If the molecules are uniformly spaced along the stacking axis (a situation which is actually met in $(TMTSF)_2ClO_4$),the unit cell contains 1/2 carrier, i.e. the conduction band is quarter-filled although the existence of a slight dimerization of the molecules along the stacks in sulfur compounds could contribute to make them half-filled band compounds (*see below*). The commensurate band filling opens a new scattering channel (Umklapp scattering) of the carriers between both sides of the Fermi surface. This leads to important modifications in the model of the gapless Luttinger liquid which in turn becomes a Wigner type insulator with a gap in the charge sector [33]subjected to the condition $K_\rho < 0.25$ for 1/4 filling although no gap exists in the spin channel. The gap in the density of states at the Fermi energy reads : $2\Delta_\rho \approx W(\frac{g_U}{W})^{1/(2-2n^2 K_\rho)}$ and g_U is the coupling constant $W(\frac{U}{W})^3$ where $n = 2$ in the Hubbard limit.

According to band calculations, the transverse overlap $t_\perp$ along the b-direction is of order 120 K and 200 K for sulfur and selenium compounds respectively. Therefore, it is quite natural to expect first 1-D theory to govern the physics of these quasi 1-D conductors, at least in the high temperature regime when $T > t_\perp$ and second, to observe a cross-over towards higher dimensionality physics below room temperature. As observed very early in the study of $(TMTSF)_2PF_6$ a plasma edge for a light polarized along the b-axis becomes observable below 100 K [34].

The striking behavior of the transport properties emerges from a comparison between the temperature dependence of longitudinal and transverse components of the resistivity [35]. The longitudinal resistivity of $(TMTSF)_2PF_6$ is metal-like down to the SDW transition at 12 K, varying like $T^{0.93}$ from 300 to 150 K once the thermal contraction is taken into account and more like T^2 below 150 K . However, the behavior of the resistance along the direction of weakest coupling i.e. along the c-axis displays an insulating character with a maximum around 120 K and becomes metallic below although still orders of magnitude above the critical Mott value. This behaviour is quite general in the TM_2X series and is independent of the shape of the anion provided the effect of thermal contraction on the transport properties is properly taken into account [36]. Similarly, the Hall constant displays a marked temperature dependence below room temperature , passing through a mimimum at the same temperature where the c-axis resitivity displays a maximum [37].

In the high temperature limit, as $T > \Delta_\rho$, we shall make the assumption that the existence of the Wigner localization can be neglected at first sight. The longitudinal transport is expected to vary according to a power law $\rho_\parallel \approx T^{4n^2 K_\rho - 3}$, [33], whereas the transport along c being incoherent should probe the density of quasiparticle states in the (a-b) planes. In the high T regime, the picture of non-coupled chains is approached. Therefore, the density of quasiparticle states should resemble the situation which prevails in a Luttinger liquid namely, $N(E) \approx \mid \omega \mid^\alpha$ where α is related to K_ρ by $\alpha = \frac{1}{4}(K_\rho + 1/K_\rho - 2)$. At energies $\omega > \Delta_\rho$, the actual density of states should be reminiscent of the density of states of the gapless Luttinger liquid. Therefore, a temperature dependence of the transport $\rho_c(T) \approx T^{-2\alpha}$ is expected in the 1-D regime ($T > T^\star$). The far infrared (FIR) conduction of $(TMTSF)_2PF_6$ exhibits also very unusual properties. There exists a FIR gap (which amounts to $\Delta_\rho = 200$ cm^{-1} in $(TMTSF)_2PF_6$) which has been taken as the signature of the (Wigner) gap. The experimental data and the theoretical power law dependence $\sigma(\omega) \approx \omega^{4n^2 K_\rho - 5}$ at $\omega > 2\Delta_\rho$ also lead to $K_\rho = 0.23$ [38]. This value for K_ρ would fit the 1/4 filled scenario with W = 12000 K (from band structure calculations and plasma edge measurements), $\Delta_\rho = 200$ cm^{-1} (from FIR data) and $U/W = 0.7$ (which is in fair agreement with the enhancement of the spin susceptibility at low temperature) [6]. Since the Wigner gap varies exponentially with K_ρ even a small variation of the ratio between the Coulomb interaction and the bandwidth under pressure would be sufficient to explain a decrease of K_ρ moving left in the diagram. Both optical and transport data lead to $2\Delta_\rho = 1000K$ in $(TMTTF)_2PF_6$ at atmospheric pressure [39]. The difference between K_ρ for selenium and sulfur compounds in which $K_\rho = 0.18$ can be accounted by the difference of their bare bandwidths. The temperature $T^\star$ corresponding to the c-resistance maximum moves up under pressure and reaches room temperature under 10 kbar. This temperature can be attributed to the beginning of a cross-over between 1-D and 2-D regimes.

The strong pressure dependence of the cross-over temperature $T^\star$ is a remarkable phenomenon of the TM_2X physics. According to the pressure data of ρ_c , the pressure dependence of $T^\star$ is about ten times larger than that of $t_\perp$ which is typically 2 $kbar^{-1}$. This feature suggests that $T^\star$ is actually a renormalized version of $t_\perp$ due to the 1-D confinement via intrastacks electron-hole interactions [6]. Admittedly, the theory of the dimensionality cross-over in a Q-1-D conductor will require some refinements taking also into account the role of the (Wigner) commensurability gap which has been overlooked so far. Some preliminary calculations are being performed towards this direction, [40].

Apparently, the finite dimerization of the intrastack bonds is not large enough to make the 1/2-filled Umklapp scattering a pertinent mechanism in the formation of the 1-D (Wigner) phase. As far as the selenium compounds are concerned this feature could be related to the transverse b-direction bandwidth ($W_b \approx 1500$K) being larger than the dimerization gap (900 K) and thus smearing out the role of the 1-D dimerization. This argument may still be valid for the sulfur series in which the dimerization gap and the transverse bandwidth are both in the same energy range.

3 Doped Spin Ladders

3.1 The $Sr_{14-x}Ca_xCu_{24}O_{41}$ Series of Spin Ladders

Undoped spin-ladders of general formula $Sr_nCu_{n+1}O_{2n+1}$, where the copper valence is exactly 2 (Cu^{2+} with spin $S = 1/2$), consist of cuprate ladder planes separated by Sr layers [41,42]. Their magnetic properties depend on the number of legs in the ladders [5]. In case of an even number of legs the formation of spin singlets on each rung leads to the opening of a spin gap in the spin excitations. In the two-leg ladder compound $SrCu_2O_3$ the spin gap can be seen by the exponential drop of the spin susceptibility [43]. This singlet ground state is not possible for an odd number of legs and a finite value of the spin susceptibility is reached at low temperature for instance in the three-leg compound $Sr_2Cu_3O_5$ [43].

The $Sr_{14-x}Ca_xCu_{24}O_{41+\delta}$ series has a more complicated structure built from the piling up of CuO_2 chains, Sr (or Ca) and Cu_2O_3 two-leg ladders layers with an orthorhombic symmetry [44], Fig. 3. The interlayer distance (along the b axis), which is 1.6 Å in $x = 0$ compound, is shortened upon Ca substitution. Moreover, there is a misfit between the lattice parameters of chains and ladders

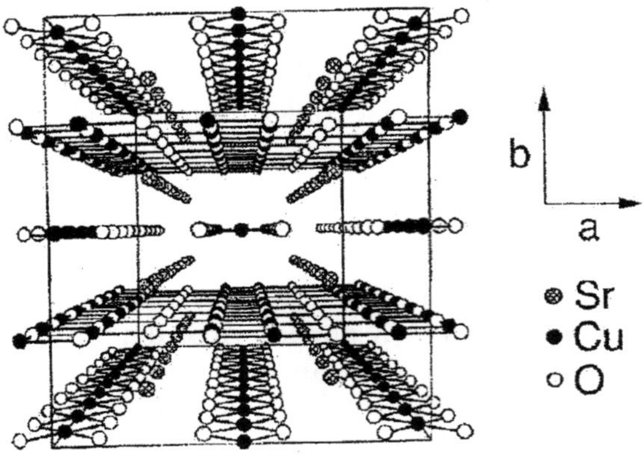

Fig. 3. Crystal structure of the $Sr_{14-x}Ca_xCu_{24}O_{41}$ series

sublattices ($10 \times c_{\text{chain}} = 7 \times c_{\text{ladder}}$). This system is inherently hole doped as the stoechiometry implies an average copper valence of 2.25. However, optical measurements [45] have shown that holes for the $x = 0$ compound are residing mainly in the chains leading in turn to an insulating behavior and to a magnetic ordering revealed by neutron scattering experiments [46] with a valency of 2.06 for the Cu ladder sites. Then, although Ca and Sr are isovalent atoms, holes are transfered from the chains to the ladders upon Ca substitution. Consequently, the copper valency in the ladders increases from 2.06 up to 2.20 (0.2 hole per Cu) as x increases from 0 to 11 [45]. At the same time, the longitudinal conductivity (along the c axis) increases, leading to quasi one dimensional (Q-1-D) metallic properties [47,48]. It has been suggested that holes doped into the ladders will share common rungs since it is the configuration which minimizes the number of damaged spin singlets [49]. As far as the transverse a-axis is concerned, a confinement of holes pairs on the ladders below a confinement temperature called $T^\star$ which is above room temperature for all samples with $x \leq 8$ and drops to below 300 K for $x = 12$. However, as for undoped ladder compounds, a spin gap is still observed in the ladder plane and a smaller spin gap ($\approx$ 140K) is also present in the chain sublattice. The existence of a spin gap on ladders was demonstrated by ^{63}Cu-NMR measurements which can easily separate the contributions coming from copper nuclei belonging to chains and ladders. The spin part of the NMR shift, which is directly proportional to the spin susceptibility, allows an accurate determination of these gaps. Upon Ca substitution, the amplitude of the spin gap in the ladders decreases from 500 K [50,51] or 430 K [52] for $x = 0$ down to 250 K for x=12 [51,52] while it remains practically unchanged in the chains (125-140 K) [50,52]. Therefore, the spin gapped structure survives the existence of a finite concentration of holes in the ladders in agreement with theoretical suggestions [49,53,54]. For samples with the largest Ca concentration, an applied pressure in the range 30-80 kbar stabilizes a superconducting ground state in the ladder planes with a transition temperature passing through a maximum at 10 K around 40 kbar [48]. The great interest of superconducting spin ladders lies in the theoretical prediction of superconductivity in the d-wave channel which is a direct consequence of the spin gapped character of the magnetic excitations [49]. The possible link between the predicted superconducting phase of isolated ladders and the (SC) phase stabilized under high pressure in $(Sr/Ca)_{14}Cu_{24}O_{41}$ is therefore a challenge in the physics of strongly correlated low dimensional fermions. The experimental results presented in this survey show that ungapped as well as gapped spin triplet excitations coexist when superconductivity is stabilized in $Sr_2Ca_{12}Cu_{24}O_{41}$ under pressure.An extended account of this work is presented in the reference [55].

3.2 Triplet Excitations in Spin Ladders Measured by the ^{63}Cu Knight Shift

The Knight shift of the Cu ladder nuclei has been studied in the whole series $(Sr/Ca)_{14}Cu_{24}O_{41}$ and also in the compound $La_5Ca_9Cu_{24}O_{41}$ with has a vanishing hole doping up to a pressure of 36 kbar. The magnetic field was aligned with the direction parallel to the packing axis of the ladders (*i.e parallel to the b axis*).

The data of $K_{\mathrm{b,s}}$ are shown in Figs. 4 and 5. For lightly doped samples, the pressure (up to 32 kbar) does not affect the Knight shift at low temperature (which remains zero). This is *at variance* with what is occurring when the Ca content overcomes x=8. Then, the spin part of the shift becomes pressure dependent at low temperature and a significant upturn of the NMR shift is observed at very low temperature although essentially under high pressure. No such effects are observed in lightly Ca-doped samples.

The pressure dependence of the zero temperature shift in Ca_{12}, Fig. 5 is indicative for the appearance of low-lying spin excitations above 20 kbar. This situation is reminiscent of the growth of zero frequency spectral weight which is observed in the undoped two-leg ladders upon substitution of Cu^{2+} by non-magnetic impurities [56,57]. A marked Curie-tail associated with the center of the spectrum is observed at low temperature under pressure in all samples with $x > 5$. Since this Curie contribution is not detected at ambient pressure in the same samples we feel confident that it cannot be attributed to impurities or instrumental shifts of the magnetic field. There is also an intermediate regime in which $K_{\mathrm{b,s}}(T)$ is activated. The scope of the present survey will be limited to the temperature regime in which Knight shifts are activated.

We shall assume that the temperature dependence of $K_{\mathrm{b,s}}$ is ascribed to the spin excitations above the spin gap Δ_{s} as for the undoped two-leg ladders. Hence, we fit the spin part $K_{\mathrm{b,s}}(T)$ using the following Troyer expression, valid in the low temperature domain [58],

$$K_{b,s}(T) = \frac{C}{\sqrt{T}} \exp \frac{-\Delta_s}{T} \tag{1}$$

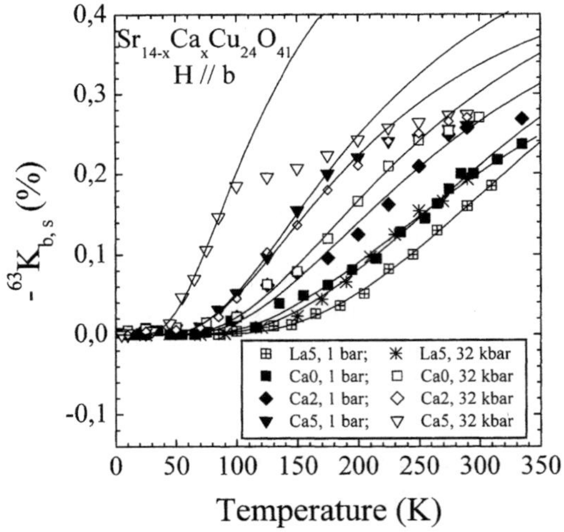

Fig. 4. T dependences of the ^{63}Cu Knight shift $^{63}K_{\mathrm{b,s}}$ in $La_5Ca_9Cu_{24}O_{41}$ (La_5) and $Sr_{14-x}Ca_xCu_{24}O_{41}$ ($x = 0, 2, 5$) for $H \parallel \mathbf{b}$ under ambient and high pressures. The lines are the fits to the data using Eq. (1)

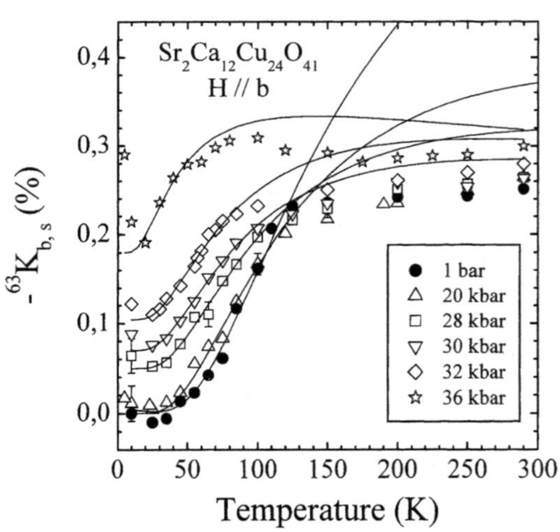

Fig. 5. T dependences of the ^{63}Cu Knight shift $^{63}K_{b,s}$ in $Sr_2Ca_{12}Cu_{24}O_{41}$ for $H \parallel \mathbf{b}$ at different pressures. The lines are the fits to the data using Eq. (1)

The values of Δ_s thus obtained via fitting the data to (1) for Ca_0, Ca_2, Ca_5, Ca_8 and Ca_{12} at ambient and under a pressure of 32 kbar are displayed in Fig. 6. The effect of pressure is large at all Ca contents. The relative reduction of Δ_s under pressure is steadily increasing with the density of holes (*see below*).We

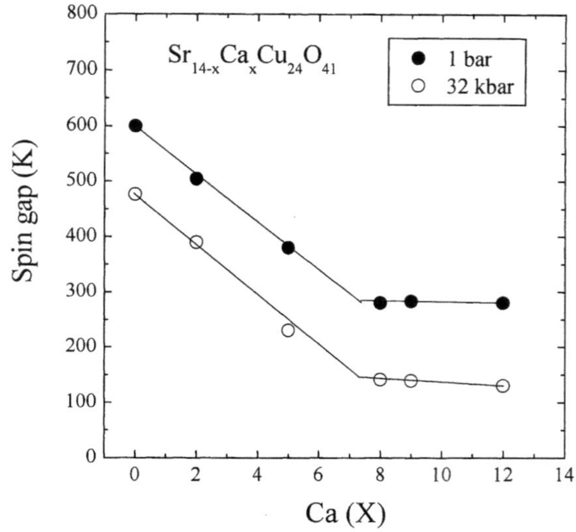

Fig. 6. Dependence of the spin gap Δ_s upon Ca substitution under ambient and high pressure. The values of Δ_s were obtained from the fits to the Knight shift data using Eq. (1). The solid lines are a guide for the eye

note that the data at ambient pressure agree fairly well with those published in Ref.[51]. The spin gap value of 840 K which is obtained for La_5 is significantly larger than the results of undoped Cu_2O_3 ladders which are obtained through the ^{17}O Knight shift in $La_6Ca_8Cu_{24}O_{41}$, ($\Delta_s = 510K$) [59] or by susceptibility in the $SrCu_2O_3$ structure [43]. However, these results fit rather well with our overall doping dependence of the spin gap in hole-doped ladders pertaining to the $(Sr/Ca)_{14}Cu_{24}O_{41}$ series which is presented in the present work.

3.3 Spin Gap and Superconductivity

A feature of the $P-T$ phase diagram for the Ca_{12} sample, Fig. 7, which deserves a special consideration is the existence of a maximum of the superconducting critical temperature when pressure drives the characteristic energy scales for the pair binding and the spin gap down to values of the order of T_c. Beyond this optimal pressure for the highest T_c, denoted as P_{opt}, these intra-ladder gaps become irrelevant and hence T_c decreases under pressure. The latter pressure domain is also characterized by a temperature dependence of resistivity along the inter-ladder a direction that becomes metallic [60,39]. Such a change of behavior in T_c can be seen as a cross-over from strong to weak coupling conditions for long-range superconducting order.

Arguments in support of this viewpoint follow from simple considerations about mechanisms of long-range superconducting order in coupled ladder systems. Let us consider in the first place the region below P_{opt} where the gaps are larger than T_c; in this domain, long-range superconducting coherence perpendicular to ladder leg direction can only be achieved through a Josephson coupling.

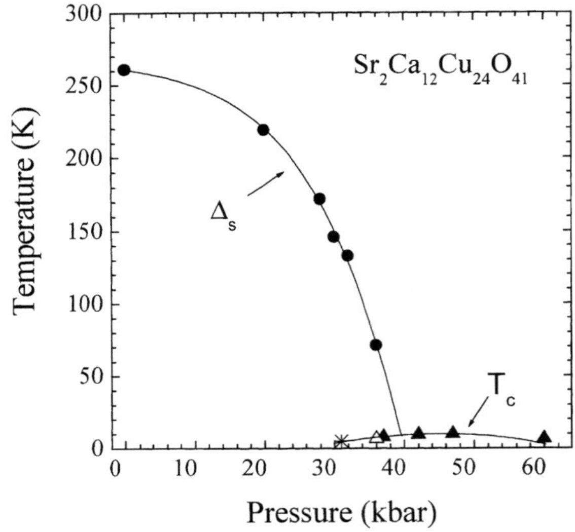

Fig. 7. Pressure dependence of the spin gap Δ_s obtained from the fits to the Knight shift data using Eq. (1) and of the superconducting transition temperature, T_c, for $Sr_{2.5}Ca_{11.5}Cu_{24}O_{41}$ (△) [4] and for $Sr_2Ca_{12}Cu_{24}O_{41}$ (*) [69]

An analysis based on the renormalized perturbation theory [61] shows that the Josephson coupling takes the form of

$$J_s \approx 2\frac{\xi_s}{c}\frac{t_{\rm L}^2}{T^*}, \tag{2}$$

where $\xi_s \approx t_{\parallel}/T^*$ stands as the size of the superconducting hole pairs within the ladders, while $t_{\rm L}$ is the inter-ladder single electron hopping ($t_{\rm L} \ll t_{\parallel}$). This expression contains the usual term $t_{\rm L}^2/T^*$ for the effective transverse motion of the pairs requiring an intermediate energy of order T^* (the confinement temperature of hole pairs on ladders, see Sec.I) and a factor ξ_s/c that accounts for the increase in probability for pair hopping due to delocalization of holes forming the pairs along the ladder direction [61]. As for pair correlations of isolated ladders, they are well known to be strongly enhanced in the presence of a spin gap. Numerical and analytical results for $t-J$ and repulsive Hubbard models for example [62,63], show that "d-wave" pair susceptibility follows a power law $\chi_{\rm s}(T) \sim (T/\Delta_{\rm s})^{-\gamma_{\rm s}}$ for temperatures below the spin gap, where $\gamma_{\rm s} \approx 1$. A molecular-field approximation of the Josepshon coupling that rigourously takes the influence of intra-ladder correlations into account, allows one to obtain at once an expression for the transition temperature. Thus from the condition $J_{\rm s}\chi_{\rm s}(T_{\rm c}) = 1$ one finds,

$$T_{\rm c} \sim t_{\rm L}^2/T^*, \tag{3}$$

which assumes that $\Delta_{\rm s} \sim T^*$. It follows that $T_{\rm c}$ increases when T^* decreases, a feature that is essentially governed by the increase of the Josephson coupling under pressure.

Hence, when pressure is further increased, one reaches the domain where $T_c \sim T^*$ and a change in the physical picture of the transition occurs. The gaps are no longer relevant for the mechanism promoting long-range order and its description evolves toward a weak coupling picture – a situation that bears close similarities with the antiferromagnetic ordering of quasi-one-dimensional organic conductors with a charge gap under pressure [6]. When the metallic character of the normal state is well developed along longitudinal and transverse directions the motion of the holes is likely to be coherent along both directions leading to Fermi liquid conditions – or deconfinement of holes, in agreement with the restoration of a metallic transverse resistivity. Thus sufficiently above $P_{\rm opt}$, the transition temperature should follow from a weak coupling BCS prescription which yields

$$T_c \approx t_L e^{-1/|g*|}. \tag{4}$$

Here g^* is the residual attractive coupling between deconfined holes. Since pressure weakens the strength of short-range superconducting correlations, it will lead to a decrease of $|g^*|$ at the energy scale t_L for deconfinement [64,61]. This in turn reduces T_c as a function of pressure. It follows then that a maximum of T_c is expected when the ladder system evolves from strong to weak coupling conditions for hole pairing at $P \sim P_{\rm opt}$.

4 Conclusion

To summarize, the picture emerging for the TM_2X phase diagram is that of a Wigner insulating phase with a correlation gap decreasing from sulfur to selenium compounds, (Fig. 2). As long as the Wigner gap is large (say larger than 300 K) the 1-D confinement is quite active and the single particle interstack hopping is meaningless despite an unrenormalized value of the transverse coupling which is still of the order of 100 K. Moving towards the right in the phase diagram, the Wigner gap decreases (because of the band broadening under pressure) and the renormalization of the transverse hopping integral become less pronouced. $(TMTSF)_2PF_6$ is a compound where T^* and $2\Delta_\rho$ are both of the same order of magnitude ($\approx 150-200$ K), (Fig.2) and T^* reaches room temperature in $(TMTSF)_2ClO_4$ under atmospheric pressure according to the data of ρ_c, [65]. Below T^* the system becomes an anisotropic conductor, first two dimensional with high energy excitations governed by the Wigner gap and ultimately an anisotropic three dimensional Fermi liquid around 5K (for $(TMTSF)_2ClO_4$ under atmospheric pressure). A recent extension of the renormalization calculations in the cross over regime [66] has shown that SDW and superconducting orders are still interfering constructively in this temperature domain although only the singlet superconducting coupling is enhanced with respect to the triplet channel. The same theory predicts the possibility of first order reentrant superconductivity within the SDW phase close to the border with superconductivity. However, recent experimental studies in $(TMTSF)_2PF_6$ have clearly identified a wide pressure domain where both superconducting and SDW regions segregate at a macroscopic scale, [67,68]. It is a pleasure to quote the words of our friend Paul Chaikin, "$(TMTSF)_2PF_6$ is the most interesting material ever discovered".

The salient result of the spin ladder study is the coexistence in the Ca_{12} sample under a 36 kbar pressure of superconductivity at 6.7 K with low lying spin excitations as obseved earlier [69] but also remainings of gapped excitations (with a gap of 80 K).The compound becomes an anisotropic 2-D conductor in the same high pressure domain where both superconductivity and gapped magnetic excitations are observed.It can be suggested that the 1-D confinement achieved at low pressure is actually related to gapped excitations whereas the deconfinement under pressure is connected to the increase of low lying excitations.

These new results are supporting the exciting prospect of superconductivity induced by the interladder tunnelling of preformed pairs [70,71,64] at least in the pressure domain where the pressure remains smaller than the pressure corresponding to the maximum of the critical temperature for superconductivity. Doped cuprate spin ladders in which the doping can be controlled by pressure are remarkable prototype compounds for the study of the mechanism for superconductivity in cuprates. The interplay between the real space pair binding and superconductivity can also be studied. Experiments at even higher pressures where a more conventional kind of superconductivity is expected would be most useful.

Acknowledgments

This article is the opportunity to ackowledge the cooperation of our co-workers who have contributed at Orsay and Copenhagen to the development of the physics and chemistry of 1D organic conductors. This field has also greatly benefited from a close cooperation with the theory group at Orsay and C.Bourbonnais at Sherbrooke, Canada. The spin ladder samples have been grown in the laboratory of A.Revcolevschi at the Université of Paris-Sud.

References

1. D. Jérome, A. Mazaud, M. Ribault, K. Bechgaard: J. Phys. Lett. **41**, L–95 (1980)
2. W.A. Little: Phys. Rev. A **134**, 1416 (1964)
3. M.Uehara et al.: J. Phys. Soc. Jpn. **656**, 2764 (1996)
4. T.Nagata et al.: Physica C **282-287**, 153 (1997)
5. E.Dagotto, T.M.Rice: Science **271**, 618 (1996)
6. C. Bourbonnais, D. Jérome: in Advances in Synthetic Metals, p 206, Elsevier, 1999
7. L.B. Coleman, M.H. Cohen, D.J. Sandman, F.G. Yamagishi, A.F. Garito, A.J. Heeger: Solid. St. Comm. **12**, 1125 (1973)
8. J. Ferraris, D.O. Cowan, V. Walatka, J.H. Perlstein: J. Am. Chem. Soc. **95**, 948 (1973)
9. J.R. Andersen, K. Bechgaard, C.S. Jacobsen, G. Rindorf, H. Soling, N. Thorup: Acta. Cryst. B **34**, 1901 (1978)
10. C.S. Jacobsen, K. Mortensen, J.R. Andersen, K. Bechgaard: Phys. Rev. B **18**, 905 (1978)

11. J.P. Pouget: Chemical Scripta **17**, 85 (1981)
12. Y. Tomkiewicz, J.R. Andersen, A.R. Taranko: Phys. Rev. B **17**, 1579 (1978)
13. A. Andrieux, C. Duroure, D. Jérome, K. Bechgaard: J. Phys. Lett. **40**, L–381 (1979)
14. A. Andrieux, P.M. Chaikin, C.D. Duroure, D. Jérome, D. Weyl, K. Bechgaard, J.R. Andersen: J. Phys. **40**, 1199 (1979)
15. G. Brun, S. Peytavin, B. Liautard, M. Maurin, E. Torreilles, J.M. Fabre, L. Giral: C.R. Acad. Sci. **284**, 211 (1977)
16. K. Bechgaard, C.S. Jacobsen, K. Mortensen, H.J. Pedersen, N. Thorup: Solid. St. Comm. **33**, 1119 (1980)
17. J.P.Pouget, S.Ravy: Synt. Met. **85**, 1523 (1997)
18. S.S. Parkin, M. Ribault, D. Jérome, K. Bechgaard: J. Phys. C **14**, L–445 (1981)
19. K. Bechgaard, K. Carneiro, M. Olsen, F.B. Rasmussen: Phys. Rev. Lett. **46**, 852 (1981)
20. W.A. Litlle: J. Polym. Sci. Part C **29** (1970)
21. L. Balicas, K. Behnia, W. Kang, E. Canadell, P. Auban-Senzier, D. Jérome, M. Ribault, J.M. Fabre: J. Phys. I **4**, 1539 (1994)
22. D. Jaccard, H.Wilhelm, D.Jérome, J.Moser, C.Carcel, J.M.Fabre: J. Phys. Condens. Matter **13**, L–1 (2001)
23. H.Wilhelm et al.: Eur. Phys. J. B **21**, 175 (2001)
24. K.Kato T.Adachi, E.Ojima, H.Kobayashi: J. Am. Chem. Soc **122**, 3238 (2000)
25. D. Jérome: Science **252**, 1509 (1991)
26. D. Jérome: in Organic Conductors, page 405, M. Dekker Inc, New-York, 1991
27. H.J. Schulz: Phys. Rev. Lett. **64**, 2831 (1990)
28. J. Voit: Rep. Prog. Phys. **58**, 977 (1995)
29. J.Solyom: Adv. Phys. **28**, 201 (1979)
30. M.Bousseau et al.: Nouveau Journal de Chimie **8**, 3 (1984)
31. E.Canadell, I.E.Rachidi, S.Ravy, J.P.Pouger, L.Brossard, J.P.Legros: J. Phys. (France) **50**, 2967 (1989)
32. C.Bourbonnais et al.: Eur. Phys. Lett. **6**, 177 (1988)
33. T. Giamarchi: Physica B **230**, 975 (1997)
34. C.S. Jacobsen, D.B.Tanner, K. Bechgaard: Phys. Rev. Lett. **46**, 1142 (1981)
35. J. Moser, M. Gabay, P. Auban-Senzier, D. Jérome, K. Bechgaard, J.M. Fabre: Eur. Phys. B **1**, 39 (1998)
36. P.Auban-Senzier et al.: to be published, 2001
37. J. Moser, J.R. Cooper, D. Jérome, B. Alavi, S.E. Brown, K. Bechgaard: Phys. Rev. Lett. **84**, 2674 (2000)
38. A. Schwartz, M. Dressel, G. Grüner, V. Vescoli, L. Degiorgi, T. Giamarchi: Phys. Rev. B **58**, 1261 (1998)
39. P. Auban-Senzier et al.: Synth. Met. **103**, 2632 (1999)
40. S. Biermann et al.: Condens. Matter, p. 0107633
41. E.M. Mc Carron et al.: Mater. Res. Bull. **23**, 1355 (1988)
42. T. Siegrist, L.F. Schneemeyer, S.A. Sunshino, J.V. Wasczak, R.S. Roth: Mater. Res. Bull. **23**, 1429 (1988)
43. M. Azuma, Z. Hiroi, M.Takano, K.Iohida, Y.Kitaoka: Phys. Rev. Lett. **73**, 3463 (1994)
44. R.S. Roth et al.: J. Am. Chem. Soc. **72**, 1545 (1989)
45. T. Osafune et al.: Phys. Rev. Lett. **78**, 1980 (1997)
46. L.P. Regnault et al.: Phys. Rev. B **59**, 1055 (1999)
47. N. Motoyama et al.: Phys. Rev. B **55**, 3386 (1997)

48. T. Nagata et al.: Phys. Rev. Lett. **81**, 1090 (1998)
49. E. Dagotto, J. Riera, D. Scalapino: Phys. Rev. B **45**, 5744 (1992)
50. K. Kumagai, S. Tsuji, M. Kato, Y. Koike: Phys. Rev. Lett. **78**, 1992 (1997)
51. K. Magishi et al.: Phys. Rev. B **57**, 11533 (1998)
52. M. Takigawa, N. Motoyama, H. Eisaki, S. Uchida: Phys. Rev. B **57**, 1124 (1998)
53. R.M. Noack, S.R. White, D.J. Scalapino: Phys. Rev. Lett. **73**, 882 (1994)
54. D. Poilblanc, H. Tsunetsugu, T.M. Rice: Phys. Rev. B **50**, 6511 (1994)
55. Y. Piskunov et al.: Eur. Phys. B submitted (2001)
56. G.B. Martins, E. Dagotto, J. Riera: Phys. Rev. B **54**, 16032 (1996)
57. N. Fujiwara et al.: Phys. Rev. Lett. **80**, 604 (1998)
58. M. Troyer, H. Tsunetsugu, D. Wurtz: Phys. Rev. B **50** , 13515 (1994)
59. T. Imai, K.R. Thurber, K.M. Shen, A.W. Hunt, F.C. Chou: Phys. Rev. Lett. **81**, 220 (1998)
60. N. Mori et al.: Physica B **239**, 137 (1997)
61. C. Bourbonnais, L.G. Caron: Int. J. Mod. Phys. B **5**, 1033 (1991)
62. C.A. Hayward, D. Poilblanc, R.M. Noack, D.J. Scalapino, W. Hanke: Phys. Rev. Lett. **75**, 926 (1995)
63. H.J. Schulz: Phys. Rev. B **53**, 2959 (1996)
64. J. Kishine, K. Yonemitsu: J. Phys. Soc. Jpn. **67**, 1714 (1998)
65. J.R. Cooper et al.: Phys. Rev B **33**A, 6810 (1986)
66. R. Duprat, C. Bourbonnais: Eur. Phys. B **21**, 219 (2001)
67. T. Vuletic et al.: Eur. Phys. B submitted (2001)
68. T. Vuletic et al.: Condens. Matter, p. 0109031
69. H. Mayaffre et al.: Science **279**, 345 (1998)
70. T.M. Rice, S. Gopolan, M. Sigrist: Europhys. Lett. **23**, 445 (1993)
71. J. Kishine K. Yonemitsu: J. Phys. Soc. Jpn. **66**, 3725 (1997)

Nucleation of Superconductivity in Low-Dimensional Systems Under Magnetic Fields

Vladimir Mineev

Commissariat à l'Energie Atomique, Département de Recherche Fondamentale sur la Matière Condensée, SPSMS, 38054 Grenoble, France

Abstract. This chapter presents a survey of the current state of art of the theory of the upper critical field in low-dimensional superconducting materials. Starting from the Ginzburg–Landau equation in the vicinity of the critical temperature we derive a new analytical expression for the angular and temperature dependence of the critical field in layered superconducting materials. This formula is shown to reproduce as limiting cases the well known results of Tinkham for an isolated film of finite thickness and of Lawrence–Doniach for the stacked array of films of zero thickness with its particular anisotropic Ginzburg–Landau limit. The well known theoretical statements of A.Lebed' about a low temperature divergence of the upper critical field parallel to the conducting layers in the layered superconductors and the reentrance of the superconductivity in ultra-high magnetic fields are reinvestigated taking into account the scattering of quasiparticles by impurities. The upper limit of the superconductor purity yielding to a finite value of the upper critical field is established. The problem of the low temperature divergence of the upper critical field in bulk materials in the quantized magnetic field is discussed. The high field behavior of the critical temperature as a function of the magnetic field for a finite but high enough crystal purity is derived.

1 Introduction

The temperature dependence of the critical field in low-dimensional superconductors differs from that observed in anisotropic superconductors. The latter in uniaxial crystals is described as [1]

$$H_{c2\perp} = \frac{\Phi_0}{2\pi\xi_{ab}^2(T)} \tag{1}$$

for the field orientation parallel to the high symmetry axis c, that is perpendicular to ab plane, and as

$$H_{c2\parallel} = \frac{\Phi_0}{2\pi\xi_{ab}(T)\xi_c(T)} \tag{2}$$

for the field directed in the ab plane. Here $\xi_{ab}(T)$ and $\xi_c(T)$ are the coherence lengths in the ab plane and along the c axis respectively. Near T_c they behave as inverse square root functions of $(T_c - T)$ and tend to finite values at zero temperature. So, near T_c the upper critical field is a linear function of $(T_c - T)$ and smoothly saturates when lowering the temperature. Such a "bulk type" behavior occurs even in the layered materials for H parallel to the conducting layers as long as the interlayer hopping integral t exceeds the critical temperature

$$t > T_c.$$

In the opposite case the "bulk" linear behavior is only preserved in the close vicinity of T_c and transforms into a square root type dependence at lower temperatures (see for example [2]). The angular dependence of the upper critical field in the layered materials also has its own peculiarities [2]. The general formula describing this temperature and the angular dimensional crossover has been recently derived in the author's paper [3] and is reproduced in the second section of this chapter. The result reproduces the critical field behaviour for the two limiting cases of single thin films (Tinkham) and anisotropic bulk materials (Lawrence–Doniach). Analytical expressions for the measurable values of the critical fields parallel and perpendicular to the conducting layers are derived. The relation established between them is used for the determination of the degree of anisotropy and thickness of superconducting layers in chemically grown compounds. The results are also applicable to artificially structured materials.

When the hopping integral exceeds the critical temperature the upper critical field in the Ginzburg–Landau region has the usual behavior typical of anisotropic bulk materials. However, at low temperatures, in quasi one-dimensional and quasi two-dimensional materials with open Fermi surfaces, the superconductivity is preserved in arbitrary large fields with direction perpendicular to the conducting filaments (quasi-1D case) or parallel to the conducting layers (quasi-2D case), as was first been pointed out by A.Lebed' [4,5]. Let us focus on quasi-two-dimensional conductors. The upper critical field divergence starts to develop at temperatures

$$T < t^2/\varepsilon_F.$$

Moreover, according to theoretical predictions [5] at high enough fields

$$\omega_{c2} = eH_{c2}v_F s/c \ > \ t$$

(here ω_{c2} corresponds to the upper critical field cyclotron frequency, s is the interlayer distance) the superconductivity is restored in the whole temperature range $T < T_c$. These statements are literally true for superconductors with triplet pairing. In the case of singlet pairing the existence of superconductivity in high fields is restricted by the paramagnetic limit H_p. So, the low temperature stability of the superconducting state under high magnetic field can be used to unravel the existence of triplet pairing superconductivity.

The tendency to a low temperature divergence of the upper critical field $H_{c2} > H_p$ in quasi-one-dimensional organic superconductor $(TMTSF)_2PF_6$ has been revealed by the group of P.Chaikin [6]. So, the observations being in correspondence with theoretical predictions [4] speak in favor of triplet type of superconductivity in this material.

Another popular layered superconductor Sr_2RuO_4 exhibits properties compatible with triplet superconductivity [7]. However, the measurements of the upper critical field in the basal plane show no tendency to divergence of $H_{c2}(T)$ down to 0.2 K [8]. Moreover, the search for reentrance of superconductivity under magnetic fields up to 33 T and temperatures down to 50 mK has given negative results [9].

The point is that the theoretical statements [4,5] are valid for absolutely clean and perfect crystalline materials. The role of crystal purity on the superconductivity reentrance in layered materials has been investigated in [6], in which the upper purity limit of the superconductor yielding a finite value of the upper critical field is established. A survey of these results is contained in the third section of this chapter.

The suppression of interlayer motion of electrons is not the only possible origin of a low temperature divergence of the upper critical field. Another source of this type of divergence is the Landau quantization of electronic orbits making the motion of the electrons in a bulk material effectively one-dimensional. In a perfect crystal the value of the upper critical field starts to grow at temperatures

$$T < \frac{T_c^2}{\varepsilon_F}.$$

The impurity scattering and the Pauli paramagnetic interaction restrict this divergence as was first pointed out in the paper of L. Gruenberg and L. Gunther [7] and discussed later by the author [8] in relationship with the calculations of the de Haas–van Alphen oscillations in the superconducting state. A brief review of these results is contained in Sect. 4 of this chapter.

2 Angular Dependence of the Critical Field in Layered Superconductors

The formulae giving the upper critical field for thin superconducting films or for bulk but anisotropic materials are well known and widely applied in the investigations of naturally or artificially grown layered superconductors [9–12]. In the case of films with thickness smaller than the coherence length $d \ll \xi$ the critical field obeys the equation proposed by M. Tinkham [13,14]

$$\left|\frac{H_{c2}(\theta)\sin\theta}{H_\perp}\right| + \left(\frac{H_{c2}(\theta)\cos\theta}{H_\parallel}\right)^2 = 1. \tag{3}$$

Here θ is the angle between the magnetic field and the film surface.

On the other hand, for uniaxial anisotropic bulk materials the critical field is given by the formula [2]

$$H_{c2}(\theta) = \frac{H_\perp}{(\sin^2\theta + a\cos^2\theta)^{1/2}}, \tag{4}$$

where

$$a = \frac{m_{ab}}{m_c} \tag{5}$$

and θ is the angle between the magnetic field and the ab plane.

The critical field in a film (3) is characterized by a cusp-like angular dependence at $|\theta| \to 0$ and a square root $\sqrt{T_c - T}$ temperature dependence for H parallel to the film surface orientation. On the contrary in anisotropic bulk materials the angular dependence of H_{c2} at $|\theta| \to 0$ is smooth and characterized by a usual linear $(T_c - T)$ behavior. For the real layered materials one is often in the region of the dimensional (2D$\to$ 3D) crossover. In absence of analytical expression, different numerical and analytical approximate treatments of the initial Ginzburg–Landau equations for layered superconductors have been applied [15–19,24,25] for a quantitative description of the upper critical field behavior. It has been proved, however, that it is possible to solve the problem analytically and to obtain a formula for H_{c2} containing both the above expressions as limiting cases as well the whole dimensional crossover in one analytic expression.

Let us start by considering a system of parallel films with thickness d and with a distance s between the middle of two adjacent films. They are coupled by Josephson tunneling and submitted to a magnetic field defined by the vector potential $\mathbf{A}(\mathbf{r}) = (0, Hx\sin\theta, -Hx\cos\theta)$. Here we have chosen the (x, y) plane as parallel to the films surfaces (ab plane). Following the Lawrence–Doniach approach [26] one can write the linearized Ginzburg–Landau equation for the order parameter ψ_n of the nth film as

$$\begin{aligned}\alpha\psi_n &- \frac{\hbar^2}{2m_{ab}}\frac{\partial^2\psi_n}{\partial x^2} + \frac{\hbar^2}{2m_{ab}}\left(\frac{2eHx\sin\theta}{\hbar c}\right)^2\psi_n \\ &+\frac{\hbar^2}{2m_{ab}}\left(-i\frac{\partial}{\partial z} + \frac{2eHx\cos\theta}{\hbar c}\right)^2\psi_n \\ &+\frac{\hbar^2}{m_c s^2}\left(1 - \cos\frac{2eHxs\cos\theta}{\hbar c}\right)\psi_n = 0,\end{aligned} \tag{6}$$

where the last term corresponds to the Josephson coupling with nearest layers with $n \pm 1$,

$$\alpha = \alpha(T) = \alpha_0\frac{T - T_c}{T_c} = -\frac{\hbar^2}{2m_{ab}\xi_{ab}^2(T)}. \tag{7}$$

It is worth mentioning that the "Ginzburg-Landau effective mass" m_c figuring in this equation is related to the electron effective mass $m_\perp$ (corresponding to a motion perpendicular to the layers) by

$$\frac{\hbar^2}{m_c s^2} \sim \frac{t^2}{\varepsilon_F} \sim \frac{m_{ab}}{m_\perp^2}. \tag{8}$$

Here t is the interlayer hopping integral, $\varepsilon_F = \hbar^2 k_F^2/2m_{ab}$. This relation has been found first by E. Kats [27]. It can also be derived from the general expression for the "effective mass" in the Ginzburg–Landau equations for a bulk anisotropic superconductor [1] in the particular case of strong uniaxial anisotropy (layered material). In contrast, for a bulk superconductor with a moderate uniaxial anisotropy, $m_c \sim m_\perp$ as follows from [1].

The solution of (6) has the following form

$$\psi_n(x,z) = \varphi_n(x) \exp\left(-\frac{2\mathrm{i}eHxz\cos\theta}{\hbar c}\right). \tag{9}$$

For films with thickness $d \ll \xi$, following [14], one can take the function φ_n as z independent. Taking the origin of the z-axis at equal distance from the film surfaces we get, after substituting (9) into (6) and averaging over the film's thickness,

$$\alpha\varphi_n - \frac{\hbar^2}{2m_{ab}}\left(\frac{\partial^2\varphi_n}{\partial x^2} - \left(\frac{eHd\cos\theta}{\sqrt{3}\hbar c}\right)^2\varphi_n\right) + \frac{\hbar^2}{2m_{ab}}\left(\frac{2eHx\sin\theta}{\hbar c}\right)^2\varphi_n$$
$$+\frac{\hbar^2}{m_c s^2}\left(1-\cos\frac{2eHxs\cos\theta}{\hbar c}\right)\varphi_n = 0. \tag{10}$$

Now, for sufficiently strong interaction between the layers, $eHs^2 \ll \hbar c\sqrt{a}$, the cosine term can be expanded. After this, the solution of (10) will be the function

$$\varphi_n(x) = \exp\left(-\frac{eHx^2}{\hbar c}(\sin^2\theta + a\cos^2\theta)^{1/2}\right) \tag{11}$$

depending on the magnetic field $H = H_{c2}(\theta)$ which has to be determined from the equation

$$\left(\frac{H_{c2}(\theta)\cos\theta}{H_\parallel^0}\right)^2 + \frac{H_{c2}(\theta)}{H_\perp}(\sin^2\theta + a\cos^2\theta)^{1/2} = 1. \tag{12}$$

Here

$$H_\perp = \frac{m_{ab}c|\alpha(T)|}{e\hbar}, \tag{13}$$

$$H_\parallel^0 = \frac{\sqrt{6m_{ab}c^2|\alpha(T)|}}{ed}. \tag{14}$$

$H_\perp$ represents the directly measurable value, whereas, unlike in the Tinkham formula (3), one needs to distinguish $H_\parallel^0$ from the measurable value $H_\parallel$. They are related by the equation resulting from (12)

$$H_\parallel^0 = \frac{H_\parallel}{\left(1 - \frac{H_\parallel}{H_\perp}\sqrt{a}\right)^{1/2}}. \tag{15}$$

It is worth noting that the temperature independent ratio

$$\frac{H_\perp}{(H_\parallel^0)^2} = \frac{ed^2}{6\hbar c} \tag{16}$$

together with (15) allows one to establish the relationship between the thickness of superconducting layers and the parameter of anisotropy a using the measured values $H_\perp$ and $H_\parallel$.

The temperature dependence of $H_\perp(T)$ and that of $H_\|(T)$ are different. While the former is determined by the usual linear $(T_c - T)$ law (13), the latter is also linear near T_c

$$H_\| \cong \frac{H_\perp}{\sqrt{a}} \tag{17}$$

as long as $H_\perp < H_\|^0\sqrt{a}$ but it transforms into a $\sqrt{T_c - T}$ dependence

$$H_\| \cong -\frac{(H_\|^0)^2\sqrt{a}}{2H_\perp} + H_\|^0, \tag{18}$$

when this condition is violated. This circumstance has been established earlier [15] and verified experimentally (see for instance [9]).

The expression (12) obviously transforms into the Tinkham formula (3) in the case of an isolated film $a \ll 1$ and into the anisotropic Ginzburg–Landau expression (4) in the case of zero films thickness $d \to 0$.

For sufficiently weak interaction between the layers $eHs^2 \gg \hbar c\sqrt{a}$ one can neglect the fast oscillating cosine term in (10) and again recover the equation for the isolated film [6] with a shifted critical temperature as $\alpha(T) \to \alpha(T) + \hbar^2/m_c s^2$. Its critical field is given by (3). The condition of weak interlayer interaction $eHs^2 \gg \hbar c\sqrt{a}$ can, with the help of (17), (5), (8) and at $s \sim d$, be rewritten as

$$t < \sqrt{T_c(T_c - T)}. \tag{19}$$

In conclusion of this section one should focus on the formula (12) which is the general analytic expression describing the temperature and angular dependent 2D - 3D dimensional crossover of upper critical fields in layered superconductors.

3 Low Temperature Divergence of the Upper Critical Field and the High Field Reentrance of Superconductivity in Layered Superconductors

We shall consider a metal with the following electron spectrum

$$\begin{aligned} \epsilon\,(\mathbf{p}) &= \frac{1}{2m}(p_x^2 + p_y^2) - 2t\cos p_z s, \\ t \ll \epsilon_F &= \frac{mv_F^2}{2}, \qquad \frac{-\pi}{s} < p_z < \frac{\pi}{s} \end{aligned} \tag{20}$$

in the magnetic field $\mathbf{H} = (0, H, 0)$, $\mathbf{A} = (0, 0, -Hx)$ parallel to the conducting layers separated by a distance s. As was shown in [5,6] the upper critical field is determined from the linear integral equation for the order parameter

$$\begin{aligned} \Delta(x) = \tilde{g} \int\limits_{|x-x_1|>a\ \sin\phi} dx_1 \int\limits_0^\pi \frac{d\phi}{2\pi v_F \sin\phi} \psi^*(\phi)\psi(\phi) \frac{2\pi T \exp\left[-\frac{|x-x_1|}{l\sin\phi}\right]}{\sinh\left[\frac{2\pi T|x-x_1|}{v_F\sin\phi}\right]} \\ \times \mathcal{I}_0\left\{\frac{2\lambda}{\sin\phi}\sin\left[\frac{\omega_c(x-x_1)}{2v_F}\right]\sin\left[\frac{\omega_c(x+x_1)}{2v_F}\right]\right\}\Delta(x_1). \end{aligned} \tag{21}$$

Here $\tilde{g} = gm/4\pi s$ is a dimensionless constant corresponding to the pairing interaction, a is the small–distance cutoff, $\omega_c = ev_F sH/c$, v_F is the Fermi velocity, $\lambda = 4t/\omega_c$, where t is the hopping integral between conducting layers, $\hbar = 1$, $\mathcal{I}_0(...)$ is the Bessel function. We neglect here the Pauli interaction with electronic spins, in order to investigate the pure orbital effect on the low temperature divergence.

The mean free path of the quasiparticles is l. We shall use also mean the free time $\tau = l/v_F$. The equation (21) is obtained within the Born approximation for the quasiparticle scattering and averaging over impurities positions. As shown in [28], the normal self-energy part produced by the impurity scattering under magnetic field has been proved to be field independent as long as

$$\frac{v_F}{\omega_c} > \frac{l}{\sqrt{k_F l}}. \tag{22}$$

The superconducting state $\Delta(\phi, x) = \psi(\phi)\Delta(x)$ is determined by the basis functions $\psi(\phi)$ of one of the irreducible representations of crystal symmetry group. They are, for instance, $\sqrt{2}(\sin^2\phi - \cos^2\phi)\Delta(x)$ for singlet pairing or $\Delta(\phi, \hat{p}_z, x) = \sqrt{2}\cos\phi\Delta(x)$ for triplet pairing and normalized as follows

$$\int_0^\pi \frac{d\phi}{\pi}\psi^*(\phi)\psi(\phi) = 1,$$

in which ϕ is the angle between $\mathbf{p}_\parallel$, the in-plane component the of momentum, and the magnetic field $\mathbf{H} \parallel \hat{y}$ (for the details see [6]). The equation (21) is valid only for superconducting states with even functions $\psi(\phi)$ and such that the anomalous self-energy part produced by impurities scattering under magnetic field is equal to zero due to the integration over ϕ (see [6]). The general case requires a more elaborate mathematical treatment which, however, does not introduce any changes at a qualitative level.

The equation (21) in the absence of a magnetic field

$$1 = \tilde{g} \int_{\frac{2\pi a T}{v_F}}^{\infty} \frac{dz}{\sinh z} \exp\left(-\frac{z}{2\pi T\tau}\right) \tag{23}$$

determines the critical temperature T_c, which is related to the critical temperature T_{c0} in a perfect crystal without impurities $l = \infty$

$$T_{c0} = \frac{v_F}{\pi a} \exp\left(-\frac{1}{\tilde{g}}\right) \tag{24}$$

by means of the well known relation

$$\ln\frac{T_c}{T_{c0}} = \psi\left(\frac{1}{2}\right) - \psi\left(\frac{1}{2} + \frac{1}{4\pi\tau T_c}\right). \tag{25}$$

For critical temperatures $T_c \sim T_{c0}$, the suppression of the superconductivity by impurities is:

$$T_c = T_{c0} - \frac{\pi}{8\tau}. \tag{26}$$

One can also point out the condition of complete suppression of superconductivity

$$\tau = \tau_c = \frac{\gamma}{\pi T_{c0}}, \tag{27}$$

where $\ln \gamma = C = 0.577...$ is the Euler constant.

The behavior of the upper critical field is determined by the relationship between the three spatial scales: $v_F/2\pi T$, l and v_F/ω_c. For temperatures close to the zero field critical temperature $T \approx T_c(H = 0)$, the upper critical field $H_{c2}(T)$, or $\omega_{c2}(T) = eH_{c2}(T)v_F d/c$ tends to zero and the inequality

$$\min\left\{\frac{v_F}{2\pi T},\ l\right\} < \frac{v_F}{\omega_{c2}(T)} \tag{28}$$

is always valid. Near T_c there is also a formal solution of (21) with $\omega_{c2} > t$ (see [5]).

As the temperature decreases the upper critical field increases, but as long as the inequality (28) takes place, the essential interval of integration over $(x - x_1)$ in (21)is determined by the $\min(v_F/2\pi T,\ l)$. Hence, one can use the smallness of $(x - x_1)v_F/\omega_c$ in the kernel of the equation (21):

$$\Delta(x) = \tilde{g} \int\limits_{|x-x_1|>a\ \sin\phi} dx_1 \int\limits_0^\pi \frac{d\phi}{2\pi v_F \sin\phi} \psi^*(\phi)\psi(\phi) \frac{2\pi T \exp\left[-\frac{|x-x_1|}{l \sin\phi}\right]}{\sinh\left[\frac{2\pi T|x-x_1|}{v_F \sin\phi}\right]}$$

$$\times \mathcal{I}_0 \left\{\frac{4t(x-x_1)}{v_F \sin\phi} \sin\frac{\omega_c x}{v_F}\right\} \Delta(x_1). \tag{29}$$

The solution of this equation gives the correct value of $\omega_{c2}(T)$ as long as the inequality (28) is true. Substituting the value of $\omega_{c2}(T)$ obtained from this equation into (28) one can determine the limit of purity below which the $H_{c2}(T)$ value derived from (29) is valid down to $T = 0$.

For pure enough samples and at low enough temperatures, the opposite condition to (28)

$$\frac{v_F}{\omega_{c2}(T)} < \min\left\{\frac{v_F}{2\pi T},\ l\right\} \tag{30}$$

can be realized, making the transformation of (21) into (29) incorrect. It can easily be shown that even in the ultraclean limit, for small enough interlayer hopping, one can still be in the range of applicability of the present theory, which is limited by the inequality (22). Hence in the temperature range

$$\frac{v_F}{2\pi l} < T < \frac{\omega_c(T)}{2\pi}$$

the magnetic field dependence of the critical temperature in (21) starts to disappear or, in other words, the tendency to the divergence of the upper critical field pointed out in [5] appears.

Solving the equation (29) for arbitrary temperature and purity is only possible numerically. Here we shall discuss the case where an analytical solution can be obtained. If the length scale ξ, on which the function $\Delta(x)$ is localized, is larger than the essential distance of integration over $(x - x_1)$

$$\xi > \min\left\{\frac{v_{\mathrm{F}}}{2\pi T},\, l\right\}, \tag{31}$$

then one can expand $\Delta(x_1) \approx \Delta(x) + \Delta'(x)(x - x_1) + \Delta''(x)(x - x_1)^2/2$ under the integral in (29). Taking into consideration that under this condition the argument of the Bessel function turns to be small even at the upper boundary of the effective interval of the integration over $(x - x_1)$:

$$\frac{t\omega_{\mathrm{c}2}(T)\xi \min\left\{\frac{v_{\mathrm{F}}}{2\pi T},\, l\right\}}{v_{\mathrm{F}}^2} \approx \frac{t \min\left\{\frac{v_{\mathrm{F}}}{2\pi T},\, l\right\}}{\epsilon_{\mathrm{F}}\, \xi} < 1, \tag{32}$$

one can also expand Bessel function $\mathcal{I}_0(x) \approx 1 - x^2/4$. As a result we get the differential equation

$$\left(\ln\frac{T_{\mathrm{c}0}}{T} - \psi\left(\frac{1}{2} + \frac{1}{4\pi\tau T}\right) + \psi\left(\frac{1}{2}\right)\right)\Delta(x) = -\frac{C\, I(\alpha)}{2}\left(\frac{v_F}{2\pi T}\right)^2 \Delta''(x)$$
$$+I(\alpha)\left(\frac{t\omega_{\mathrm{c}2}(T)x}{\pi v_{\mathrm{F}} T}\right)^2 \Delta(x), \tag{33}$$

where $\alpha = (2\pi T\tau)^{-1}$,

$$I(\alpha) = \int\limits_0^\infty \frac{z^2 dz}{\sinh z}\exp(-\alpha z) = 4\sum_{n=0}^{\infty}\frac{1}{(2n+1+\alpha)^3}, \tag{34}$$

$$C_\psi = \int\limits_0^\pi \frac{d\phi}{\pi}\psi^*(\phi)\psi(\phi)\sin^2\phi. \tag{35}$$

In the pure case, $\alpha \approx \alpha_c = (2\pi T_{\mathrm{c}}\tau)^{-1} \ll 1$, the inequality (31) is valid only in vicinity of the critical temperature. Putting $T = T_{\mathrm{c}}$ in the right hand side and taking its lowest eigenvalue we obtain

$$\ln\frac{T_{\mathrm{c}0}}{T} - \psi\left(\frac{1}{2} + \frac{1}{4\pi\tau T}\right) + \psi\left(\frac{1}{2}\right) = \frac{\sqrt{C_\psi} I(\alpha_c) t\omega_{\mathrm{c}2}(T)}{2\sqrt{2}\pi^2 T_{\mathrm{c}}^2}. \tag{36}$$

Summing this equation with (25) and keeping only the terms linear in $T - T_{\mathrm{c}}$ and in the impurity concentration we get

$$\omega_{\mathrm{c}2}(T) = \frac{ev_{\mathrm{F}}s}{c}H_{\mathrm{c}2}(T) = \frac{4\sqrt{2}\pi^2}{7\zeta(3)\sqrt{C_\psi}t}\left(T_{\mathrm{c}0} - \beta\frac{\pi}{8\tau}\right)(T_{\mathrm{c}} - T). \tag{37}$$

Here the coefficient

$$\beta = 2 - \frac{90\zeta(4)}{7\pi^2\zeta(3)} \approx 0.83, \tag{38}$$

shows that the slope of $H_{c2}(T)$ at $T = T_c$ decreases with increasing impurity concentration somewhat slower than T_c itself (see (26)) .

The equation (29) does not contain any divergence of $H_{c2}(T)$ and the expression (37) which is valid in the Ginzburg–Landau region can be used at arbitrary temperature as an upper estimation of upper critical field keeping in mind that the formal Ginzburg–Landau value of upper critical field at any temperature is always larger than its genuine value. Hence, to establish the limits of sample purity at which equation (29) works, one may substitute (37) at zero temperature into the inequality (28). Omitting the numerical factor of the order of unity we have

$$l < \frac{t}{T_c}\xi_{ab}, \tag{39}$$

where $\xi_{ab} = v_F/2\pi T_c$ is the basal plane coherence length. We see that there is a wide range of sample purity within which one cannot expect low temperature divergence of the upper critical field. In Sr_2RuO_4 the mean free path should be approximately 10 times larger than the basal plane coherence length ξ_{ab} to go out of the range (39).

Let us now consider the dirty case: $T_c \ll T_{c0}$, $\alpha \gg 1$ and $I(\alpha) \approx \alpha^{-2}$ allowing an analytical solution for $H_{c2}(T)$ at arbitrary temperature. Taking the lowest eigen value of (33) we get

$$\ln\frac{T_{c0}}{T} - \psi\left(\frac{1}{2} + \frac{1}{4\pi\tau T}\right) + \psi\left(\frac{1}{2}\right) = \sqrt{2C_\psi}\tau^2 t\omega_{c2}(T). \tag{40}$$

Combining this equation with (25) yields

$$\omega_{c2}(T) = \frac{\ln\frac{T_c}{T} - \psi\left(\frac{1}{2} + \frac{1}{4\pi\tau T}\right) + \psi\left(\frac{1}{2} + \frac{1}{4\pi\tau T_c}\right)}{\sqrt{2C_\psi}\tau^2 t}. \tag{41}$$

This expression is correct at any temperature. One can rewrite it approximately in the more simple form

$$\omega_{c2}(T) = \frac{\sqrt{2}\pi^2}{3\sqrt{C_\psi}t}(T_c^2 - T^2), \tag{42}$$

where

$$T_c = \frac{1}{\pi\tau}\left(\frac{3}{2}\ln\frac{\pi T_{c0}\tau}{\gamma}\right)^{1/2}. \tag{43}$$

In conclusion of this section we can stress once more that as long as the purity of the samples does not exceed a high threshold level (39), the upper critical field has no tendency to diverge. The layered organic materials with their intrinsic chemical purity are certainly the most appropriate candidates for the realization of a low temperature upper critical field divergence and high field superconductivity restoration.

4 Quantum Corrections to the Upper Critical Field at Low Temperatures

The critical temperature as a function of H at $T \ll T_c$ has been derived by L. Gruenberg and L. Guenther [7]. It is determined by the equation

$$H = H_{c2o}\left(1 - \frac{T^2}{T_c^2}S_0 + 2\pi^{3/2}\left(\frac{\omega_c}{\mu}\right)^{1/2} S_1 + 2^{3/2}\pi^{1/2}\frac{\omega_c}{\mu}S_2\right), \quad (44)$$

where S_0 is a slow (logarithmic) function of H with numerical values of the order of unity,

$$S_1 = \frac{2\pi T}{\omega_c}\Re\sum_{n=1}^{\infty}\sum_{\nu=0}^{\infty}\exp\left(-\frac{4\pi n(\tilde{\omega}_\nu + \mathrm{i}\mu_e H)}{\omega_c}\right), \quad (45)$$

$$S_2 = \frac{2\pi T}{\omega_c}\Re\sum_{n=1}^{\infty}\sum_{m>n}^{\infty}\sum_{\nu=0}^{\infty}(-1)^{m+n}\exp\left(-\frac{2\pi(m+n)(\tilde{\omega}_\nu + i\mu_e H)}{\omega_c}\right)$$
$$\times\frac{\cos\left(2\pi|n-m|\frac{\mu}{\omega_c} - \frac{\pi}{4}\right)}{|n-m|^{1/2}}, \quad (46)$$

$$H_{c2o} = \frac{\pi^2 e^2}{2\gamma}\frac{T_c^2}{v_F^2}. \quad (47)$$

Here, μ is the chemical potential, v_F is the Fermi velocity, Γ is the Landau level width, μ_e is the electron's magnetic moment and $N_0 = m^* k_F/2\pi^2$ is the normal state electron density of states per one spin direction.

The first two terms of (44) reproduce of the quasiclassical result found by L. Gor'kov [29]

$$H = H_{c2o}(1 - \frac{T^2}{T_c^2}S_0) \quad (48)$$

for the upper critical field at $T \to 0$. The third and the fourth terms in (44) give, respectively non-oscillating and oscillating corrections to Gor'kov's expression due to the quantization of the quasiparticle levels. For further considerations it will be convenient to represent the functions S_1 and S_2 in more simple analytical forms [8].

The summation over ν in (45) yields

$$S_1 = \frac{\pi T}{\omega_c}\Re\sum_{n=1}^{\infty}\frac{\exp\left(-\frac{4\pi n a}{\omega_c}\right)}{\sinh\frac{4\pi^2 nT}{\omega_c}}, \quad (49)$$

where

$$a = \Gamma + \mathrm{i}\eta\omega_c, \quad (50)$$

and

$$\eta\omega_c = \mu_e H - \frac{q}{2}\omega_c, \quad (51)$$

is the Zeeman energy deviation from a half-integer number $q/2$ of energies ω_c between Landau levels. For a non-interacting electron gas in a perfect crystal $a = 0$ and

$$S_1 \approx \frac{1}{4\pi} \ln \frac{\omega_c}{4\pi^2 T} \tag{52}$$

diverges at $T \to 0$. This result formally states the existence of superconductivity for a non-interacting electron gas in a perfect crystal at $T = 0$ in an arbitrary large field [7].

At the opposite limit, when, $\pi T \ll |a|$

$$S_1 = \Re \sum_{n=1}^{\infty} \left(\frac{1}{4\pi n} - \frac{2\pi^3 n}{3} \left(\frac{T}{\omega_c} \right)^2 \right) \exp\left(-\frac{4\pi n a}{\omega_c} \right). \tag{53}$$

At $|a| \ll \omega_c/4\pi$ it is

$$S_1 = \frac{1}{4\pi} \ln \frac{\omega_c}{4\pi|a|} - \frac{\pi}{24} \left(\frac{T}{|a|} \right)^2. \tag{54}$$

For $4\pi\Gamma \sim \omega_c$ one can approximate the summation (49) by its first term

$$S_1 = \frac{\pi T}{\omega_c \sinh \frac{4\pi^2 T}{\omega_c}} \exp\left(-\frac{4\pi\Gamma}{\omega_c} \right) \cos 4\pi\eta. \tag{55}$$

Substituting (55) into (44) provides us with the behavior of the upper critical field at low temperatures (averaged over oscillations)

$$\bar{H}_{c2}(T) = H_{c2o} \left\{ 1 + \frac{1}{2} \sqrt{\frac{\pi\omega_c}{\mu}} \left(1 - \frac{8\pi^4}{3} \left(\frac{T}{\omega_c} \right)^2 \right) \exp\left(-\frac{4\pi\Gamma}{\omega_c} \right) \cos 4\pi\eta \right\} \tag{56}$$

Let us denote the number of Landau levels at $H = H_{c2}$ lying below the Fermi level ε_F as

$$n_{c2} = \frac{\varepsilon_F}{\omega_c} = \frac{1}{2\pi^2} \left(\frac{\varepsilon_F}{T_c} \right)^2. \tag{57}$$

Here $\omega_{c2} = H_{c2}/m^*$ is the cyclotron frequency at the upper critical field. The number n_{c2} in typical materials with small Fermi energy and relatively high T_c, in which dHvA effect in the mixed state has been observed, is of the order of one hundred. So, the upper critical field at zero temperature is noticeably larger ($\sim n_{c2}^{-1/2}$) than its quasiclassical value (48). The decrease of H_{c2} with temperature follows a T^2 law. This parabolic dependence in the region

$$T \ll \frac{\omega_c}{2\pi^2} \sim \frac{T_c^2}{\varepsilon_F}$$

is much faster than Gorkov's parabola (48). For S_2 the summation over ν in (16) produces

$$S_2 = \frac{\pi T}{\omega_c} \Re \sum_{n=1}^{\infty} \sum_{m>n}^{\infty} (-1)^{m+n} \frac{\exp\left(-\frac{2\pi(m+n)(\Gamma + i\mu_e H)}{\omega_c} \right)}{\sinh \frac{2\pi^2(m+n)T}{\omega_c}} \frac{\cos\left(2\pi(m-n)\frac{\mu}{\omega_c} - \frac{\pi}{4} \right)}{(m-n)^{1/2}}. \tag{58}$$

Changing the summation variables to n and $m-n=l$ we get

$$S_2 = \frac{\pi T}{\omega_c}\Re\sum_{l=1}^{\infty}\frac{(-1)^l}{\sqrt{l}}\exp\left(-\frac{2\pi l(\Gamma+\mathrm{i}\mu_e H)}{\omega_c}\right)\cos\left(2\pi l\frac{\mu}{\omega_c}-\frac{\pi}{4}\right)$$
$$\sum_{n=1}^{\infty}\frac{\exp\left(-\frac{4\pi n(\Gamma+\mathrm{i}\mu_e H)}{\omega_c}\right)}{\sinh\frac{2\pi^2(l+2n)T}{\omega_c}}. \tag{59}$$

As for S_1 for a non-interacting electron gas in a perfect crystal ($2\mu_e H=\omega_c$, $\Gamma=0$) the sum over n in (59) diverges at $T\to 0$ (see also [7]).

In the more realistic case $4\pi\Gamma\sim\omega_c$, one can keep only the first term in the sum over n:

$$S_2 = \frac{\pi T}{\omega_c}\sum_{l=1}^{\infty}\frac{(-1)^l\exp\left(-\frac{2\pi\Gamma(l+2)}{\omega_c}\right)}{\sqrt{l}\,\sinh\frac{2\pi^2(l+2)T}{\omega_c}}\cos\left(2\pi l\frac{\mu}{\omega_c}-\frac{\pi}{4}\right)\cos\left(\frac{2\pi\mu_e H(l+2)}{\omega_c}\right). \tag{60}$$

Substitution of (55) and (60) into (44) and the solution of this equation with respect to T gives the oscillating behavior of critical temperature. A simple estimation shows that the amplitude of oscillations at $T\sim\omega_{c2}\sim 2\pi^2T_c^2/\varepsilon_F$ is of the order of

$$\frac{\delta T_{\mathrm{osc}}}{T_c}\sim\left(\frac{T_c}{\varepsilon_F}\right)^2 \tag{61}$$

At lower temperatures the amplitude of the oscillations is somewhat larger. The period of one oscillation δH is easy to find from

$$1=\delta n=\delta\frac{\varepsilon_F}{\omega_c}=-\frac{\varepsilon_F}{\omega_c}\frac{\delta H}{H}$$

So,

$$\frac{\delta H}{H_{c2}}\approx 2\pi^2\left(\frac{T_c}{\varepsilon_F}\right)^2. \tag{62}$$

As has been shown in [8], this value coincides with the critical region in the vicinity of the upper critical field at low temperatures. Hence the critical temperature oscillating behavior as a function of H exists only in the framework of mean field approximation and has poor chances of being observable. (For more details see Sect. 6 of the paper [8]).

To conclude this section we note that, due to the Landau level quantization, in ultra high purity superconductors in the ultra low temperature range

$$T<a<\omega_c \tag{63}$$

there is a parabolic upturn of the upper critical field with a noticeably larger value ($\propto\sqrt{\omega_c/\varepsilon_F}$) than the quasiclassical one (47). The effect is more pronounced in the ultraclean superconductors with both small Fermi energy and high value of $H_{c2}(T=0)$. These conditions are the same as those for the observation of de Haas–van Alphen oscillations in the superconducting state [30] which is accessible in the compounds with A-15 structure (V_3Si, Nb_3Sn...), borocarbides and heavy fermionic materials[31–34].

References

1. L.P. Gorkov, T.K. Melik-Barkhudarov: Zh. Eksp. Teor. Fiz. **45**, 1493 (1963) [Sov.Phys.-JETP **18**, 1031 (1964)]
2. M.Tinkham: *'Introduction to superconductivity'*, Second Edition, McGraw-Hill, Inc., 1996
3. V.P. Mineev: Phys.Rev. B **65**, 012508 (2002)
4. A.G. Lebed': Pis'ma Zh. Eksp. Teor. Fiz. **44**, 89 (1986) [JETP Lett. **44**, 114 (1986)]
5. A.G. Lebed', K. Yamaji: Phys. Rev. Lett. **80**, 2697 (1998)
6. I.J. Lee, P.M. Chaikin, M.J. Naughton: Phys. Rev. B **62** R14669 (2000)
7. Y. Maeno, T.M. Rice, M. Sigrist: Phys. Today **54**, 42 (2001)
8. T. Akima, S. Nishizaki, Y. Maeno: J. Phys. Soc. Jpn. **68**, 694 (1999)
9. E. Ohmichi, Y. Maeno, S. Nagai, Z.Q. Mao, M.A. Tanatar, T. Ishiguro: Phys. Rev. B **61**, 7101 (2000)
10. V.P. Mineev: J. Phys. Soc. Jpn. **69**, 3371 (2000)
11. L.W. Gruenberg, L. Gunther: Phys. Rev. **176**, 606 (1968)
12. V.P. Mineev: Phil. Mag. **80**, 307 (2000)
13. S.T. Rugigiero, T.W. Barbee, Jr., M.R. Beasley: Phys. Rev. Lett. **45**, 1299 (1980)
14. B.Y. Jin, J.B. Ketterson, E.J. McNiff, Jr., S. Foner, I. Schuller: J. Low Temp. Phys. **69**, 39 (1987)
15. M.J. Naughton, R.C. Yu, P.K. Davies, J.E. Fisher, R.V. Chamberlin, Z.Z. Wang, T.W. Jing, N.P. Ong, P.M. Chaikin: Phys. Rev. B **38**, 9280 (1988)
16. R. Marcon, E. Silva, R. Fastampa, M. Giura: Phys. Rev. B **46**, 3612 (1992)
17. M. Tinkham: Phys. Rev. **129**, 2413 (1963)
18. F.E. Harper, M. Tinkham: Phys. Rev. **172**, 441 (1968)
19. G. Deutscher, O. Entin-Wohlman: Phys. Rev. B **17**, 1249 (1978)
20. L.I. Glazman: Zh. Eksp. Teor. Fiz. **193**, 1373 (1987) [Sov. Phys. JETP **66**, 780 (1987)]
21. K. Tanaka: J. Phys. Soc. Jpn. **56**, 4245 (1987)
22. B.Y. Jin, J.B. Ketterson: Adv. Phys. **38**, 189 (1989)
23. V. Gvozdikov: Fiz. Nizk. Temp. **16**, 5 (1990) [Sov. J. Low Temp. Phys **16**, 1 (1990)]
24. T. Schneider, A.Schmidt: Phys. Rev. B **47**, 5915 (1993)
25. N. Takezawa, T. Koyama, M. Tachiki: Physica C **207**, 231 (1993)
26. W.E. Lawrence, S. Doniach, E.Kanda eds.: Proc. 12th Int. Conf. Low. Temp. Phys. (Kyoto, 1970; Keigaku, Tokyo 1971), p.361
27. E.I. Kats: Zh. Eksp. Teor. Fiz. **56**, 1675 (1969) [Sov.Phys.-JETP **29**, 897 (1969)]
28. Yu.A. Bychkov: Zh. Eksp. Teor. Fiz. **39**, 1401 (1960) [Sov. Phys. JETP **12**, 977 (1961)]
29. L.P. Gor'kov: Zh. Eksp. Teor. Fiz. **193**, 1373 (1987) [Sov. Phys. JETP **66**, 780 (1987)]
30. M.G. Vavilov, V.P. Mineev: Zh. Éksp. Teor. Fiz. **112**, 1873 (1997) [Sov. Phys. JETP **85**, 1024 (1997)]
31. T.J.B.M. Janssen, C. Haworth, S.M. Hayden, P. Meeson, M. Springford: Phys. Rev. B **57**, 11698 (1998)
32. H. Ohkuni, T. Ishida, Y. Inada, Y. Haga, E. Yamamoto, Y. Onuki, S.Takahashi: J. Phys. Soc. Jpn. **66**, 945 (1997)
33. T. Terashima, C. Haworth, H. Takeya, S. Uji, H. Aoki: Phys. Rev. **56**, 5120 (1997)
34. C. Bergemann, S.R. Julian, G.J. McMullan, B.K. Howard, G.G. Lonzarich, P. Lejay, J.P. Brison, J. Flouquet: Physica B **230-232**, 438 (1997)

Superconductivity Under High Magnetic Fields in Low-Dimensional Organic Salts

Takehiko Ishiguro

[1] Department of Physics, Kyoto University, Kyoto 606-8502, Japan
[2] CREST Japan Science and Technology Corporation, Kawaguchi 332-0012, Japan

Abstract. The organic superconductors characterized by low-dimensionality and prepared in high crystalline quality are suitable materials for the study of superconductivity in the high-magnetic-field limit. Under a selected direction with respect to the crystalline axes so as to depress the orbital pair breaking effect, the effect of spin-polarization on the superconducting state comes to the fore. The upper critical field down to low temperatures is reviewed and the breakthrough of the BCS Pauli paramagnetic limit is discussed.

1 Introduction: Low-Dimensional Superconductors in Oriented Magnetic Fields

The structure of the organic superconductors is characterized by stacking of planar molecules. This results in either a column structure or a layer structure, depending on the molecular arrangement, and provides a low-dimensional electronic structure. Although many atoms are involved in the constituent molecules, the electronic structures are notably simplified, because the electronic states near the Fermi level are determined principally by the correlated electrons in the highest occupied and/or the lowest unoccupied molecular orbitals [1–3].

Most of organic superconductors are prepared as single crystals. This enables us to study the anisotropic properties intrinsic to the low-dimensionality, for example, by applying an oriented high magnetic field with respect to the crystalline axes. The upper critical field H_{c2} is rather low, due to the low critical temperature T_c. This allows us to cover the whole temperature-versus-magnetic-field $(T-H)$ phase diagram with conventional magnets.

Either the anisotropic Ginzburg–Landau (G-L) model or the Lawrence–Doniach (L-D) model can describe layered superconductors, depending on the interlayer interaction [4]. When the interlayer interaction is negligible so that the layered structure is represented by stacks of independent sheets, the L-D model can be used. With the increase in the interlayer interaction, the anisotropic G-L model can approach the system more properly. A crossover between L-D to anisotropic G-L regimes appears with a boundary represented by the comparison of the out-of-plane component of the coherence length $\xi_\perp(T)$ to the interlayer spacing s, as

$$\xi_\perp(T^*) = s/\sqrt{2} \tag{1}$$

where T^* is the crossover temperature. With the shorter $\xi_\perp(T)$, the system can be represented more properly in terms of the L-D model. In the two-dimensional

(2D) limit, H_{c2} diverges when the magnetic field is applied exactly parallel to the superconducting plane, provided that the spin polarization effect is not taken into account.

The upper critical field can be formulated as a nucleation point of superconductor on decreasing the magnetic field from a high field. H_{c2} is given as the maximum magnetic field providing non-trivial eigen value of the G-L equation [5–8]. The destruction of superconductivity in a high magnetic field is due to the time-reversal symmetry breaking. Meanwhile, for superconductors with singlet pairing, superconductivity can be destroyed by a magnetic field due to the spin polarization effect, according to Clogston and Chandraskhar [9,10]: the condensation energy is compensated by Zeeman energy for the normal electrons. In this case the limiting field at $T = 0$ caused by the Pauli paramagnetic effect, H_{p}, is given by

$$H_{\mathrm{p}} = \Delta_0/\sqrt{2}\mu_0 \tag{2}$$

where μ_0 is the magnetic moment of electrons. For a weak-coupling BCS superconductor, the relation can be simplified as

$$H_{\mathrm{p}}^{\mathrm{BCS}} = 1.84 T_{\mathrm{c}} \tag{3}$$

where $H_{\mathrm{p}}^{\mathrm{BCS}}$ is given in units of Tesla and T_{c} in Kelvin. Hereafter, we call $H_{\mathrm{p}}^{\mathrm{BCS}}$ as the BCS Pauli paramagnetic limit field. The relationship between H_{c2} determined by the orbital effect (H_{c2o}) and H_{p} can be evaluated as

$$\frac{H_{c2o}}{H_{\mathrm{p}}} \sim \left(\frac{g}{2}\right)\left(\frac{m^*}{m_0}\right)\left(\frac{1}{p_{\mathrm{F}}\xi}\right) \tag{4}$$

where g is the effective g-value, m^* is the effective mass, m_0 is the electron mass, p_{F} is the Fermi momentum and ξ is the coherence length [19].

In the organic layered superconductors, the in-plane coherence length ($\xi_{\parallel}$) is longer than the interlayer spacing and hence $1/p_{\mathrm{F}}$. Provided that $g \sim 2$ and $m^* \sim m_0$, H_{p} becomes much larger than H_{c2o}. This means that H_{c2} is determined by the orbital magnetic effect. For layered superconductors, the situation corresponds to the case with the magnetic field applied perpendicular to the plane, where the magnetic field with a finite perpendicular component to the superconducting plane destroys superconductivity by the orbital magnetic effect [5–8]. On the other hand when the direction of the magnetic field is in proximity to the parallel direction, the magnetic field effect to the kinetic energy is depressed because of the weak electron transfer between layers. This can be represented by a very large m^* in eq. (4). When the magnetic field direction is rotated toward the in-plane direction, H_{c2} shoots out. In this case, H_{c2o} becomes very high so that H_{c2} is determined by H_{p}.

For layered organic superconductors, it is intriguing that the upper critical field often exceeds $H_{\mathrm{p}}^{\mathrm{BCS}}$ when the magnetic field is aligned parallel to the superconducting plane [11]. In this article, we discuss the origin briefly. It is interesting that the periodic superconductivity order parameter state [12–16,19,17,18] has been pointed out for ET (or BEDT-TTF) salts with the 2D nature, while the boundless tendency of H_{c2} for $(TMTSF)_2PF_6$ [20,21] is related to the survival of superconductivity in high magnetic fields which may be understood in terms of a triplet pairing [22–24].

2 Upper Critical Field Parallel to the Conducting Axis

In 2D organic superconductors such as $(ET)_2X$ and $(BETS)_2X$ (X represents counter anion), H_{c2} shows a robust change with respect to the inclination angle from the conducting plane and exhibits a sharp maximum in the direction parallel to the plane [4]. For quasi one-dimensional (Q1D) superconductors $(TMTSF)_2X$, H_{c2} exhibits similar anisotropy with respect to the most conducting plane. It has been intriguing then whether the observed value of H_{c2} under the parallel magnetic field exceeds H_p^{BCS} or not [11,25,26]. Also, the pattern of the temperature dependence is to be noted, since it should reflect the underlying pair-breaking mechanism.

Under the dominance of the orbital pair breaking effect, H_{c2} is proportional to $(T_c - T)$ near T_c, while proportional to (1 - $(T/T_c)^2$) at low temperature [7]. When the orbital effect is not effective in the pair-breaking, superconductivity is broken by the spin polarization due to the Zeeman effect. In this case, the temperature dependence of H_{c2} is approximated by $(T_c - T)^{1/2}$. Typical temperature dependences obtained under parallel magnetic field in organic superconductors are compiled in Fig. 1 and Fig. 2 for salts with T_c higher than 4 K (with and without pressure), and in Fig. 3 for salts with T_c lower than 2 K. The absolute values of H_{c2} obtained under the magnetic field parallel to the superconducting plane ($H_{c2\parallel}(0)$) and the BCS Pauli paramagnetic limit (H_p^{BCS}) are listed in Table I together with T_c and $dH_{c2\parallel}/dT$.

The patterns of the temperature dependence are of wide variety. In Fig. 1, $(ET)_4Hg_{2.89}Br_8$, $(ET)_2Cu[N(CN)_2]Br$, $(ET)_2I_3$, and $(BETS)_2GaCl_4$ show that H_{c2} tends to saturate at low temperatures, while H_{c2} of $(ET)_2Cu(NCS)_2$ changes almost linearly with temperature down to 0 K. It is interesting that the pressurized salts of $(ET)_2Cu[N(CN)_2]Br$, $(ET)_2Cu[N(CN)_2]Cl$, $(ET)_2Cu[N(CN)_2]I$ and $(ET)_2Cu(NCS)_2$ shown in Fig. 2 also tend to show the linear temperature dependence although the slope is lowered compared to that for $(ET)_2Cu(NCS)_2$ at ambient pressure. For $(TMTSF)_2ClO_4$ and $(TMTSF)_2PF_6$ in Fig. 3, the $dH_{c2\parallel}/dT$ exhibits divergent behavior near 0 K.

On the other hand, for $(ET)_2Cu[N(CN)_2]Br$, $(ET)_2Cu(NCS)_2$ and for $(ET)_4Hg_{2.89}Br$, H_{c2} starts to increase with a low slope from T_c followed by a high slope on decreasing temperature. The change is ascribed to the 3D to 2D crossover represented by eq. (1), due to the temperature dependence of the coherence length, which becomes divergent in the proximity to T_c. In the high slope region, $dH_{c2\parallel}/dT$ decreases continuously as $(T_c - T)^{1/2}$. The data can be understood in terms of the spin polarization effect coming to the front under the suppression of the orbital effect. This is in agreement with the saturation of H_{c2} at low temperatures. The values of $dH_{c2\parallel}/dT$ in Table I denote the slope appearing at the onset of the high slope region. It is noteworthy that

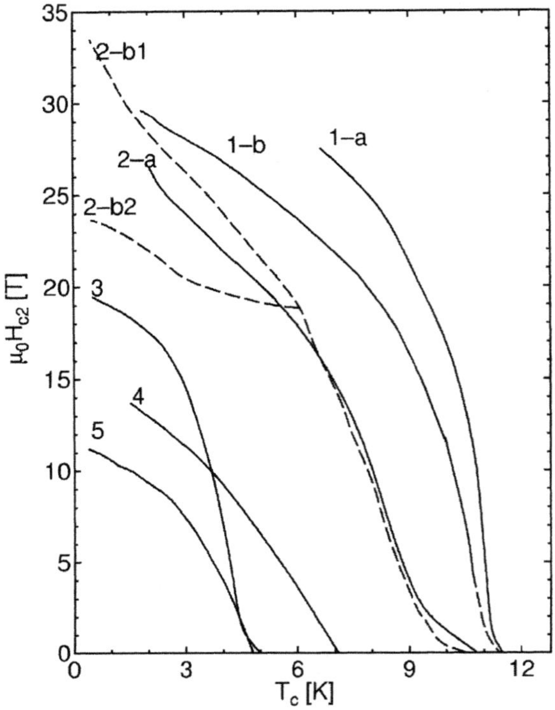

Fig. 1. The temperature dependence of H_{c2} under magnetic fields applied exactly parallel to the superconducting plane. 1-a: κ-$(ET)_2Cu[N(CN)_2]Br$ [28], 1-b: κ-$(ET)_2Cu[N(CN)_2]Br$ [29], 2-a: κ-$(ET)_2Cu(NCS)_2$ [29], 2-b: κ-$(ET)_2Cu(NCS)_2$ [30], 3: κ-$(ET)_4Hg_{2.89}Br_8$ [27], 4: β_H-$(ET)_2I_3$ [25], 5: λ-$(BETS)_2GaCl_4$ [31]

the reported data for $(ET)_2IBr_2$ and $(ET)_2AuI_2$ quoted in Table I may not be intrinsic, since in early days the applied field might not be set so exactly parallel to the superconducting plane.

The possibility that H_{c2} exceeds H_p^{BCS} in the organic superconductors was first pointed out for $(ET)_2I_3$ by Laukhin et al. [25]. Their evaluation, giving the value of $H_{c2}(0)$ corresponding to 1.2 times of H_p^{BCS}, was subject to the experimental uncertainties in determining H_{c2}, for example, due to the width in the resistive transition. Meanwhile, H_{c2} of $(ET)_4Hg_{2.89}Br_8$, reaching more than twice the value of H_p^{BCS} at 1.5 K, appealed the transcendence over H_p^{BCS} [26,27].

For $(ET)_2Cu[N(CN)_2]Br$, the measurement by a pulse magnetic field showed that H_{c2} does not saturate but the slope decreases on approaching 0 K [29]. For this salt, it is remarkable that the low temperature properties including T_c change depending on the cooling procedure. These changes are due to the modification of the ordering in the CH_2-groups existing at the terminals of ET molecules [38]. The difference in the temperature dependence of H_{c2} , shown as 1-a and 1-b in Fig. 1 is ascribed to the heat treatment. The saturating behavior is ascribed partly to the effect of disorder. For $(ET)_4Hg_{2.89}Br_8$, the incommensurate anion

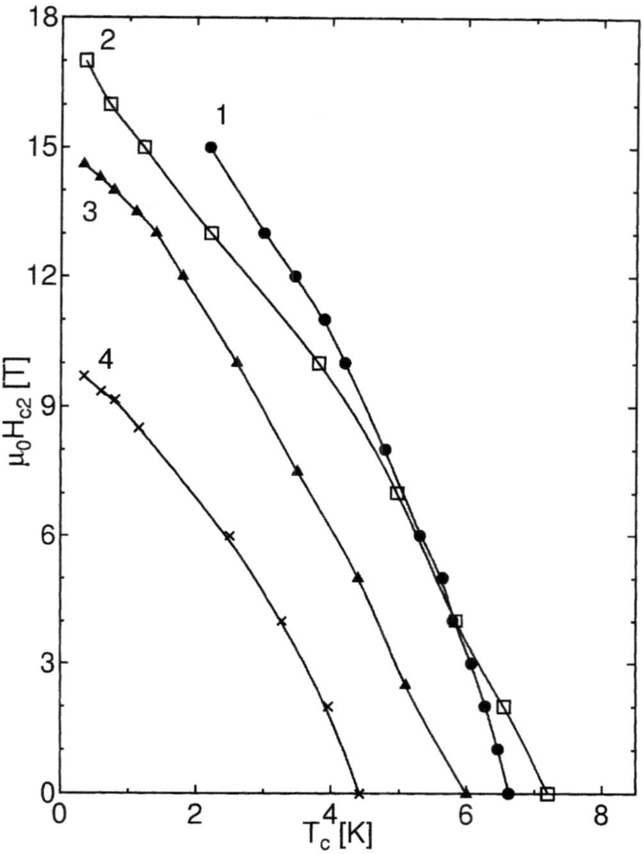

Fig. 2. The temperature dependence of H_{c2} under magnetic fields applied exactly parallel to the superconducting plane. 1: κ-$(ET)_2Cu[N(CN)_2]Cl$ under 1.9 kbar [28], 2: κ-$(ET)_2Cu[N(CN)_2]I$ under 1.2 kbar [33], 3: κ-$(ET)_2Cu[N(CN)_2]Br$ under 1.9 kbar [34], 4: κ-$(ET)_2Cu(NCS)_2$ under 1.5 kbar [34]

content accompanies the disorder resulting in the saturating behavior at low temperatures.

H_{c2} of $(ET)_2Cu(NCS)_2$ is remarkable not only in its absolute value which well exceeds H_p^{BCS}, as determined by the resistive measurement but also in the temperature dependence near 0 K [30], showing no tendency of saturation. The resistive measurement carried out independently [37,29] showed a similar behavior in the low temperature region. It is to be noted that H_{c2} evaluated from the ac magnetic susceptibility is depressed substantially at low temperatures as shown by 2-b2 [30].

The family of $(ET)_2CuX$ (X = $[N(CN)_2]Cl$, $[N(CN)_2]Br$ and $[N(CN)_2]I$), provides the highest T_c among the low dimensional organic superconductors but has to be weakly pressurized to stabilize superconductivity. Application of pressure is useful also to cover the whole T-H phase diagram since T_c and hence

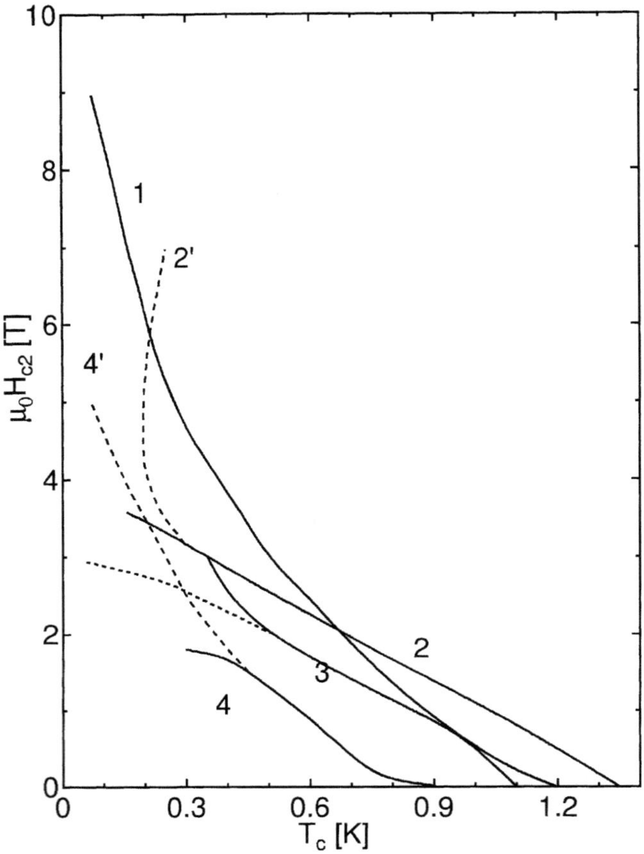

Fig. 3. The temperature dependence of H_{c2} under magnetic fields applied exactly parallel to the superconducting plane. 1: $(TMTSF)_2PF_6$ [21], 2: $(TMTSF)_2ClO_4$ (mid point of transition to zero resistance) [35], 2': $(TMTSF)_2ClO_4$ (onset) [35], 3: $(DMET\text{-}TSF)_2AuI_2$ [36], 4: α-$(ET)_2NH_4Hg(SCN)_4$ (mid point of transition to zero resistance) [32], 4': α-$(ET)_2NH_4Hg(SCN)_4$ (onset of transition) [32]

H_{c2} are reduced. The results obtained under pressure of $1 \sim 2$ kbar together with those for $(ET)_2Cu(NCS)_2$ are compiled in Fig. 2. It is remarkable that the temperature dependence is almost linear in these cases but tends to show a slight decrease in the slope on approaching 0 K. A typical example is shown in Fig. 2 for $(ET)_2Cu(NCS)_2$ in contrast to that at ambient pressure (Fig. 1). It is interesting also that when pressurized, H_{c2} near 0 K region for $(ET)_2Cu[N(CN)_2]Br$ and $(ET)_2Cu(NCS)_2$ decreased substantially to the level of H_p^{BCS} [34].

The temperature dependence of H_{c2} in $(TMTSF)_2PF_6$, pressurized to induce superconductivity, shows a divergent behavior on approaching 0 K [21]. The quoted data were obtained under 5.9 kbar, in the proximity to the antiferromagnetic phase, with the field exactly oriented so that the carrier orbits along the open Fermi surface are confined into the layer plane. On the other

Table 1. $H_{C2\parallel}(0)$ and H_{C2}^{BCS} in organic superconductors

Material	T_C [K]	$H_{C2\parallel}(0)$ [T]	H_{C2}^{BCS} [T]	$dH_{C2\parallel}/dT$ [T/K]	Ref.
$(TMTSF)_2ClO_4$	1.35	>7.0	2.5	3.1	[35]
@H‖b′					
$(TMTSF)_2PF_6$					
@under 5.9 kbar	1.2	>9.0	2.2	3.2	[21]
@H‖b′					
@under 6 kbar	1.14	>4.0	2.1	4.8	[20]
@H‖a					
$(DMET\text{-}TSF)_2AuI_2$	1.1	4.0	2.0	2.7	[36]
$\beta\text{-}(ET)_2I_3$	7.1	~15.0	13.0	3.3	[25]
$\beta\text{-}(ET)_2IBr_2$	2.3	~3.7	4.3	1.4	[2]
$\beta\text{-}(ET)_2AuI_2$	4.0	~6.3	7.4	1.7	[2]
$\kappa\text{-}(ET)_2Cu(NCS)_2$	10.4	34.0	19.0	4.5	[30]
$\kappa\text{-}(ET)_2Cu(NCS)_2$	5.2	12.5	9.6	3.1	[34]
@under 1.5 kbar					
$\kappa\text{-}(ET)_2Cu[N(CN)_2]Br$	11.2	>30.0	20.7	14.0	[28]
$\kappa\text{-}(ET)_2Cu[N(CN)_2]Br$	6.0	15.0	11.1	2.7	[34]
@under 1.9 kbar					
$\kappa\text{-}(ET)_2Cu[N(CN)_2]Cl$	7.1	17.0	13.0	2.7	[32]
@under 1.9 kbar					
$\kappa\text{-}(ET)_2Cu[N(CN)_2]I$	6.7	15.5	12.3	4.3	[33]
@under 1.2 kbar					
$\kappa\text{-}(ET)_4Hg_{2.89}Br_8$	5.8	25.0	10.7	12.0	[27]
$\alpha\text{-}(ET)_2NH_4Hg(SCN)_4$	1.3	5.5	2.4	4.0	[32]
		2.0			
$\lambda\text{-}(BETS)_2GaCl_4$	4.7	12.0	8.7	5.1	[31]

hand, H_{c2} for $(TMTSF)_2ClO_4$ is shown with two branches as represented with solid and broken lines. At low fields, zero resistance was observed and H_{c2} was evaluated from the mid point of the resistive transition as shown by the solid line. At higher fields, the resistivity decreased showing a peak in the temperature dependence. In this case the transition to the superconducting state is not as sharp as for $(TMTSF)_2PF_6$. If we assume that the decrease is due to superconductivity, we obtain the data shown by the broken line. The up-turn behavior is particularly interesting in relation to the recurrence of superconductivity [22] although the experimental results are subject to further invstigation with regard to the mechanism of the resistive transition. It is interesting also to know whether the divergence of H_{c2} is specific to the Q1D Fermi surface of $(TMTSF)_2X$. H_{c2} of $(DMET\text{-}TSF)_2AuI_2$ with a similar Fermi surface, however, showed a linear temperature dependence in the low temperature region ,but not a divergent behavior [36] although $H_{c2}(0)$ is twice the value of H_p^{BCS}.

The data for $(ET)_2NH_4Hg(NCS)_4$ are depicted with two branches shown by the solid and broken lines in Fig. 3. H_{c2} determined from the transition to zero resistance shows a saturation behavior as depicted by the solid line. If we define H_{c2} from the onset of the resistive decrease, assuming that the decrease is due to superconductivity, H_{c2} continues to increase by cooling to very low temperature as extended by the broken line [32]. In this case, the origin of the residual resistance is open to further investigation.

3 Breaking Through the BCS Pauli Limit

There are several factors which can be pointed out as possible sources for breaking through H_p^{BCS}. First, the spin-orbit scattering effect decreasing the effective magnetic moment μ_0 increases H_{c2} (eq. (2)). This was studied for transition-metal dichalcogenide superconductors showing a high $H_{c2}(0)$ which exceeds H_p^{BCS} [40]. In these compounds the transition-metal atoms lying on the superconducting layer serve the spin-orbit interaction. For organic superconductors, the heavy atoms, like Cu, I and Hg, stay mostly in the insulating layers and are kept away from the itinerant electrons. In $(ET)_2Cu[N(CN)_2]Br$ and $(ET)_2Cu(NCS)_2$, the evaluated scattering time of the electrons τ from the Dingle temperature of the quantum oscillation [41,42], which can be shorter than the spin-orbit scattering time τ_{so}, is not low enough to satisfy the condition for the dominance of the spin-orbit scattering time, $\tau eH/m_0c \ll 1$ [39].

Alternatively, the effective magnetic moment in the normal state is depressed by the many body effects or electron correlation effects [43,44]. It is known that the superconducting state of $(ET)_2Cu[N(CN)_2]Br$ is adjacent to the antiferromagnetic state. The effective g-factor of $(ET)_2Cu[N(CN)_2]Br$ from the angle-dependent magneto-quantum oscillation is evaluated to be 1.5 [41]. This should increase the field intensity to satisfy the paramagnetic limit condition. The similar effective g-factor is reported for $(ET)_2Cu(NCS)_2$ and can be more or less common to the low-dimensional organic superconductors with strong electron correlation.

Secondly, the effect of strong coupling giving a high Δ_0 value is pointed out (eq. (2)). Several reports claim that the coupling is stronger than the weak-coupling BCS value: for example, tunneling junction experiments [45], the specific heat jump at T_c [46]. Concerning $(ET)_2Cu[N(CN)_2]Br$, the experimentally evaluated $H_{c2}(0)/H_p^{BCS}$ of 1.7 can be explained by taking into account the decreased g-factor and the enhanced coupling intensity [34]. The factor of 1.31 for $(ET)_2Cu[N(CN)_2]Cl$ and 1.26 for $(ET)_2Cu[N(CN)_2]I$ can be understood along this line.

Interestingly, as possibilities to break through the Pauli paramagnetic effect, the involvement of either the inhomogeneous order-parameter state or the triplet pairing mechanism can be considered. The former works for the singlet pairing resulting in the enhancement of H_{c2} in the low temperature region below $0.56T_c$. It is to be noted that both cases are easily destroyed by disorder. For $(ET)_2Cu[N(CN)_2]Br$, disorder in the terminal ethylene-group arrangement of ET molecules can work to suppress these mechanisms. For $(ET)_2Hg_{2.89}Br_8$ which has substantial disordered structure due to the incommensurate counter anion configuration, these mechanisms are hard to work, although the value of $H_{c2}(0)/H_p^{BCS} = 2.3$ is extraordinary.

3.1 Possibility of Inhomogeneous Order Parameter State

Superconductivity becomes unstable when in a high magnetic field the Zeeman energy of the normal state exceeds the condensation energy in a high magnetic field. Provided that the destruction of superconductivity is controlled by the paramagnetic effect, the self-consistent gap equation is modified and a new type of depaired superconducting ground state can emerge. The state called Fulde-Ferrell-Larkin-Ovchinikov (FFLO or LOFF) was predicted theoretically by Fulde and Ferrell [47] and Larkin and Ovchinikov [48] (in 1964). In relation to the BCS state with singlet pairing electrons, a first-order transition from BCS to FFLO states is expected to take place by increasing an applied magnetic field in a certain condition. With further increase of the field, the FFLO state turns to be the normal one through a second-order transition. The momentum of such pair is not equal to zero and the order parameter determining the spectrum of one-particle excitation turns out to be spatially periodic and inhomogeneous.

A number of theoretical reports assert that the layered clean superconductor with weak interlayer interaction under a magnetic field aligned exactly parallel to the conducting plane is a suitable system to realize the FFLO state [19,12,15,17,18]. The Q2D superconductor like $(ET)_2Cu(NCS)_2$ can be the candidate, since the mean free path in the normal state is very large according to the clear magneto-quantum oscillation. When the inhomogeneous order-parameter state is formed, H_{c2} increases with the decreasing temperature, transcending the paramagnetic limit. The degree of transcendence depends on the dimensionality of the Fermi surface. For 3D superconductors, the enhanced paramagnetic limit is of the order of 10 % [47], while for 2D one the enhancement factor can reach 1.4 and increases with increasing the nesting tendency in the Fermi surface [51]. For 1D one the paramagnetic limit even diverges at 0 K [15].

The FFLO state can be detected through the electrodynamic properties such as the magnetic field penetration, which becomes anisotropic and depends on the direction of the magnetic field. Also, since the one-particle excitation spectrum is expected not to have a gap, due to the periodic order parameter distribution, thermodynamic properties like the heat capacity vary with temperature as a power-law instead of exponential.

The difference in the H_{c2} for $(ET)_2Cu(NCS)_2$ detected through the resistivity and the magnetic susceptibility, represented by two lines denoted 2-b1 and 2-b2 in Fig. 1, was ascribed to the FFLO state [30]. The line 2-b1 detected by the resistivity was understood as the border line between the FFLO state and the normal state, while the line 2-b2 was assigned to that between the BCS and FFLO states. From the similarity of a theoretical phase diagram [19], it was claimed that 2-b2 line corresponds to the transition line to the FFLO state.

The sensitivity to the disorder provides another clue to find the FFLO state. Recently the thermal conductivity of $(BETS)_2GaCl_4$ in the parallel magnetic field was found to show a distinctive behavior in the H_{c2} region depending on the sample quality [49]. The thermal conductivity in disordered samples showed an abrupt transition to the normal state, ascribable to the first-order transition to the normal state, as expected for Pauli paramagnetic limiting case. On the other hand, the thermal conductivity in pure crystals started to show a change at the lower field but reached the normal state in the higher field. The former can be interpreted in terms of the formation of the FFLO state. The advantage of the thermal conductivity probing the superconducting state is that it does not have direct concern with magnetic vortices, in contrast to the resistive and magnetic measurements.

In order to derive a definite conclusion on the FFFF state, the observation of the modulated order parameter state, for example, by a scanning tunneling spectroscopy is required[19].

3.2 Relation to Triplet Pairing

The value of H_{c2} of $(TMTSF)_2PF_6$ reaches 9 T, $\sim 4H_{c2}^{BCS}$ [21] (Fig. 3). The superconductivity is sensitive to nonmagnetic defects. To interpret the divergent behavior of H_{c2}, far exceeding H_{c2}^{BCS}, the scenario ascribing to the the FFLO state can be considered. In the FFLO state of a pure 1D system, H_{c2} can be infinite. In a realistic system with non-negligible dispersion as represented by a Q1D band structure, however, H_{c2} extended to higher field by the FFLO mechanism is bounded. The value is estimated to be in the order of 3-4 T for $(TMTSF)_2PF_6$ [24].

We note that in a triplet superconductor the decrease in T_c with increasing field can be recovered at high fields, provided that the magnetic field is aligned exactly along the plane on which the parallel spins are lying. This is because the orbital destructive effect is suppressed under the aligned high magnetic field while the pair with parallel spins is not. As a result, reentrance of superconductivity in a very high magnetic field is expected for triplet-pairing superconductors. The data of $(TMTSF)_2PF_6$ and $(TMTSF)_2ClO_4$ shown in Fig. 3 remind us the feature connected to the triplet pairing, but the definite reentrance is not yet obtained. It is noteworthy that superconductivity reentrance can be suppressed even by extremely weak impurity potential[55]

Superconductivity of $(TMTSF)_2X$ ($X = PF_6$, ClO_4) has been considered to be unusual due to the sensitivity to nonmagnetic impurities [50]. Pulsed NMR Knight shift measurements of ^{77}Se indicate that no observable change between the metallic and superconducting states. Since the Knight shift is linearly dependent on the electron spin susceptibility, it means that the spin polarization is not changed at the superconducting transition [35]. However, the experimental results seem to be in controversy around the conditions of the measurement. Thus the relevance to triplet pairing is still open for further study.

4 Concluding Remarks

In this review paper, we presented the T-H phase diagrams placing emphasis to cover the whole regime. We showed then how the effect of the spin polarization effect comes to the front in the high field limit.

The recent report on the magnetic-field-induced superconductivity in the material $(BETS)_2$-$FeCl_4$ [56,57] suggests that the internal magnetic field by magnetic anions such as $FeCl_4$ existing in the insulating layer provides a new realm for the study of superconductivity in a high magnetic field.

Acknowledgment

The author would like to express sincere thanks to S. Kamiya, M. V. Kartsovnik, V. Kresin, A. G. Lebed, Y. Maeno, K. Maki, V. P. Mineev, E. Ohmichi, H. Shimahara, Y. Shimojo and M.A. Tanatar for helpful discussion.

References

1. J.M. Williams, J.R. Ferraro, R.J. Thorn, K.D. Carlson, U. Geiser, H.H. Wang, A.M. Kini, M.-H. Whangbo: *Organic Superconductors (Including Fullerenes)* (Printece Hall, Eaglewood Cliff, NJ, 1992)
2. T. Ishiguro, K. Yamaji, G. Saito: *Organic Superconductors, Second Edition* (Springer, Heidelberg, 1998)
3. S. Kagoshima, R. Kato, H. Fukuyama, H. Seo, H. Kino: in *Advances in Synthetic Metals: Twenty Years of Progress in Science and Technology* eds. P. Bernier, S. Lefrand and G. Bidan (Elsevier, Amsterdam, 1999) p.262
4. M. Tinkham: *Introduction to Superconductivity* 2nd ed. (McGraw-Hill, New York, 1996) p. 139
5. A.A. Abrikosov: Sov. Phys. JETP **5**, 1174 (1957)
6. L.P. Gor'kov: Sov. Phys. JETP **10**, 593 (1960)
7. N.R. Werthmer, E. Helfand, P.C. Hohenberg: Phys. Rev. **147**, 295 (1966)
8. K. Maki, T. Tsuneto: Prog. Theor. Phys. **31**, 945 (1964)
9. A.M. Clogston: Phys. Rev. Lett. **9**, 266 (1962)
10. B.S. Chandraskhar: Appl. Phys. Lett. **1**, 7 (1962)
11. T. Ishiguro: J. Phys. IV France **10**, Pr3-139 (2000)
12. L.N. Bulaevskii: Sov. Phys. JETP **38**, 634 (1974)
13. L.W. Gutenberg, L. Gunter: Phys. Rev. Lett. **16**, 996 (1966)

14. K. Aoi, W. Dieterich, P. Fulde: Z. Phys. **267**, 223 (1974)
15. A.I. Buzdin, J.P. Brison: Europhys. Lett. **35**, 707 (1996)
16. N. Dupuis: J. Supercon. **12**, 475 (1999) N. Dupuis, G. Montambaux: Phys. Rev. B **49**, 8993 (1994) N. Depuis, G. Montambaux, C.A.R. Sa de Melo: Phys. Rev. Lett. **70**, 2613 (1993)
17. H. Burkhardt, D. Rainer: Ann. Physik **3**, 181 (1994)
18. C.A.P. Sa de Melo: J. Supercon. **12**, 459 (1999)
19. H. Shimahara: J. Supercon. **12**, 469 (1999) Phys. Rev. B **50**, 12760 (1994)
20. I.J. Lee, M.J. Naughton, G.M. Danner, P.M. Chaikin: Phys. Rev. Lett. **78**, 3555 (1997)
21. I.J. Lee, P.M. Chaikin, M.J. Naughton: Phys. Rev. B **62**, R14669 (2000)
22. A.G. Lebed: JETP Lett. **44**, 144 (1986) J. Supercond. **12**, 453 (1999)
23. A.G. Lebed: Phys. Rev. B **59**, R721 (1999)
24. A.G. Lebed, K. Machida, M. Ozaki: Phys. Rev. B **62**, R795 (2000)
25. V.N. Laukhin, S.I. Pesozkii, E.B. Yagbuskii: JETP Lett. **45**, 392 (1987)
26. R.B. Lyubovskii, R.N. Lyubovskaya, O.A. D'yachenko: J. Phys. I France **6**, 1609 (1996)
27. E. Ohmichi, T. Ishiguro, T. Sakon, T. Sasaki, M. Motokawa, R.B. Lyubovskii, R.N. Lyubovskaya: J. Supercon. **12**, 505 (1999)
28. Y. Shimojo, A.E. Kovalev, S. Kamiya, E. Ohmichi, T. Ishiguro, H. Yamochi, G. Saito, A. Ayari, P. Monceau: Physica B **294-295**, 427 (2001)
29. E. Ohmichi: private communication
30. J. Singleton, J.A. Symington, M.-S. Nam, A. Ardavan, M. Karmoo, P. Day: J. Phys.: Condens. Matter **12**, L641 (2000)
31. M.A. Tanatar, T. Ishiguro, H. Tanaka, A. Kobayashi, H. Kobayashi: J. Supercond. **12**, 511 (1999)
32. Y. Shimojo, T. Ishiguro, H. Yamochi, G. Saito: unpublished
33. M.A. Tanatar, T. Ishiguro, N.D. Kushch, E.B. Yagubskii: unpublished
34. Y. Kamiya, Y. Shimojo, M.A. Tanatar, T. Ishiguro, G. Saito, J. Yamada: unpublished
35. I.J. Lee, A.P. Hope, M.J. Leone, M.J. Naughton: Synth. Met. **70**, 747 (1995)
36. Y. Shimojo, M.A. Tanatar, T. Ishiguro, R. Kato: unpublished
37. F. Zou, J.S. Brooks, R.H. MaKenzie, J.A. Schlueter, J.M. Williams: Phys. Rev. B **61**, 750 (2000)
38. See for example, M.A. Tanatar, T. Ishiguro, T. Kondo, G. Saito: Phys. Rev. B **59** 3841 (1999) and the references.
39. A.E. Kovalev, T. Ishiguro, T. Kondo, G. Saito: Phys. Rev. B **62**, 103 (2000)
40. R.A. Klemm, A. Luther, M.R. Beasley: Phys. Rev. B **12**, 877 (1975)
41. H. Weiss, M.V. Kartsovnik, W. Biberacher, E. Balthes, A.G.M. Jansen, N.D. Kushch: Phys. Rev. B **60** R16259 (1999)
42. J. Wosnitza: Fermi surfaces of low-dimensional organic metals and superconductors, Springer Tracts in Modern Physics vol. 134, Springer, Berlin, 1996
43. M. Schossmann, J.P. Garbotte: Phys. Rev. B **39**, 4210 (1989)
44. T.P. Orlando, M.R. Beaseley: Phys. Rev. Lett. **46**, 1589 (1981)
45. See for example, M.E. Howley, K.E. Gray, B.D. Terris, H.H. Wang, K.D. Carlson, J.M. Williams: Phys. Rev. Lett. **57**, 629 (1986), more recently, T. Arai, K. Ichimura, K. Nomura et al.: Phys. Rev. B **63**, 104518 (2001)
46. B. Andraka, J.S. Kim, G.R. Stewart et al.: Solid State Commun. **79** 57 (1991) I.I. Elsinger, J. Wosnitza, S. Wanka et al.: Phys. Rev. Lett. **84** 6098 (2000)
47. P. Fulde, R.A. Ferrell: Phys. Rev. **135**, A550 (1964)

48. A.I. Larkin, Yu. N. Ovchinikov: Sov. Phys. JETP **20**, 762 (1965)
49. M.A. Tanatar, T. Ishiguro, A.G. Lebed, H. Tanaka, H. Kobayashi: unpublished
50. D. Jérome, H.J. Schultz: Adv. Phys. **31**, 299 (1982)
51. H. Shimahara: J. Phys. Soc. Jpn. **68**, 3069 (1999)
52. I.J. Lee, D.S. Chow, W.G. Clark, J. Strouse, M.J. Naughton, P.M. Chaikin, S.E. Brown: Phys. Rev. Lett. **88**, 17004 (2002)
53. A.G. Lebed, K. Yamaji: Phys. Rev. Lett. **80**, 2697 (1998)
54. M. Miyazaki, K. Kishigi, Y. Hasegawa: J. Phys. Soc. Jpn. **67**, L2618 (1998)
55. V.P. Mineev: J. Phys. Soc. Jpn. **69**, 3371 (2000)
56. S. Uji, H. Shinagawa, T. Terashima, T. Yakabe, Y. Terai, M. Tokumoto, A. Kobayashi, H. Tanaka, H. Kobayashi: Nature **410**, 908 (2001)
57. L. Balicas, J.S. Brooks, K. Storr, S. Uji, M. Tokumoto, H. Tanaka, H. Kobayashi, A. Kobayashi, V. Barzykin, L.P. Gor'kov: Phys. Rev. Lett. **87** 067002-1 (2001)

Vortex Phases

T. Giamarchi[1] and S. Bhattacharya[2]

[1] Laboratoire de Physique des Solides, CNRS-UMR8502, UPS Bât 510, 91405 Orsay, France
[2] NEC Research Institute, Princeton, New Jersey 08540, USA

Abstract. This chapter provides a pedagogical introduction, and presents the latest theoretical and experimental developments on the physics of vortices in type II superconductors.

1 Background

The discovery of high T_c superconductors has shattered the comforting sense of understanding that we had of the phase diagram and physical properties of type II superconductors, and in particular of the mixed (vortex) phase in such systems. Indeed, it was well known since Abrikosov [1] that above H_{c1} the magnetic field penetrates in the form a vortex, made of a filament of radius ξ (the coherence length) surrounded by supercurrent screening the external field running over a radius λ (the penetration depth). Because of the repulsion between vortices due to supercurrents (see Fig. 1), the naive idea is that vortices will form a perfect triangular crystal (the Abrikosov lattice). This has led to the phase diagram shown in Fig. 1, that has been the cornerstone of our understanding of all type II superconductors for more than three decades [2,3]. However in high T_c, one could reach much higher temperatures, and it was soon apparent that some of the physics linked to the existence of the thermal fluctuations and disorder had been overlooked. This led to a burst of investigations, both theoretical and experimental, to understand the physical properties of such vortex matter. Of course, high T_c were not the only field of investigations and low T_c superconductors were reexamined as well, now that we knew what to look for in them.

Indeed the vortex phase provides an excellent system for both the fundamental researcher and one in search of physics with useful applications. From the fundamental point of view, vortex matter provides a unique system where one can study a crystal, in which one can vary the density (the lattice spacing) at the turn of a knob (simply by varying the magnetic field). In addition, because this crystal is embedded in a "space" with a much finer lattice constant (the real atomic crystal), it can be submitted to various perturbations such as disorder, difficult to investigate in normal crystals. This provides thus a unique opportunity to study the combined effect of disorder and thermal fluctuations on a crystal. From a more practical point of view, if one has vortices in a superconductor and passes a current I through it, the current will act as a force on the magnetic tubes that are the vortices, and they will start to slide. The sliding

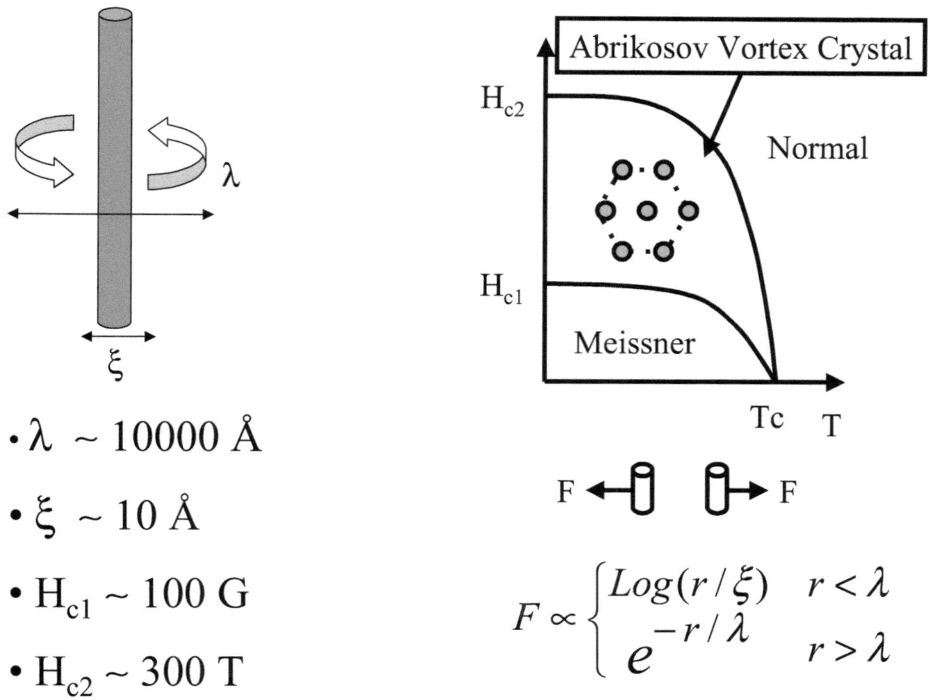

Fig. 1. The structure of a vortex, with a core size ξ and supercurrent running over a radius λ the penetration length. Typical values for high T_c are given. Due to supercurrents vortex repel with a force F. This leads to the naive phase diagram where a crystal of vortex exists between H_{c1} and H_{c2}.

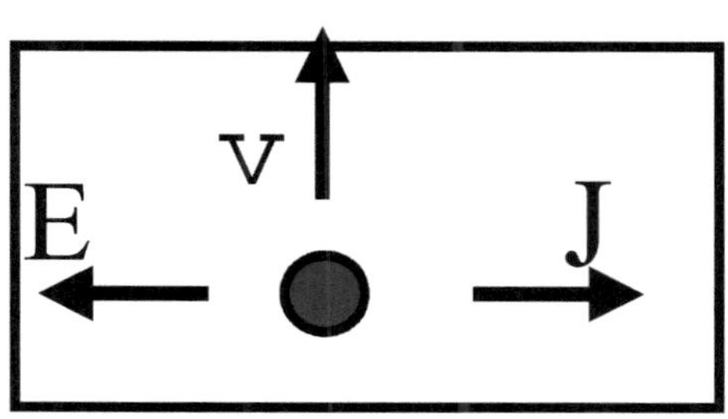

Fig. 2. A current density J driven in a superconductor exerts a force on the vortices, making them slide sideways with a velocity v. Since the vortices are flux tubes this generates an electric field E in the direction of the current and thus a finite (and rather high) resistance.

in turn generates an electric field E, which means that the superconductor is not superconducting any more, due to the motion of vortices (see Fig. 2). The resistance is rather large and is simply $\rho = \rho_n(H/H_{c2})$ (Bardeen–Stephens resistance). In order to get a good superconductor, it is thus necessary to prevent the vortices from moving by pinning them. Hence the strong practical incentive to understand the properties (both static and dynamics) of vortices in the presence of disorder.

A study of vortices prompts several questions, that we will try to address in these notes:

- What is the effect of disorder on the Abrikosov vortex crystal?
- How is one to describe the vortex phase? Does one need the full Ginzburg-Landau description or can one use a simplified description modeling vortices as elastic spaghettis?
- What is the phase diagram and what are the static physical properties of the vortices in presence of disorder and thermal fluctuations?
- What are the dynamical properties? Is there a linear response and more generally what is the $I-V$ characteristics?
- What is the nature of the vortex system when it is in motion?
- Are there links with other physical systems exhibiting the same competition between crystal order and disorder?

Of course, these questions have been examined intensively in the last 30 years, and represent an impressive body of research work. So in these few pages we have to make a choice. Reviews already exist on vortices [4–7] so we will try to present in these notes the basic ideas enabling the reader to understand the concept behind the variety of studies, and then bring the reader up to date with the recent theoretical and experimental developments that have been left out of the previous reviews. The plan of this chapter is as follows: in Section 2 we discuss an elastic description of the vortices, and the issue of lattice melting. In Section 3 we discuss the effects of disorder, introduce the basic lengthscales and physical concepts and discuss the limitations of the previously proposed solutions to tackle this problem. In Section 4 we discuss the recent theory of the Bragg glass and compare it with the host of experimental data on the statics of the vortex lattice. In Section 5 we discuss the more complicated issue of the dynamics. Finally conclusions, perspectives and contact with other physical systems can be found in Section 6.

2 Elastic Description of Vortices

A way to get a tractable description of the vortex lattice is to ignore the microscopic aspects of the superconducting state and the Ginzburg-Landau description of the vortex, and simply consider the vortices as an elastic object. The core is like a piece of string and the supercurrents provide the repulsive (elastic) forces. Of course such a description is a simplification and, depending on the problem, it will be necessary to check that important physics has not been left out in the process. However, such a description has the advantage of being simple enough so that additional effects such as disorder can be included, and retains in fact most of the interesting physics for a macroscopic description of the vortex lattice (phase diagram, imaging, transport). Another advantage is that this allows us to make contact with a large body of related problems as will be discussed in section 6.

We thus describe the vortex system as objects having an equilibrium position R_i^0 (on a triangular lattice for the vortices, but this is of course general) and a displacement u_i compared to this equilibrium position. u_i is a vector with a certain number n of components. The vortices being lines $n = 2$, since displacements are on the plane perpendicular to the z axis. The elastic Hamiltonian is

$$\mathcal{H} = \frac{1}{2\Omega} \sum_{\alpha\beta} \sum_{q} c_{\alpha\beta}(q) u_\alpha(q) u_\beta(-q) \tag{1}$$

where $\alpha = x, y, z$ are the spatial coordinates. The $c_{\alpha\beta}(q)$ are the elastic constants. The fact that they have a non trivial dependence on q comes from the long range nature of the forces between the vortices. The c can be computed from the microscopic forces between vortices. Standard elasticity corresponds to $c(q) = cq^2$. Such a behavior will always be correct at large distance (small q) since the forces have a finite range λ. In (1) various physical process have in principle to be distinguished and correspond to different elastic constants. This corresponds to bulk, shear and tilt deformations of the vortex lattice as shown on Fig. 3. Although these different elastic constants can be widely different in magnitude (for example bulk compression is usually much more expensive than shear), this is a simple practical complication that does not change the quadratic nature of the elastic Hamiltonian. Such a description is of course also valid for anisotropic superconductors (such as the layered High T_c ones), provided that the anisotropy is not too large. If the material is too layered then it is better to view the vortices as pancakes living in each plane and coupled by Josephson or electromagnetic coupling between the planes as shown in Fig. 4. For moderately anisotropic materials viewing the stack of pancakes as a vortex line is however enough. In this notes we will stick to this description.

The melting of the vortex lattice can easily be extracted from the elastic description. Although a detailed theory of melting is still lacking, one can use a basic criterion, known as Lindemann criterion that states that the crystal melts when the thermally induced displacements of a particle in the crystal becomes some sizeable fraction of the lattice spacing. On a more formal level the melting

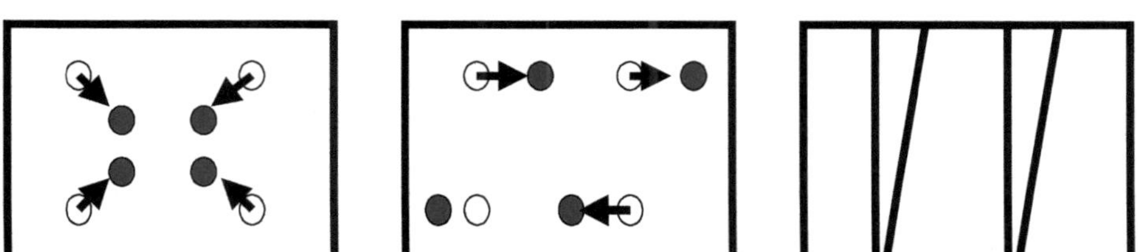

Fig. 3. Three different deformations of the vortex lattice (compression (bulk), shear and tilt) correspond to three different elastic constants, respectivelly called c_{11}, c_{66} and c_{44}.

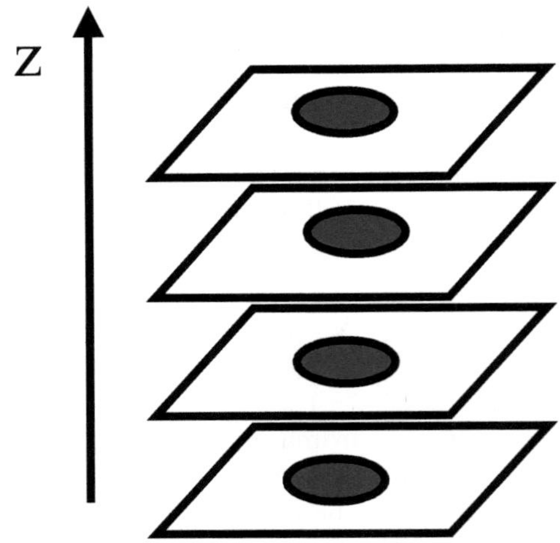

Fig. 4. In very anisotropic materials, it is best to see a vortex line as a stack of pancakes vortices coupled by electromagnetic or Josephson coupling. For most cases however a line description will be sufficient.

is given by

$$\langle u^2 \rangle = l_{\mathrm{T}}^2 = C_{\mathrm{L}}^2 a^2 \tag{2}$$

which defined the "effective" (thermal) size of the particle (also known as Lindemann length). The proportionality constant that reproduces correctly the melting is empirically determined to be $C_{\mathrm{L}} \sim 0.1 - 0.2$. A simple calculation based on the elastic description thus gives

$$\frac{T_m}{c} = C_{\mathrm{L}}^2 a^2 \tag{3}$$

showing that the melting temperature goes down as the lattice spacing goes down (or the magnetic field up). Of course the full quantitative study for the vortex lattice should be done with the full fledged elastic Hamiltonian (including bulk, shear, tilt), but the main conclusions remain unchanged [8,9]. This leads to the first modification of the naive Abrikosov phase diagram taking into account the melting shown in Fig. 5. Close to H_{c1} the elastic constants drop down (since the vortices get separated by more than λ the force between them is exponentially small), and from (3) the crystal also melts, leading the reentrant behavior shown in Fig. 5.

Early experimental studies [10–13] of the melting transition were based on transport measurements. The sudden onset of an ohmic resistance was argued to signify melting and its location in the (H, T) space is the locus of the melting phase boundary. Typical experimental results are shown in Fig. 6, where the onset is characterized by a pronounced knee in the resistance. With increasing external field, the onset moves to lower temperatures and eventually broadens considerably. This implies that at sufficiently large fields, the sharp first order transition crosses over to a more continuous second order like transition, an effect expected to be the result of disorder (see later). At lower fields and higher

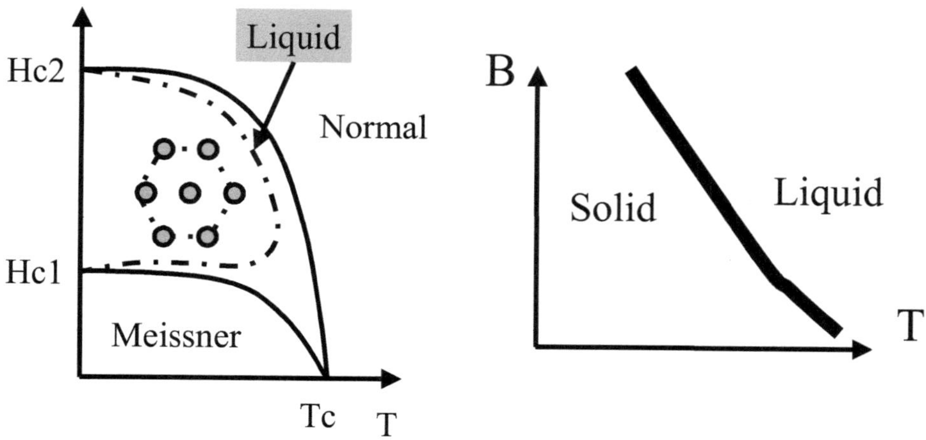

Fig. 5. Thermal fluctuations induce melting of the vortex lattice. On the left the full melting curve is shown. For high T_c, in practice one is often far from both H_{c1} and H_{c2} leading to the apparent phase diagram shown on the right.

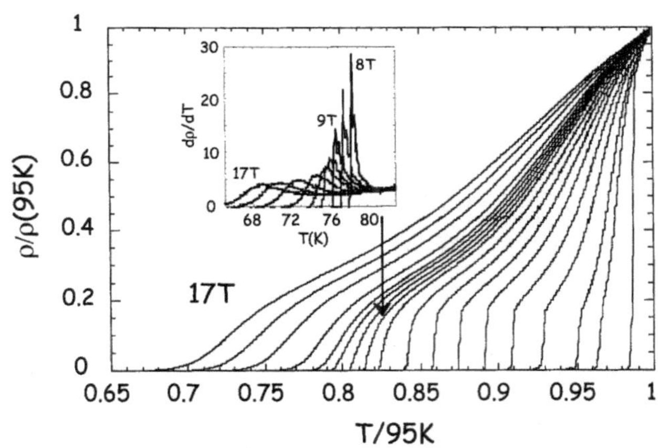

Fig. 6. T-dependence of ohmic resistance at various field values. The sudden onset marked by a pronounced knee marks the melting transition. The inset shows a plot of the temperature derivative of the resistance which, surprisingly mimics the specific heat jump across the transition [13]

temperatures, the melting transition is closely approximated by the disorder-free case, discussed above. Later experiments performed on thermodynamic quantities such as the magnetization and specific heat confirmed the first order nature of the melting transition, at least for weak disorder. Fig. 7 shows an experimental phase diagram of BSCCO obtained by local magnetization using a novel hall-bar technique [14]. Figure 8 shows typical measurements of the jump in the local induction across this transition, by either changing temperature at a fixed field or changing field at a fixed temperature. In both cases, a positive

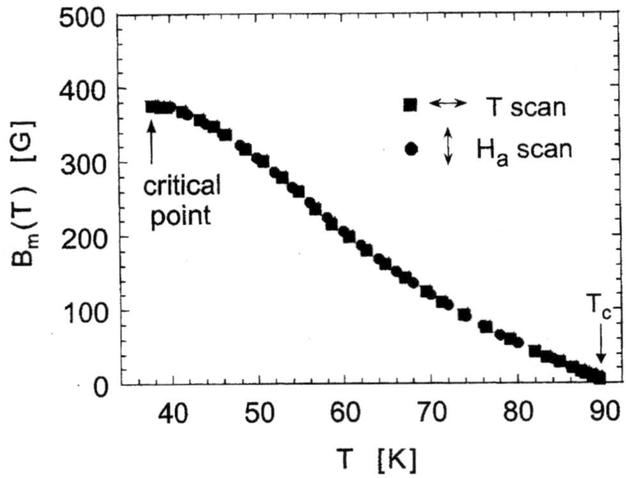

Fig. 7. Melting phase diagram obtained from the jump in local induction for both isothermal and isofield data [14]. Note the similarity of the phase boundary with theoretical expectations in Fig. 5 above.

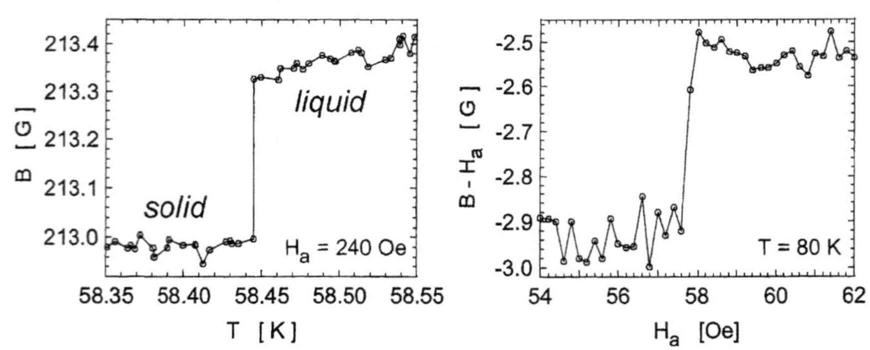

Fig. 8. Typical jumps in local induction for isothermal and isofield measurements from Hall bar method [14]. Both the solid and liquid phases have no measurable pinning in the bulk and the data are reversible.

jump in B is observed in going from the solid to the liquid phase. Using the Clapeyron-equation:

$$\frac{dH_{\mathrm{m}}}{dT} = -\frac{\Delta S}{\Delta M} \tag{4}$$

where ΔS is the entropy change and ΔM is the change in magnetization (density of vortices). The experimental phase diagram is consistent with theoretical expectations as shown in Fig. 5 above. The negative slope of the melting curve

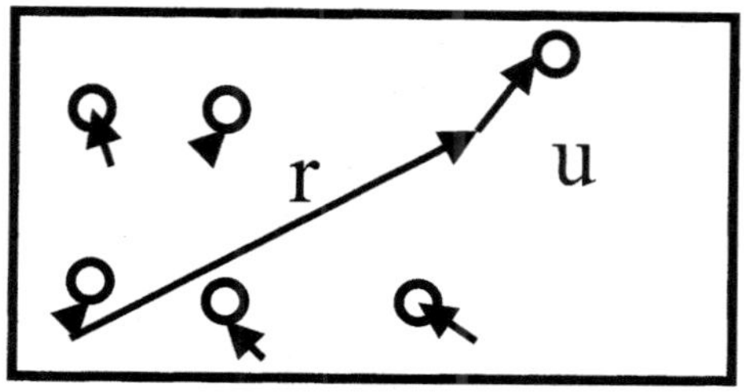

Fig. 9. The relative displacement correlation function $B(r)$ measuring the displacements between two points separated by a distance r is directly measured in experiments such as decoration experiments.

is consistent with an increase of density in the liquid phase, curiously akin to a "water-like" melting phenomenon. Very little experimental work is available on the low field reentrant branch of the melting phase boundary. At low fields the intervortex interaction is weak and the effects of disorder dominate. Reentrant phenomena in peak effect (see Section 4.4) have been somewhat widely observed and is thought to be dominated by effects of disorder, rather than thermal fluctuations.

Finally, besides the phase diagram, what are the physical quantities that one can in principle compute and that are directly connected to experiments? The first important information is the relative displacements correlation function

$$B(r) = \overline{\langle [u(r) - u(0)]^2 \rangle} \tag{5}$$

where $\langle\rangle$ is the average over thermal fluctuations and $\overline{\cdots}$ is the average over disorder (if need be). (5) indicates how the displacements between two points in the system separated by a distance r grow (see Fig. 9). Decoration experiments provide a direct measure of this correlation function as we will see. In a perfect crystal $B(r) = 0$, whereas both thermal fluctuations and disorder will make the displacements grow. How $B(r)$ grows tells us whether the system is well ordered or not. In a good crystal $B(r)$ will saturate to a finite value whereas it will grow unboundedly if the perfect positional order of the crystal is destroyed.

Another important quantity, directly measures in diffraction experiments is the structure factor

$$S(q) = \overline{\langle \rho(q)\rho(-q) \rangle} \tag{6}$$

In a perfect crystal this consists of Bragg peaks at the vectors K of the reciprocal lattice of the crystal. If one considers one such peak its shape (as shown on Fig. 10) is the Fourier transform of the positional correlation function

$$C(r) = \overline{\langle e^{iKu(r)} e^{-iKu(0)} \rangle} \tag{7}$$

Thus in a perfect crystal $C(r) = 1$ and the Fourier transform is a $\delta(q)$ Bragg peak. If there are only thermal fluctuations $C(r \to \infty) \to$ Cste (in fact $C(\infty) = e^{-K^2 l_{\mathrm{T}}^2/2}$). The Fourier transform is still a δ peak but with a reduced weight which is simply the Debye Waller factor. The faster $C(r)$ decreases the more

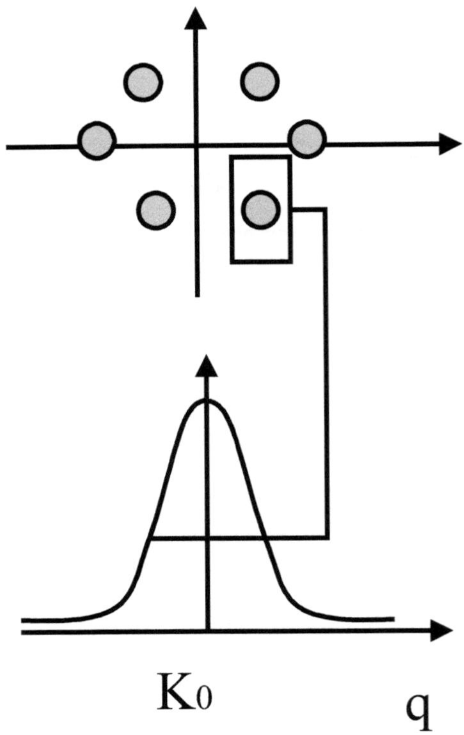

Fig. 10. The structure factor measured in diffraction experiments such as neutrons or X-rays. The shape of a peak is given by the Fourier transform of the positional correlation function $C(r)$ (see text).

disordered is the crystal. If $C(r)$ decreases exponentially to zero with a characteristics lengthscale R_a, the peak in the structure factor is some broad (lorentzian like) shape with a width R_a^{-1} indicating that the perfect translational order is lost. Although it is not always true (it is only true for gaussian fluctuations such as thermal fluctuations) a rule of thumb is

$$C(r) \sim e^{-K^2 B(r)/2} \tag{8}$$

showing quite logically that the faster the relative displacements grow the more positional order (measured by the Bragg peaks) is destroyed in the system.

3 Disorder, Basic Lengths and Open Questions

The next task is to consider the effects of disorder. In real systems, disorder exists in all varieties in the underlying atomic crystal: vacancies, interstitials, lattice dislocations, grain boundaries, twin boundaries, second-phase precipitates, etc. In high quality single crystals it is possible to limit dominant disorder to point

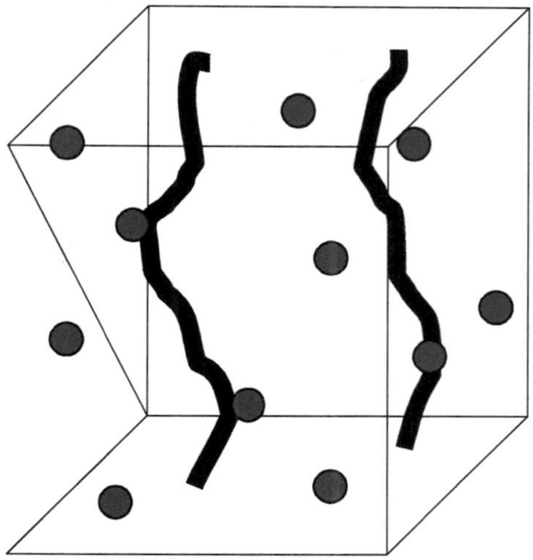

Fig. 11. Point like impurities (small circles) act as pinning centers for the vortices.

like impurities. Additionally, point like disorder can be, and sometimes is, intentionally added to systems in the form of electron irradiation in the case of the cuprates and/or substitutional atomic impurities of various kinds in low T_c systems. More artificial disorder can be introduced for example by heavy ions irradiation that produce columns of defects in the material. We will briefly mention the consequences of such artificial disorder in Section 6, but most of these notes will be devoted to the effects of point like impurities.

Such impurities (as shown on Fig. 11) can be modeled by a random potential $V(r)$ coupled directly to the density $\rho(r) = \sum_i \delta(r - R_i^0 - u_i)$ of vortex lines

$$\mathcal{H}_{\text{dis}} = \int d^d r V(r) \rho(r) \tag{9}$$

In principle one has "just" to add (9) to (1) and solve. Unfortunately the coupling of the displacements to disorder is highly non linear (since it occurs inside a δ function), and thus this is an horribly complicated problem. Physically this traduces the fact that there is a competition between the elastic forces that want the system perfectly ordered and the disorder that let the lines meander. This competition is bound to lead to complicated states where the system tries to compromise between these two opposite tendencies.

In order to understand the basic physics of such problem a simple (but ground breaking !) scaling argument was put forward by Larkin [15]. To know whether the disorder is relevant and destroys the perfect crystalline order, let us assume that there exists a characteristic lengthscale R_a for which the relative displacements are of the order of the lattice spacing $u(R_a) - u(0) \sim a$. If the displacements vary of order a over the lengthscale R_a the cost in elastic energy from (1) is

$$\frac{c}{2} R_a^{d-2} a^2 \tag{10}$$

by simple scaling analysis. Thus in the absence of disorder minimizing the energy would lead to $R_a = \infty$ and thus to a perfect crystal. In presence of disorder the fact that displacements can adjust to take advantage of the pinning center on a volume of size R_a^d allows to gain some energy. Since V is random the energy gained by adapting to the random potential is the square root of the potential over the volume R_a^d, thus one gains an energy from (9)

$$-VR_a^{d/2}\rho_0 \tag{11}$$

Thus minimizing (10) plus (11) shows that below four dimensions the disorder is always relevant and leads to a finite lengthscale

$$R_a \sim a\left(\frac{c^2 a^d}{V^2\rho_0^2}\right)^{1/(4-d)} \tag{12}$$

at which the displacements are of order a. The conclusion is thus that even an arbitrarily weak disorder destroys the perfect positional order below four dimensions, and thus no ordered crystal can exist for $d \leq 4$. This is an astonishing result, which has been rediscovered in other context (for charge density waves R_a is known as Fukuyama-Lee [16] length and for random field Ising model this is the Imry-Ma length [17]). Of course it immediately prompts the question of what is the resulting phase of elastic system plus disorder?

Since solving the full problem is tough another important step was made by Larkin [15,18]. For small displacements he realized that (9) could be expanded in powers of u leading to the simpler disorder term

$$H_{\text{larkin}} = \int d^d r f(r) u(r) \tag{13}$$

where $f(r)$ is some random force acting on the vortices. Because the coupling to disorder is now linear in the displacements the Larkin Hamiltonian is exactly solvable. Taking a local random force $\overline{f(r)f(r')} = \Delta\delta(r-r')$ gives for the relative displacements correlation function and structure factor

$$B(r) = B_{\text{thermal}}(r) + \frac{\Delta}{c^2} r^{4-d} \tag{14}$$

$$C(r) = e^{-K^2 B(r)/2} \simeq e^{-r^{4-d}} \tag{15}$$

where B_{thermal} are the displacements in the absence of disorder due to thermal fluctuations (which remain bounded in $d > 2$ and are thus negligible at large distance compared to the disorder term). Thus the solution of the Larkin model confirms the scaling analysis: (i) displacements do grow unboundedly (as a power law) and thus perfect positional of the crystal is lost; (ii) the lengthscale at which the displacements are of the order of the lattice is similar to the one given by the scaling analysis. In addition the Larkin model tells us how fast the positional order is destroyed: the displacements grow as power law thus the positional is destroyed exponentially fast, leading to peaks in the structure factor of width

R_a^{-1}. However all these conclusions should be taken with a grain of salt. Indeed the Larkin model is an expansion in powers of u, and thus cannot be valid at large distance (since the displacements grow unboundedly the expansion has to break down at some lengthscale). What is this characteristic lengthscale? A naive expectation is that the Larkin model ceases to be valid when the displacements are of order a i.e. at $r = R_a$. In fact this is too naive as was noticed by Larkin and Ovchinikov. To understand why, let us rewrite the density of vortices in a more transparent way than the original form

$$\rho(x) = \sum_i \delta(r - R_i^0 - u_i) \tag{16}$$

Taking the continuum limit for the displacements in (16) should be done with care since one is interested in variations of the density at scales that can be *smaller* than the lattice spacing. A very useful way to rewrite the density is [19,20] (see also [21,22]):

$$\rho(r) = \rho_0 - \rho_0 \nabla \cdot u(r) + \rho_0 \sum_K e^{iK(r-u(r))} \tag{17}$$

which is a decomposition of the density in Fourier harmonics determined by the periodicity of the underlying perfect crystal as shown on Fig. 12. The sum over the reciprocal lattice vectors K obviously reproduces the δ function peaks of the density (16). If one considers particles with a given size (for vortices it is the core size ξ) then the maximum K vector in the sum should be

$$K_{\max} \sim 2\pi/\xi \tag{18}$$

in order to reproduce the broadening of order ξ of the peaks in density. This immediately allows us to reproduce the Larkin model by expanding (9) using (17)

$$\rho_0 \int d^d r \sum_K e^{iK(r-u(r))} V(r) \tag{19}$$

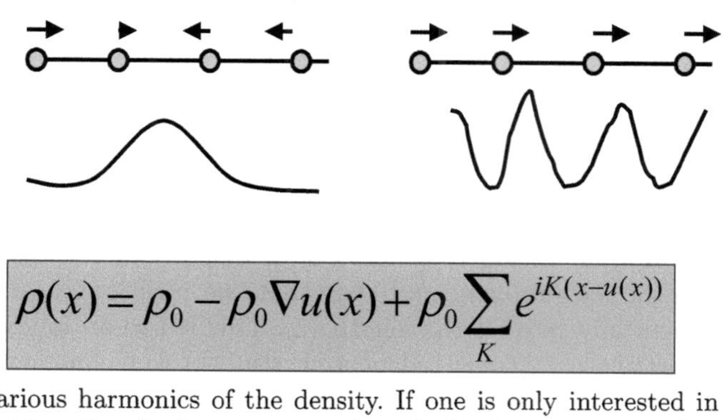

Fig. 12. Various harmonics of the density. If one is only interested in variations of the density at lengthscales large compared to the lattice spacing one has the standard "elastic" expression of the density in terms of the displacements. In the vortex system however on has to consider variations of density at lengthscales smaller than the lattice spacing and higher harmonics are needed [19,20].

in powers of u. Clearly the expansion is valid as long as $K_{\max}u \ll 1$ This will thus be valid up to a lengthscale R_c such that $u(R_c)$ is of the order of the *size of the particles* ξ. Note that this lengthscale is different (and quite generally smaller) than the lengthscale R_a at which the displacements are of the order of the lattice spacing. The Larkin model ceases to be valid well before the displacements become of the order of a and thus *cannot be used* to deduce the behavior of the positional order at large length scale. In addition it is easy to check that because the coupling to disorder is linear in the Larkin model, this model does not exhibit any pinning. Any addition to an external force leads to a sliding of the vortex lattice. It thus seems that this model is not containing the basic physics needed to describe the vortex lattice. In a masterstroke of physical intuition Larkin realized that the lengthscale at which this model breaks down is precisely the lengthscale at which pinning appears [18]. The lengthscale R_c is thus the lengthscale above which various chunks of the vortex system are collectively pinned by the disorder. A simple scaling analysis on the energy gained when putting an external force

$$\mathcal{H} = \int d^d r F_{\text{ext}} u(r) \tag{20}$$

allows to determine the critical force needed to unpin the lattice. Assuming that the critical force needed to unpin the lattice is when the energy gained by moving due to the external force is equal to the balance of elastic energy and disorder $\frac{c}{2}\xi^2 R_c^{d-2}$, one obtains

$$J_c \propto \frac{c\xi}{R_c^2} \tag{21}$$

This is the famous Larkin-Ovchinnikov relation which allows to relate a dynamical quantity (the critical current at $T = 0$ needed to unpin the lattice) to purely static lengthscales, here the Larkin-Ovchinikov length at which the displacements are of the order of the size of the particle. Let us insist again that this lengthscale controling pinning is quite different from the one R_a at which displacements are of the order of the lattice spacing at that controls the properties of the positional order.

The lack of efficiency of the Larkin model to describe the behavior of displacements beyond R_c still leaves us with the question of the nature of the positional order at large distances. However, extrapolating naively the Larkin model would give a power law growth of displacements. Such behavior is in agreement with exact solutions of interface problems in random environments (so called random manifold problems) and solutions in one spatial dimension. It was thus quite naturally assumed that an algebraic growth of displacements was the correct physical solution of the problem, and thus that the positional order would be destroyed exponentially beyond the length R_a. This led to an image of the disordered vortex lattice that consisted of a crystal "broken" into crystallites of size R_a due to disorder. To reinforce this image (incorrect) "proofs" were given [23] to show that due to disorder dislocations would be generated at the lengthscale R_a (even at $T = 0$) further breaking the crystal apart and leaving no hope of keeping positional order beyond R_a. A summary of this (incorrect) physical image is shown on Fig. 13.

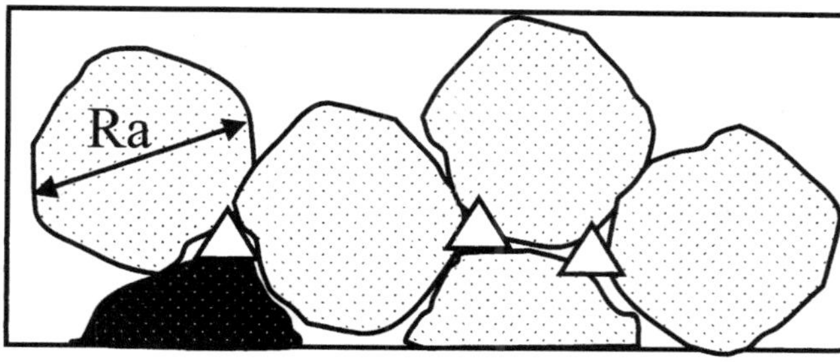

Fig. 13. The (incorrect) physical image that was the commonly accepted view of what a disordered elastic system would look like. The crystal would be broken into crystallites of size R_a by the disorder. Dislocations would be generated by the disorder at the same lengthscale.

If one believes that the positional order is lost and dislocations are spontaneously generated one can wonder whether an elastic description of the vortex lattice is a good starting point. An intermediate attitude is to consider that such a description is useful at intermediate lengthscale and can be used to obtain the pinning properties (since they are controlled by lengthscales below R_a) or as a first step in absence of a better description [4]. A more radical view is to consider that since positional order would be lost at large length scale it is best to ignore it from the start and that an elastic description of the vortex lattice is a bad starting point: it is much better to ignore positional order altogether and to focus on the phase of a vortex [24,23]. The effect of disorder is thus introduced by a random gauge field destroying the phase coherence between the vortices. The system is then described by a random phase energy

$$\mathcal{H} = \sum_{ij} \cos(\phi_i - \phi_j - A_{ij}) \tag{22}$$

With certain assumptions on the properties of the gauge field (essentially that $\lambda = \infty$) the idea is that the solid vortex phase will be transformed into a glassy Gauge glass (called the vortex glass), leading to the phase diagram shown in Fig. 14. The vortex glass phase has a continuous transition, with a divergent lengthscale, towards the liquid. It thus exhibit scaling at the transition. It was also suggested that inside the glass phase there should be no linear response to an applied current [24]. We will come back to this point in Section 5.

Although this description of the vortex lattice/ vortex glass phase was very successful in the beginning, it started to run into serious problems as both the experiments and the theory were refined. Among the experimental problems one would notice (the corresponding data will be presented in the next sections)

- The transition between the solid (vortex glass?) and the liquid was shown to be discontinuous by various measurements. Specific heat measurements have now proved that this transition is first order.

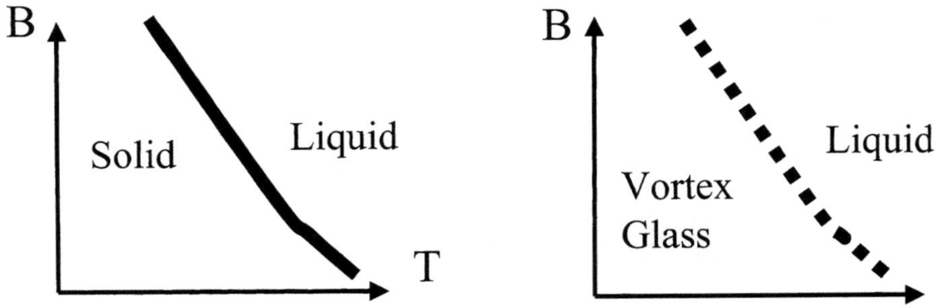

Fig. 14. In a vision where positional order is ignored, and vortices are described by their phases, the solid phase is transformed into a vortex glass phase, having a continuous transition (with scaling) towards the liquid phase [24,23].

- Decoration experiments were seeing very large regions free of dislocations and showing a very good degree of positional order. This did not seem to fit well with the idea that disorder would strongly affect the positional order.
- Neutron scattering was exhibiting quite good Bragg peaks, again showing stronger positional order than naively anticipated.
- The phase diagram seemed more complicated than the one shown in Fig. 14.

On the theoretical front two main points were raised: (i) the gauge glass model was shown to have no glass transition in $d = 3$ for realistic values of the parameters (such as a finite λ) [25]. (ii) The important suggestion was made, using scaling arguments [22,26] and then firmly established in detailed calculations [27,19] that *provided dislocations were ignored* displacements in vortex lattices were growing much more slowly than a power law (logarithmically).

These experimental facts and theoretical points suggested that the effect of disorder on the vortex lattice could be less destructive than naively anticipated. They thus strongly prompted for an understanding of the physical properties, such as the positional order, stemming from the elastic description. They also made it mandatory to resolve the issue of the asserted [26,23] ever presence of disorder induced dislocations, which would invalidate the elastic results and always destroy the positional order above R_a.

4 Statics: Experimental Facts and Bragg Glass Theory

To get a quantitative theory of the disordered system, and go beyond the simple scaling analysis, it is necessary to solve the full Hamiltonian (1) plus (9). Fortunately the theoretical "technology" had developed tools allowing to obtain a rather complete solution of this problem [19,20]. We describe the solution here and examine the consequences for experimental systems.

4.1 Bragg Glass

The problem one needs to solve is (using the decomposition of density (17))

$$\mathcal{H} = \frac{c}{2}\int d^d r (\nabla u)^2 + \rho_0 \sum_K \int d^d r e^{iK(r-u(r))} V(r) \tag{23}$$

Although we have written here the simplified form of the elastic hamiltonian the full one has to be considered but this does not change the method. One then gets rid of the disorder using replicas. After averaging over disorder the problem to solve becomes

$$\mathcal{H} = \frac{c}{2}\sum_{a=1}^{n}\int d^d r (\nabla u_a)^2 - D\rho_0^2 \sum_K \sum_{a,b=1}^{n} \int d^d r \cos(K(u_a(r) - u_b(r))) \tag{24}$$

where $\overline{V(r)V(r')} = D\delta(r-r')$. One has thus traded a disordered problem for a problem of n interacting fields. The limit $n \to 0$ has to be taken at the end for the two problems to be identical. So far the mapping is exact, but (24) is still too complicated to be solved exactly. Two methods are available to tackle it: (i) a variational method; (ii) a renormalization group method around the upper critical dimension $d = 4$ (a $4 - \epsilon$ expansion). The renormalization method is relatively involved and we refer the reader to the various reviews and to [20,28] for more details and discussions. The variational method is simple in principle [29], and has the advantage to give the essential physics. One looks for the best quadratic Hamiltonian

$$\mathcal{H}_0 = \sum_{ab}\int d^d q G_{ab}^{-1}(q) u_a(q) u_b(-q) \tag{25}$$

that approximate (24). H_0 leads to a variational free energy

$$\mathcal{F}_{\text{var}} = \mathcal{F}_0 + \langle \mathcal{H} - \mathcal{H}_0 \rangle_{\mathcal{H}_0} \tag{26}$$

that has to be minimized with respect to the variational parameters. The unknown Green's function $G_{ab}(q)$ are thus determined by

$$\frac{\partial \mathcal{F}_{\text{var}}}{\partial G_{ab}(q)} = 0 \tag{27}$$

This is nothing but the well known self consistent harmonic approximation. The technical complication here consists in taking the limit $n \to 0$ [30]. The best variational parameters are the ones that break the replica symmetry, in a similar way than what happens in spin glasses. This is very comforting since we expect on physical grounds that the competition between the elasticity and the disorder causes a strong competition where the system has to find its ground state. It is thus quite natural that such a competition leads to glassy properties. This is what the solution of the problem confirms. A similar effect appears in the

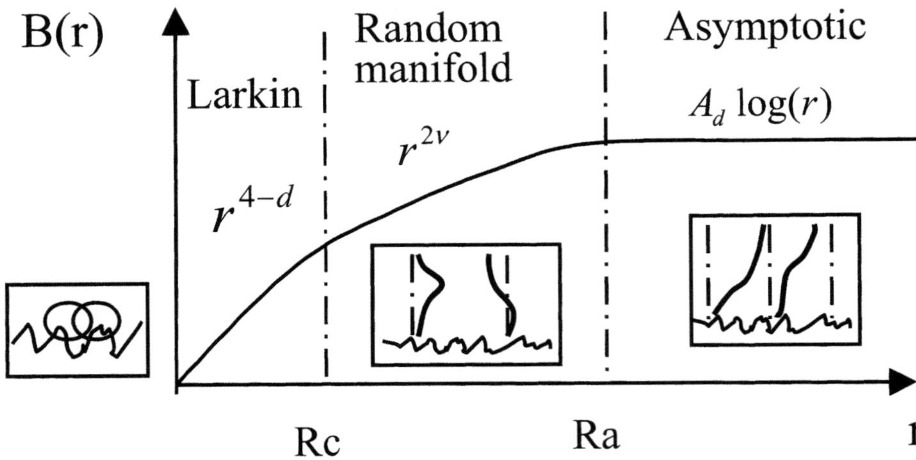

Fig. 15. The relative displacement correlation function $B(r)$ as a function of distance. The lengthscales R_c and R_a define three regimes. Below R_c one recovers the Larkin behavior. Between R_c and R_a the displacements still grow algebraically albeit with a modified exponent. Above R_a the growth becomes logarithmic.

renormalization solution where a non-analyticity appears, signaling again glassy properties. The two methods thus agree quite well (not only qualitatively but also quantitatively).

Let us now describe the full solution given in [19,20]. One finds for the relative displacements correlation function the one shown on Fig. 15. Three regime can be distinguished, separated by the characteristic lengthscales R_c and R_a. Below R_c, one is in the Larkin regime, and the displacements grow as $r^{(4-d)/2}$. Then between R_c and R_a (i.e. when the displacements are between ξ and a), each line wanders around its equilibrium position and the problem is very much like the one of a single line in a disordered environment, i.e. a random manifold problem. The growth of displacements is still algebraic, albeit with a different exponent ν. Above R_a however the displacements grow much more slowly and $B(r) = A_d \log(r)$. The physics is easy to understand: because of the periodic nature of the system, each line can take care of the disorder around its equilibrium position. There is thus no interest for one line to make displacements much larger than the lattice spacing to pass through a particularly favorable region of disorder (this would be the case for a single line). For many lines it is better to let the line closeby take care of this region. This ensures that the total energy of the *whole system* is the lowest. As a result the displacements do not need to grow much above a. The variational method and RG techniques allow to compute the prefactor A_d and to address the issue of the positional order. The positional correlation function is simply given by [19,20]

$$C(r) \sim \left(\frac{1}{r}\right)^{\eta} \tag{28}$$

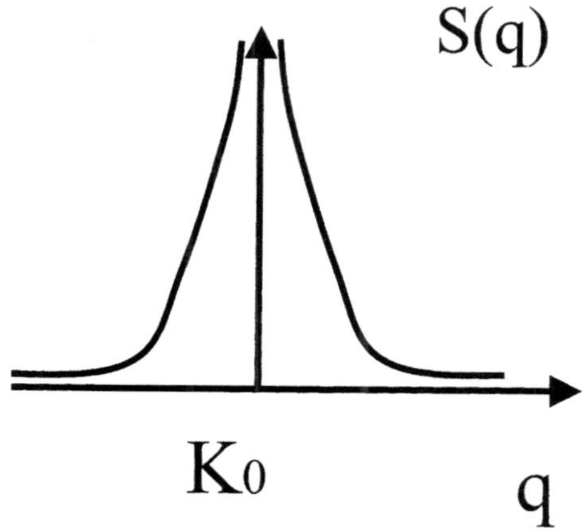

Fig. 16. The disordered system still has quasi-long range positional order and algebraically divergent Bragg peaks. It is thus nearly as ordered as a perfect solid. Such a phase, which is a glass when one looks at its dynamics properties, and in addition possesses perfect topological order (no defects such as dislocations) is the Bragg glass phase [20].

where η is an exponent independent of the strength of disorder or temperature (for example $\eta_{\text{var}} = 1$ in $d = 3$). The physics described above, and obtained from the variational ansatz is totally generic. The alternate RG approach in $4 - \epsilon$ does indeed recover identical physics, as was shown first in the Larkin and asymptotic regimes [19,20] and more recently for the full $B(r)$ [28]. The variational approach even gives quite accurate values of the exponents themselves as can be checked by comparing with the RG. The quite striking consequence is that far from having the positional order destroyed by the disorder in an exponential fashion, a quasi-long range order (algebraic decay of positional order) exists in the system. Algebraically *divergent* Bragg peaks still exist as shown on Fig. 16. It is to be noted that this phase, which is thus practically as ordered as a perfect solid, is a glass when one looks at its dynamical properties. This is seen in the analysis by the presence of replica symmetry breaking in the variational approach or the existence of non-analyticities in the renormalization solution. From a physical point of view this means that the system has many metastable states separated from its ground state by divergent barriers. As a result it exhibits pinning and non-linear dynamics (creep motion), as we will discuss in more details in Section 5.

What about the argument that dislocations should always be generated at lengthscale R_a? In fact it was shown [20] that this argument which ignores the fact that the coupling of the displacements to the disorder is non linear is simply incorrect and that in fact in $d > 2$ dislocations are *not* generated by disorder provided the disorder is moderate (i.e. R_a large enough compared to a). This means that the results given above corresponds to the *true thermodynamic* ground state of the system. Thus there exists a thermodynamically stable *glassy* phase, the Bragg glass, with quasi long range order (algebraically divergent Bragg peaks), perfect topological order (absence of defects such as dislocations). This is a surprising result since one naively associate a glass with a very scrambled system. The Bragg glass shows that this is not the case and that one has to distinguish the positional properties from the energy landscape (or the dynamical ones).

The existence of this Bragg glass phase has of course many consequences for the vortex systems, consequences that we now investigate and compare with the available experimental data on vortices.

4.2 Positional Order: Decorations and Neutrons

The first consequence is the existence of the perfect topological order and the algebraic Bragg peaks.

On the theoretical side, after the initial proposal [20] the existence of algegraic bragg peaks and the absence of dislocations have been confirmed by further analytical results [31–33] and numerical simulations [34,35]. On the experimental side, direct structural information has been obtained from both real space studies such as magnetic decoration, scanning tunneling microscopy, Lorentz microscopy and reciprocal space studies by neutron diffraction, summarized in Fig. 17. The upper two panels show a decoration micrograph of a fairly ordered lattice in $NbSe_2$ with a few defects [36] and an STM micrograph [37] of the same material with no defects. The Lorentz micrograph in a thin film of Nb [38] produces a nearly amorphous vortex assembly while the neutron diffraction data on a single crystal sample of Nb [39] show Bragg peaks up to third order reflections, suggesting a very high degree of order. These results show that defect-free phases with Bragg reflections are experimentally observed, while highly defective or even amorphous or liquid like phases also exist. The task is to find which parts of the phase space are occupied by each and what controls the phase transformations among them.

The Fourier transforms of the real space data [40,36,41] show very large regions free of dislocations and yield Bragg peaks routinely, suggesting a much stronger solid like order than would be expected naively from a vortex glass model. The situation with neutron diffraction is similar as shown on Figure 17.

Recently neutron data have provided a direct evidence of the presence of the Bragg glass phase. Indeed the power-law Bragg peaks shown in Fig. 16 gives when convolved with the experimental resolution the result shown in Fig. 18. The width at half maximum of the observed peak is constant and determined by the experimental resolution, and the height is fixed by the position correlation length R_a. Thus if disorder (or magnetic field – see the next section) is increased the Bragg glass predicts that observed neutron peaks should collapse without broadening. This behavior, has been quantitatively tested on the compound $(K,Ba)BiO_3$ [42] (see Fig. 18) which has the advantage of being totally isotropic and thus avoid all complications associated with anisotropy such as possible 2D-3D crossovers. Peaks are seen to collapse without any broadening thus providing a direct evidence of the Bragg glass phase and its algebraic positional order.

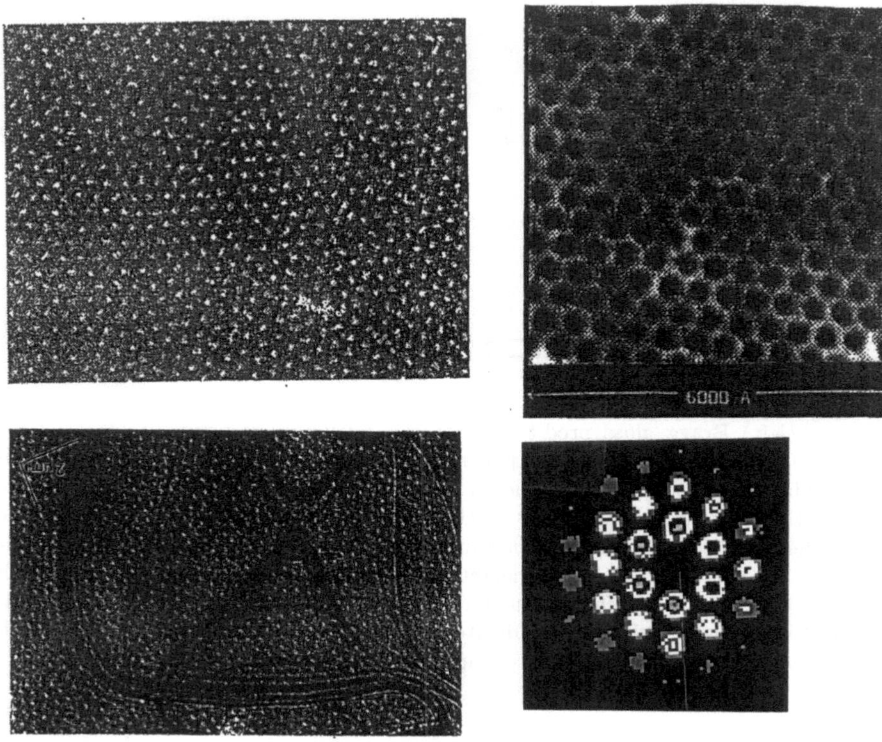

Fig. 17. Vortex phase structure using various methods. The upper two panels show a decoration data on the left [36] and an STM data on the right in $NbSe_2$ [37]. The lower two panels show a Lorentz holography micrograph of a thin film Nb [38] and a neutron diffraction picture of a single crystal Nb [39]

4.3 Unified Phase Diagram

Another striking consequences of the existence of the Bragg glass, is that it imposes a *generic* phase diagram for all type II superconductors [20]. Indeed the Bragg glass has no dislocations, thus if either thermal fluctuations *or* disorder are increased the Bragg glass should "melt" to a phase that contains defects . If the thermal fluctuations increase, the standard melting occurs and leads to the liquid phase. Because the Bragg glass is nearly as ordered as a perfect solid one can expect it to melt though a first order phase transition. More importantly this "melting" of the Bragg glass can also occur because the disorder is increased in the system. For vortices increasing the field has a similar effect. Indeed the effective disorder in (9) is $V\rho_0$, thus increasing the average density makes the disorder term stronger compared to the elastic term (1). Indeed for moderate fields the change in elastic constants due to the field is quite small. Thus increasing the field is like increasing disorder. One should thus have a *"melting"* transition of the Bragg glass (induced by the disorder) as a function of the field.

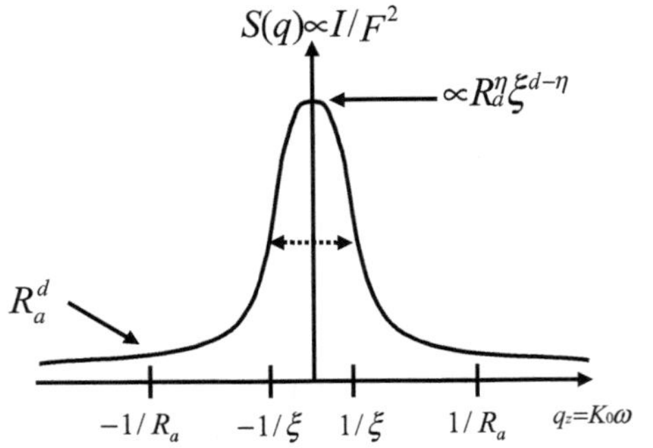

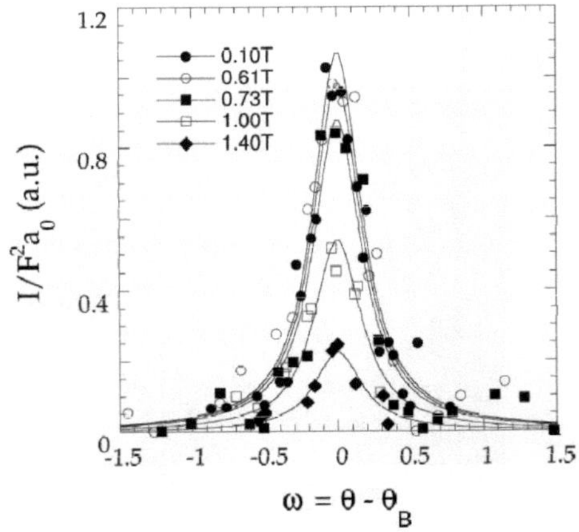

Fig. 18. (left) Bragg glass predictions for the angular dependence of the neutron diffracted intensity when a finite experimental resolution is taken into account. The arrows indicate the values of $S(q)$ for $q = 0$ and $1/R_a$ respectively. If the disorder increases (Ra decreases), the height of the peak decreases as R_a^η but the peak does not broaden since the half width at half maximum is always given by the experimental resolution $1/\xi$ and not $1/R_a$. The height of the peaks gives a direct measure of the characteristic length R_a.(right) Angular dependence of neutron intensity diffracted by a (K,Ba)BiO_3 crystal [42] at the indicated applied fields. Those data show that the diffracted intensity (rocking curves) collapse without any broadening above 0.7 T.

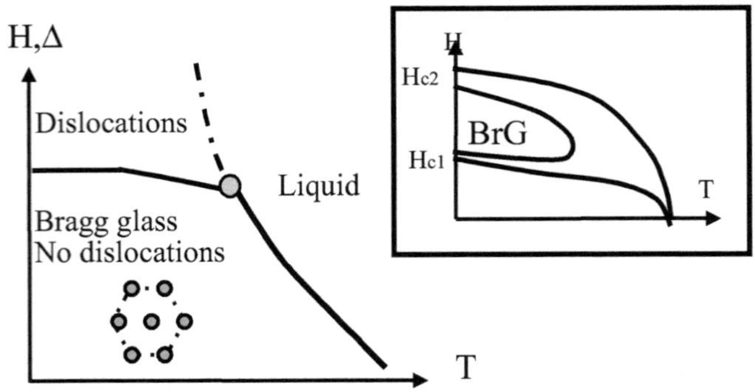

Fig. 19. Universal phase diagram for type II superconductors. The Bragg glass "melts" due to thermal fluctuations of disorder induced fluctuations. This lead to transitions as a function of magnetic field. The left diagram is far from H_{c1} (more adapted to High T_c) whereas the insert shown the full diagram. The melting towards the liquid is expected to be first order.

Close to H_{c1} because the change of elastic constants is then drastic, one expects a similar transition. This leads to the universal phase diagram shown in Fig. 19. Numerical simulations [34,35,43–45] are in agreement with this phase diagram and allowed a detailed study of the disorder induced melting of the Bragg glass and of the high field phase.

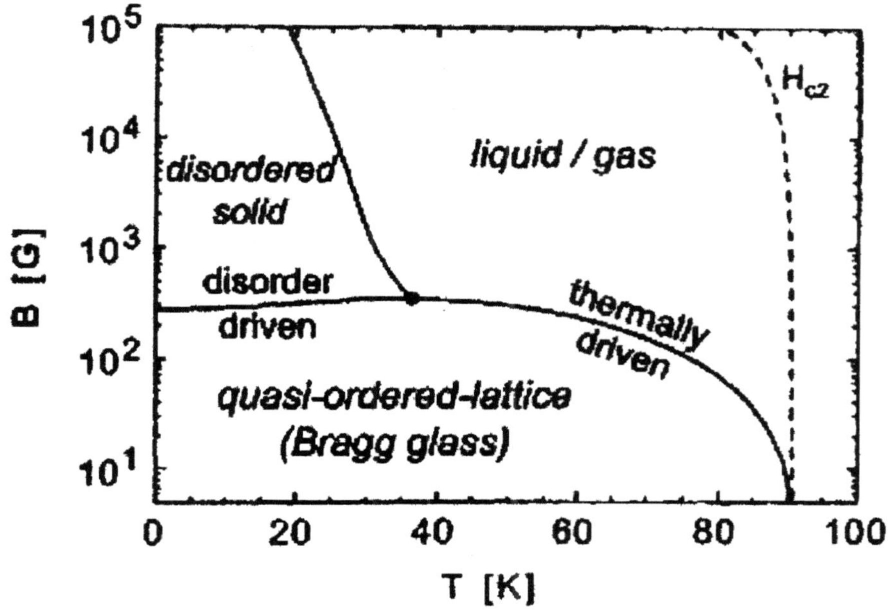

Fig. 20. Experimental phase diagram for the cuprate system BSCCO [14,46]. At the thermally induced melting transition one sees thermodynamic anomalies as in Fig. 8. The disorder-driven transition is accompanied by the second peak or peak effect, shown later in Fig. 26.

We first focus on the experimental results of the loss of Bragg glass order in the cuprate systems. A remarkable experimental determination of the phase diagram of BSCCO is shown in Fig. 20. Three phases are clearly identified in it: a quasi-ordered Bragg glass phase which "melts" by thermal fluctuations into a liquid and also "amorphizes" or "melts" by quenched disorder into a disordered solid. First we focus on structural evidence of the two phase transitions from neutron diffraction [47]. In the Bragg glass phase one clearly observes the resolution limited Bragg peaks. Upon increasing temperature across the thermally driven transition the Bragg peaks lose intensity and become unobservable in the putative liquid phase. A similar loss of Bragg peak intensity is observed when the magnetic field is increased at fixed T across the Bragg glass to the disordered or vortex glass phase. Fig. 21 and Fig. 22 show the results of neutron diffraction of BSCCO as the two phase boundaries are crossed, one from the ordered Bragg glass to the liquid and the other from Bragg glass to the disordered solid phase [47]. In both cases the Bragg reflections lose intensity and disappear at the phase boundary, in a way similar to the one discussed in Fig. 18. Due to limited neutron intensity, detailed studies of the disordered phase have not been performed in the cuprates. But recent neutron diffraction studies of low T_c system Nb [48,39] with a short penetration depth have directly shown a transformation of bragg peaks to a ring of scattering, implying liquid like (amorphous) order in the disor-

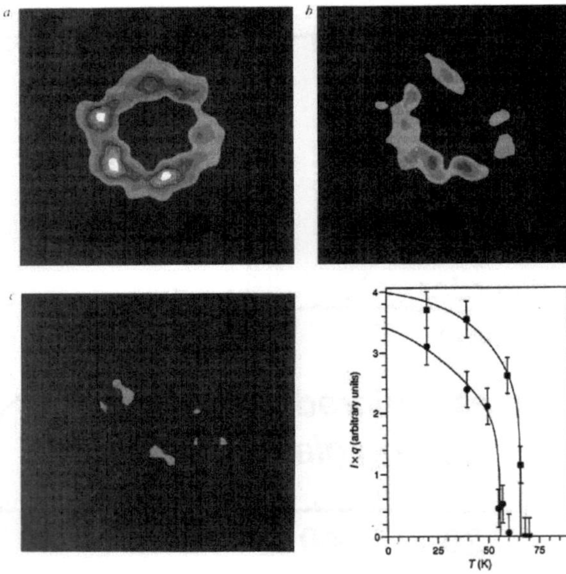

Fig. 21. Evolution of the neutron bragg peaks in the ordered phase (upper left panel) with increasing temperature. The lower right panel shows the loss of Bragg intensity with increasing T

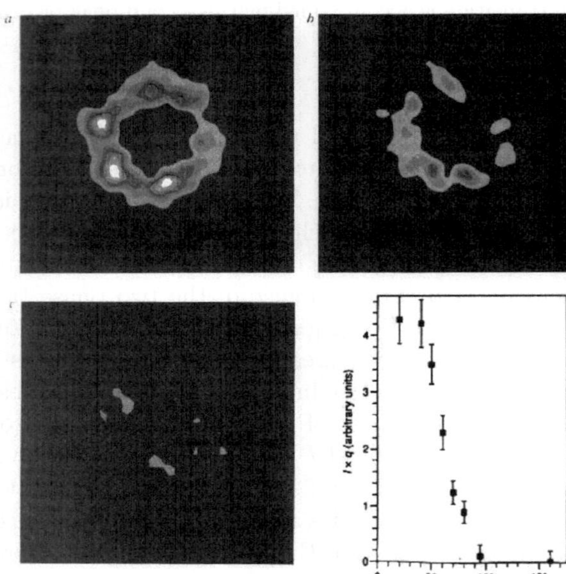

Fig. 22. Loss of Bragg intensity upon increasing the magnetic field at a fixed temperature

dered phase. The experimental phase behavior is thus entirely compatible with theoretical expectations in Fig. 19. The comparison with the theoretical phase diagram identifies the quasi-ordered solid phase with the Bragg glass phase. The phase with dislocations is expected to correspond to the disordered solid phase. For BSCCO the position of the field melting line has been computed by Lindemann argument or similar cage arguments [49–52] and the value of the "melting" field is in good agreement with the observed experimental value. The distinction between the disordered solid phase and the liquid phase remains an experimentally open question for different systems with different types of disorder. For the very anisotropic BSCCO system there is also the question of the existence of additional phases. Structural results of the same kind are not available for the other common cuprate system, namely YBCO. However, thermodynamic data on the magnetization jump and entropy jump were measured. A composite of the data is shown in Fig. 23 (see also [53]) which demonstrates close agreement within the Clapeyron equation confirming the first order nature of the thermally driven melting transition.

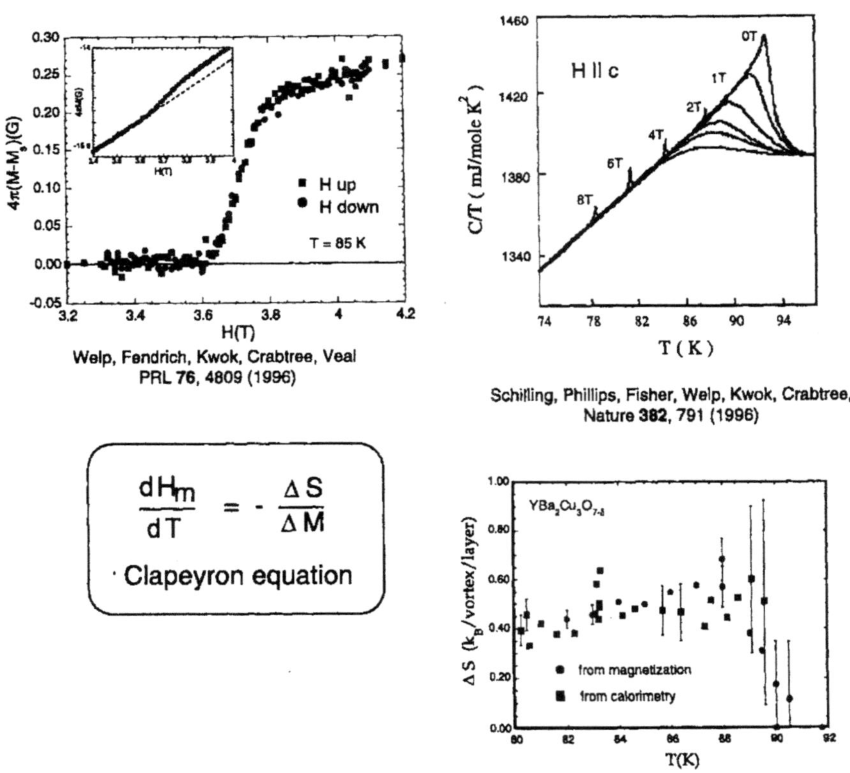

Fig. 23. Experimental data [54,55] showing thermodynamic measurements from magnetization and calorimetry of the first order transition in YBCO

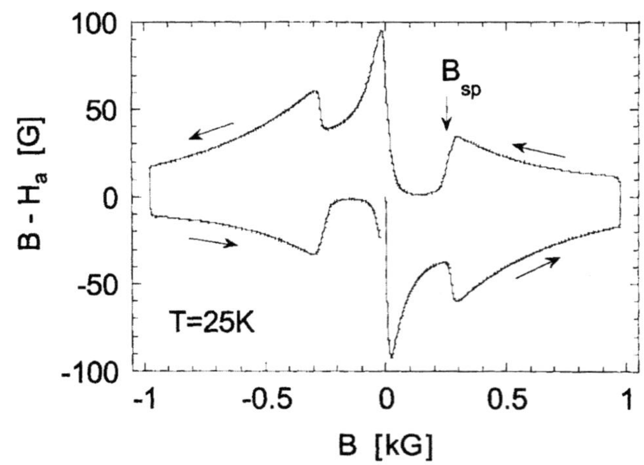

Fig. 24. Typical second magnetization peak data from the magnetic hysteresis loop in BSCCO[46], the sudden enhancement of the diamagnetic signal at B_{sp} marks the enhancement of critical current across the Bragg glass to vortex glass phase transition.

4.4 Second Peak and Peak Effect

Now we focus on the disorder driven phase transition shown above. The magnetization hysteresis loop across this transition manifests a typical second peak effect, often called the fishtail effect due to its shape, as shown in Fig. 24 for BSCCO [46]. The sharp jump is magnetization at B_{sp} marks a sudden increase in the irreversibility, i.e., a jump in critical current J_{c}. In a naive view of the Larkin scenario, this marks a sudden decrease in the correlation volume, i.e., a sharp loss of order, consistent with the neutron diffraction data shown above. Similar results were obtained for YBCO also, yielding a qualitatively similar phase diagram [56–58].

Peak effects are ubiquitous in low T_{c} systems and have been known for nearly four decades and provided the primary motivation for the Larkin-Ovchinnikov scenario of collective pinning. In these systems the peak effect [59] usually occurs very close to the normal phase boundary, unlike in the cuprate systems where the fishtail anomaly occurs very far from it. Recent resurgence of activity on the peak effect phenomenon in low T_{c} systems also provide a phase diagram of ordered and disordered vortex phases not dissimilar to the cuprates. Due to the smallness of the thermal fluctuation effects, however, the peak effect transition often occurs in close proximity to the melting transition, or even coincident with it. Separating the two effects remains a matter of considerable controversy in the low T_{c} systems. In the popular low T_{c} system $NbSe_2$ the peak effect phenomenon has been studied extensively in recent years [60]. Fig. 25 shows a typical set of data for the resistive detection of the peak effect where a rapid drop in the resistance at the peak effect boundary marks a sudden increase in the critical current analogous to the magnetization jump shown above. Direct

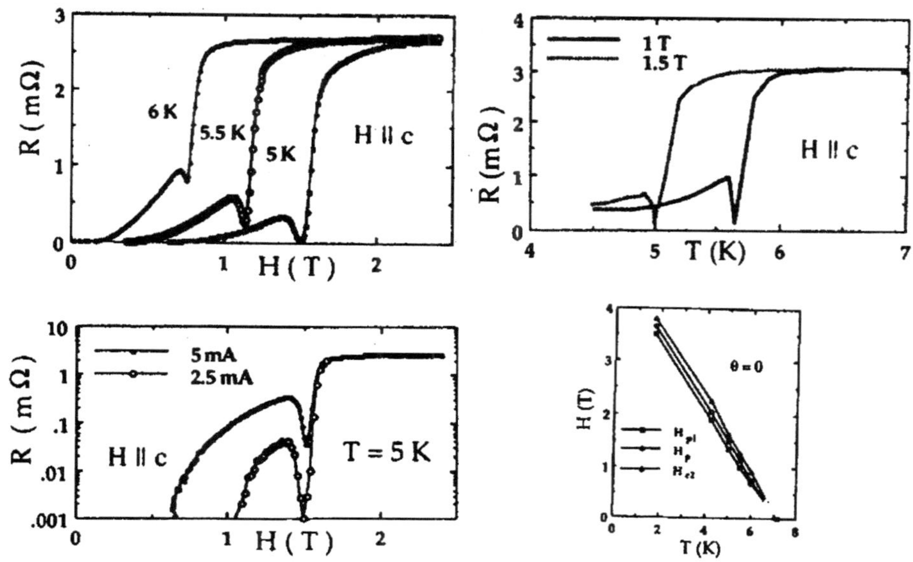

Fig. 25. Typical observation of peak effect in transport [61]. The upper two panels show the transition in isothermal and isofield measurements. The lower left panel shows the independence of the locus on driving current suggesting a thermodynamic origin of the anomaly. The right panel shows the locus of the peak region bounded by the onset at H_{pl} and the peak at Hp. The close proximity to the upper critical field is typical of low T_{c} materials..

structural evidence also clearly shows the amorphization of the Bragg glass phase with six fold symmetric Bragg spots to a ring of scattering at the peak in the elemental system Nb [39]. In $NbSe_2$ the same peak effect is accompanied by a sharp change in the line shape as seen in the asymmetry parameter [62]. These results are shown in Fig. 26.

Of special importance is the clear experimental observation of a reentrant phase behavior for the $NbSe_2$ system [63,64]. From the Meissner phase an increase in field shows two anomalies, first from a disordered phase into an ordered phase and then a reentrance into a disordered phase just below the upper critical field. Especially significant is the pronounced shift in the order-disorder phase boundary with varying quenched disorder (addition of magnetic dopants). Figure 27 shows the progressive shrinkage of the Bragg glass phase from both high field side as well as from the low field side as disorder is increased from sample A through sample C. These results are in excellent agreement with the theoretical discussions above. Direct structural evidence of amorphization in this system was obtained through muon spin relaxation experiments [62] that are entirely analogous to the results [65] in the cuprate systems.

Several questions remain open for the peak effect from the phase behavior point of view. In addition to the second magnetization peaks, a peak effect

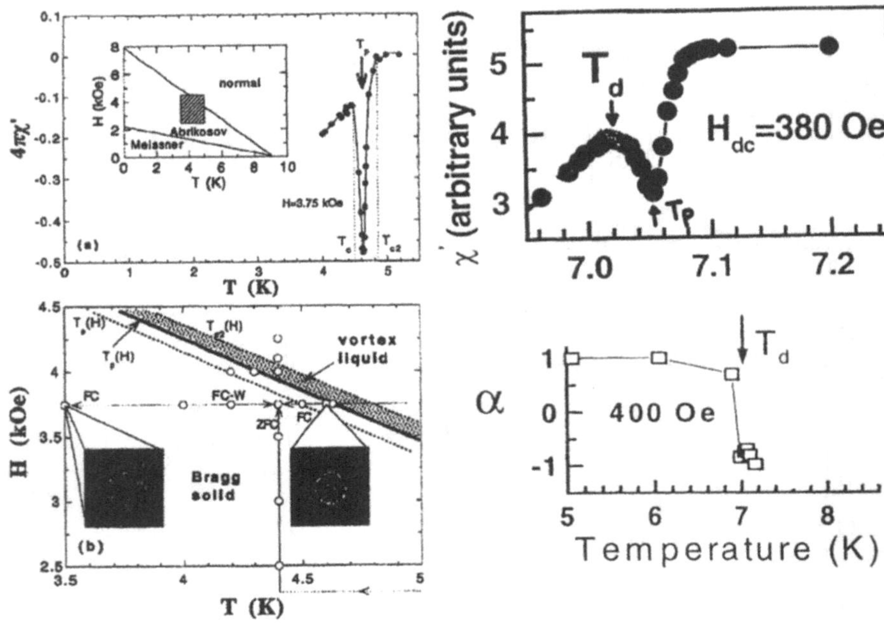

Fig. 26. Experimental data on Nb [39] on the left and $NbSe_2$ [62] on the right. The upper panel shows the peak effect from ac susceptibility for both samples. The lower left panel shows the structure from neutron diffraction in Nb showing the Bragg spots disappear into a ring of scattering. Muon spin relaxation data in $NbSe_2$ on the right shows a sudden drop in the skewness parameter at the peak effect that marks the first order structural disordering of the lattice.

is often observed in YBCO very close to or even coincident with the melting transition [66]. This suggests that the disordered solid phase may protrude as a sliver all around the Bragg glass phase [67] or that there are two types of peak effects, one associated with disorder induced melting and the other with the thermally induced melting transition. In what follows, we indeed show that the melting of the Bragg glass provides a natural explanation for the peak effect.

How the melting of the Bragg glass signals itself? To understand it let us look at the $V-I$ characteristics, shown on Fig. 28. The Bragg glass is collectively pinned so it has a small critical current J_1 but very high barriers leading to practically no motion (hence no V) in the pinned phase (below J_1). On the other hand in the high field phase (with dislocations) or the liquid it is more easy to pin small parts leading to higher critical currents, but the pinning is not collective hence a much more linear response below J_2. The $V-I$ characteristics thus cross at the melting of the Bragg glass [50]. This crossing leads for an apparent increase of the critical current close to the melting and thus to a peak effect in the transport measurements or to a second peak in the magnetization measurements.

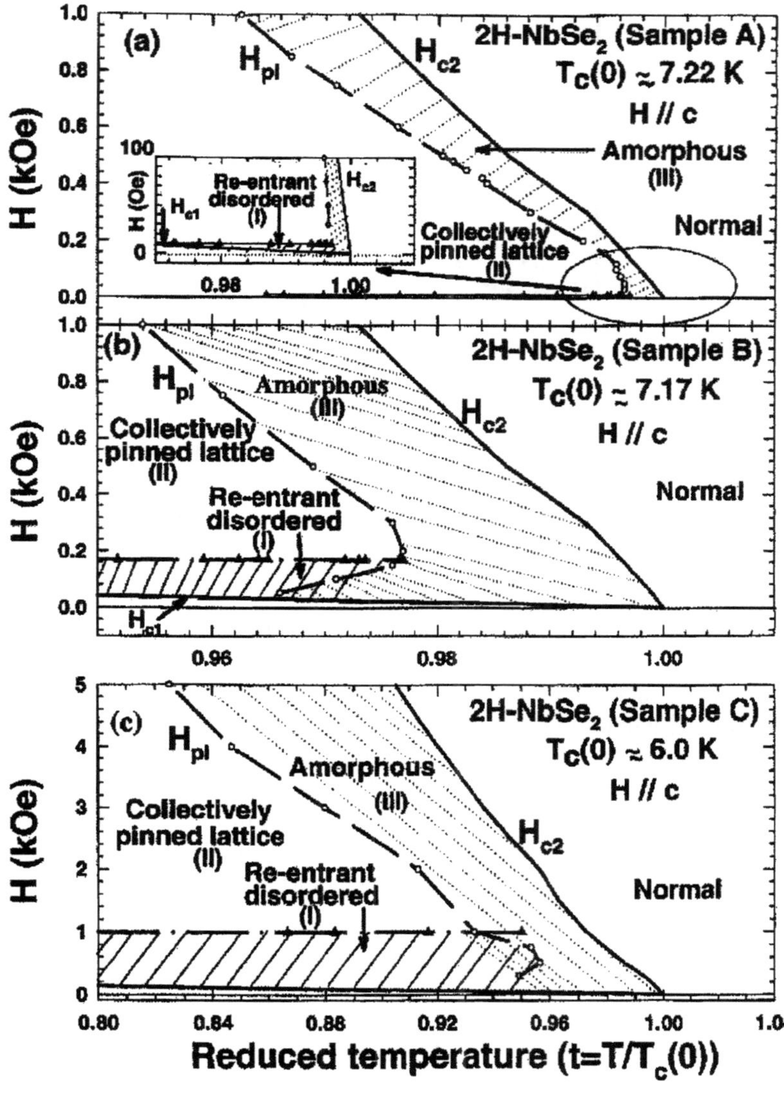

Fig. 27. Evolution of the phase diagram [64] with increased disorder from sample A to C. As disorder increases, the collectively pinned lattice, i.e., the Bragg glass phase, shrinks from both above and below.

Recent simulations and reexamination of older experimental data [68,61] are in excellent agreement with this $I - V$ crossing scenario at the peak effect near melting or the second magnetization peak.

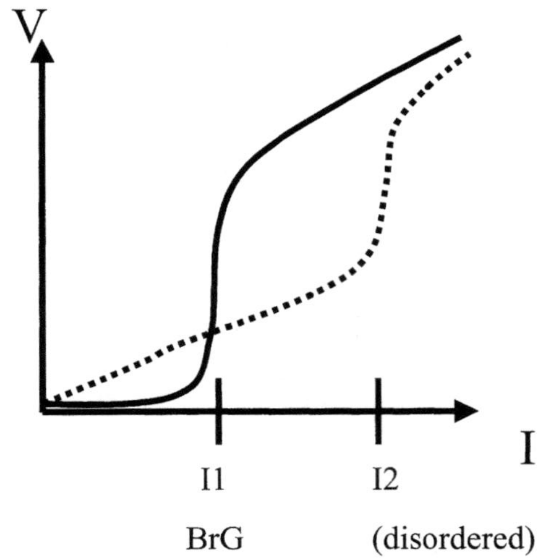

Fig. 28. The $V-I$ characteristics for the Bragg glass (full line) and the high field phase or the liquid (dashed line) [50].

5 Dynamics of Vortices

The competition between disorder and elasticity manifests also in the dynamics of such systems, and if any in a more dramatic manner. When looking at the dynamics, many questions arise. Some of them can be easily asked (but not easily answered) when looking at the $V-I$ characteristics shown in Fig. 29. The simplest question is prompted by the $T=0$ behavior. Since the system is pinned the velocity is zero below a certain critical force F_c. For $F > F_c$ the system moves. What is F_c? We saw that the scaling theory of Larkin and Ovchinikov relates it directly to the static characteristic length R_c. Can one extract this critical force directly from the solution of the equation of motion of the vortex lines

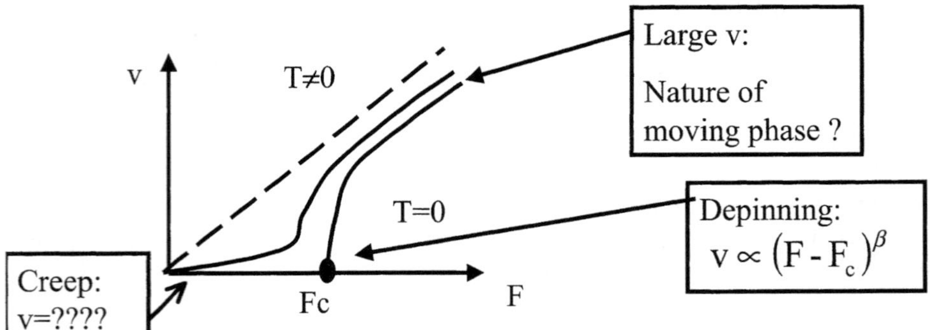

Fig. 29. The $v-F$ characteristics for a vortex system. This is in fact the velocity of the system v, in response to an external force F.

$$\eta \frac{du_i(t)}{dt} = -\frac{\delta \mathcal{H}}{\delta u_i} + F + \zeta_i(t) \tag{29}$$

This equation is written for overdamped dynamics, but can include inertia as well. η is the friction coefficient taking into account the dissipation in the cores, F the externally applied force, and ζ a thermal noise. Can one obtain from this equation the velocity above F_c? The $v - F$ curve at $T = 0$ is reminiscent of the one of an order parameter in a second order phase transition. Here the system is out of equilibrium so no direct analogy is possible but this suggests that one could expect $v \sim (F - F_c)^\beta$ with a dynamical critical exponent β. We will not investigate these questions here, because of lack of space and refer the reader to the above mentioned reviews and to [69] for an up to date discussion of this issue and additional references.

The second question comes from the $T \neq 0$ curve. Well below threshold $F \ll F_c$ the system is still expected to move through thermal activation. What is the nature of this motion and what is the velocity? If the system has a glassy nature one expects it to manifest strongly in this regime since thermal activation will have to overcome barriers between various states. We address this question in Section 5.1.

Finally, there are many questions beyond the simple knowledge of the average velocity. One of the most interesting is the nature of the moving phase. If one is in the moving frame where the system looks motionless, how much this moving system resembles or not the static one. This concerns both the positional order properties and the fluctuations in velocity such as the ones measured in noise experiments. This is discussed in Section 5.2.

5.1 Creep

Let us first examine the response of the vortex system to a very small external force. For usual systems we expect the response to be linear (leading to a finite resistivity for the system). Indeed earlier theories of such a motion found a linear response. The idea is to consider that a blob of pinned material has to move in an energy landscape with barriers Δ as shown in Fig. 30. The external force F tilts the energy landscape making forward motion possible. The barriers are overcome by thermal activation (hence the name: Thermally Assisted Flux Flow) with an Arrhenius law. If the minima are separated by a distance a the velocity is

$$v \propto e^{-\beta(\Delta - Fa/2)} - e^{-\beta(\Delta + Fa/2)} \simeq e^{-\beta\Delta} F \tag{30}$$

The response is thus linear, but exponentially small. One thus recovers that pinning drastically improves the transport qualities of superconductors. For old superconductors β was small enough so that this formula was not seriously challenged. However with high T_c one could reach values of β where it was clear that the TAFF formula was grossly overestimating the motion of vortex lines.

The reasons is easy to understand if one remembers that the static system is in a vitreous state. In such states a "typical" barrier Δ does not exist, but barriers are expected to diverge as one gets closer to the ground state of the

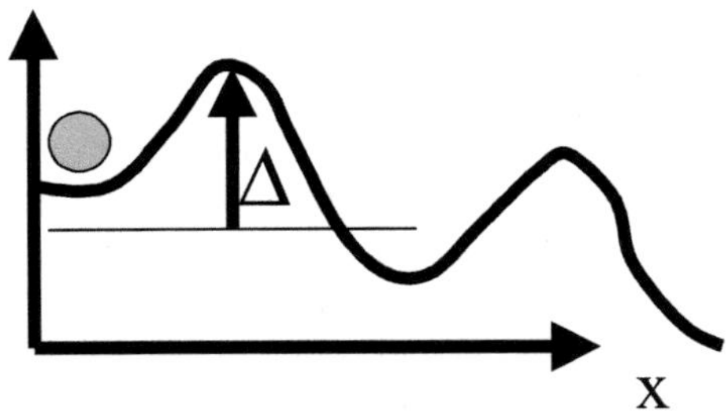

Fig. 30. In the Thermally Assisted Flux Flow (TAFF) [70] a region of pinned material is considered as a particle moving in an energy landscape with barriers. This leads to an exponentionally small but linear response.

system. The TAFF formula is thus valid in system where the glassy aspect is killed. This is the case in the liquid where various parts of the system are pinned individually. When the glassy nature of the system persists up to arbitrarily large lengthscales the theory should be accommodated to take into account the divergent barriers [24,23]. This can be done quantitatively within the framework of the elastic description. In fact such a theory was developed before for interfaces [71,72] and then adapted for periodic systems such as the vortex lattice [73,22]. The basic idea is beautifully simple. It rests on two quite strong but reasonable assumptions: (i) the motion is so slow that one can consider at each stage the lattice as motionless and use the *static* description; (ii) the scaling for barriers which is quite difficult to determine is the same than the scaling of the minimum of energy (metastable states) that can be extracted again from the static calculation. If the displacements scale as $u \sim L^\nu$ then the energy of the metastable states (see (1)) scale as

$$E \sim L^{d-2+2\nu} \tag{31}$$

on the other hand the energy gained from the external force over a motion on a distance u is

$$E_F = \int d^d x F u(x) \sim F L^{d+\nu} \tag{32}$$

Thus in order to make the motion to the next metastable state one needs to move a piece of the pinned system of size

$$L_{\mathrm{min}} \sim \left(\frac{1}{F}\right)^{\frac{1}{2-\nu}} \tag{33}$$

The size of the minimal nucleus able to move thus grows as the force decrease. Since the barriers to overcome grow with the size of the object the minimum barrier to overcome (*assuming* that the scaling of the barriers is *also* given by (31))

$$\Delta(F) \sim \left(\frac{1}{F}\right)^{\frac{d-2+2\nu}{2-\nu}} \tag{34}$$

leading to a velocity

$$v \propto e^{-\beta\left(\frac{1}{F}\right)^{\frac{d-2+2\nu}{2-\nu}}} \tag{35}$$

This is a remarkable equation. It relates a dynamical property to *static* exponents, and shows clearly the glassy nature of the system. The corresponding motion has been called creep since it is a sub-linear response. Of course the derivation given here is phenomenological, but it was recently possible to directly derive the creep formula from the equation of motion of the system [74,69]. This proved the two underlying assumptions behind the creep formula and in particular that the scaling of the barriers and metastable states is similar. More importantly this derivation shows that although the formula for the velocity given by the phenomenological derivation is correct, the actual motion is more complicated than the naive phenomenological picture would suggest. Indeed the phenomenological image is that a nucleus of size $L_{\rm min}$ moves through thermal activation over a length $L_{\rm min}^{\nu}$, and then the process starts again in another part of the system. In the full solution, one finds that the motion of this small nucleus triggers an avalanche in the system over a much larger lengthscale [69]. Checking for this two scales process in simulations or actual experiments if of course a very challenging problem.

The creep formula is quite general and will hold for interfaces as well as periodic systems. For periodic systems, dislocations might kill the collective behavior by providing an upper cutoff to the size of the system that behaves collectively (as if the system was torn into pieces). In the Bragg glass the situation is clear. Since there are no dislocations the creep behavior persists to arbitrarily large lengthscales. Since $\nu = 0$ in the Bragg glass the creep exponent in (35) is $\mu = 0.5$ [22,20]. What becomes the creep when dislocations are present is still an open [75] and very challenging question. What is sure is that one can expect a weakening of the growth of the barriers or even their saturation, when going from the Bragg glass phase to the “melted” phase [50]. This is at the root of the crossing of the $I - V$ shown in Fig. 28. Experimental verification of the creep effects postulated above have proved difficult due to the functional form of (35) where the power law appears in the exponentiated factor and requires data spanning many decades in the drive to determine the exponents with adequate precision. Transport experiments [76] as well as magnetic relaxation experiments [77] have reported creep exponents compatible with the Bragg glass prediction, as well as weakening of the barrier growth when going to the disordered phase. But this is clearly a very challenging and difficult issue that would need more investigations.

5.2 Dynamical Phase Diagram

Let us now turn to the problem of the nature of the moving phase.

This problem was directly prompted by experimental observations. Indeed early measurements of the dynamics showed dramatic effects near the peak effect (see below), led to the construction of an experimental dynamical “phase diagram” [60] for the moving phases shown in Fig. 31. In addition dramatic evidence of the evolution of vortex correlations with driving force was obtained

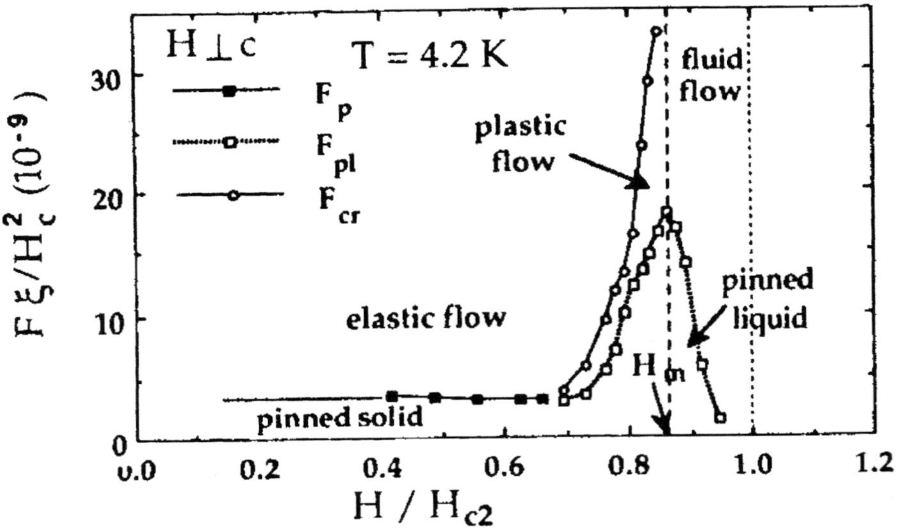

Fig. 31. A typical dynamic phase diagram in $NbSe_2$. Below the transition region a direct transition into a moving ordered phase is seen while in the peak regime a pronounced intermediate plastic flow regime is seen. At large drives above F_{cr}, the moving ordered phase is established. In the equilibrium pinned liquid phase, the crossover current is immeasurably large.

many years ago by neutron diffraction studies [78]. The Bragg peak in the pinned phase broadened significantly at the onset of motion showing a loss of order (or appearance of defects) and a subsequent healing of the Bragg peak at large drives showing a reentrance into an ordered moving phase. It was thus necessary to understand the nature of the "phases" once the lattice was set into motion.

One regime in which one could think to tackle this problem is when the lattice is moving at large velocities. Indeed in that case it is possible to make a large velocity expansion. Such expansion was performed with success to compute the corrections to the velocity due to pinning [79,80], and get an estimate of the critical current. An important step was to use the large v method to compute the displacements [81]. It was found that due to the motion the system averages over the disorder. As a result the system do not feel the disorder any more above a certain lengthscale and recrystalizes. The memory of the disorder would simply be kept in a shift of the effective temperature seen by this perfect crystal. This picture was consistent with what was shown to be true for interfaces, even close to threshold [82] (as shown on Fig. 32) and thus provided a nice explanation for the recrystallization observed at sufficiently large velocities. Perturbation approach along those lines has been extended in [83]. However the peculiarities of the periodic structure manifest themselves again, and they do not follow this simple scenario, the way that the interfaces would. The crucial

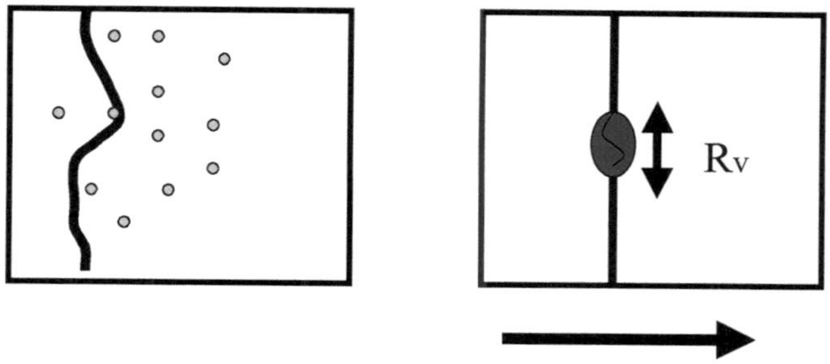

Fig. 32. When an interface is in motion, it averages over the disorder. As a result the interface does not feel the disorder above a certain lengthscale R_v and becomes flat again.

ingredient present in periodic system is the existence of a periodicity *transverse* to the direction of motion. Because of this, the transverse components of the displacements still feel a disorder that is non averaged by the motion. As a result the large-v expansion always breaks down, even at large velocities and cannot be used to determine the nature of the moving phases. To describe such moving phase the most important components are the components transverse to the direction of motion (this is schematized in Fig. 33). The motion of these components can be described by a quite generic equation of motion, as explained in [84,85]. The transverse components still experience a static disorder, and as a result the system in motion remains *a glass* (moving glass).

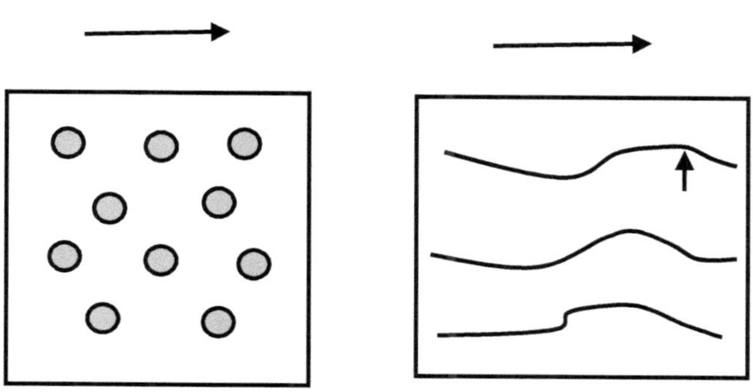

Fig. 33. When in motion a periodic structure averages over the Fourier components of disorder along the direction of motion whereas the component perpendicular to motion remains unaffected. As a result the components of the displacements transverse to the direction of motion are the essential ingredients to describe a moving system.

In the moving glass the motion of the particles occurs through elastic channels as shown in Fig. 34. The channels are the best compromise between the elastic forces and the static disorder still experienced by the moving system. Like lines submitted to a static disorder the channels themselves are rough and can meander arbitrarily far from a straight line (displacements grow unbound-

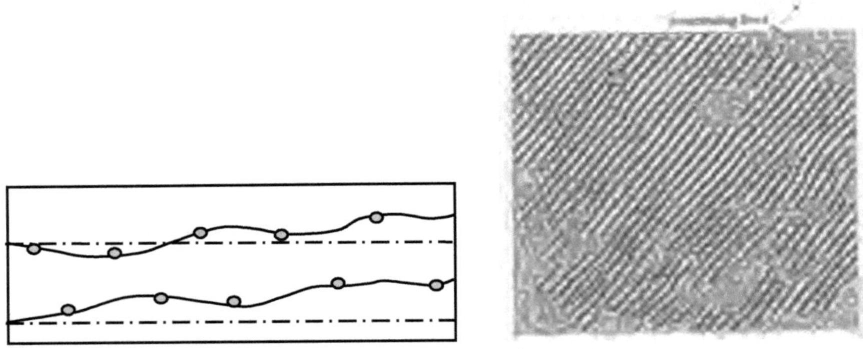

Fig. 34. In a moving glass the motion occurs through elastic channels. The channels themselves are rough and meander arbitrarily far from a straight line, but all particles follow on these channels like cars on a highway. These channels have been observed in decoration experiments, as shown on the right [86]

edly). However although the channels themselves are rough, the particles of the system are bound to follow these channels and thus follow exactly the *same* trajectory when in motion. Needless to say the moving system (moving glass) is thus very different than a simple solid with a modified temperature where the particles would just follow straight line trajectories (with a finite thermal broadening).

When does this picture breaks down? Clearly this should be the result of defects appearing in the structure. For example close to depinning, it was shown experimentally [60] (see Fig. 31) that some of the regions of the system can remain pinned while other parts of the system flow, leading to a plastic flow with many defects. One thus need to check again for the stability of the moving structure to defects. Fortunately the very existence of channels provide a very natural framework to study the effect of such defects: they will lead quite naturally to a coupling and a decoupling of the channels [84,87,85,83].

The various phases that naturally emerges in $d = 3$ are thus the ones shown in Fig. 35 (a similar study can be done for $d = 2$). At large velocity the channels are coupled and the system possesses a perfect topological order (no defects). The moving glass system is thus a moving Bragg glass. The structure factor has six Bragg peaks (with algebraic powerlaw divergence) showing that the system has quasi-long range positional order. If the velocity is lowered a Lindemann analysis shows that defects that appear first tend to decouple the channels. This means that positional order along the direction of motion is lost, but since the channel structure still exist a smectic order is preserved (channels become then the elementary objects). As a result the structure factor now has only two peaks. This phase is thus a moving smectic (or a moving transverse glass, as first found in [88]). It is important to note that in these two phases the channel structure is preserved and described by the moving glass equation. Both these phases are thus a moving glass. A quite different situation can occur if the velocity is lowered further. In that case the channel structure can be destroyed altogether,

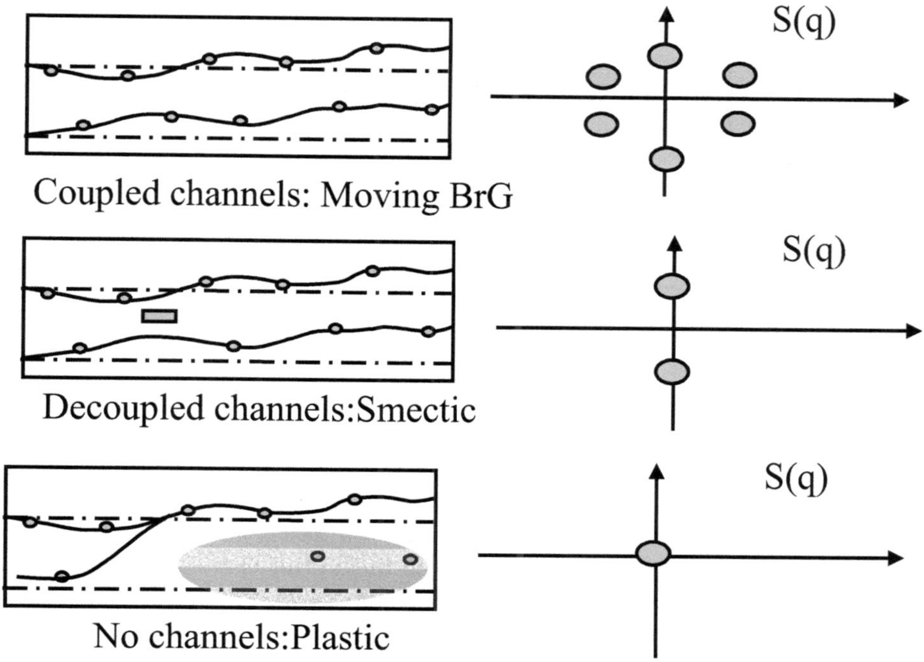

Fig. 35. The various dynamical phases of a moving periodic system.

leading to a plastic phase. Note that depending on the amount of disorder in the system this may or may not occur, and in $d = 3$ a purely elastic depinning could be possible in principle (in $d = 2$ the depinning is always plastic). These various phases lead to the dynamics phase diagram shown in Fig. 36. These various phases as well as the dynamical phase diagram have been confirmed in numerous simulations (see e.g. [88–91]).

Let us now turn to the experimental analysis of the dynamics. Early experiments on the dynamics of the vortex phases near the peak effect [60] showed that not only does the critical current rise sharply at the transition, but the $I-V$ curves also change in character. This is shown schematically in fig. 37. The top panel shows the peak effect at a fixed T and varying H. Three distinct types of $I-V$ curves are observed in the regions marked I, II and III, shown in the middle panel. In the peak region the curves show opposite curvature to that in the other regions, the voltage grows convex upwards. The lower panel illustrates this behavior through the differential resistance R_{d} $(= dV/dI)$for each region. In the peak region it shows a pronounced maximum above which it rapidly decreases to a terminal asymptotic value of the Bardeen-Stephen flux flow resistance above a crossover current. A comparison with simulations [92] shown in Fig. 38 suggests that the peak signifies a plastic flow region where the vortex matter moves incoherently and a coherent flow is recovered at high drives. This behavior contrasts with that in region I below the peak effect where the pinned vortex matter is in the Bragg glass phase. In this case the flux flow resistance is approached

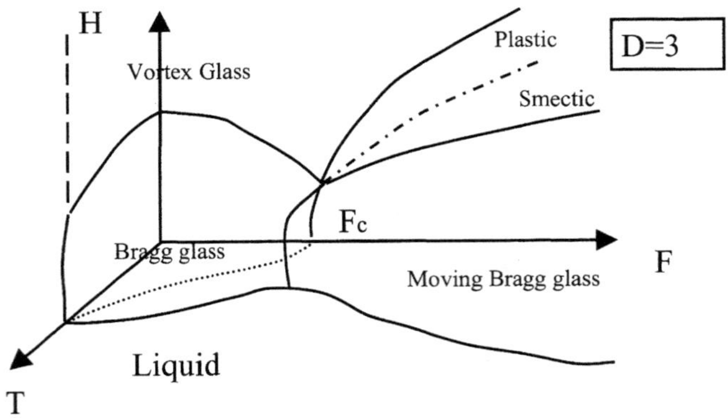

Fig. 36. The dynamical phase diagram [85] as a function of the temperature T, the magnetic field (far from H_{c1}) and the applied force F. Both the Bragg glass and the Moving Bragg glass have perfect topological order.

monotonically without a measurable signature of plastic flow. The plastic flow regime is also accompanied with pronounced fingerprint effect: a repeatable set of features in R_d as the current is ramped up and down, signifying a repeatable sequence of chunks of pinned vortex assembly depinning and joining the main flow. This clearly establishes the defective nature of the moving vortex matter in this regime. The contrast with this behavior in region I is thought to represent a depinning of the Bragg glass phase directly into a moving Bragg glass phase.

A striking result is seen in the behavior of the flux flow noise [93], summarized in Fig. 39. The noise is larger in the peak regime by orders of magnitude (seen in the lower panel) and is restricted to current values between the depinning current and the crossover current (shown in the upper panel). The noise is, therefore, associated with an incoherent flow of a defective moving phase and a coexistence of moving and pinned vortex phases as was seen in detailed studies [94,95]. The qualitative behavior observed in the experiment is consistent with many simulations of the flux flow noise characteristics. However, recent work suggests that an edge contamination mechanism in the peak regime is responsible for triggering much of the defective plastic flow in the bulk and is a subject under active investigation [96,97]. There have also been reports of narrow band noise [98], i.e., noise peaks at the so-called washboard frequency. The observation of such noise would be compatible with the moving Bragg glass phase.

More experimental evidence in favor of the moving glass is also found in recent decoration and STM studies of the moving vortex assembly [99,86,100].

5.3 Metastability and History Dependence

A current focus on experimental vortex phase studies is a renewed interest in history effects that have long been known to occur in vortex matter. In order to understand the equilibrium phase behavior of the system, we need to ascertain that we have indeed reached equilibrium in an experiment. However, most experiments in low T_c systems and at low temperatures in high T_c systems as well,

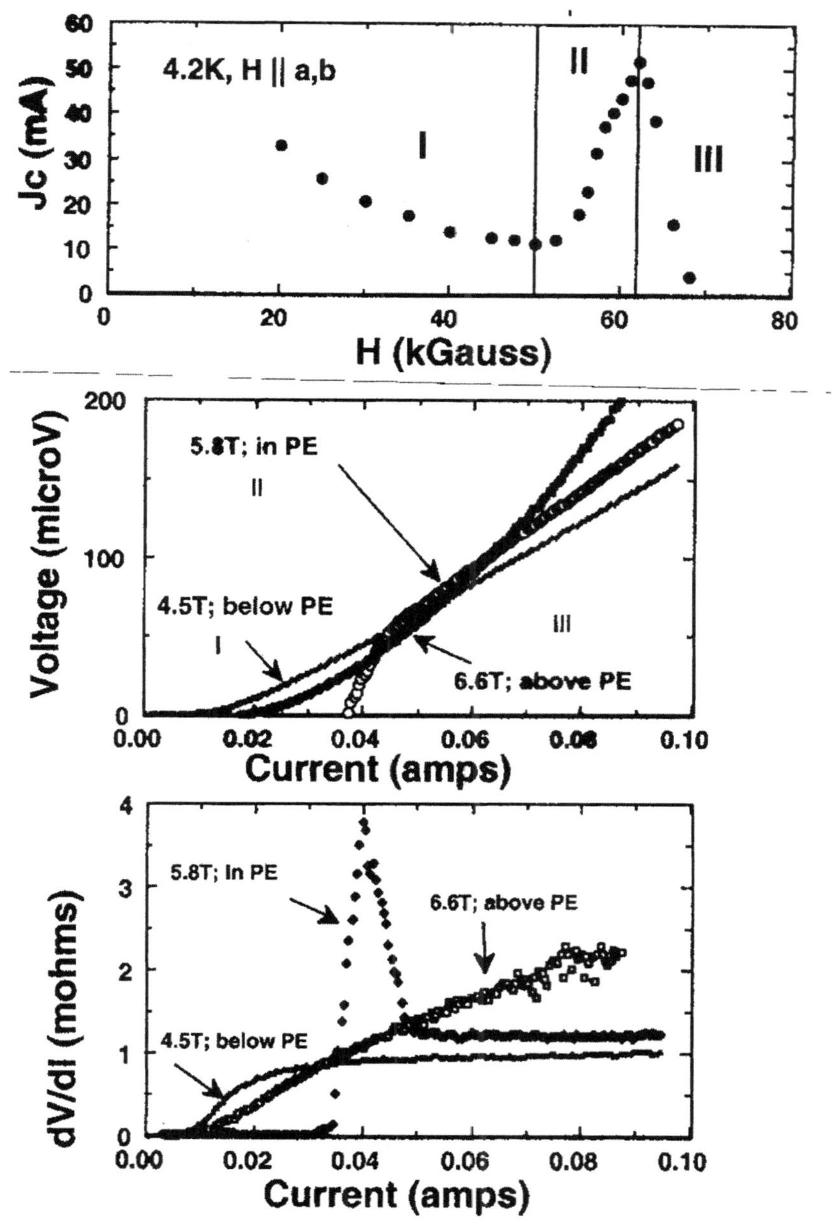

Fig. 37. Summary of the variation in the $I - V$ characteristic across the peak effect, i.e., ordered to disordered phase transition [60]. The upper panel shows the peak effect transition and marks the three regimes of flow. The middle and lower panels show the $I - V$ curves and the $dV/dI - I$ curves for the three regimes.

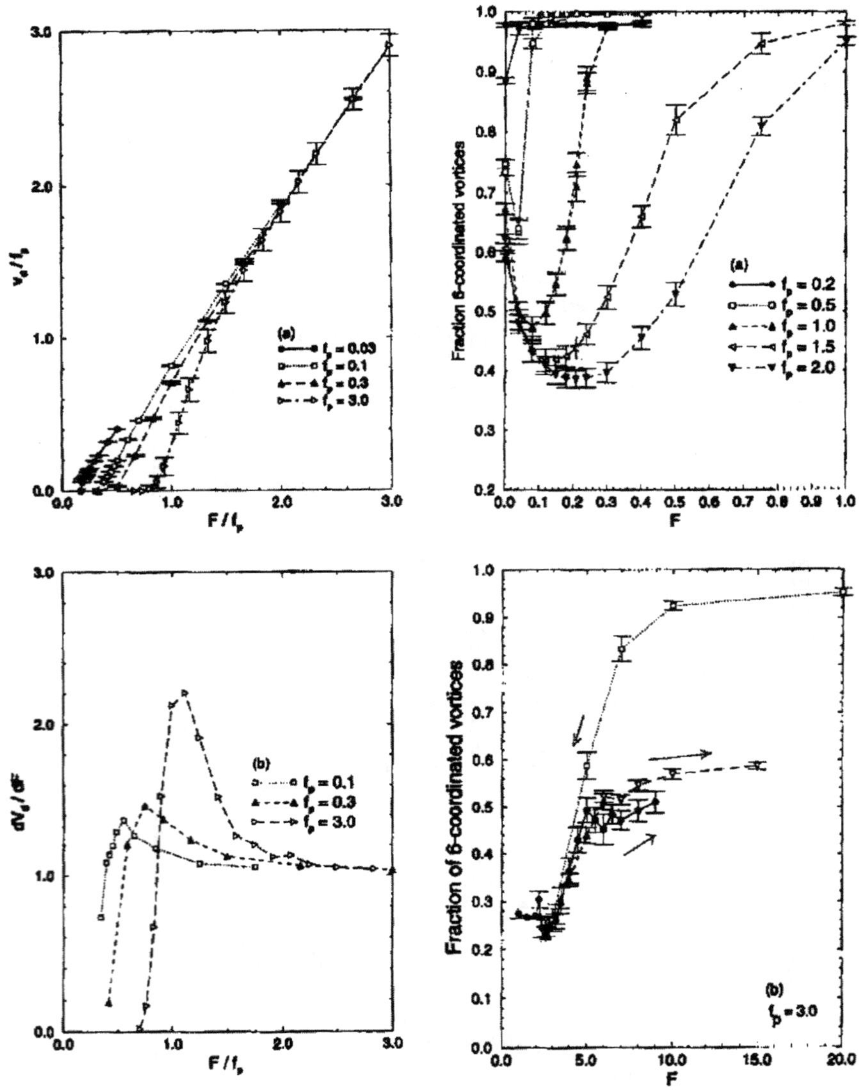

Fig. 38. Simulation data [92] of the force-velocity curves together with the variation of defect density with driving force. Compare the simulation data with the experimental data for plastic flow.

show a pronounced dependence of the vortex correlations on the thermomagnetic history of the system. A striking example is shown in Fig. 40 where the magnetic response [101,102], and transport critical current [103] are measured for field cooled (FC) and zero field cooled(ZFC) cases. In the latter case one sees a pronounced peak effect but not in the former. In other words, the FC state yields a highly disordered vortex glass phase and the latter yields an ordered

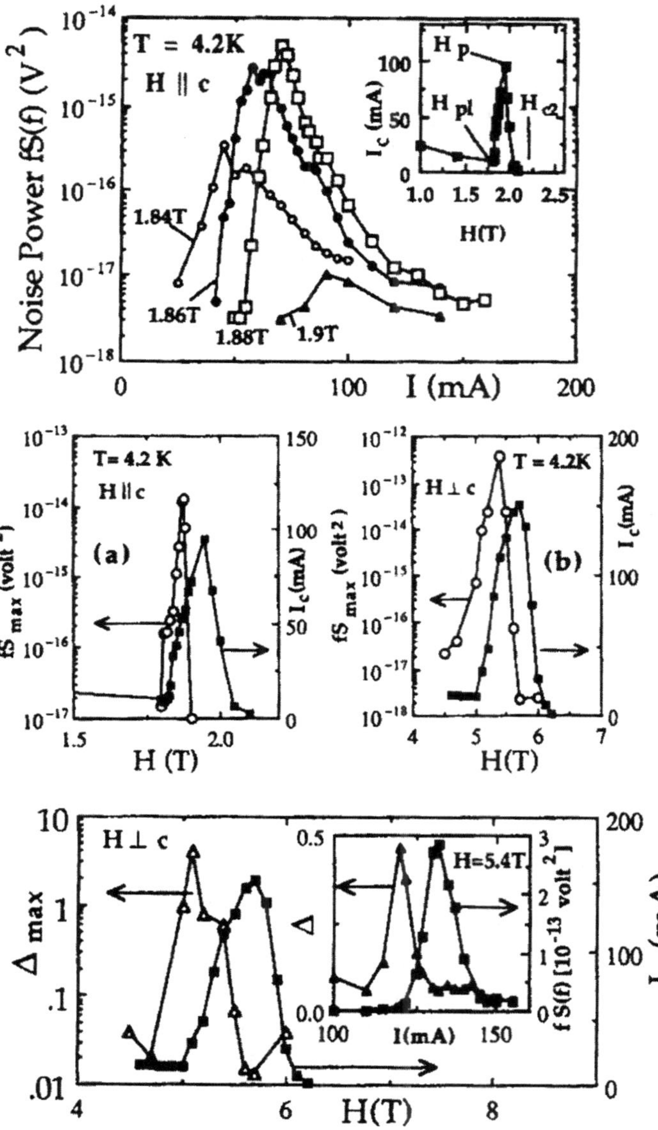

Fig. 39. Noise characteristics in the plastic flow regime [93]. The upper panel shows the current dependence of the noise which turns on at the critical current and depletes at large current when a moving ordered phase is recovered. The middle panels show the field dependence of the noise; it is restricted only to the peak regime. The lowest panel show a measure of the non-gaussianity of the noise which is maximum (i.e.,few noise sources) at the onset of motion supporting the moving "chunk" scenario in the plastic flow regime.

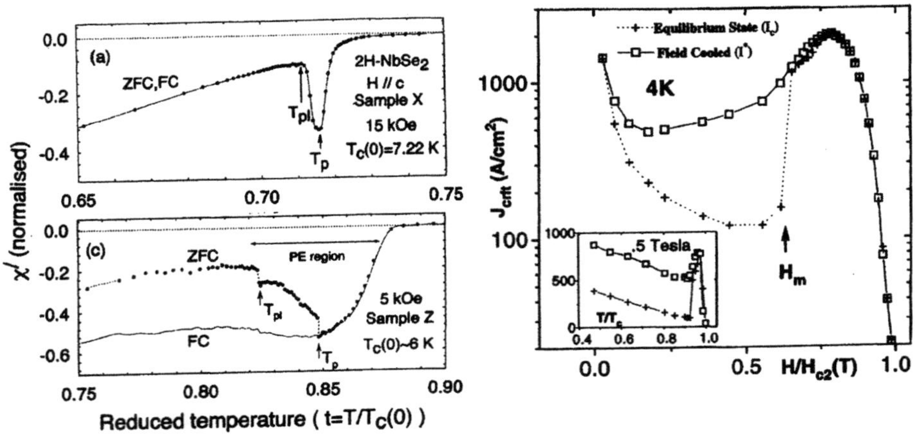

Fig. 40. History dependence of the critical current seen with magnetic susceptibility(left) [101], and transport(right) [103]. For a very clean sample on the top panel on the left, no history effect is seen unlike the bottom panel for a dirty sample. The peak of the peak effect marks the onset of an equilibrium disordered phase.

Bragg glass phase. The question then is: which is the stable and equilibrium state of the system and how does one find it? A similar question arose in the interpretation of neutron experiments [104] where it was suggested [105] that the observed broadening of the lines that disappeared after a cycling above the critical current, was due to the presence of *out of equilibrium* dislocations (on the top of the equilibrium Bragg glass phase).

One possible resolution of the problem comes from a "shaking experiment" where the FC state is subjected to a large oscillatory magnetic field and the system evolves to the ZFC state [106]. Once in the ZFC ordered state, below the peak effect, the system cannot be brought to the FC disordered state regardless of any external perturbation. It is thus reasonable to conclude that the ordered (Bragg glass) state is the equilibrium state below the peak effect. The FC disordered (vortex glass) phase is simply supercooled from the liquid phase above. When pinning sets in, the system fails to explore the phase space due to the pinning barrier and stays frozen into glassy, disordered phase. Shaking with an ac field then provides an annealing mechanism to bring the system in the true ground state which is the Bragg glass. On the other hand, shaking fails to produce an ordered state above the peak and thus the disordered phase is indeed the ground state there. Curiously then the ZFC state is formed by vortices entering the system at high velocity, thereby ignoring pinning and forming a moving Bragg glass phase from which the pinned Bragg glass phase evolves easily. Recent aging experiments [107] have provided compelling evidence in support of these conclusions. Further evidence of a thermodynamic nature of the transition is obtained also from magnetization anomalies from annealed vortex states [108].

5.4 Edge Effects

Yet another phenomenon has begun to be explored in experimental studies [96,97]. This relates to the observation that edges of a sample provide nucleation centers for the disordered phase. The net results are (1) the order-disorder phase transition becomes spatially non-uniform and leads to phase coexistence that marks the width of the peak effect region and (2) an external driving current flows non-uniformly in the system. The contamination of the ordered phase by the disordered phase from the edge and the subsequent annealing back to the ordered phase at larger drives occur in a non-uniform manner leading to a variety of unusual time and frequency dependent phenomena observed earlier. Differentiating the effects of these processes from the bulk response of the system, usually assumed in interpreting data and in simulations, remains a subject of current study.

5.5 Transverse Critical Force

One of the unexpected consequences of the presence of channels in the moving glass phase, is the existence of transverse pinning [84]. Let us examine what would happen if one tried to push the moving system sideways by applying a force F_y in a direction perpendicular to motion. This is depicted in Fig. 41. The

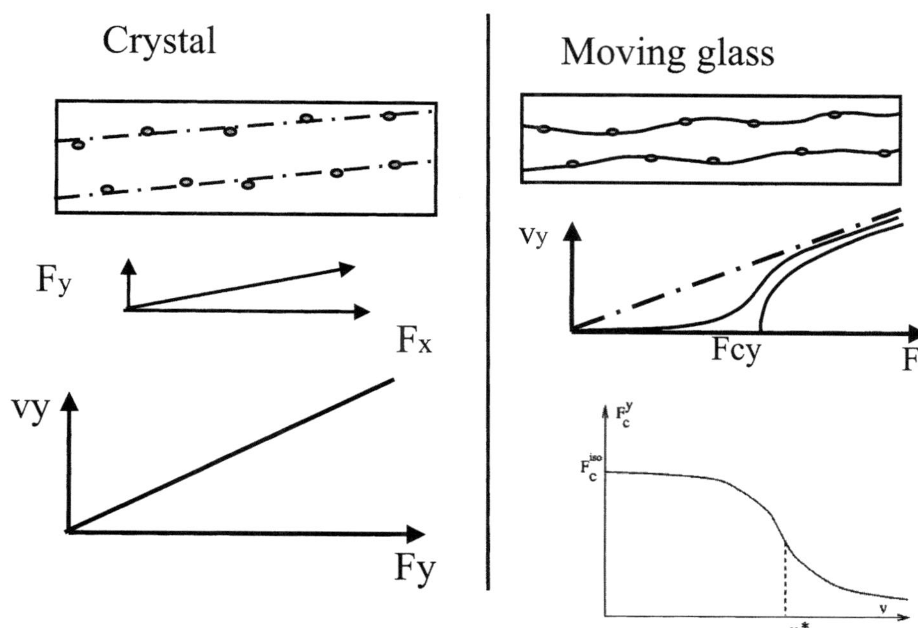

Fig. 41. Response of the system to a force F_y transverse to the direction of motion. Although a crystal would respond linearly, the moving glass remains pinned in the transverse direction although the particles are moving along the direction x [84,85].

naive answer would be that the system is submitted to a total force $F = F_x + F_y$ and because the modulus of this force is larger than the threshold ($F_x > F_c$ since the system is already sliding), the system will slide along the total force. This means that there is a *linear* response v_y to the applied transverse force F_y. This is what would occur if the moving system was a crystal. However in the moving glass although the particles themselves move along the channels, the channels are submitted to a static disorder and thus pinned. It means that if one applies a transverse force, the channels will have to pass barriers before they can move. As a results the system is transversally pinned even though it is moving along the x direction. The existence of this transverse critical force is a hallmark of the moving glass. It is a fraction of the longitudinal transverse force and decreases as the longitudinal velocity increases as shown on Fig. 41. The existence of such transverse force has been confirmed in many simulations (see e.g. [88,109,110]). Its experimental observation for vortex systems is still an experimental challenge.

6 Conclusions and Perspectives

It is thus clear from the body of work presented in these notes that the field of vortex matter has considerably matured in the last ten years or so. Many important physical phenomena have been unravelled, and a coherent picture starts to emerge. As far as the statics of the problem is concerned we start now to have a good handle of the problem. The Bragg glass description has allowed to build a coherent picture both of the phase diagram and of most of the previously poorly understood striking phenomena such as imaging, neutrons and peak effect or more generally transport phenomena. Important issues such as the contamination by the edge and the necessity to untangle these effects to get the true thermodynamics properties of the system are now understood, allowing to get reliable data. Although, understanding the dynamics is clearly much more complicated both theoretically and experimentally, here also many progress have been made. The vortex matter has allowed to introduce and check many crucial concepts such as the one of creep motion. For the dynamics also it is now understood that periodicity is crucial and that strong effects of disorder can persist even when a lattice is fast moving.

Of course the progress realized make only more apparent the many exciting issues not yet understood and that are waiting to be solved. For the statics the nature of the high field phase above the melting field of the Bragg glass is still to be understood. Is this phase a distinct thermodynamic phase from the liquid? is it a true glass in the dynamical sense? Many difficult questions that will need the understanding of a system in which disorder *and* defects (dislocations, etc.) play a crucial role. Similar question occur for the dynamics whenever a plastic phase occurs. There is thus no doubt that understanding the role of defects and disorder is now one of the major challenge of the field. The glassy nature of the various phases also certainly needs more detailed investigations, in particular the "goodies" such as aging that have been explored in details for systems such as spin glasses would certainly prove useful to investigate. Finally, both theorists and experimentalists have mostly focussed for the moment on the steady state dynamics. Clearly the issue of noise, out of equilibrium dynamics, and history dependence are challenging and crucial problems.

In addition, the material covered in this chapter is only a part of the many interesting phenomena related to the physics of vortex lattices, and more generally disordered elastic systems. Because of space limitation, it is be impossible to cover here all these interesting aspects but we would like to at least mention few of them.

In addition to point like impurities there is much interest, both theoretical and practical, to introduce artificial disorder. The most popular is the one produced by heavy ion irradiation [111–113], leading to the so called columnar defects and to the Bose Glass phase [114]. Other types of disorder such as splay [115] or regular pinning arrays have also been explored. We refer the reader to the above mentioned reviews for more details on these issues.

Quite interestingly, the description of the vortices in term of elastic objects allows to make contact with a large body of other physical systems who share in fact the same effective physics. This ranges from classical systems such as magnetic domain walls, wetting contact lines, colloids, magnetic bubbles, liquid crystals or quantum systems: charge density waves, Wigner crystals of electrons, Luttinger liquids. These systems share the same basic physics of the competition between elastic forces that would like to form a nice lattice or a flat interface and disorder. This makes the question of exploring the connections with the concepts useful for the vortex lattices (similarities/differences) particularly interesting and fruitful. For example, magnetic domain walls are ideal systems to quantitatively check the creep formula [116], Wigner crystals [117] or charge density waves [118] have offered the possibility to test the existence of a transverse critical force. The question of the observation of the noise in these various systems is also a very puzzling question. Of course, these systems have also their own particularities and present their own challenging problems. The interested reader is again directed to the review papers for the classical problems and a short review along those lines for quantum systems can be found in [119].

There is thus no doubt that the field is now growing and permeating many branches of condensed matter physics, providing a unique laboratory to continue developing unifying concepts such as the ones born and grown for vortices since more than half of a century. To what new developments this extraordinary richness of physical situations will lead, only time will tell.

Acknowledgements

We have benefitted from many invaluable discussions with many colleagues, too numerous, to be able to thank them all here. We would however like to especially thank G. Crabtree, P. Kes, M. Konczykowski, W. Kwok, C. Marchetti, A. Middleton, C. Simon, K. van der Beek, F.I.B Williams, E. Zeldov and G.T. Zimanyi. T.G. acknowledges the many fruitful and enjoyable collaborations with P. Le Doussal, and with P. Chauve, D. Carpentier, T. Klein, S. Lemerle, J.

Ferre. S.B. acknowledges the special contributions of M. Higgins and thanks E. Andrei, S. Banerjee, P. deGroot, A. Grover, M. Marchevsky, Y. Paltiel, S. Ramakrishnan, T.V.C Rao, G. Ravikumar, M. Weissman, E. Zeldov and A. Zhukov for fruitful collaborations. TG would like to thank the ITP where part of these notes were completed for hospitality and support, under the NSF grant PHY99-07949. S.B. thanks the Tata Institute of Fundamental Research for hospitality while materials for the lectures were prepared.

References

1. A.A. Abrikosov: Sov. Phys. JETP **5**, 1174 (1957)
2. M. Tinkham: *Introduction to Superconductivity* (Mc Graw Hill, New York, 1975)
3. P.G. De Gennes: *Superconductivity of Metals and Alloys* (W.A. Benjamin, New York, 1966)
4. G. Blatter et al.: Rev. Mod. Phys. **66**, 1125 (1994)
5. H. Brandt: Rep. Prog. Phys. **58**, 1465 (1995)
6. T. Giamarchi, P. Le Doussal: in *Spin Glasses and Random fields*, ed. A.P. Young (World Scientific, Singapore, 1998), p.321, cond-mat/9705096
7. T. Nattermann, S. Scheidl: Adv. Phys. **49**, 607 (2000)
8. A. Houghton, R. Pelcovits, A. Sudbo: Phys. Rev. B **40**, 6763 (1989)
9. D. Nelson: Phys. Rev. Lett. **69**, 1973 (1988)
10. M. Charalambous, J. Chaussy, P. Lejay: Phys. Rev. B **45**, 5091 (1992)
11. H. Safar et al.: Phys. Rev. Lett. **70**, 3800 (1993)
12. H. Safar et al.: Phys. Rev. B **52**, 6211 (1995)
13. W. Kwok et al.: Phys. Rev. Lett. **72**, 1088 (1994)
14. E. Zeldov et al.: Nature **375**, 373 (1995)
15. A.I. Larkin: Sov. Phys. JETP **31**, 784 (1970)
16. H. Fukuyama, P.A. Lee: Phys. Rev. B **17**, 535 (1978)
17. Y. Imry, S.K. Ma: Phys. Rev. Lett. **35**, 1399 (1975)
18. A.I. Larkin, Y.N. Ovchinnikov: J. Low Temp. Phys. **34**, 409 (1979)
19. T. Giamarchi, P. Le Doussal: Phys. Rev. Lett. **72**, 1530 (1994)
20. T. Giamarchi, P. Le Doussal: Phys. Rev. B **52**, 1242 (1995)
21. F.D.M. Haldane: Phys. Rev. Lett. **47**, 1840 (1981)
22. T. Nattermann: Phys. Rev. Lett. **64**, 2454 (1990)
23. D.S. Fisher, M.P.A. Fisher, D.A. Huse: Phys. Rev. B **43**, 130 (1990)
24. M.P.A. Fisher: Phys. Rev. Lett. **62**, 1415 (1989)
25. H.S. Bokil, A.P. Young: Phys. Rev. Lett. **74**, 3021 (1995)
26. J. Villain, J.F. Fernandez: Z. Phys. B **54**, 139 (1984)
27. S.E. Korshunov: Phys. Rev. B **48**, 3969 (1993)
28. T. Emig, S. Bogner, T. Nattermann: Phys. Rev. Lett. **83**, 400 (1999)
29. R.P. Feynman: *Statistical Mechanics* (Benjamin Reading, MA, 1972)
30. M. Mezard, G. Parisi: J. de Phys. I (Paris) **4**, 809 (1991) E.I. Shakhnovich, A.M. Gutin: J. Phys. A **22**, 1647 (1989)
31. D. Carpentier, P. Le Doussal, T. Giamarchi: Europhys. Lett. **35**, 379 (1996)
32. J. Kierfeld, T. Nattermann, T. Hwa: Phys. Rev. B **55**, 626 (1997)
33. D.S. Fisher: Phys. Rev. Lett. **78**, 1964 (1997)
34. M.J.P. Gingras, D.A. Huse: Phys. Rev. B **53**, 15193 (1996)
35. A.V. Otterlo, R. Scalettar, G. Zimanyi: Phys. Rev. Lett. **81**, 1497 (1998)
36. M. Marchevsky: Ph.D. thesis, University of Leiden, 1998

37. H. Hess et al.: Phys. Rev. Lett. **62**, 214 (1989)
38. K. Harada et al.: Nature **360**, 51 (1992)
39. X. Ling et al.: Phys. Rev. Lett. **86**, 126 (2001)
40. D.G. Grier et al.: Phys. Rev. Lett. **66**, 2270 (1991)
41. P. Kim, Z. Yao, C.A. Bolle: Phys. Rev. B **60**, R12589 (1999)
42. T. Klein et al.: Nature **413**, 404 (2001)
43. S. Ryu, A. Kapitulnik, S. Doniach: Phys. Rev. Lett. **77**, 2300 (1996)
44. Y. Nonomura, X. Hu: Phys. Rev. Lett. **86**, 5140 (2001)
45. P. Olsson, S. Teitel: Phys. Rev. Lett. **87**, 137001 (2001)
46. B. Khaykovich et al.: Phys. Rev. Lett. **76**, 2555 (1996)
47. R. Cubbit et al.: Nature **365**, 407 (1993)
48. P.L. Gammel et al.: Phys. Rev. Lett. **80**, 833 (1998)
49. D. Ertas, D.R. Nelson: Physica C **272**, 79 (1996)
50. T. Giamarchi, P. Le Doussal: Phys. Rev. B **55**, 6577 (1997)
51. J. Kierfeld: Physica C **300**, 171 (1998)
52. A.E. Koshelev, V.M. Vinokur: Phys. Rev. B **57**, 8026 (1998)
53. F. Bouquet et al.: Nature **411**, 448 (2001)
54. U. Welp et al.: Phys. Rev. Lett. **76**, 4908 (1996)
55. A. Schilling, R.A. Fisher, G.W. Crabtree: Nature **382**, 791 (1996)
56. K. Deligiannis et al.: Phys. Rev. Lett. **79**, 2121 (1997)
57. T. Nishizaki et al.: Phys. Rev. B **61**, 3649 (2000)
58. S. Kokkaliaris, A.A. Zhukov, P.A.J. de Groot: Phys. Rev. B **61**, 3655 (2000)
59. R. Wordenweber, P. Kes, C. Tsuei: Phys. Rev. B **33**, 3172 (1986)
60. S. Bhattacharya, M.J. Higgins: Phys. Rev. Lett. **70**, 2617 (1993)
61. M.J. Higgins, S. Bhattacharya: Physica C **257**, 232 (1996)
62. T.V.C. Rao et al.: Physica C **299**, 267 (1998)
63. K. Ghosh et al.: Phys. Rev. Lett. **76**, 4600 (1996)
64. S.S. Banerjee et al.: Europhys. Lett. **44**, 91 (1998)
65. C.M. Aegerter et al.: Phys. Rev. B **57**, 1253 (1998)
66. T. Ishida et al.: Phys. Rev. B **56**, 5128 (1997)
67. G.I. Menon: 2001, cond-mat/0103013
68. A. van Otterloo et al.: Phys. Rev. Lett. **84**, 2493 (2000)
69. P. Chauve, T. Giamarchi, P. Le Doussal: Phys. Rev. B **62**, 6241 (2000)
70. P.W. Anderson, Y.B. Kim: Rev. Mod. Phys. **36**, 39 (1964)
71. T. Nattermann: Europhys. Lett. **4**, 1241 (1987)
72. L.B. Ioffe, V.M. Vinokur: J. Phys. C **20**, 6149 (1987)
73. M. Feigelman, V.B. Geshkenbein, A.I. Larkin, V. Vinokur: Phys. Rev. Lett. **63**, 2303 (1989)
74. P. Chauve, T. Giamarchi, P. Le Doussal: Europhys. Lett. **44**, 110 (1998)
75. J. Kierfeld, H. Nordborg, V.M. Vinokur: Phys. Rev. Lett. **85**, 4948 (2000)
76. D. Fuchs et al.: Phys. Rev. Lett. **80**, 4971 (1998)
77. C.J. van der Beek et al.: Physica C **341**, 1279 (2000)
78. R. Thorel et al.: J. Phys. (Paris) **34**, 447 (1973)
79. A.I. Larkin, Y.N. Ovchinnikov: Sov. Phys. JETP **38**, 854 (1974)
80. A. Schmidt, W. Hauger: J. Low Temp. Phys **11**, 667 (1973)
81. A.E. Koshelev, V.M. Vinokur: Phys. Rev. Lett. **73**, 3580 (1994)
82. T. Nattermann, S. Stepanow, L.H. Tang, H. Leschhorn: J. Phys. (Paris) **2**, 1483 (1992)
83. S. Scheidl, V. Vinokur: Phys. Rev. B **57**, 13800 (1998)
84. T. Giamarchi, P. Le Doussal: Phys. Rev. Lett. **76**, 3408 (1996)

85. P. Le Doussal, T. Giamarchi: Phys. Rev. B **57**, 11356 (1998)
86. M. Marchevsky et al.: Phys. Rev. Lett. **78**, 531 (1997)
87. L. Balents, C. Marchetti, L. Radzihovsky: Phys. Rev. B **57**, 7705 (1998)
88. K. Moon et al.: Phys. Rev. Lett. **77**, 2378 (1997)
89. C.J. Olson et al.: Phys. Rev. Lett. **81**, 3757 (1998)
90. A. Kolton et al.: Phys. Rev. Lett. **83**, 3061 (1999)
91. H. Fangohr, S.J. Cox, P.A.J. de Groot: Phys. Rev. B **64**, 064505 (2001)
92. A.A.M.M.C. Faleski, M.C. Marchetti: Phys. Rev. B **54**, 12427 (1996)
93. A.C. Marley, M.J. Higgins, S. Bhattacharya: Phys. Rev. Lett. **74**, 3029 (1995)
94. R.D. Merithew et al.: Phys. Rev. Lett. **77**, 3197 (1996)
95. M. Rabin et al.: Phys. Rev. B **57**, R720 (1998)
96. Y. Paltiel et al.: Nature **403**, 398 (2000)
97. M. Marchevsky, M.J. Higgins, S. Bhattacharya: Nature **409**, 591 (2001)
98. Y. Togawa et al.: Phys. Rev. Lett. **85**, 3716 (2000)
99. F. Pardo et al.: Nature **396**, 348 (1998)
100. A.M. Troyanovski et al.: Nature **399**, 665 (1999)
101. S.S. Banerjee et al.: Phys. Rev. B **58**, 995 (1998)
102. G. Ravikumar et al.: Phys. Rev. B **57**, R11069 (1998)
103. W. Henderson et al.: Phys. Rev. Lett. **77**, 2077 (1996)
104. U. Yaron et al., Phys. Rev. Lett. **73**, 2748 (1994)
105. T. Giamarchi, P. Le Doussal: Phys. Rev. Lett. **75**, 3372 (1995)
106. S.S. Banerjee et al.: Appl. Phys. Lett. **74**, 126 (1999)
107. F. Portier et al.: 2001, cond-mat 0109077.
108. G. Ravikumar et al.: Phys. Rev. B **63**, 0240505 (2001)
109. C.J. Olson et al.: Phys. Rev. B **61**, R3811 (1999)
110. H. Fangohr, P.A.J. de Groot, S.J. Cox: Phys. Rev. B **63**, 064501 (2001)
111. M. Konczykowski et al.: Phys. Rev. B **44**, 7167 (1991)
112. L. Civale et al.: Phys. Rev. Lett. **67**, 648 (1991)
113. C.J. van der Beek et al.: Phys. Rev. B **51**, 15492 (1995)
114. D.R. Nelson, V.M. Vinokur: Phys. Rev. B **48**, 13060 (1993)
115. T. Hwa, P. Le Doussal, D.R. Nelson, V.M. Vinokur: Phys. Rev. Lett. **71**, 3545 (1993)
116. S. Lemerle et al.: Phys. Rev. Lett. **80**, 849 (1998)
117. F. Perruchot et al.: Physica B **256**, 587 (1998)
118. N. Markovic, N. Dohmen, H. van der Zant: Phys. Rev. Lett. **84**, 534 (2000)
119. T. Giamarchi, E. Orignac: in *New Theoretical Approaches to Strongly Correlated Systems*, Vol.23 of *NATO SCIENCE SERIES: II: Mathematics, Physics and Chemistry*, ed. A.M. Tsvelik (Kluwer Academic Publishers, Dordrecht, 2001), cond-mat/0005220

Colossal Magnetoresistive Oxides in High Magnetic Fields

Tsuyoshi Kimura[1] and Yoshi Tokura[1,2]

[1] Department of Applied Physics, University of Tokyo, Tokyo 113-8656, Japan
[2] Joint Research Center for Atom Technology, Tsukuba, Ibaraki 305-0046, Japan

Abstract. The perovskite-type manganese oxides are overviewed in the light of the mechanism of colossal magnetoresistance (CMR). The essential ingredient of the CMR physics is not only the double-exchange interaction but also other competing interactions, such as ferromagnetic/antiferromagnetic superexchange interactions and charge/orbital ordering instabilities as well as their strong coupling with the lattice deformation. In particular, the orbital degree of freedom of the conduction electrons in the nearly-degenerate $3d$ e_g state plays an essential role in producing the unconventional metal-insulator phenomena in the manganites via strong coupling with spin, charge, and lattice degrees of freedom. Insulating or poorly conducting states arise from the long or short-range correlations of charge and orbital, but can be mostly melted or turned into the orbital-disordered conducting state by application of a magnetic field, producing CMR or ane insulator−metal transition.

1 Introduction

The particular magnetoresistance (MR) phenomena described here are the gigantic decrease of resistance by application of a magnetic field that is observed in the manganese oxides (manganites) with perovskite or related structures [1]. Such a gigantic negative MR is now termed "colossal magnetoresistance" (CMR) to distinguish it from the giant magnetoresistance (GMR) observed in transition-metal systems in multilayer or granular forms. The MR in the perovskite manganites near the Curie temperature (T_C) seems to have already been known at the very early stage of the study on transition metal oxides. For example, the paper in 1969 by Searle and Wang [2] reported thoroughly the magnetic field dependence of the resistivity for a flux-grown crystal of $La_{1-x}Pb_xMnO_3$, in particular the large MR near T_C, as well as the phenomenological analysis. Soon after, Kubo and Ohata [3] have given a theoretical account for this phenomenon using the so-called double-exchange Hamiltonian (or the $s-d$ model with the on-site ferromagnetic exchange interaction), that includes the essential ingredient of the double-exchange mechanism elaborated by Zener [4], Anderson and Hasegawa [5], and de Gennes [6]. Interest in the MR of those manganites has revived more lately since the rediscovery of the CMR or even more astonishing magnetic field induced insulator−metal [7–11] and/or lattice-structural [12] transitions. On the one hand, the highly sensitive and electrically readable magnetic-field sensors have recently been in industrial demand for the read-head of magnetic memories and it is anticipated that the CMR oxides may be one such candidate material.

The "renaissance" of the CMR manganites was due partly to the revived interest in the barely metallic state of the transition metal oxides with strong electron–electron and/or electron–lattice interaction since the discovery of the high-temperature superconductivity in copper oxide (cuprate) compounds. Thanks to the experience gained in the research fever for superconducting cuprates, there has been much progress in the preparative method of oxide specimens, e.g., growth of single crystals and fabrication of epitaxial thin films, as well as in the comprehensive understanding of the electronic and magnetic properties of such correlated-electron systems [13]. This circumstance has expedited the research on CMR oxides, up to now. In perovskite-related structures, the two important parameters, i.e., the band filling (or doping level) and the bandwidth (or electron hopping interaction), can be controlled to a considerable extent by modifying the chemical composition of the perovskite. Both parameters control the kinetic energy of the conduction electrons which not only governs the metal–insulator phenomena but also the competing magnetic interactions, i.e. ferromagnetic vs. antiferromagnetic, in the perovskite manganites. We will see ample examples of the chemical control of the electronically important parameters in the following.

2 Fundamental Electronic and Lattice Features

2.1 Electronic Features

The manganese (Mn) ion in the CMR manganites is surrounded by the oxygen octahedron, as shown in the left panel of Fig.1. The $3d$ orbitals on the Mn-site placed in such an octahedral coordination are subject to the partial lifting

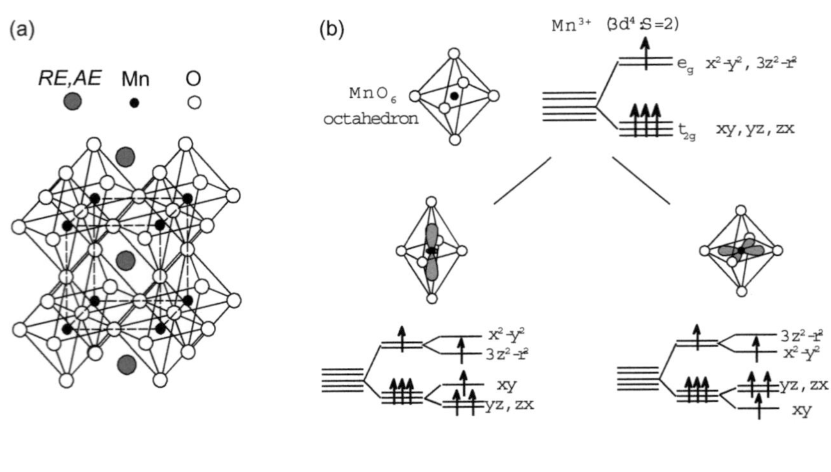

Fig. 1. (*left*): Schematic structure of perovskite-type manganese oxide. (*right*): Schematic electronic structure of Mn^{3+} ion in MnO_6 octahedron with JT distortion

of the degeneracy into lower-lying t_{2g} states and higher-lying e_g sates. In the Mn^{3+}-based compounds, the Mn site shows the $t_{2g}^3 e_{g^1}$ configuration (total spin number $S = 2$). The t_{2g} electrons, less hybridized with $2p$ states and stabilized by the crystal field splitting, are viewed as always localized by strong correlation effects and as forming a local spin ($S = 3/2$) even in the metallic state. Even the e_g state electrons, hybridized strongly with O $2p$ states, are affected by such a correlation effect, and tend to localize in the "carrier-undoped" or the parent all-Mn^{3+}-based compound, forming the so-called Mott insulator. However, the e_g electrons can be itinerant and hence play a role of conduction electrons, when electron vacancies or holes are created in the e_g orbital states of the crystal. The latter hole-doping procedure corresponds to creation of mobile Mn^{4+} species on the Mn sites.

The important consequence of the apparent separation into spin and charge sectors in the $3d$ orbital states are the effective strong coupling between the e_g conduction electron spin ($S = 1/2$) and the t_{2g} electron local spin ($S = 3/2$). This on-site ferromagnetic coupling follows the Hund's rule. The exchange energy J_H (Hund's-rule coupling energy) is as large as 2-3 eV for the manganites and exceeds the intersite hopping interaction t_{ij}^0 of the e_g electron between the neighboring sites, i and j. In the case of the strong coupling limit with $J_H/t_{ij} \gg 1$, the effective hopping interaction t_{ij} can be expressed in terms of the Anderson-Hasegawa relation [5], by neglecting the Berry phase term,

$$t_{ij} = t_{ij}^0 \cos(\theta_{ij}/2). \tag{1}$$

Namely, the absolute magnitude of the effective hopping depends on the relative angle (θ_{ij}) between the neighboring (classical) spins. The ferromagnetic interaction via the exchange of the (conduction) electron is termed the double-exchange interaction after the naming by Zener [4]. By creating electron-vacancy sites (or hole-doping) the e_g electron can hop depending on the relative configuration of the local spins. The ferromagnetic metallic state is stabilized by maximizing the kinetic energy of the conduction electrons ($\theta_{ij} = 0$). When the temperature is raised up to near or above the ferromagnetic transition temperature (T_C), the configuration of the spin is dynamically disordered and accordingly the effective hopping interaction is also subject to disorder and reduced on average. This would lead to an enhancement of the resistivity near and above T_C. Therefore, a large MR can be expected around T_C, since an external field relatively easily aligns the local spins and hence the randomness of the e_g hopping interaction is reduced. This is the intuitive explanation of the MR observed for the manganites around T_C in terms of the double-exchange (DE) model.

The physics of the colossal magnetoresistance (CMR) is obviously more complex. There are other important factors than in the above simplest DE scenario, which are necessary to interpret important features observed experimentally. Those are, for example, electron-lattice interaction, ferromagnetic/antiferromagnetic superexchange interaction between the local spins, intersite exchange interaction between the e_g orbitals (orbital ordering tendency), intrasite and intersite Coulomb repulsion interactions among the e_g electrons, etc. These interactions/instabilities occasionally compete with the ferromagnetic DE interaction, producing complex but intriguing electronic phases as well as the gigantic response of the system to an external field, such as CMR or the field-induced insulator-metal transition. Among the above interactions other than the DE interaction, the important electron-lattice interaction stems from the Jahn-Teller type coupling of the conduction e_g electrons with oxygen displacement. In the crystal, such a Jahn-Teller distortion is collective and a coherent distortion of Mn-O network is realized, as typically seen in $LaMnO_3$ [14]. The itinerant e_g holes (or mobile Mn^{4+} species) are obviously destructive to the static distortion, yet the subsisting dynamical Jahn-Teller coupling has been argued as one of the major factors relevant to the CMR phenomena.

2.2 Perovskite Structures

Most of the CMR phenomena have been found out and investigated for the manganese oxide (manganite) compounds with perovskite-type structure. The perovskite-type magnanites have the general formulas $RE_{1-x}AE_x\mathrm{MnO}_3$, where RE stands for a trivalent rare earth element such as La, Pr, Nd, Sm, Eu, Gd, Ho, Tb,Y, and etc. or for Bi^{3+}, and AE for a divalent alkaline earth ions such as Sr, Ca, and Ba or for Pb^{2+}. The (RE, AE) site (so-called perovskite A-site) can, in most cases, form homogeneous solid solution. Perovskite-based structures occasionally show lattice distortion as modifications from the cubic structure. One of the possible origin in the lattice distortion is the deformation of the MnO_6 octahedron arising from the Jahn-Teller effect that is inherent to the high-spin ($S = 2$) Mn^{3+} with double degeneracy of the e_g orbital. Another lattice deformation comes from the connective pattern of the MnO_6 octahedra in the perovskite structure, forming rhombohedral or orthorhombic (so-called $GdFeO_3$-type) lattice. In those distorted perovskites, the MnO_6 octahedra show alternating buckling. Such a lattice distortion of the perovskite in the form of ABO_3 (here $A = RE_{1-x}AE_x$ and $B =$ Mn for the present manganites) is governed by the so-called tolerance factor f [15], which is defined as

$$f = (r_B + r_O)/(r_A + r_O). \tag{2}$$

Here, r_i ($i = A, B$ or O) represents the (averaged) ionic size of each element. The tolerance factor f measures, by definition, the lattice-matching of the sequential AO and BO_2 planes. When f is close to 1, the cubic perovskite structure is realized. As r_A or equivalently f decreases, the lattice structure transforms to the rhombohedral ($0.96 < f < 1$) and then to the orthorhombic structure ($f < 0.96$), in which the B-O-B bond angle is bent and deviated from 180°. In the case of the orthorhombic lattice, the bond angle θ varies continuously with f. The bond angle distortion decreases the one-electron bandwidth W, since the effective d electron transfer interaction between the neighboring B-sites is governed by the supertransfer process via O $2p$ states. For example, let us consider the hybridization between the $3d$ e_g state and the $2p\sigma$ state in the $GdFeO_3$ type lattice which is composed of the quasi-right BO_6 octahedra tilting alternatively. In the strong ligand field approximation, the $p-d$ transfer interaction

t_pd is scaled as $t^0_{pd} \cos\theta$, t^0_{pd} being for the cubic perovskite. Thus, W for the e_g electron state is approximately proportional to $\cos^2\theta$.

Another important feature in the perovskite and related structures is that the compounds are quite suitable for the carrier doping procedure (filling control) since the structure is very robust against the chemical modification on the A-site. For example, in $La_{1-x}Ca_xMnO_3$, the solid solution can extend from $x = 0$ to 1. For AE=Sr the x can be extended up to $x = 0.7$ under normal synthetic condition, but up to $x = 1$ under high pressure. In the formula unit of $RE_{1-x}AE_xMnO_3$, the averaged Mn valence varies as 3+x. Namely, the x produces the vacancy in the e_g electron state at a rate of x per Mn-site, and hence is referred to as the hole-doping. To be exact, the filling (n) of the e_g electron conduction band is given as $n = 1 - x$.

3 Magnetoresistance of $La_{1-x}Sr_xMnO_3$

The $La_{1-x}Sr_xMnO_3$ is the most canonical double-exchange (DE) system, since it show the largest one-electron bandwidth W and accordingly less significantly affected by the electron-lattice and Coulomb correlation effects. Nevertheless, the end compound $LaMnO_3$ ($x = 0$) is strongly affected by the collective Jahn-Teller effect as well as by the electron correlation effect due to the $n = 1$ filling of the e_g band. The collective Jahn-Teller distortion present in $LaMnO_3$ reflects the orbital ordering such as the alternating $d_{x^2-r^2}$ and $d_{3y^2-r^2}$ orbitals on the ab plane [14]. In this Jahn-Teller-distorted and orbital-ordered state, $LaMnO_3$ undergoes an antiferromagnetic transition at 120 K. The spin ordering structure is the layered type or so-called A-type, in which the ferromagnetic ab planes are coupled antiferromagnetically along the c-axis.

With hole-doping by substitution of La with Sr, the ordered spins are canted toward the c-axis direction. Such a feature was first reported by Wollan and Koehler for $La_{1-x}Ca_xMnO_3$ [16] and was interpreted in terms of the DE mechanism by de Gennes [6]: The extension of the wave-function of the hole along the c-axis can mediate the DE-type ferromagnetic coupling, producing the spin canting. As the doping level x increases, the spin-canting angle is continuously increased to the nearly ferromagnetic spin structure. In $La_{1-x}Sr_xMnO_3$ [17,18], the canted antiferromagnetic phase appears to persist up to $x = 0.15$, although for $x > 0.10$ the spin-ordered phase is almost ferromagnetic. With further doping, the ferromagnetic phase appears below T_C which steeply increases with x up to 0.3 and then saturates. Another important consequence of the hole-doping is that the static collective Jahn-Teller distortion present in $LaMnO_3$ is diminished with increase of the doping level, perhaps beyond $x = 0.10$. This is manifested by the structural transformation between two types of orthorhombic forms [18].

The temperature dependence of the resistivity and the magnetic field effect are exemplified in Fig. 2 (left panel) for the $x = 0.175$ crystal [1,17]. The resistivity shows a steep decrease around T_C (indicated by an arrow), indicating that the relatively high resistivity above T_C is dominantly due to the scattering of the conduction electrons by thermally-disordered spins. The magnetic field greatly

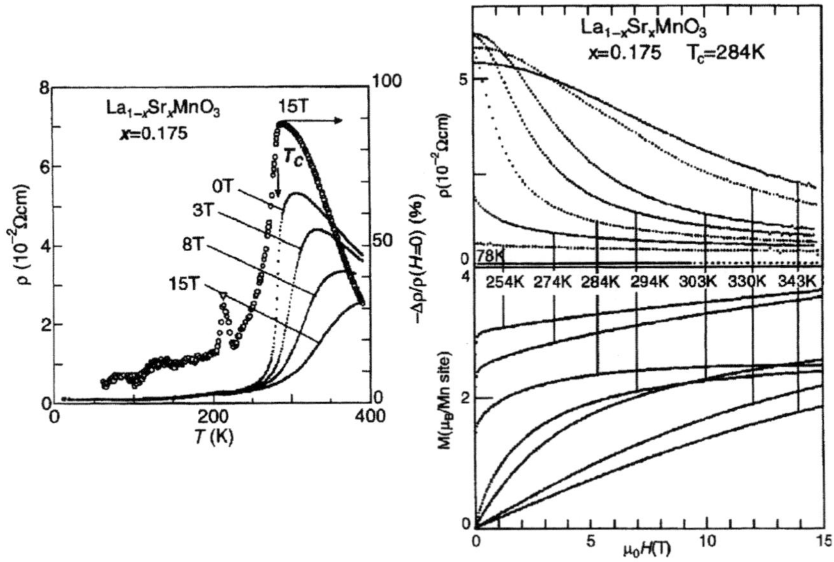

Fig. 2. (*left*): Temperature dependence of the resistivity for a $La_{1-x}Sr_xMnO_3$ ($x = 0.175$) crystal at various magnetic fields. Open circles represent the magnitude of the negative magnetoresistance $-[\rho(H) - \rho(0)]/\rho(0)$ with a magnetic field of 15 T. (*right*): The resistivity (upper panel) and magnetization (lower panel) as a function of magnetic field for a $La_{1-x}Sr_xMnO_3$ ($x = 0.175$) crystal

reduces the resistivity near T_C due to the suppression of the spin-scattering of the e_g-state carriers. Such a negative magentoresistance (MR) is nearly isotropic with respect to the direction of the applied field.

4 Compositional Tuning of CMR and Its Implication in the CMR Mechanism

4.1 Variation of Electronic Phase Diagrams

We show in Fig. 3 electronic phase diagrams in the temperature (T) vs. hole concentration (x) plane for prototypical compounds [19]: (a) $La_{1-x}Sr_xMnO_3$, (b) $Nd_{1-x}Sr_xMnO_3$ and (c) $Pr_{1-x}Ca_xMnO_3$. As the tolerance factor or equivalently the averaged ionic radius of the perovskite A-site decreases from (La,Sr) to (Pr,Ca) through (Nd,Sr), the orthorhombic distortion of $GdFeO_3$-type increases, resulting in the bending of the Mn-O-Mn bond and hence in the decrease of the one-electron bandwidth (W) of the e_g-state carriers. This means that other electronic instabilities, such as the charge/orbital-ordering and super-exchange interactions which compete with the double-exchange (DE) interaction, may become dominant in specific regions of x and temperature.

The phase diagram for $La_{1-x}Sr_xMnO_3$ with maximal W is canonical as the DE system, as described in the former section. The decrease of W complicates

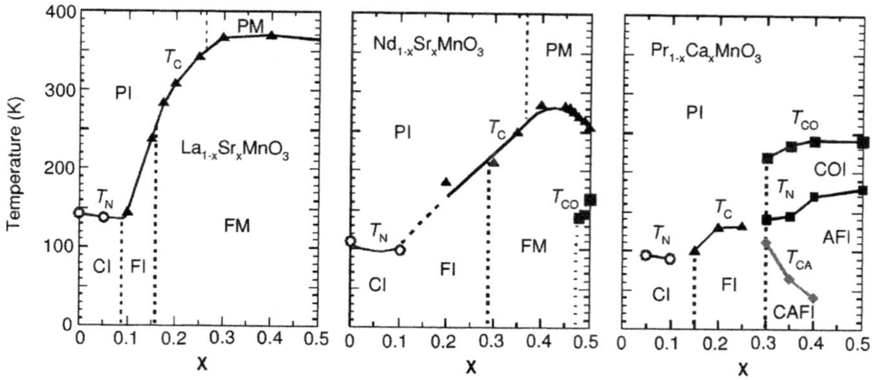

Fig. 3. The magnetic as well as electronic phase diagrams of $La_{1-x}Sr_xMnO_3$ (*left*), $Nd_{1-x}Sr_xMnO_3$ (*middle*) and $Pr_{1-x}Ca_xMnO_3$ (*right*). The PI, PM and CI denote the paramagnetic insulating, paramagnetic metallic and spin-canted insulating states, respectively. The FI, FM, and AFM denote the ferromagnetic insulating and ferromagnetic metallic, and antiferromagnetic (A-type) metallic states, respectively. At $x = 0.5$ in $Nd_{1-x}Sr_xMnO_3$, the charge-ordered insulating (COI) phase with CE-type spin ordering is present. For $0.3 \leq x \leq 0.5$ in $Pr_{1-x}Ca_xMnO_3$, the antiferromagnetic insulating (AFI) state exists in the charge-ordered insulating (COI) phase. The canted antiferromagnetic insulating (CAFI) state also shows up below the AFI state in the COI phase for $0.3 \leq x \leq 0.4$

the phase diagram. In particular, when the doping x is close to the commensurate value $x = 1/2$, the charge-ordering instability occurs. In the case of $Nd_{1-x}Sr_xMnO_3$ (middle panel of Fig. 3), the ferromagnetic metallic (FM) phase shows up for $x > 0.3$, yet in the immediate vicinity of $x = 1/2$ the FM state changes into the charge-ordered insulating (COI) state with decrease of temperature below $T_{CO} = 160$ K [20]. This COI state accompanies the concomitant antiferromagnetic ordering of orbitals and spins, the so-called CE type [16,21–23], as shown in Fig. 4. The nominal Mn^{3+} and Mn^{4+} species with a 1:1 ratio show a real-space ordering on the (001) plane of the orthorhombic lattice ($Pbnm$), while the e_g orbital shows the 1×2 superlattice on the same (001) plane. Reflecting such an orbital ordering, the ordered spins form a complicated sublattice structure extending over a larger unit cell as shown in the figure. Although a similar concomitant charge- and spin-ordering is widely seen in $3d$ electron transition metal oxides [13], the most notable feature of the manganites is the magnetic field induced melting of the CO state, which shows up as the field-induced phase transition from an antiferromagnetic insulator to a ferromagnetic metal [20].

With the hole concentration of $x > 0.52$, the A-type (layer-type) AF ordering turns up again [24–26]. Taking into account the change in lattice parameters toward the A-type AF phase [i.e., the a, b-axes expanding and the c-axis shrinking.], the uniform ordering of $d_{x^2-y^2}$ orbitals appears to be realized in the A-type AF phase in such an overdoped region ($0.52 < x < 0.62$). Note the difference in the orbital ordering in the present overdoped metallic manganites

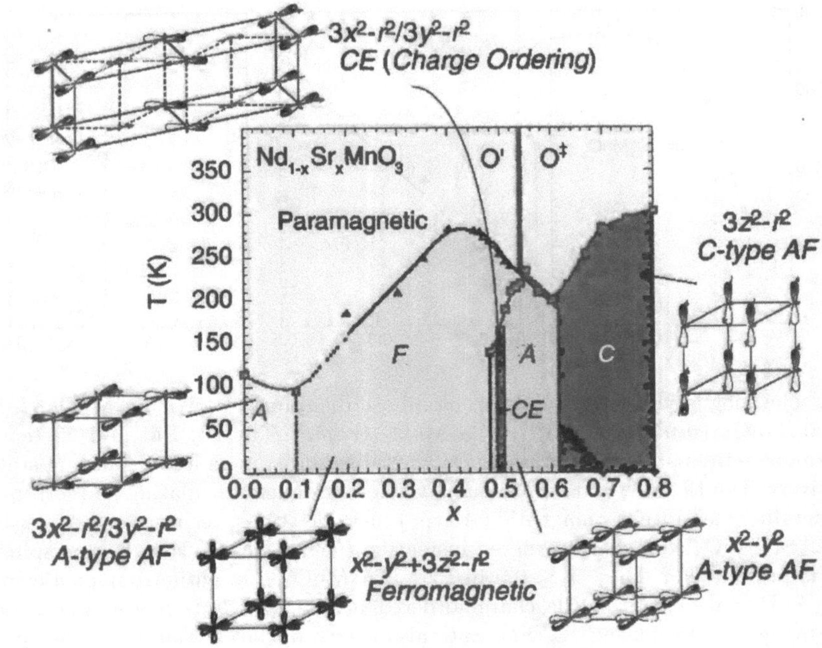

Fig. 4. Electronic phase diagram of $Nd_{1-x}Sr_xMnO_3$ crystal. The abbreviations mean paramagnetic (P), ferromagnetic (F), *CE*-type antiferromagnetic and charge-ordered (CE), *A*-type antiferromagnetic (A), *C*-type antiferromagnetic (C). O' and O‡ indicate the orthorhombic lattice structures with $a \approx b > c$ and $a \approx b < c$, respectively

$RE_{1-x}Sr_xMnO_3$ (RE =La, Pr, and Nd; $0.5 < x < 0.6$) and the insulating parent material $REMnO_3$ ($x = 0$), in the latter of which the same *A*-type AF structure appears with the orbital ordering of $3x^2 - r^2$ and $3y^2 - r^2$ in the *ab* plane. The uniform ordering of $x^2 - y^2$ orbitals makes the phase metallic but with a high 2D character ($\rho_{ab}/\rho_c \approx 10^4$ at 35 mK) [25], since the inter-plane (*c*-axis) hopping of the fully spin-polarized $d_{x^2-y^2}$ electron is almost forbidden due to the in-plane confinement in the layered-AF structure as well as to the minimal supertransfer interaction between the $x^2 - y^2$ orbitals along the z-direction. The mean-field theory for the electronic phase diagram of pseudo-cubic manganites also predicts that the layered *A*-type AF state near $x = 0.5$ accompanies the $x^2 - y^2$ orbital-ordering [27]. As x is further increased, the rod type (*C*-type) AF ordering turns up at $x > 0.62$ (Fig. 4). Considering the elongation of a lattice parameter along the *c*-axis, the $3z^2 - r^2$ orbitals are mostly occupied in the *C*-type AF phase.

For $Pr_{1-x}Ca_xMnO_3$ with a further reduced tolerance factor, the CO phase is present below 220-240 K for $x = 0.3$ and no FM phase shows up in any x-region at zero field. In the CO phase of this compound, there are successive magnetic transitions to the CO antiferromagnetic insulating (AFI) phase and to the CO spin-canted insulating (CAFI) phase [22,23,28,29]. An observed variation of the transition temperatures T_N and T_{CA} with x can be interpreted in terms of the partial revival of the DE carriers in the CO phase [30]: The CO pattern is always $(\pi,\pi,0)$ in the pseudo-cubic setting irrespective of x and hence the deviation of x from $x = 0.5$ may produce extra carriers which mediate the DE interaction along the c-axis and play a role of modifying the antiferromagnetic spin structure towards the ferromagnetic one, as observed. Such a discommensuration of the nominal hole concentration seems to affect not only the temperature-dependent spin structure but also the metal–insulator phase diagram in the $T-H$ plane.

4.2 Compositional Tuning of CMR

The ferromagnetic transition temperature T_C is a good measure for the kinetic energy of the double-exchange (DE) carriers, which would scale linearly with the e_g electron hopping interaction or the conduction bandwidth (W) in the case of the simple DE ferromagnet. The T_C in the perovskite manganties, $RE_{1-x}Sr_xMnO_3$ (RE =La,Pr,Nd,Sm and their solid solution), are in fact very sensitive to the doping level and the tolerance factor (degree of lattice distortion) which are both closely relevant to the carrier kinetic energy [30,31]. The left panel of Fig. 5 shows T_C for various doping levels (x) as a function of the tolerance factor f, defined by Eq.(2) [31]. The temperature of the charge ordering transition is also indicated for the $x = 0.5$ crystals. The ferromagnetic transition temperature T_C decreases with decease in f or W irrespective of x. In particular, the rate of the reduction in T_C is enhanced below some critical value of $f \sim 0.975$ for $x = 0.5$, $\sim$0.970 for $x = 0.45$, and $\sim$0.964 for $x = 0.4$.

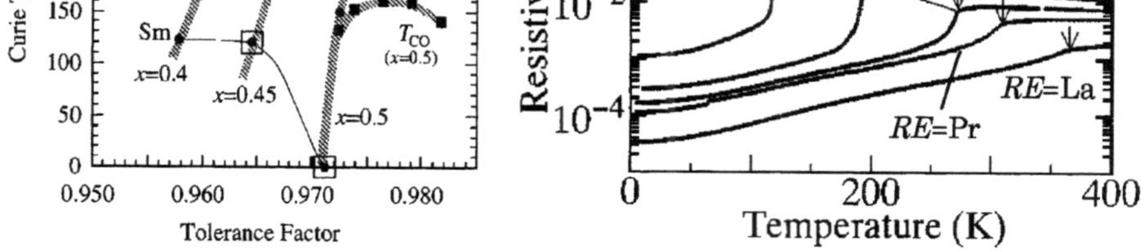

Fig. 5. (*left*): The critical temperatures of the ferromagnetic transition (solid circles) and the charge-ordering transition for $x = 0.5$ (solid squares) in $RE_{1-x}Sr_xMnO_3$ (RE=La, Pr, Nd, Sm and their solid solution) crystals as a function of the tolerance factor. The tolerance factor (f) is defined as $f = (r_A + r_O)/\sqrt{2}(r_B + r_O)$ where r_A, r_B and r_O are ionic radii of the perovskite A, B (Mn) site cations and oxygen, respectively. The hatched thick lines and the solid thin ones are guides to the eye, connecting the identical x or RE compounds. (*right*): The temperature dependence of the resistivity for crystals of $RE_{1-x}Sr_xMnO_3$ ($x = 0.4$)

Let us first see the case of $x = 0.5$. For $f > 0.976$ (RE=Nd) the T_C is almost constant and the charge-ordered state exists at lower temperatures (see the curves for $Nd_{0.5}Sr_{0.5}MnO_3$ in the left panel of Fig. 5). For $f < 0.976$, however, the reduction of T_C with the decrease of f is pronounced. Since T_C should depend almost linearly on W (and hence f in this region) in the simple DE model, such a pronounced suppression of T_C suggests that other instabilities which compete with the DE interaction become important for $f < 0.976$. In this region, the hysteresis (first-order nature) of the ferromagnetic transition and the deviation of the inverse susceptibility from the Curie-Weiss law are also pronounced [31]. For $f < 0.971$, the ferromagnetic metallic state disappears and in turn the charge-ordered state is present below some critical temperature, e.g., T_{CO} =230 K for $Pr_{1/2}Ca_{1/2}MnO_3$ (see also Figs. 3 and 6). In the left panel of Fig. 5, similar suppression in T_C is seen also for $f < 0.970$ in $x = 0.45$ and for $f < 0.964$ in $x = 0.4$, although no (long-range) charge-ordered state is present even at low temperatures. In such a f-region showing anomalously low T_C, the restoration of the DE interaction by the magnetic field is most conspicuous and hence the magnetoresistance around T_C is maximized.

We show in the right panel of Fig. 5 the temperature dependence of the resistivity of $x = 0.4$ crystals with different tolerance factors [32]. The most enhanced MR is observed around T_C for $Sm_{0.6}Sr_{0.4}MnO_3$ with other instabilities

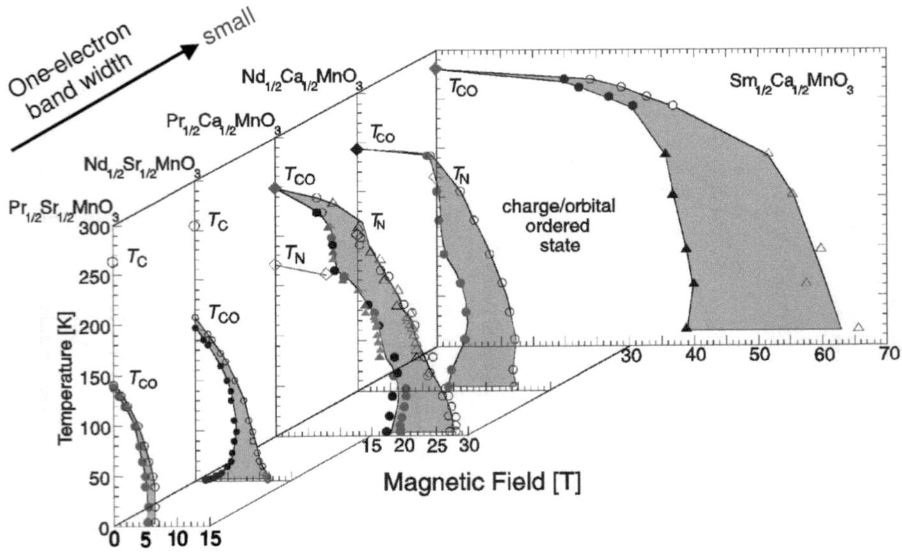

Fig. 6. The charge/orbital-ordered phase of various $RE_{1/2}AE_{1/2}MnO_3$ plotted on the magnetic field-temperature plane. The phase boundaries have been determined by measurements of the magnetic field dependence of the resistivity ($\rho - H$) and magnetization ($M - H$) at fixed temperatures [25,34], and those for $RE_{1/2}Ca_{1/2}MnO_3$ (RE = Pr, Nd and Sm) have been obtained by measurements using pulsed magnetic fields up to 40 T [35]

which compete with the DE interaction. The anomalous increase of the resistivity above T_C (and its critical field suppression) for $Sm_{1-x}Sr_xMnO_3$ ($x = 0.45$) must be relevant to such an antiferromagnetic interaction competing with the ferromagnetic DE interaction [31]. The colossal feature of the MR is relevant to the switching of magnetism above T_C as well as lattice strain. The anisotropic feature of the magnetostriction coupled to the gigantic MR for the $Sm_{1-x}Sr_xMnO_3$ ($x = 0.45$) crystal implies that the relevant dynamical Jahn-Teller distortions are strongly spatially correlated [33]. Such a collective nature of the Jahn-Teller distortion above T_C likely arises from the anisotropic short-range orbital ordering or fluctuation. In particular, the e_g orbitals are already aligned perpendicular to the c-axis and show a directional order (i.e., with little $3z^2 - r^2$ component), while the in-plane ordering remains short-range. Raman spectroscopy and X-ray diffuse scattering have provided some experimental supports to this view [33]. The situation may be viewed as a "liquid-crystal" state (like smectic phase) for the orbital degree of freedom.

5 Magnetic-Field-Induced Melting of Charge/Orbital-Ordered States

5.1 Charge/Orbital Ordering at $x = 1/2$

As seen in Fig. 4, in $Nd_{1/2}Sr_{1/2}MnO_3$, the ferromagnetic and metallic state due to the double-exchange (DE) interaction is seen below T_C ~255 K, and subsequently the transition to the antiferromagnetic charge/orbital-ordered state occurs at T_{CO} (T_N) =160 K [25]. However, the antiferromagnetic spin-ordering occurs not concurrently but at a lower temperature, T_N ~170 K, and no ferromagnetic state is present at zero field. In the reduced-bandwidth systems, such as $RE_{1/2}Ca_{1/2}MnO_3$ (RE = Pr, Nd, Sm,..), in which no ferromagnetic or metallic state is realized, a split between T_{CO} and T_N is generally observed. In $Nd_{1/2}Sr_{1/2}MnO_3$, an antiferromagnetic spin-ordering of the CE-type (Fig. 4) with the magnetic unit cell expanding to 2×2×1 of the original orthorhombic lattice has been confirmed below T_{CO} by a neutron diffraction study [24]. The charge/orbital ordering at $x = 1/2$ is thus characterized by the antiferromagnetic CE-type structure which has originally been revealed for $La_{1-x}Ca_xMnO_3$ ($x \sim 1/2$) [16].

In the magnetization curves for the charge/orbital ordered state, we can see a metamagnetic transition. In accord with this, the resistivity also shows a steep decrease by several orders of magnitude. Thus, the transition from the antiferromagnetic charge/orbital-ordered insulating state to the ferromagnetic metallic one is caused by application of an external magnetic field. From a thermodynamic point of view, both states are energetically almost degenerate, but the free energy of the ferromagnetic (FM) state can be decreased by the Zeeman energy $-M_sH$ (M_s: the spontaneous magnetization) so that the magnetic field induced transition to the FM metallic state takes place in an external magnetic field. Figure 6 shows the charge/orbital ordering phase diagrams for various $RE_{1/2}AE_{1/2}MnO_3$

crystals which are presented on the magnetic field-temperature ($H - T$) plane. The phase boundaries have been determined by the measurements of the magnetic field dependence of the resistivity ($\rho - H$) and magnetization ($M - H$) at fixed temperatures [20,34], and those for $RE_{1/2}Ca_{1/2}MnO_3$ (RE = Pr, Nd and Sm) have been obtained by measurements using pulsed magnetic fields up to 40 T [35]. In this figure, the critical field to destruct the charge/orbital-ordered state in $Nd_{1/2}Sr_{1/2}MnO_3$ is about 11 T at 4.2 K, while that in $Pr_{1/2}Ca_{1/2}MnO_3$ increases to about 27 T. In the case of $Sm_{1/2}Ca_{1/2}MnO_3$, the charge/orbital ordering is so strong that the critical field becomes as large as about 50 T at 4.2 K. Figure 6 thus demonstrates that the robustness of the charge/orbital ordering at $x = 1/2$ critically depends on W, which is understood as a competition between the DE interaction and the ordering of Mn^{3+}/Mn^{4+} with a 1:1 ratio accompanied by the simultaneous ordering of Mn^{3+} e_g-orbitals.

To be further noted in Fig. 6 is the large hysteresis of the transition (hatched in the figure) which is characteristic of a first order transition coupled with the change of crystal lattice. The change in lattice parameters originates from the field-destruction of orbital ordering. As seen in Fig. 6, the hysteresis region (hatched area) expands with decreasing temperature especially below about 20 K. In the case of a first order phase transition, the transition from the metastable to the stable state occurs by overcoming a potential barrier. Since the thermal energy is reduced with decrease in temperature, a larger (smaller) field than the thermodynamic value is needed to induce the transition from (to) the AF charge/orbital-ordered to (from) the FM metallic state. Thus the hysteresis between the field increasing and decreasing runs increases with the decrease of temperature.

5.2 Effect of Discommensuration of Doping Level

The charge/orbital-ordering phenomena tend to be most stabilized when the band filling coincides with a rational number for the periodicity of the crystal lattice. In fact, the charge/orbital-ordered state in $Nd_{1-x}Sr_xMnO_3$ emerges only around $x = 0.5$, as shown in Fig. 4, and disappears for $x < 0.48$ and $x > 0.52$. In perovskite manganites for which W is further reduced, such as in $Pr_{1-x}Ca_xMnO_3$, however, the charge/orbital ordering of the similar CE-type appears in a much broader range of x [22]. We have already shown in Fig. 3 the electronic phase diagram of $Pr_{1-x}Ca_xMnO_3$ ($0 < x < 0.5$) [29]. In Fig. 3, the ferromagnetic and metallic state is not realized at zero magnetic field and at ambient pressure due to the reduced W, while the ferromagnetic but insulating phase appears for $0.15 < x < 0.3$. With further increase in x, the charge/orbital-ordered state with 1:1 ordering of Mn^{3+}/Mn^{4+} appears for $x > 0.3$. As an earlier neutron diffraction study [22] reported, the charge/orbital ordering exists in a broad range of x ($0.3 < x < 0.75$), where the pattern of spin, charge and orbital ordering is basically described by that of $x = 0.5$. The coupling of spins along the c-axis is antiferromagnetic at $x = 0.5$, while the spin arrangement is not antiferromagnetic (or collinear) but canted for $x < 0.5$. As the doping level deviates from the commensurate value of 0.5, extra electrons are doped on the

Mn^{4+} sites in a naive sense. To explain the modification of the arrangement of spins along the c-axis, Jirak et al. [22] postulated that the extra electrons hop along the c-axis mediating the ferromagnetic double exchange interaction. (This is analogous to the lightly hole-doped case of $LaMnO_3$, where the canted antiferromagnetic spin structure is realized due to the hole motion along the c-axis [6].) Such an effect of extra electrons on the magnetic structure should be enhanced with decrease in x from 0.5. In fact, neutron diffraction studies [22,23] have revealed that the coupling of spins along the c-axis becomes almost ferromagnetic at $x = 0.3$ in spite of that the CE-type ordering is maintained within the orthorhombic ab-plane.

In $Pr_{1-x}Ca_xMnO_3$, the dependence of the field induced transition (or collapse of the charge/orbital-ordered state in a magnetic field) on the doping level has systematically been investigated for $0.3 < x < 0.5$ [29,35]. The x-dependent features are well demonstrated by the charge/orbital ordering phase diagram displayed on the magnetic field-temperature ($H - T$) plane as shown in Fig. 7. In this figure, the phase boundaries have been determined by the measurements of the magnetic field dependence of the resistivity ($\rho - H$) and magnetization ($M - H$) at fixed temperatures. The charge/orbital ordering for $x = 0.5$ is so strong that the critical field to destroy the charge/orbital-ordered state becomes as large as about 27 T at 4.2 K, and a similar feature is also seen for $x = 0.45$ [35]. For $x < 0.4$, by contrast, the charge/orbital-ordered phase-region becomes remarkably shrunk, in particular at low temperatures. The averaged value (H_{av})

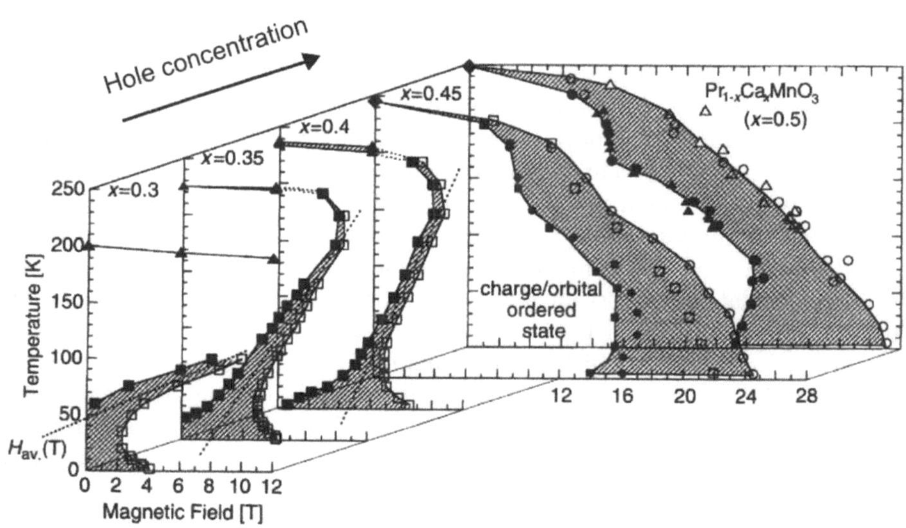

Fig. 7. The charge/orbital-ordered state of $Pr_{1-x}Ca_xMnO_3$ ($x = 0.5$, 0.45, 0.4, 0.35 and 0.3), which is plotted on the magnetic field-temperature plane [29,35]. The hatched area indicates the hysteresis region. In the cases of $x = 0.3$, 0.35 and 0.4, the lines of $H_{av}(T)$ at which the free energies of both the charge/orbital-ordered and the FM metallic states are supposed to be equal are indicated by dashed lines

of the critical fields in the H-increasing and -decreasing runs at a constant temperature rather shows a decrease with decrease in temperature below ≈175 K ($dH_{av}/dT < 0$). In the case of $x = 0.3$, the collapse of the charge/orbital-ordered state (i.e., appearance of the FM metallic state) is realized by applying an external magnetic field of only several Tesla when the temperature is set below 50 K. Another noteworthy aspect in Fig. 7 is the expansion of the field-hysteresis with decreasing temperature, which is characteristic of first order phase transition as mentioned above [20]. Such a variation of the charge/orbital-ordered phase with the doping level has been observed similarly for a further W-reduced system, $Nd_{1-x}Ca_xMnO_3$ [35,37]. The common feature for the modification of the phase diagram with x seems to be correlated with the action of the extra electron-type carriers in the CE-type structure [23]. Thermodynamically, excess entropy may be brought into the charge/orbital-ordered state by the extra localized carriers and their related orbital degrees of freedom, which are pronounced as x deviates from 0.5. The excess entropy may reduce the stability of the charge-ordered state with decreasing temperature and cause the reduction in the critical magnetic field ($dH_{av}/dT < 0$) as observed in the case of $x = 0.4$ in Fig. 7. In reality, however, the microscopic phase separation into the $x \approx 0.5$ charge-ordered state and the $x < 0.5$ ferromagnetic state should be taken into account for a deeper understanding of the present features for $x < 0.5$ systems.

6 Summary

Fundamental features of the colossal magnetoresistive (CMR) manganites with perovskite-related structures have been described. One of the advantage of these perovskites is the possibility of controlling the band filling and the bandwidth, which are both important physical parameters for the strongly correlated electron system or strongly coupled electron-lattice system like the CMR manganites. First, we have seen a canonical example of the double-exchange (DE) manganite, that is the case of $La_{1-x}Sr_xMnO_3$, whose MR near T_C can well be described by the canonical DE model. Even in that case, however, in particular near the metal-insulator compositional (x) phase boundary, we see large deviations from the simplest DE model. Those are ascribed partly to the Jahn-Teller type electron-lattice coupling, which causes dynamical Jahn-Teller distortion or small polaron conduction above T_C. In other words, the hole-doping level (concentration of nominal mobile Mn^{4+} species in the Jahn-Teller Mn^{3+}-ion backgrounds) is not large enough to extinguish the orbital "moment". In narrower-bandwdith manganite systems than $La_{1-x}Sr_xMnO_3$, even more complex but intriguing features emerge due to the competition between the DE and other generic instabilities, such as antiferromagnetic superexchange, orbital-ordering, charge-ordering, and orbital-lattice (Jahn-Teller type) interactions. Complex electronic phase diagrams are presented for several $RE_{1-x}AE_xMnO_3$ (RE = La, Pr, Nd, and Sm; AE = Sr and Ca) as a function of the doping level, where the charge/orbital/spin ordering occasionally shows up near the doping level of $x = 1/2$. Conspicuous phase transition from a charge/orbital ordered state to a ferromagnetic metal is caused by application of an external magnetic field. More lately, it has been demonstrated that other external stimuli than magnetic field, such as impurity (e.g. Cr^{3+})-doping on the Mn site [38], current-injection [39] and photo-irradiation [40] or X-ray irradiation [41] can drive such a phase transformation, although the transitions remain within a spatially local area. These phenomena have been attracting a great interest in the context of unconventional phase control of the magnetic and electronic states in magnetic oxides.

Acknowledgment

The authors would like to thank A. Asamitsu, Y. Tomioka, and H. Kuwahara for their help in preparation of the manuscript.

References

1. For a more comprehensive review, see for example, *Colossal Magnetoresistive Oxides*, ed. Y.Tokura, Godon&Breach Science Publishers, 1999
2. C.W. Searle, S.T. Wang: Can. J. Phys. **47**, 2023 (1969)
3. K.Kubo, N.Ohata: J. Phys. Soc. Jpn. **33**, 21 (1972)
4. C.Zener, Phys. Rev. **82**, 403 (1951)
5. P.W. Anderson, H. Hasegawa: Phys. Rev. **67**, 1000 (1955)
6. P.-G. de Gennes, Phys. Rev. **118**, 141 (1960)
7. R. M. Kusters, D. A. Singleton, D. A. Keen, R. McGreevy, W. Hayes, Physica B **155**, 362 (1989)
8. K. Chabara, T. Ohno, M. Kasai, Y. Kozono, Appl. Phys. Lett. **63**, 1990 (1993)
9. R. von Helmolt, J. Wocker, B. Holzapfel, M. Schultz, K. Samwer, Phys. Rev. Lett. **71**, 2331 (1993)
10. S. Jin, T. H. Tiefel, M. mcCormack, R. A. Fastnacht, R. Ramesh, L. H. Chen, Science **264**, 413 (1994)
11. Y. Tokura, A. Urushibara, Y. Moritomo, T. Arima, A. Asamitsu, G. Kido, N. Furukawa: J. Phys. Soc. Jpn. **63**, 3931 (1994)
12. A. Asamitsu, Y. Moritomo, Y. Tomioka, T. Arima, Y. Tokura: Nature **373**, 407 (1995)
13. M. Imada, A. Fujimori, Y. Tokura: Rev. Mod. Phys. **70**, 1039 (1998)
14. J. Kanamori, J. Appl. Phys. **31**, 14S (1960)
15. For a review, see J.B. Goodenogh, J.M. Longon: Magnetic and Other Properties of Oxides and Related Compounds, eds. K.-H. Hellwege, O. Madelung, Landolt-Bornstein, New Series, Group III, Vol.4, Pt. a (Springer-Verlag, Berlin, 1970)
16. E.O. Wollan, W.C. Koehler: Phys. Rev. **100**, 54 (1955)
17. A. Urushibara, Y. Moritomo, T. Arima, A. Asamitsu, G. Kido, Y. Tokura: Phys. Rev. B **51**, 14103 (1995)
18. H. Kawano, R. Kajimoto, M. Kubota, H. Yoshizawa: Phys. Rev. B **53**, R14709 (1996); 2202 (1996)
19. Y. Tokura, Y. Tomioka, H. Kuwahara, A. Asamitsu, Y. Moritomo, M. Kasai: J. Appl. Phys. **79**, 5288 (1996)
20. H. Kuwahara, Y. Tomioka, A. Asamitsu, Y. Moritomo, Y. Tokura: Science **270**, 961 (1995)
21. J.B. Goodenough: Phys. Rev. **100**, 564 (1955)

22. Z. Jirak, S. Krupicka, Z. Simsa, M. Dlouha, Z. Vlatislav: J. Magn. Magn. Mater. **53**, 153 (1985)
23. H. Yoshizawa, H. Kawano, Y. Tomioka, Y. Tokura: Phys. Rev. B **52**, R1345 (1995)
24. H. Kawano, R. Kajimoto, H. Yoshizawa, Y. Tomioka, H. Kuwahara, Y. Tokura: Phys. Rev. Lett. **78**, 4253 (1997)
25. H. Kuwahara, T. Okuda, Y. Tomioka, A. Asamitsu, Y. Tokura: Phys. Rev. Lett. **82**, 4316 (1999)
26. Y. Moritomo, T. Akimoto, A. Nakamura, K. Hirota, K. Ohoyama, M. Ohashi: Phys. Rev. B **58**, 5544 (1998)
27. R. Maezono, S. Ishihara, N. Nagaosa: Phys. Rev. B **58**, 11583 (1998)
28. Y. Tomioka, A. Asamitsu, Y. Moritomo, Y. Tokura: J. Phys. Soc. Jpn. **64**, 3626 (1995)
29. Y. Tomioka, A. Asamitsu, H. Kuwahara, Y. Tokura: Phys. Rev. B **53**, R1689 (1996)
30. H.Y. Hwang, S.-W. Cheong, P.G. Radaelli, M. Marezio, B. Batlogg: Phys. Rev. Lett. **75**, 914 (1995)
31. Y. Tomioka, H. Kuwahara, A. Asamitsu, M. Kasai, Y. Tokura: Appl. Phys. Lett. **70**, 3609 (1997)
32. E. Saitoh, Y. Okimoto, Y. Tomioka, T. Katsufuji, Y. Tokura: Phys. Rev. B **60**, 10362 (1999)
33. E. Saitoh, Y. Tomioka, T. Kimura, Y. Tokura: J. Phys. Soc. Jpn. **69**, 2403 (2000)
34. Y. Tomioka, A. Asamitsu, Y. Moritomo, H. Kuwahara, Y. Tokura: Phys. Rev. Lett. **74**, 5108 (1995)
35. M. Tokunaga, N. Miura, Y. Tomioka, Y. Tokura: Phys. Rev. B, **57**, 5259 (1998)
36. H. Yoshizawa, H. Kawano, Y. Tomioka, Y. Tokura: J. Phys. Soc. Jpn. **65**, 1043 (1996)
37. K. Liu, X.W. Wu, K.H. Ahn, T. Sulchek, C.L. Chien, J.Q. Xiao: Phys. Rev. B **54**, 3007 (1996)
38. B. Raveau, A. Maignan, C. Martin: J. Solid State Chem. **130**, 162 (1997)
39. A. Asamitsu, Y. Tomioka, H. Kuwahara, Y. Tokura: Nature **388**, 50 (1997)
40. K. Miyano, T. Tanaka, Y. Tomioka, Y. Tokura: Phys. Rev. Lett. **78**, 4257 (1997)
41. V. Kiryukhin, D. Casa, J.P. Hill, B. Keimer, A. Vigliante, Y. Tomioka, Y. Tokura: Nature **386**, 813 (1997)

Half-Metallic Ferromagnets

J.M.D. Coey, M. Venkatesan and M.A. Bari

Physics Department, Trinity College, Dublin 2, Ireland

Abstract. A broad classification is proposed for half-metallic ferromagnets, and examples of each type are given. Most half-metals are transition metal oxides, sulfides or Heusler alloys. The problems of defining and measuring spin polarization are discussed, and the principles of the various methods available are outlined. Some half-metallic characteristics other than integral spin moment are presented.

1 Introduction

A half-metal is a solid with an unusual electronic structure. For electrons of one spin orientation it is a metal, but for the other spin orientation it is a semiconductor or insulator. This definition assumes a magnetically-ordered state with an axis to define the spin quantization. For clarity, we consider the situation at zero temperature, where there is no spin mixing.

Normal ferromagnets, even strong ferromagnets, are not half-metals. Cobalt and nickel, for example, have fully spin-polarized d-bands with a filled $\uparrow$ $3d$ band and only $\downarrow$ d electrons at the Fermi level E_F, but the Fermi level also crosses the $4s$ band which is almost unpolarized so there is a density of both $\uparrow$ and $\downarrow$ electrons there (Fig. 1) [1]. The $4s$ electrons carry most of the current. In order to arrive at a situation where there are only $\uparrow$ or $\downarrow$ electrons at E_F, it is necessary to reorder the $3d$ and $4s$ bands of the ferromagnetic transition element. This can be achieved pushing the bottom of the $4s$ band up above E_F or by depressing the Fermi level below the bottom of the $4s$ band. Otherwise a hybridization gap might be introduced at E_F for one spin orientation. In any case, it is necessary to pass from a pure element to an alloy or compound; all half-metals contain more than one element. Some are stoichiometric compounds, others are solid solutions.

Two possible situations which can arise when there is a single spin orientation at E_F are illustrated in Fig. 2. The type I_A half-metal in Fig. 2a is metallic for $\uparrow$ electrons, but semiconducting for $\downarrow$ electrons, whereas the opposite is true for the type I_B half-metal in Fig. 2b. Half-metallic oxides where the $4s$ states are pushed above E_F by $s-p$ hybridization are of type I_A when there are less than five d-electrons, but those with more than five d-electrons are of type I_B. Likewise, half-metallic Heusler alloys with heavy p-elements like Sb tend to have the $3d$ levels depressed below the $4s$ band edge by $p-d$ hybridization.

In a second class of half-metals, type II, the carriers at the Fermi level are in a band that is sufficiently narrow for them to be localized. The heavy carriers

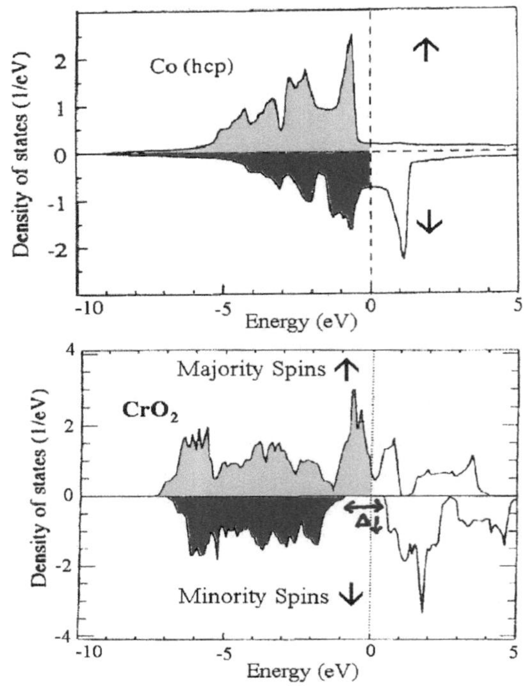

Fig. 1. Spin polarization of the electronic density of states for a strong ferromagnet (Co) and a half-metallic ferromagnet (CrO_2)

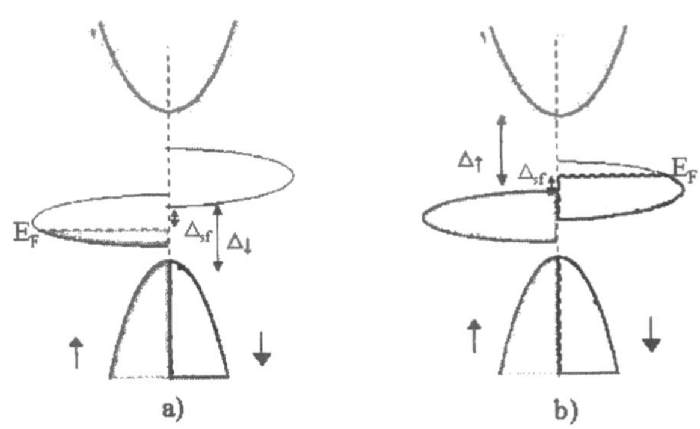

Fig. 2. Schematic density of states for a half-metal with **(a)** type I_A with only ↑ electrons at E_F and **(b)** type I_B with only ↓ electrons at E_F. In narrow d-bands, the states at E_F may be localized

form polarons and conduction is then by hopping from one site to another with the same spin [11]. Otherwise, in the presence of disorder, the states in the band tails tend to be localized. There is a mobility edge, and the conduction

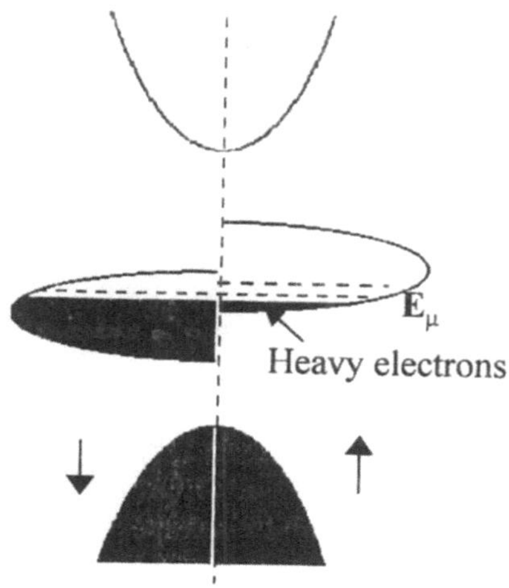

Fig. 3. Schematic density of states for a type III_A half-metal, where electrons of one spin direction are itinerant and the others are localized

process involves excitation to that edge. In either case, the heavy carriers give an activated conduction $\rho \approx \rho_\infty \exp(E_a/kT)$ where $E_a \approx 0.1eV$

A third class of half-metals, known as "transport half-metals", have localized ↑ carriers and delocalized ↓ carriers or vice versa [3]. A density of states exists for both sub-bands at E_F, but the carriers in one band have a much larger effective mass than those in the other. So far as electronic transport properties are concerned, only one sort of carriers contributes significantly to the conduction. Schematic band structure of a half-metal of type III is shown in Fig. 3. The heavy carriers give an activated conduction $\rho \approx \rho_\infty \exp(E_a/kT)$ with $E_a \approx 0.1$ eV, whereas the light carriers are metallic with a resistivity given by Matthiessen's rule $\rho \approx \rho_0 + \rho(T)$.

Normally, there is no connection or confusion between a half-metal and a semimetal. A semimetal, of which bismuth is the textbook example, has small and equal numbers of electrons and holes (≈ 0.01 per atom) due to a fortuitously small overlap between valence and conduction bands (Fig. 4). However, when the semimetal is magnetically ordered with a great difference in effective mass between electrons and holes, it is possible that it could resemble the previous type of half-metal. We refer to these as type IV half-metals, rather than semimetallic type III half-metals. An example is given in the next section. Our classification is summarized in Table 1. For completeness, we have included magnetic semiconductors in the table as type V half-metals, although we will not discuss them here.

Half-metallicity is not an easy property to detect experimentally. There is no intrinsic behaviour that constitutes the unmistakable signature of a half-metal.

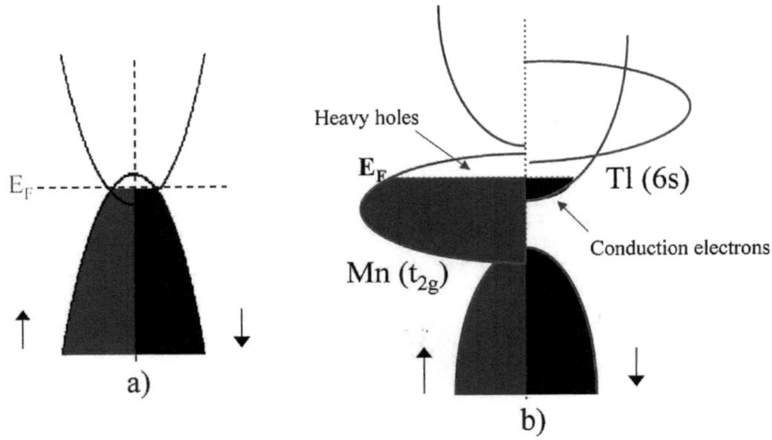

Fig. 4. Schematic density of states for **(a)** a semi-metal and **(b)** type IV_A half-metal

Table 1. Summary of the classification of half-metals

Type	Density of states	Conductivity	↑ electrons at E_F	↓ electrons at E_F
I_A	half-metal	metallic	itinerant	none
I_B	half-metal	metallic	none	itinerant
II_A	half-metal	nonmetallic	localized	none
II_B	half-metal	nonmetallic	none	localized
III_A	metal	metallic	itinerant	localized
III_B	metal	metallic	localized	itinerant
IV_A	semimetal	metallic	itinerant	localized
IV_B	semimetal	metallic	localized	itinerant
V_A	semiconductor	semiconducting	few, itinerant	none
V_B	semiconductor	semiconducting	none	few, itinerant

The best indication of a type I or type II half-metal is metallic conduction in a material with a spin moment at $T = 0$ which is *precisely an integral number of Bohr magnetons per unit cell.* In a stoichiometric compound, the number of electrons per unit cell $n = n_\uparrow + n_\downarrow$ is an integer. On account of the gap in one of the spin-polarized bands, $n_\uparrow$ or $n_\downarrow$ is also an integer. It follows that both $n_\uparrow$ and $n_\downarrow$ are both integers, and so then is the difference $n_\uparrow - n_\downarrow$ which is the spin moment in units of the Bohr magneton. The integer spin moment criterion, or an extension of it to cover the case of a solid solution, is a necessary but not a sufficient condition for half-metallicity.

The methods of measuring spin polarization discussed in Sect. 3 all somehow involve extracting electrons at the surface of the material, with the chance that their inherent polarization will be modified in the process. It has therefore been customary to rely on electronic structure calculations to identify half-metals. The first compound to be so identified was NiMnSb in the 1983 calculation of de Groot et al [4]. These authors coined the term 'half-metal'. The first oxide to be identified as a half-metal in this way was CrO_2 [5]. Nevertheless, schematic spin-polarized density of states diagrams of the type shown in Fig. 2 were introduced by Goodenough and his many followers from the 1970s [6]. For example, both CrO_2 [7] and Fe_3O_4 [8] had been identified as half-metals in this way before the term was actually coined.

Electronic structure calculations usually deal with an ideally-ordered bulk stoichiometric compound. In some cases, there are electronic structure calculations of the surface [9,10]. Variations in stoichiometry, antisite defects and nonstoichiometry are usually ignored in the calculations. Sometimes the half-metallic character is robust with respect to these variations, but more often it is not.

Perfect spin polarization is potentially very useful. The emerging science of spin electronics [11] seeks to exploit the two separate spin channels in increasingly-sophisticated electronic devices. The idea is that half-metallic electrodes can act as sources of spin-polarized electrons, and as magnetically-controllable spin filters. So far these expectations have not been fully realized, for reasons which are discussed later.

We first review some materials that are claimed to be half-metals, and then consider the experimental methods of determining spin polarization and other half-metallic characteristics before discussing the evidence that different materials belong to this select, but potentially useful class.

2 Materials

Table 2 is a list of representative materials which are claimed to be half-metals, together with the category to which they belong, and references to electronic structure calculations. The density of states at E_F is taken from the calculations, and the localized or metallic conduction is based on the Ioffe-Regel criterion [11]. The predicted spin moment per formula unit and the Curie temperature are included in each case.

Table 2. Some half-metallic materials.

Material	Type	↑ electrons	↓ electrons	T_C(K)	$(n_\uparrow - n_\downarrow)z$	Ref.
CrO_2	I_A	Cr (t_{2g})	-	396	2	[12]
Sr_2FeMoO_6	I_B	-	Mo(t_{2g})	421	4	[15]
$(Co_{1-x}Fe_x)S_2$	I_A	Co(e_g)	-	≈100	(1-x)	[22]
NiMnSb	I_A	-	Ni(e_g)	730	4	[4]
Mn_2VAl	I_B	Mn(t_{2g})	-	760	2	[24]
Fe_3O_4	II_B	-	Fe(t_{2g})	860	4	[19]
$(La_{0.7}Sr_{0.3})MnO_3$	III_A	Mn(e_g)	Mn(t_{2g})	390	< 3.7	[3]
$Tl_2Mn_2O_7$	IV_B	Mn(t_{2g})	Tl(6s)	120	6	[23]

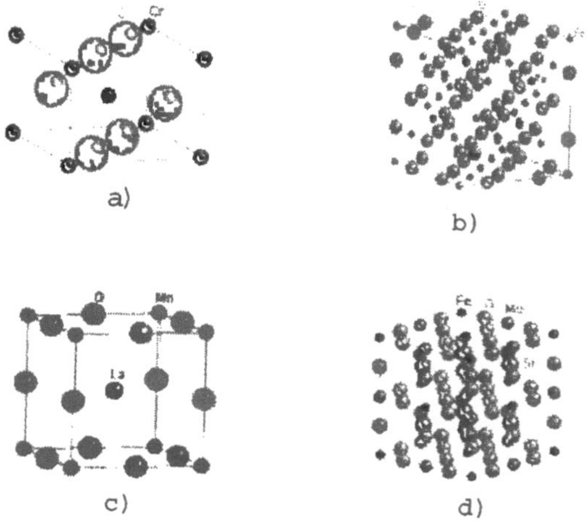

Fig. 5. Crystal structures of oxide and sulfide half-metals: **(a)** rutile **(b)** spinel **(c)** perovskite and **(d)** double perovskite

Materials listed in the table are oxides and sulfides on the one hand, and Heusler alloys on the other. Some are stoichiometric compounds, others are solid solutions based on a nonmetallic end-member (FeS_2, $LaMnO_3$). The prospects for finding new half-metallic compounds are quite limited, but prospects are better for finding new solid solutions with robust half-metallicity.

We now discuss the electronic structure of these representative materials:

CrO_2 Chromium dioxide is the only stoichiometric binary oxide that is a ferromagnetic metal (T_C = 392 K). The compound has the rutile structure illustrated in Fig. 5 with a formal electronic configuration of $(t_{2g}^2)^{\uparrow}$ for the Cr^{4+}, although there is strong mixing of oxygen holes and chromium electron states at E_F [12]. The compound is thermodynamically metastable, but it has proved possible to grow tiny crystals and good-quality films which have a residual resistivity ρ_0 as low as 4 x 10^{-8} Ω m. [13]. The measured moment is the expected 2.0 μ_B/ formula [14]. Properties of CrO_2 are reviewed in Ref. [48].

Sr_2FeMoO_6 This compound has the double perovskite structure with a supercell with NaCl-type order of Fe and Mo. Formal electronic configurations are $Fe^{3+}(t_{2g}^3e_g^2)^{\uparrow}$ and $Mo^{5+}(t_{2g}^1)^{\downarrow}$, although the Mo and Fe $t_{2g}^{\downarrow}$ orbitals are strongly mixed and a half-metallic structure is predicted [15]. It is a type I_B half-metal. The compound is ferrimagnetic with an expected spin moment of 4 μ_B and T_C = 420 K. Some good-quality films have moments approaching this value [16], but moments measured in single crystals are generally lower [17] which is usually attributed to antisite disorder of Fe and Mo.

Fe_3O_4 Magnetite, the most famous magnetic mineral, is the half-metal with the highest Curie point (860 K). The B-sites of the spinel structure are populated by an equal mixture of Fe^{3+} and Fe^{2+}, so the average B-site configuration is $(t_{2g}^3 e_g^2)^{\uparrow}(t_{2g}^{0.5})^{\downarrow}$. The A-sites contain oppositely magnetized Fe^{3+} $(t_{2g}^3 e_g^2)^{\downarrow}$ cores. The ↓ B-site electrons form small polarons which hop among the B-sites [18]. Resistivity at 120 K, the lowest temperature before the Verwey transition where the B-site charges order, is 10^{-4} Ωm. High-quality films and crystals have a spin moment of 4.0 μ_B at this temperature, reflecting the ferrimagnetic structure of A and B sites. Magnetite has a spin gap in the majority density of states [19]; it is a type II_B half-metal.

$(La_{0.7}Sr_{0.3})MnO_3$ The optimally-doped ferromagnetic manganite has a Curie temperature of 390 K, and a rhombohedrally-distorted perovskite structure. The substitution of Sr for La means that there is a mixture of Mn^{3+} $(t_{2g}^3 e_g)^{\uparrow}$ and $Mn^{4+}(t_{2g}^3)^{\uparrow}$ on the B-sites of the structure [20]. The hopping $e_g^{\uparrow}$ electron produces ferromagnetic coupling by double exchange. Residual resistivity is as low as 4×10^{-7} Ω m. Electronic structure calculations on this [3] and the similar manganite $(La_{0.67}Ca_{0.33})MnO_3$ [21] place the Fermi level slightly *above* the bottom of the $t_{2g}^{\downarrow}$ band, and the ferromagnetic moment is consistently reported as slightly less than the 3.7 μ_B expected for a type I half-metal. $(La_{0.7}Sr_{0.3})MnO_3$ is a type III_A half-metal, with both spin states present at E_F, but very different mobilities for the two spins.

$(Co_{1-x}Fe_x)S_2$ Pyrite (fools' gold), FeS_2, is a diamagnetic material with the Fe^{2+} in a low-spin state $(t_{2g}^3)^{\uparrow}(t_{2g}^3)^{\downarrow}$. The extra electron associated with the cobalt then enters an $e_g^{\uparrow}$orbital. The solid solutions are ferromagnetic with $T_C \propto (1-x)$ and $m \approx (1-x)\mu_B$ for $x > 0.1$. This seems to be a robust half-metal of type I_A [22].

$Tl_2Mn_2O_7$ This ferromagnetic pyrochlore-structure oxide has an unusual semimetallic band structure with approximately 0.01 electrons and holes per formula [23]. There are a small number of heavy holes at the top of a narrow ↑ band of $Mn(t_{2g})$ character and an equal number of ↓ electrons in a broad band of mixed Tl(6s), O(2p) and Mn(3d) character, making this a type IV_A half-metal.

Mn_2VAl This ordered Heusler alloy (Fig. 6a) has a ferrimagnetic structure with oppositely directed moments on Mn and V, and a net moment of 1.94 μ_B. It is not a robust half-metal because incomplete atomic order and substoichiometry in manganese can eliminate the spin gap, introducing light ↓ holes at the Fermi level. The ideal compound is predicted to be a type I_B half-metal with $Mn(t_{2g})$ ↓ electrons at E_F.

NiMnSb The original half-metallic Heusler alloy [4] has atoms ordered on three of the four sublattices of the fcc structure, with the fourth remaining vacant (Fig. 6b). NiMnSb is a type I_A half-metal with ↑ electrons of $Ni(e_g)$ character at E_F. Again modest amounts of atomic disorder can destroy the half-metallicity [26] that appears in this, and other Heusler alloys with 22 valence electrons [27].

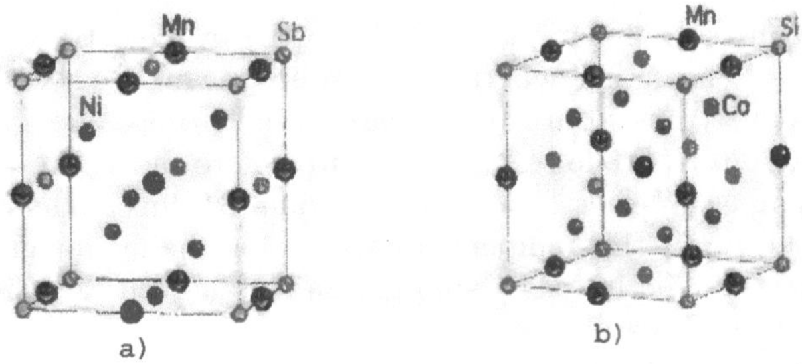

Fig. 6. Crystal structures of **(a)** Heusler and **(b)** half-Heusler alloy

3 Spin Polarization

Unlike solids with other electronic structures–metallic, semiconducting, superconducting – there is no distinctive property or syndrome that allows us to identify a material as a half-metal. The difficulty of actually measuring the spin polarization is a key problem. The methods available are spin-polarized photoemission, and transport measurements in point contacts and tunnel junctions, either with two ferromagnetic electrodes, or with one ferromagnetic and one superconducting electrode, as summarized in Fig. 7.

The straightforward definition of spin polarization is

$$P_0 = (N^{\uparrow} - N^{\downarrow})/(N^{\uparrow} + N^{\downarrow}) \tag{1}$$

where $N^{\uparrow,\downarrow}$ are the spin polarized densities of states at the Fermi level, but this is not necessarily what is measured [28]. In an experiment involving ballistic transport, the densities of states must be weighted by the averaged Fermi velocity of electrons passing through:

$$P_1 = (\langle N^{\uparrow} v_{\mathrm{F}}^{\uparrow} \rangle - \langle N^{\downarrow} v_{\mathrm{F}}^{\downarrow} \rangle)/(\langle N^{\uparrow} v_{\mathrm{F}}^{\uparrow} \rangle + \langle N^{\downarrow} v_{\mathrm{F}}^{\downarrow} \rangle) \tag{2}$$

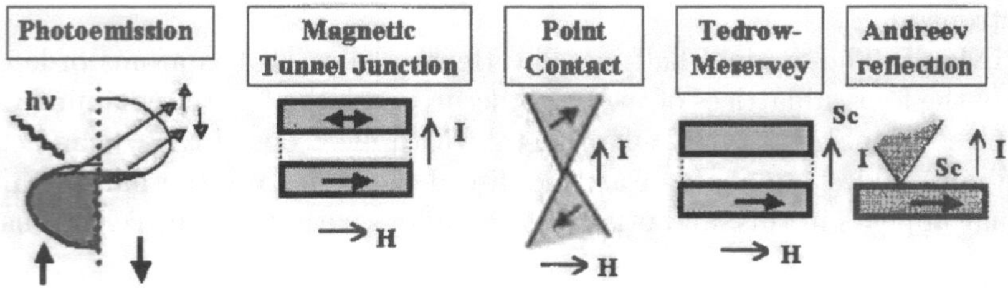

Fig. 7. Schematic of five methods of measuring P. Sc indicates a superconductor

P_1 will be large for transport half-metals (types III and IV), but P_0 can be small for the same materials. In the diffusive limit, the densities of states must be weighted by the square of the Fermi velocity

$$P_2 = (\langle N^{\uparrow} v_F^{\uparrow 2} \rangle - \langle N^{\downarrow} v_F^{\downarrow 2} \rangle)/(\langle N^{\uparrow} v_F^{\uparrow 2} \rangle + \langle N^{\downarrow} v_F^{\downarrow 2} \rangle) \tag{3}$$

In tunnelling experiments, the density of states should be weighted by the appropriate spin-dependent tunnelling matrix element to give P_T. Mazin shows that P_T is equal to P_2 in the case of a specular barrier with low transparency [28].

When the electrons are fully spin polarized, $N^{\uparrow}$ or $N^{\downarrow} = 0$ and $P_0 = P_1 = P_2 = 100$ %, but for any other polarization they will be different. For example, for $(La_{0.67}Ca_{0.33})MnO_3$ [21] $N^{\uparrow}$ and $N^{\downarrow}$ are 0.58 and 0.27 states eV^{-1}, respectively, and $v_F^{\uparrow}$ and $v_F^{\downarrow}$ are 0.76×10^6 and 0.22×10^6 ms^{-1}, hence $P_0 = 36\%$, but $P_2 = 92$ %.

The magnetoresistance of a contact or junction, defined as

$$\Delta R/R_{\uparrow\downarrow} = (R_{\uparrow\downarrow} - R_{\uparrow\uparrow})/R_{\uparrow\downarrow} \tag{4}$$

cannot exceed 100 %. Since the conductance is $G = 1/R$, $\Delta R/R$ is also equal to $(G_{\uparrow\uparrow} - G_{\uparrow\downarrow})/G_{\uparrow\uparrow}$. It is often used in conjunction with the Julière formula [29]

$$\Delta R/R = 2PP'/(1 + PP') \tag{5}$$

to infer the spin polarization if the electrodes are identical ($P = P'$) or if the polarization of one of them is known. This expression is based on the transport of classical spins which have strictly no probability of travelling from a $\uparrow$ electrode ($\theta = 0$) to one where the direction of magnetization is opposite ($\theta = 180°$). In fact, for quantum spins s=1/2 which hop or tunnel between ion cores with spin S', there is the possibility of a $|S = S' + 1/2; M_S = S' - 1/2\rangle$ final state in the other electrode. The probability of transmission when there is no electron depolarization in the barrier or interface is [31]

$$f(\theta) = t[\cos^2 \theta/2 + [P(T, V_b)/(2S' + 1)] \sin^2 \theta/2] \tag{6}$$

where $P(T, V_b)$ reflects the probability of exciting a spin wave at a given temperature and bias voltage, and t is the transmission probability for fully polarized electrodes with parallel magnetization. This reduces to $\cos^2 \theta/2$ for classical spins $S' \gg 1$, hence the probability is 1 for parallel alignment of the electrodes, and 0 for antiparallel alignment. In the $S' = 1/2$ quantum limit, the probability is again 1 for parallel alignment, but may be as high as 1/2 for antiparallel alignment. The conductance is proportional to the transmission probability; so in that case the magnetoresistance can not be lower than 50% for fully spin polarized electrodes.

Another complication arises when the transport process across the barrier does not occur in one step. Co-tunneling processes via intermediate states in the barrier are possible, and if the state involves a magnetic impurity, depolarization may occur [30]. Otherwise the polarization in a tunnel junction may be enhanced by multiple reflection in the barrier, or by transit via an intermediate state in a Coulomb blockaded region [31].

Table 3 lists a number of measurements of spin polarization for the materials listed in Table 2, as well as the ferromagnetic elements Fe, Co and Ni. The methods used to measure P are now discussed briefly.

An alternative definition of magnetoresistance $\triangle R/R_{\uparrow\uparrow} = (G_{\uparrow\uparrow} - G_{\uparrow\downarrow})/G_{\uparrow\downarrow}$ is sometimes known as the "headline" or "optimistic" definition. It is infinite for fully polarized electrodes and classical spins because there is no possibility of electron transfer in the antiparallel configuration.

Table 3. Some measurements of Spin Polarization.

Material	Method	Temperature(K)	P(%)	Reference
CrO_2	Andreev	L He	94-98	[41][46]
	PMR	L He	82	[48]
Fe_3O_4	Point contact	RT	84	[36]
$(La_{0.7}Sr_{0.3})MnO_3$	Andreev	L He	58-92	[3]
	Photoemission	L He	~100	[33]
	Tunnel Junction	L He	85	[34]
	Tedrow-Meservey	L He	72	[47]
NiMnSb	Andreev	L He	58	[41]
	Tedrow-Meservey	L He	28	[40]
Fe	Andreev	L He	43	[41]
	Tedrow-Meservey	L He	45	[39]
Co	Andreev	L He	40	[41]
	Tedrow-Meservey	L He	42	[39]
Ni	Andreev	L He	42	[41]
	Tedrow-Meservey	L He	31	[39]

3.1 Spin-Polarized Photoemission

The spin polarization of ejected photoelectrons can be measured for different incident photon energies. This method provides a rather direct image of the spin-polarized density of states near E_F, but the photoelectrons which carry the information are coming from a thin layer at the surface. The measured polarization is weighted by the absorption cross-section for $\uparrow$ and $\downarrow$ electrons. Typical data for $(La_{0.7}Sr_{0.3})MnO_3$ are shown in Fig. 8 [32].

3.2 Tunnel Junctions

Spin-polarized tunneling in ferromagnetic junctions has been recently reviewed by Moodera and Mathon [33]. These are usually made with thin films as the ferromagnetic electrodes, separated by a thin barrier layer of insulating oxide such as Al_2O_3 or $SrTiO_3$. For best results, the junction area should be small, so

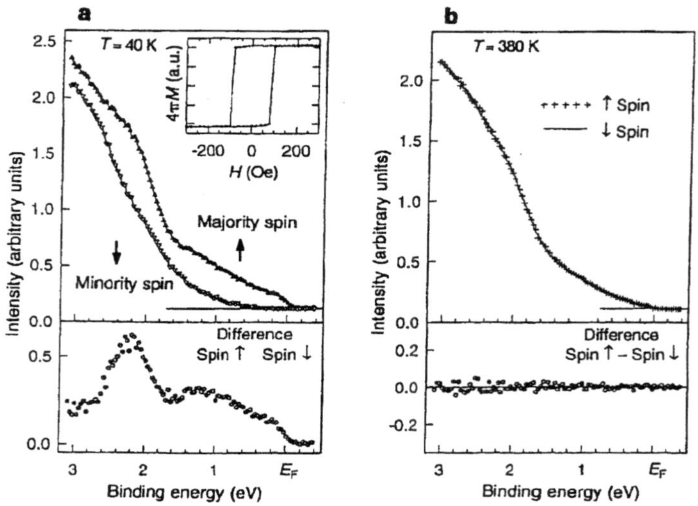

Fig. 8. Spin-polarized photoemission data for $(La_{0.7}Sr_{0.3})MnO_3$ [32]

the ferromagnetic electrodes can be deposited as two perpendicular stripes using shadow masks, or else the films are patterned using a lithographic technique. The electrode shapes are chosen so that they have different switching fields, and the magnetoresistance is measured from the difference in resistance between the parallel and antiparallel configurations. The simplest analysis relates the magnetoresistance (4) to the spin polarization P via the Julière formula (5). A typical magnetoresistance loop for $(La_{0.7}Sr_{0.3})MnO_3$(25 nm)/ $SrTiO_3$(3 nm)/ $(La_{0.7}Sr0._3)MnO_3$(33 nm) is shown in Fig. 9 [34], from which a polarisation of 85 % at 4.2 K can be deduced. However, this spin polarisation declines very rapidly with increasing temperature.

The magnetoresistance is also sensitive to applied voltage, falling rapidly with increasing bias. More serious from the point of view of deducing the spin polarization is the dependence of magnetoresistance on the barrier oxide, and its thickness [35]. The spin polarization of the tunneling electrons is governed by the bonding at the electrode/barrier interface.

3.3 Point Contacts

Here a stable conducting contact is established between two crystallites of the ferromagnetic material, and the magnetoresistance of the contact is measured. Normally, there is no method of controlling the magnetization directions in the initial contact before applying the magnetic field, so the experiment involves making many contacts and selecting those with the largest magnetoresistance. Some data for Fe_3O_4 are illustrated in Fig. 10 [36]. A variant of the point contact method is powder magnetoresistance, discussed further in the next section.

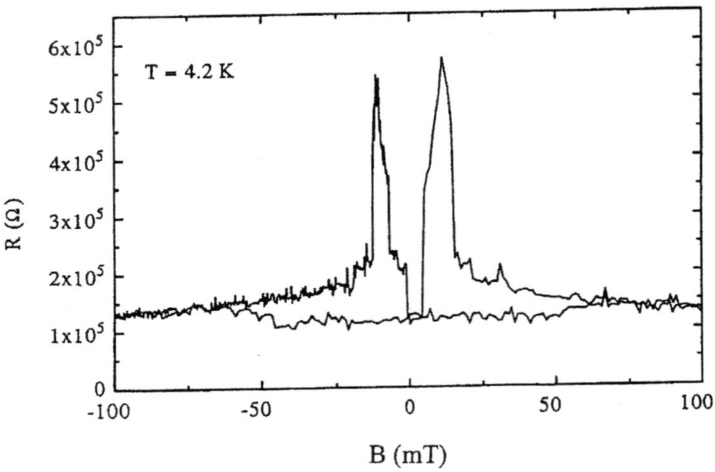

Fig. 9. Magnetoresistance loop for a $(La_{0.7}Sr_{0.3})MnO_3/SrTiO_3/(La_{0.7}Sr_{0.3})MnO_3$ tunnel barrier [34]

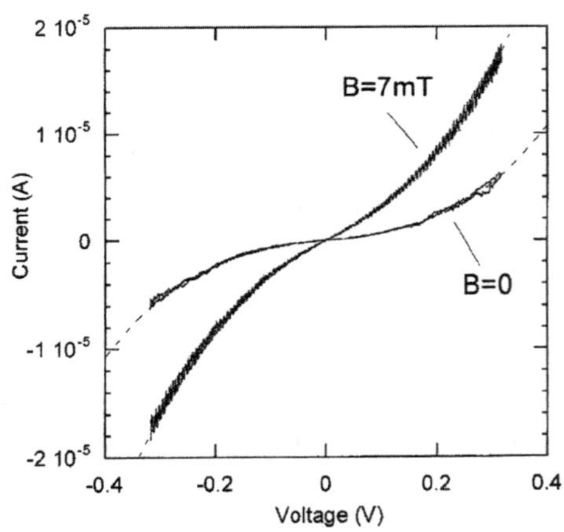

Fig. 10. Magnetoresistance effect at room temperature in an Fe_3O_4 point contact [36]

A related structure is a nanowire between two spin-polarized electrodes. Provided the transfer of the electron across the contact does not interfere with its spin (the transport may be ballistic, by hopping or by tunneling) the magnetoresistance for classical spins is still given by Eq. 5 [37]. For fully-polarized electrodes at low bias and zero temperature, $P = P' = 1$, and the magnetoresistance is 100%.

3.4 Tedrow–Meservey Experiment

This elegant experiment [38] involves a tunnel barrier where the second electrode is a thin layer of superconducting aluminum. The superconducting transition of Al is normally $T_{sc} = 1.2$ K, but there is a high critical field because the penetration depth in a type I superconductor can be much greater than the film thickness. The normal tunneling characteristic from a ferromagnet to a superconductor in zero field is illustrated in Fig. 11. It depends on the convolution of the ferromagnetic and superconducting densities of states, and the twin peaks are separated by the superconducting energy gap; for Al, $2\Delta \approx 0.25$ meV. In an applied field, there is a spin splitting of the density of states, and the curve becomes asymmetric, as shown in the figure. The spin polarization is usually inferred from the conductivity at the four points labelled σ_1 to σ_4, using the

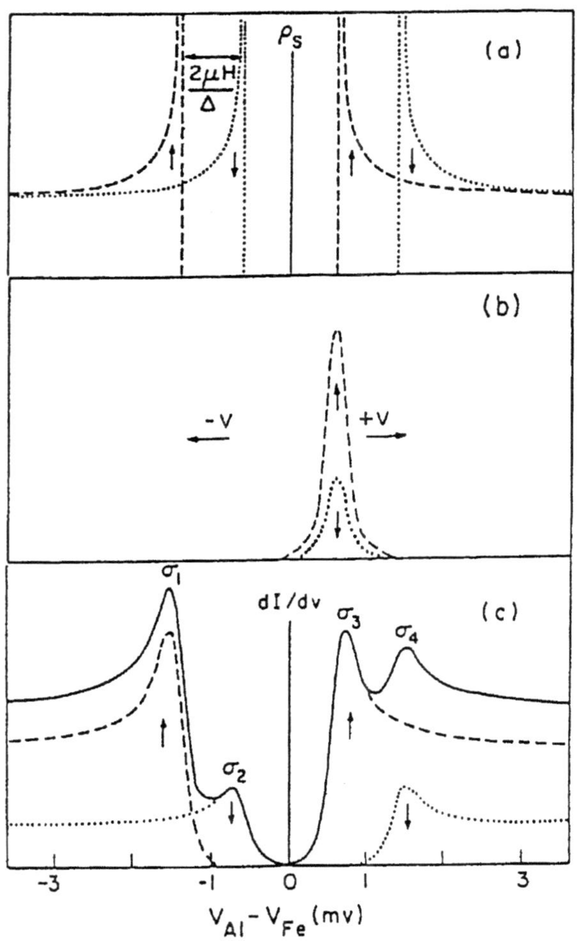

Fig. 11. Schematic explanation of the Tedrow-Meservey experiment [38]

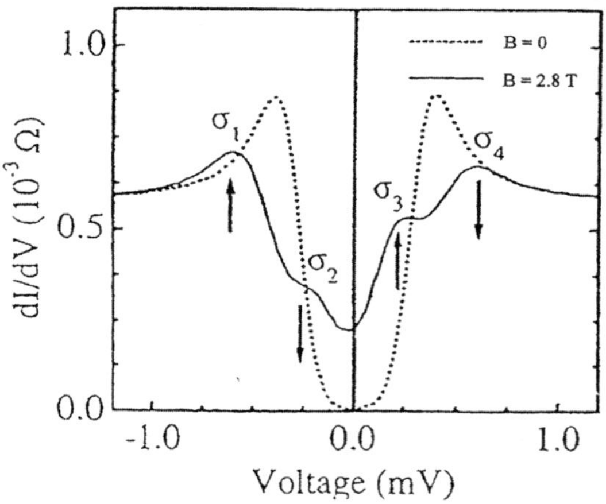

Fig. 12. Field-dependence of the dI/dV vs V characteristic of an $NiMnSb/Al_2O_3/Al$ tunnel barrier [40]

formula

$$P = [(\sigma_1 - \sigma_3) - (\sigma_2 - \sigma_4)]/[(\sigma_1 - \sigma_3) + (\sigma_2 - \sigma_4)] \tag{7}$$

Some data for NiMnSb are shown in Fig. 12 [40].

3.5 Andreev Reflection

This also involves a ferromagnet and a superconductor, but with a ballistic point contact, rather than a tunnel junction [41][42]. The ↑ electrons injected from the normal metal into the superconductor must form Cooper pairs, and this is achieved by the simultaneous injection of a ↓ hole from the superconductor back into the metal. The injected current is doubled in this way for a normal metal, when the junction is biased within the gap, but if the metal is a half-metallic ferromagnet, there are no states available, and the current is then zero. In general, the effect depends on the degree of spin polarization, which may be deduced from data at $T = 0$ using

$$P_1 = 1 - G_0/2G_n \tag{8}$$

where G_0 is the conductance at zero bias and G_n is the conductance when the applied voltage is much greater than the energy gap. Some typical data for Cu and CrO_2 are shown in Fig. 13.

All the methods of measuring P are somehow surface sensitive, since electrons are removed from the ferromagnet for spin analysis in an external detector. The surface or interface states may have a critical influence on the result. In particular, different crystallographic surfaces can yield up electrons with different values of P.

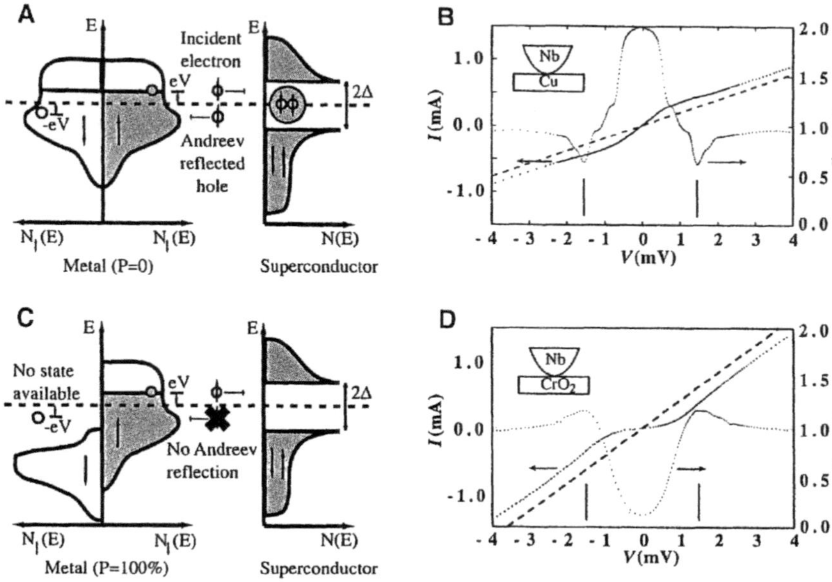

Fig. 13. Schematic explanation of Andreev reflection, with data for a normal metal (Cu) and a half-metal (CrO_2) [41]

4 Other Half-Metallic Characteristics

In view of the difficulties in measuring P directly, it is worth asking what other methods can be used. Apart from an integral spin moment and metallic conduction, what else can be used to identify a compound experimentally as a half-metal?

4.1 Powder Magnetoresistance (PMR)

This is a variant of the point contact experiment which is rather easy to perform [43]. A powder of a supposedly half-metallic material is compressed into a compact pellet, which typically has 55–60 % of the X-ray density. The resistance of the pellet is governed by contacts between the grains of powder, and it is several orders of magnitude greater than that of the fully dense polycrystalline material. The magnetic orientations of the grains will be statistically random in the unmagnetized state, so all angles are present between the crystallites magnetic axes at first, but on cycling the field, a typical butterfly magnetoresistance curve is traced out, with maximum resistance at the coercive field, as shown in Fig. 14.

The contacts between the powder grains are typically a mixture of tunnel barriers and ballistic point contacts, and the resistance is determined by the most conducting paths through the maze. It is helpful to dilute the conducting magnetic powder with a nonconducting, nonmagnetic powder of similar particle

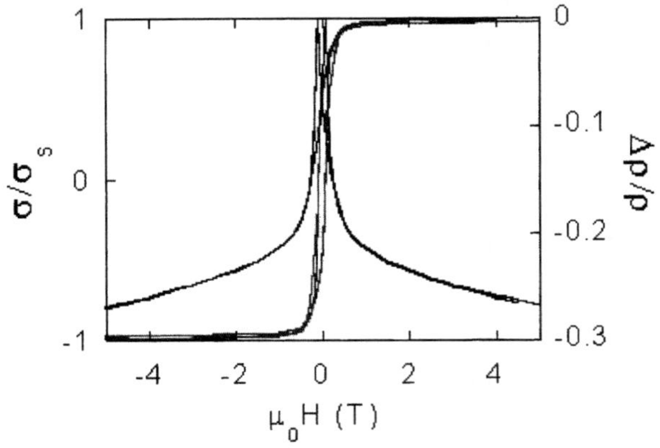

Fig. 14. Correlation of magnetic hysteresis and hysteresis of the magnetoresistance of a pressed powder of CrO_2

size distribution. Close to the percolation threshold, the conduction paths are attenuated and the largest magnetoresistance effects are observed. Some data are shown in Fig. 15 including CrO_2. The PMR effect is large in CrO_2 and the other oxides, but it is more than 100 times smaller in NiMnSb.

4.2 Two-Magnon Scattering

Spin waves can be excited in half-metals just as they can in other ferromagnetic metals or insulators. All that is needed is the presence of ferromagnetically-aligned atomic moments. However, an important electron-scattering process in a normal ferromagnetic metal is one-magnon scattering, where an electron undergoes a spin-flip in an inelastic process involving creation or annihilation of a magnon. This gives rise to a T^2 term in the resistivity [44]. Such processes are suppressed in a half-metal, at least at low temperature, because there are no states near the Fermi energy into which the electron can be scattered. The first available magnetic scattering processes involve two magnons, which give rise to a term in the resistivity varying as $T^{9/2}$ [45]. Unfortunately, such a term is very difficult to disentangle from regular scattering processes involving phonons.

4.3 Chemical Potential Shift

A characteristic feature of a half-metal is the small shift in chemical potential in an applied magnetic field. The shift, $g\mu_B sB$ is about 60 μV T^{-1}, so it should be possible to create potential differences of this magnitude by subjecting a half-metal to an appropriate magnetic field gradient. So far, there has been no report in the literature of such an effect.

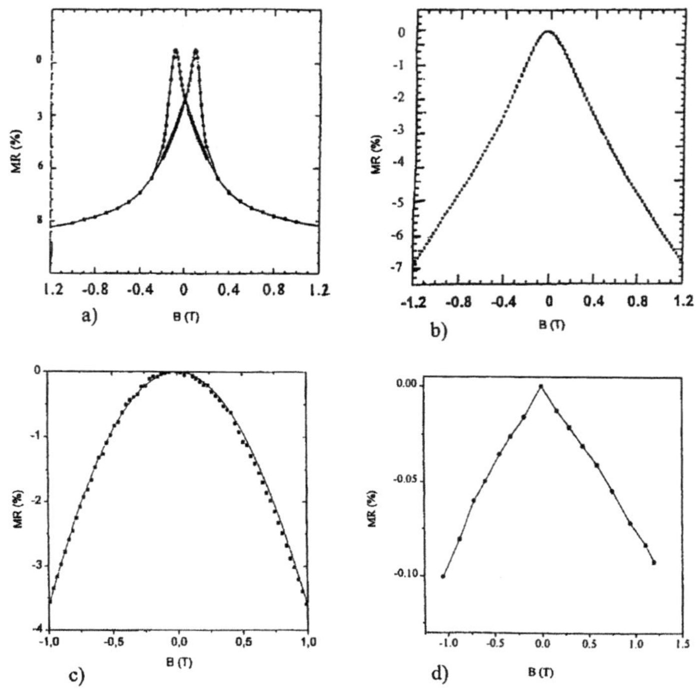

Fig. 15. Powder magnetoresistance measurements at 77 K on **(a)** CrO_2 **(b)** $(La_{0.7}Sr_{0.3})MnO_3$ **(c)** $Tl_2Mn_2O_7$ and **(d)** Mn_2VAl

5 Discussion

The complete spin-polarization predicted for a half-metal is rarely measured in practice. Results from the literature for the candidate materials identified as half-metals from band-structure calculations were listed in Table 3. Some notable results are 100 % [32] and 82 % [34] deduced for $(La_{0.7}Sr_{0.3})MnO_3$ from spin-polarized photoemission, and a tunnel junction, respectively, both at 4.2 K, 94-98 % for CrO_2 from Andreev reflection [41] [42] and 82 % from PMR at 4.2 K [48], and 84 % for Fe_3O_4 point contacts at room temperature [36].

Some insight into these data is achieved by comparing with results obtained on Fe, Co and Ni, the classic weak and strong ferromagnets. Surprisingly, all three yield values of P around 40 % in Andreev [41] and Tedrow-Meservey [38] experiments. The sign of P determined in the Tedrow Meservey experiment is positive for all three, even though the density of states for $\downarrow$ electrons in iron exceeds the density of states for $\uparrow$ electrons by a factor of three. This strongly suggests that it is principally the spin polarization of the 4s electrons that counts, on account of their high Fermi velocity (Eq. (2) and (3)). The contribution of 3d electrons to P_1 or P_2 is small. Although the 4s band has little net polarization, a substantial polarization of the $4s$ states at the Fermi level results from $4s-3d$ hybridization.

CrO_2 is probably the best studied example of a half-metal [48], but P falls off very rapidly with temperature. The calculated spin structure of the surface shows that it is fully spin polarized [9]. It is nevertheless possible that soft surface spin waves tend to destroy the polarization of emitted electrons.

$(La_{0.7}Sr_{0.3})MnO_3$ shows very high spin polarization in photoemission, although both band calculations and Andreev studies as a function of residual resistance [3] indicate it is a transport half-metal of type III_A.

Fe_3O_4 has the highest Curie temperature of any half-metal, and measurements of point contacts suggest that a high room-temperature spin polarization is not just a dream [36]. At present, it seems to offer the best prospects for applications.

In materials such as Sr_2FeMoO_6 and the Heusler alloys, the spin polarization is very sensitive to site disorder. The Heusler alloys, in particular, have yielded disappointing values of P. Here the electronic structure of the surface may be to blame, as it is quite different from the bulk.

6 Conclusions

- The extended classifications scheme proposed for half-metals takes account of both itinerant and localised electrons.
- We depend heavily on band calculations to identify potential half-metals. There is no experimental "smoking gun" test.
- Measurements of the spin polarization P always involve electrons that have come from the surface, not the bulk. The density of states and exchange interactions at the surface may be very different from those of the bulk and these factors influence the value of P.
- P is weighted by the high Fermi velocity of $4s$ electrons, whenever these are present.
- The measured temperature-dependence of P is associated with spin wave excitation. It seems to be largely a surface effect.
- Interface engineering will be critical for device operation, for example making use of multiple reflection as in spin tunnel junctions [31].
- There is a need for new half-metals with higher Curie temperatures.

References

1. R.C. O'Handley: Modern Magnetic Materials Wiley, New York 2000
2. N.F. Mott: Metal-Insulator Transitions, 2nd Edition, Taylor and Francis, London 1990
3. B. Nadgorny, I.I. Mazin, M. Osofsky, R.J. Soulen, P.Broussard, R.M. Stroud, D.J. Singh, V.G. Harris, A. Arsenov, Ya Mukovskii: Phys. Rev. B **63**, 18 4433 (2001)
4. R.A. de Groot, F.M. Mueller, P.G. van Engen, K.H.J. Buschow: Phys. Rev. Lett. **50**, 2024 (1983)

5. K. Schwarz: J. Phys.F **16**, L211 (1986)
6. J.B. Goodenough: Magnetism and the Chemical Bond Wiley-Interscience (1976)
7. J.B. Goodenough: Prog. Solid State Chem. **5**, 235 (1971)
8. D.L. Camphausen, J.M.D. Coey, B.K. Chakraverty: Phys. Rev. Lett. **29**, 657 (1972)
9. H. van Leuken, R.A. de Groot: Phys. Rev. B **51**, 7176 (1995)
10. S. Ishida, T. Masaki, S. Fujii, S. Asano: Physics B **245**, 1 (1998)
11. Spin Electronics (M Ziese, M Thompson, eds.) Springer, Berlin 2001
12. S.P. Lewis, P.B. Allen, T. Sasaki: Phys. Rev. B **55**, 10253 (1997) M.A. Korotin, V.I. Anisimov, D.I. Khomski, G.A. Sawatzky: Phys. Rev. Lett. **80**, 4305 (1998)
13. S.M. Watts, S. Wirth, S. von Molnar, A. Barry, J.M.D. Coey: Phys. Rev. B **61**, 149621 (2000)
14. B.L. Chamberland: Mater. Res. Bull. 2,827 (1967)
15. K.I. Kobayashi, T.Kimura, H. Sawada, K.Terakura, Y.Tokura: Nature **395**, 677 (1998)
16. W. Westerberg, D. Reisinger, G. Jakob: Phys. Rev. B **62**, R767(2000)
17. Y. Tomioka, T. Okuda, Y. Okimoto, R. Kumai, K.I. Kobayashi, Y. Tokura: Phys. Rev. B **61**, 422 (2000)
18. V.A.M. Brabers, in Ferromagnetic Materials vol. 8 (K.H.J. Buschow, ed. Elsevier, Amsterdam, p.189, 1995
19. M. Penicaud, B. Silberchiot, C.B. Sommers, J. Kubler: J. Magn. Magn. Mater. **103**, 212 (1992)
20. J.M.D. Coey, M.Viret, S. von Molnar: Adv. Phys. **48**, 167 (1999)
21. W.E. Pickett, D.J. Singh: Phys. Rev. B **53**, 1146 (1996)
22. I.I. Mazin: Appl. Phys. Lett. **77**, 3000 (2000)
23. D.J. Singh: Phys. Rev. B **55**, 313 (1997)
24. R Weht, W E Pickett: Phys. Rev. B **60**, 13006 (1999)
25. C. Jiang, M. Venkatesan, J.M.D. Coey: Solid State Commun. **118**, 513 (2001)
26. D. Orgassa, H. Fujiwara, T.C. Schulthess, W.H. Butler: Phys. Rev. B **60**, 13237 (1999)
27. J. Tobola , J. Pierre: J. Phys. CM **10**, 1013 (1998)
28. I.I. Mazin: Phys. Rev. Lett. **83**, 1427 (1999)
29. M. Julière: Phys. Lett. A **54**, 225 (1975)
30. P. Guinea: Phys. Rev. B **58**, 9212 (1998)
31. S. Takahashi, S. Maekawa: Phys. Rev. Lett. **81**, 2799 (1998)
32. J.H. Park, E. Vescovo, H.J. Kim, C. Kwon, R. Ramesh, T. Venkatesan: Nature **392**, 794 (1998)
33. J.S. Moodera, G. Mathon: J.Magn. Magn. Mater. **200**, 248 (1999)
34. M.Viret, M. Drouet, J. Nassar, J.P Contuor, C. Fermon, A. Fert: Europhys. Lett. **39**, 545 (1997)
35. J.M. De Teresa, A.Barthelémy, A. Fert, J.P. Contuor, R. Lyonnet, F. Montaigne, P. Seneor, A.Vaurès: Phys. Rev. Lett. **82**, 4288 (1999)
36. J.J. Versluijs, M. Bari, J.M.D. Coey: Phys. Rev. Lett. **87**, 026601 (2001)
37. N. Garcia: Appl. Phys. Lett. **77**, 131 (2000)
38. R.Meservey, P.M. Tedrow: Physics Reports **238**, 173 (1994)
39. D.J. Monsma, S.S.P. Parkin: Appl. Phys. Lett. **77**, 720 (2000)
40. C.T. Tanaka, J. Nowak, J.S. Moodera: J. Appl. Phys. **86**, 6239 (1999)
41. R.J. Soulen, J.M. Byers, M.S. Osofsky, B. Nadgorny, T. Ambrose, S.F. Chong, P.R. Broussard, C.T. Tanaka, J.S. Moodera, A. Barry, J.M.D. Coey: Science **282**, 88 (1998)

42. S.K. Upadhyay, A. Palanisami, R.N. Louie, R.A. Burman: Phys. Rev. Lett. **81**, 3247 (1998)
43. J.M.D. Coey: J. Appl. Phys. **85**, 5576 (1999)
44. G.K. White, S.P. Woods Phil: Trans. Roy Soc. A **251**, 273 (1958)
45. K.Kubo, N.A. Ohata: J Phys. Soc. Japan **33**, 21 (1972)
46. Y. Ji, G.J. Strijkers, F.Y. Yang, C.L. Chien, J.M. Byers, A. Anguelouch, G. Xiao, A. Gupta: Phys. Rev. Lett. **86**, 5585 (2001)
47. D.C. Worledge, T.H. Geballe: Appl. Phys. Lett. **76**, 900 (2000)
48. J.M.D. Coey, M. Venkatesan: J. Appl. Phys. **91**, (2002)

Effects of Electron–Electron Interactions Near the Metal-Insulator Transition in Indium-Oxide Films

Zvi Ovadyahu

The Racah Institute of Physics, The Hebrew University,
Jerusalem 91904, Israel

Abstract. The metallic phase is theoretically distinguished from the insulating one by having a finite conductance. This distinction however applies strictly at absolute zero. Near the transition and at finite temperatures, the two phases often exhibit similar transport behavior. The difference between the metal and the insulator may be more readily recognized in systems driven far from equilibrium. When electron–electron interactions are dominant, the insulating phase is expected to exhibit glassy properties. Some manifestations of these effects will be shown and discussed

The metal insulator transition (MIT) is currently viewed as an example of a quantum phase transition. The latter term refers to a phase transition that takes place at the absolute zero of temperature as one or more parameters of the system are varied. The disorder driven MIT (the Anderson transition) is a competition between the potential disorder experienced by the charge carriers and their kinetic energy [1]. Once a critical amount of disorder is exceeded, the conductivity σ of the system vanishes at $T = 0$ while σ is finite for smaller disorder.

In the absence of electron–electron interactions, it is generally accepted today that this transition is continuous [2]. Strictly speaking, a true MIT is believed to occur only for a three-dimensional system while a $1D$ and $2D$ systems are insulating for any degree of disorder. On the other hand, one can still distinguish between a *diffusive* transport regime and an *activated* one even in $1D$ and $2D$ systems. Here we shall adopt the simple-minded view that diffusive transport is essentially a metallic behavior.

In the presence of interactions, the situation is less clear, and there are currently experimental results that are interpreted as evidence for a true metallic phase in a $2D$ system. This controversial issue will not be treated here. On the other hand, it will be stressed that interactions can modify considerably the out of equilibrium properties of the insulating phase, and lead to glassy behavior.

The distinction between the insulating (σ=0) versus a metallic phase ($\sigma\geq0$) is theoretically well defined. Experiments however are performed at finite temperatures where σ is never really zero. The common procedure to determine the transition point is to rely on extrapolating $\sigma(T)$ to zero based on data taken at low enough temperatures. Extrapolation always entails an uncertainty, especially when carried over an infinite distance (from a finite T to zero). The procedure of obtaining $\sigma(0)$ by extrapolation is thus never certain. It may be plausible only if the temperature is much smaller than any relevant energy *and* if the law for $\sigma(T)$ has physical basis. An experimental definition of the MIT thus must involve knowing the properties of the insulating versus metallic phases at *finite* temperatures. So the question is what are the properties that distinguish a metal from an insulator. When $\sigma(T \to 0) = 0$, the conductance must decrease upon cooling and therefore $d\sigma/dT$ is positive at low temperatures. Thus, the TCR (temperature coefficient of the resistance), is perforce negative for the insulating phase. The first point we wish to make is that the converse is not necessarily true. In fact, a diffusive system should show a negative TCR at sufficiently low temperatures unless something like superconductivity intervenes.

The reason that a negative TCR is sometimes associated with an insulator is presumably based on intuition nurtured by the semi-classical, Boltzmann transport theory for a diffusive system. According to this picture, the resistivity ρ of a metal is related to the two characteristic scattering rates $\tau_{\rm imp}^{-1}$ and $\tau_{\rm in}^{-1}$:

$$\rho(T) \propto \frac{1}{\tau_{\rm imp}(0)} + \frac{1}{\tau_{\rm in}(T)} \tag{1}$$

The former is the elastic scattering rate associated with static disorder (e.g., impurities, lattice imperfections etc.) and involves just a change in momentum $\mathbf{k}$. $\tau_{\rm in}^{-1}$ on the other hand involves scattering events that include exchange of energy between the particle and some other entity (namely, a change in k^2) . This expression hinges on the assumption that the two types of scattering rates are independent of one another. It is further assumed that $\tau_{\rm imp}$ is independent of temperature while $\tau_{\rm in}$ is temperature dependent and has the general form:

$$\tau_{\rm in}(T) \propto \frac{1}{T^P} \tag{2}$$

with P a positive number that depends on the particular mechanism of scattering (*i.e.*, electron-phonon, electron–electron etc.), and the system dimensionality. This picture has two immediate implications:

- $\rho(T) = \rho(0) + \rho_{\rm in}(T)$
- $\frac{\partial \rho}{\partial T} \geq 0$

The first says that $\rho(T)$ may be decomposed into a component that is controlled just by the amount of disorder $\rho(0)$, which is temperature independent, and another component that depends on T but not on disorder. This means that samples with different degrees of disorder should exhibit "parallel" $\rho(T)$ curves (the Mathiessen rule). The second result means that the TCR of a metallic (diffusive) system is positive for *any finite* temperature. Both implications are well obeyed in clean metallic systems when measured at not-too-low temperatures. Significant deviations from the behavior expected by the Mathiessen rule are observed once the MIT is approached from the metallic side [2]. These appear as a contribution to the resistivity $\Delta\rho$, which has a negative TCR. The magnitude of $\Delta\rho$ increases with disorder, and below a certain (disorder dependent) temperature it is the dominant contribution to the resistivity.

Today we recognize two distinct mechanisms that may account for this excess-ρ. Both are quantum mechanical in nature, and each addresses an issue ignored by the semi-classical model. It turns out that neglecting these correlations is justified only in the limit $\tau_{\text{in}} \ll \tau_{\text{imp}}$. Note that this condition agrees with the statement made above, namely that the predictions of the Boltzmann picture hold for weak disorder (large τ_{imp}) and high T (small τ_{in}, c.f. (2)). When this is not the case, quantum interference effects, not addressed by the Boltzmann model, need be taken into account. These, so called, weak-localization (WL) effects reflect on the failure of the assumption that elastic and inelastic scattering rates are independent. In $2D$ diffusive systems WL corrections lead to $\Delta\rho$ that diverges logarithmically as $T \rightarrow 0$. Such a behavior was confirmed in many experiments. An example for such a behavior in a thin In_2O_{3-x} film is illustrated in Fig. 1. However, it soon became clear that $\Delta\rho \propto -\log(T)$ cannot be unambiguously ascribed to WL effects. As was shown by Aronov and Altshuler, electron–electron interactions (EEI) effects may lead to a similar temperature dependence [1]. These effects result from the Coulomb interaction between the charge carriers. In "clean" metals, it has been traditionally believed that screening weakens the effects of the Coulomb interaction and thus e-e (electron-electron) effects are rather weak. What Aronov and Altshuler showed is that in the presence of static disorder, screening is impaired, and effects of interactions can be observed even for relatively clean (but still diffusive) systems.

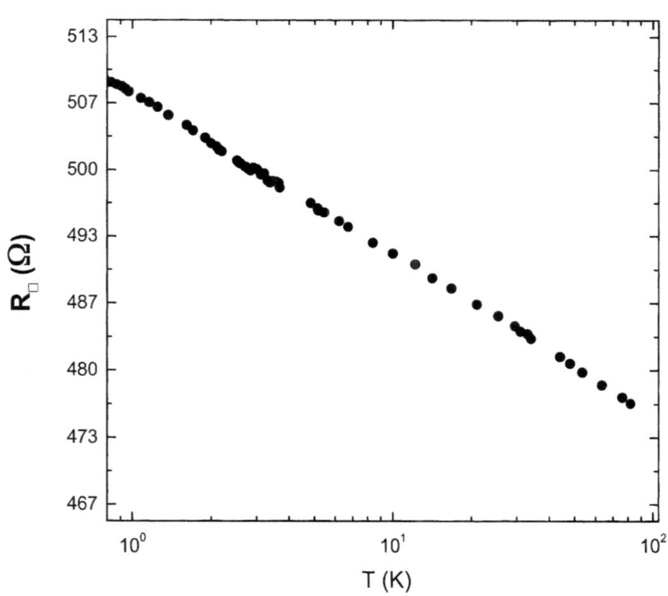

Fig. 1. The dependence of the sheet resistance $R_{\square}$ on temperature for a 190 Å In_2O_{3-x} film showing the logarithmic law expected by both WL and EEI effects in a $2D$ system

Since in terms of $\rho(T)$, the contributions of WL versus those due to EEI turn out to be similar, to distinguish between them other properties of the system need to be probed. The following table summarizes the effects expected by either mechanism.

	Weak-localization	Electron–electron interactions
Resistance versus T	$\Delta\rho \propto -\log(T)$ (in $2D$)	$\Delta\rho \propto -\log(T)$ (in $2D$)
Density of states (DOS)	Regular	Cusp in $N(0)$ around E_F
Magneto-resistance	Orbital	Zeeman, negligible at small H
Hall coefficient R_H	T-independent	$\frac{\delta R_H}{R_H} = 2\frac{\delta R}{R}$

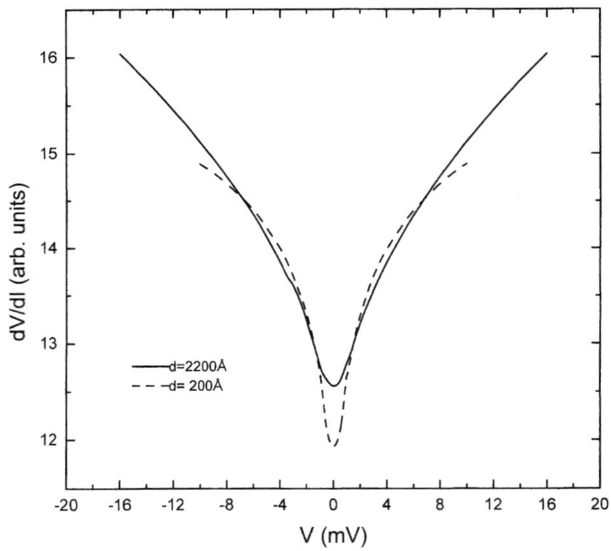

Fig. 2. Tunneling characteristics of a $2D$ (thickness 190 Å and $3D$ In_2O_{3-x} samples $[dI/dV(N(\varepsilon)]$. The measurements were carried out at T=1.2 K. See reference [3] for details

An important difference between the WL corrections and those of EEI is the modulation produced by the latter in the single particle DOS. Such an effect, characteristic of many-body interaction situations, has been seen in experiments and the results agree in detail with the theory. Examples of the characteristic cusp in $N(\varepsilon)$, the tunneling-DOS of $2D$ and a $3D$ system are shown in Fig. 2. The figure illustrates that the DOS has indeed a cusp centered at the chemical potential, and it has precisely the energy dependence predicted by the EEI picture for both $2D$ and $3D$. Namely, $N(\varepsilon)$ scales like $\ln(\varepsilon)$ and $\varepsilon^{1/2}$ in $2D$ and $3D$ respectively. Furthermore, a crossover from $2D$ to $3D$ behavior is observed as function of the tunneling bias as expected by theory [3]. Clearly then, interaction effects are quite prominent in these disordered systems.

On the other hand, magneto-resistance measurements taken at small fields H reveal equally good agreement to the predictions of the WL picture. The data in Fig. 3 compares the magneto-resistance of two In_2O_{3-x} films. One of these samples has been intentionally doped with 2% Au. This is known to lead to a strong spin-orbit scattering, which is weak in this system in its pure (un-

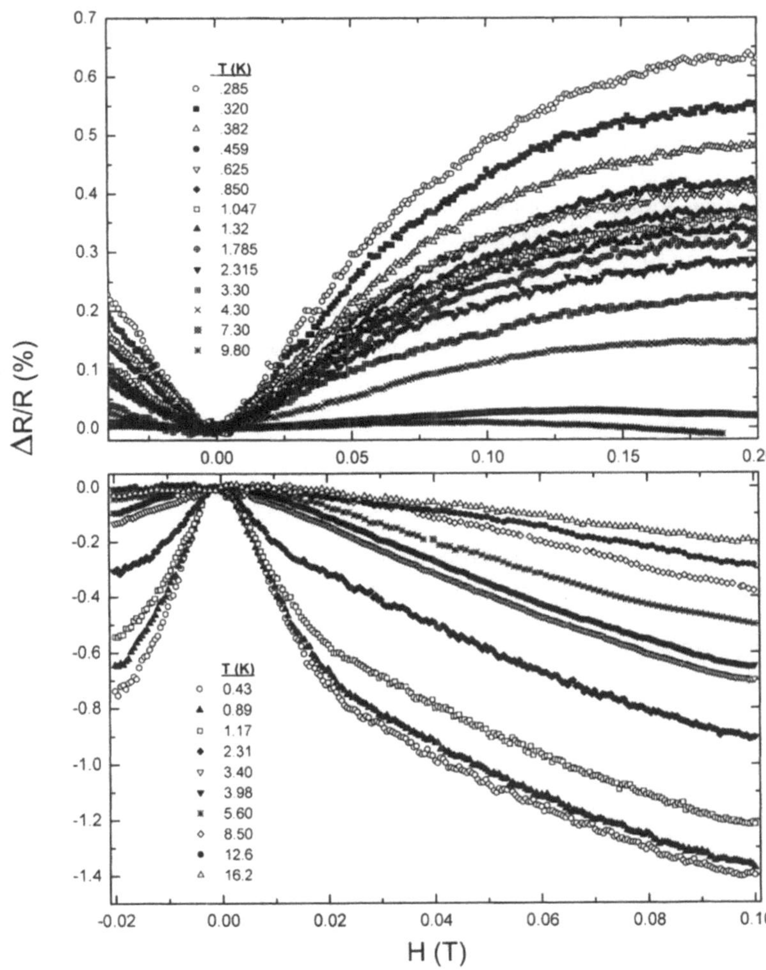

Fig. 3. Magneto-resistance of a typical pure In_2O_{3-x} sample (bottom graph) versus a gold doped In_2O_{3-x} sample (top graph) at several temperatures. Both were measured with the same electric field and had similar $R_\square$. Note that the sign of the magneto-resistance is different

doped) form. As illustrated in the figure the magneto-resistance in positive for the Au-doped sample and negative for the pure one. This is in agreement with the property expected by the WL picture – the sign of the magneto-resistance should depend on whether the spin-orbit scattering is strong or weak. This is in contrast with the EEI prediction that anticipates the magneto-resistance to be always positive, independent of spin-orbit scattering strength.

Moreover, there is clear evidence that the magneto-resistance is orbital in nature. This can be tested by changing the orientation of the field H relative to the sample plane as shown in Fig. 4. In this case, the sample has a thickness d of about 600 Å and the figure shows magneto-resistance for both orientations of the field (H parallel and perpendicular to the sample plane). Note that the relative change of the resistance due to H depends on the orientation, and for small H this difference is quite large. The magneto-resistance is due to the suppression of the quantum interference, which is the underlying mechanism responsible for the WL corrections. This quantum interference involve trajectories that extend on scale $L_\phi = (D\tau_{\text{in}})^{1/2}$ – the inelastic diffusion length (D is the diffusion constant). The electron maintains phase coherence over this scale. The effect of the field H is to modify the interference through the Aharonov–Bohm effect. The latter

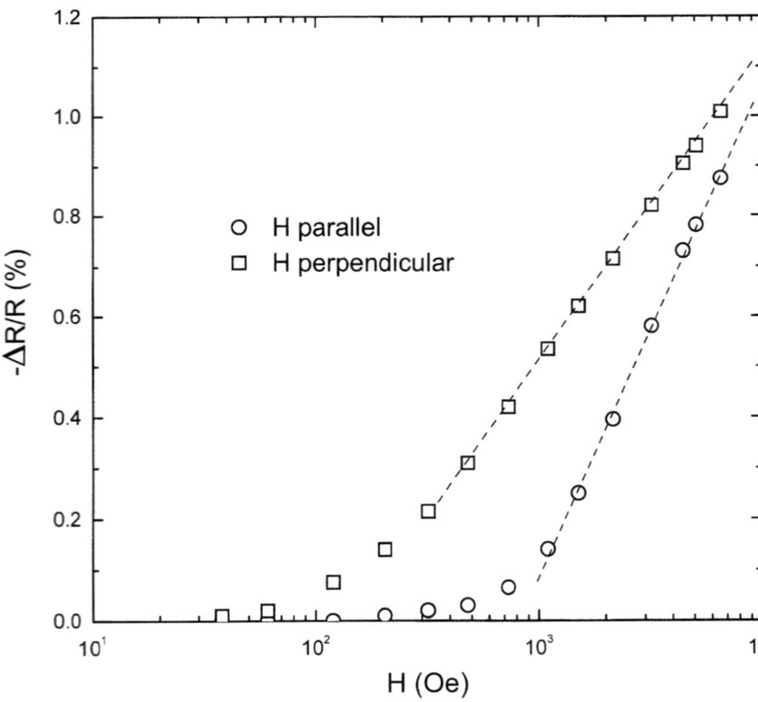

Fig. 4. Magneto-resistance of a 600 Å pure In_2O_{3-x} sample for perpendicular and parallel field orientation. Note that the curves tend to meet at large H

depends on the magnetic flux threading these loops, *i.e.*, HL_ϕ^2. Once L_ϕ is larger than the sample thickness d, the relevant flux is $L_\phi dH$ hence smaller when H is oriented parallel to the sample plane than for the perpendicular orientation. The magneto-resistance is thus smaller for parallel H. This anisotropy is a characteristic signature of an *orbital* mechanism. For large values of H we see that the magneto-resistance curves tend to merge and the anisotropy vanishes (Fig. 4). This is another feature expected by the theory. The characteristic length associated with a field H is $L_H = (h/eH)^{1/2}$ defined by $HL_H^2 = \phi_0 = h/e$ – the flux quantum. L_H,called the magnetic length, is the scale over which the Aharonov–Bohm effect causes an appreciable phase shift. Namely, at a field H all loops with scale $L < L_H$ are still unaffected while the phases for loops with $L > L_H$ are fully scrambled. When $L_H < d$(*i.e.*, for large enough H) there is no distinction between parallel and perpendicular field in terms of the magneto-resistance. In other words, a crossover from a two-dimensional to a three-dimensional behavior is expected when H becomes so large that $L_H < d$ which is precisely what is observed in Fig. 4.

So we now may conclude that both EEI and WL effects play a role in the transport properties of the system and manifest themselves by different measurements. It seems natural then to assume that the excess resistivity depicted in e.g., Fig. 1 is the combined effect of both mechanisms. The question is what part of $\Delta\rho(T)$ is due to EEI and what part is due to WL effects. It is a common practice to use the magneto-resistance measurements to de-convolute the $R(T)$ data and obtain a quantitative answer for this question. For the In_2O_{3-x} system used here for illustration, it turns out that about 80% of $\Delta\rho(T)$ comes from EEI. This seems to suggest that the effects of WL and EEI may be just superimposed in an additive way. Indeed this conclusion appears to be consistent with experiments involving $\rho(T)$ and magneto-resistance measurements on many different systems. However, when we try the next test of the theories – the behavior of the Hall coefficient R_H, we are led to review this conclusion.

The WL picture anticipates that R_H is temperature *independent* just as in a "classical metal". The EEI model on the other hand has the non-trivial prediction that the relative change in R_H will be twice as big as the corresponding change in ρ over the same temperature range. Given the observation reached above namely, that $\Delta\rho(T)$ is mostly due to EEI, we expect to find significant temperature dependence for R_H in these samples. As shown in Fig. 5 however, no such dependence is found when R_H is measured at small fields. Interestingly, when the Hall-effect measurements are carried out at larger values of fields, R_H does show temperature dependence that asymptotically approaches the prediction of the EEI theory (Fig. 6). This anomalous behavior has been seen also in Si-inversion samples [4]. It is important to realize that the theoretical treatment of the Hall effect assumes linear response. Strictly speaking, the behavior of R_H can be compared with theory only in the limit $H \to 0$. In this limit it seems that $\delta R_H = 0$ *as if no EEI effects are present* in contrast with the data for the tunneling DOS. The field above which the EEI prediction does hold correlates with the field that suppresses the WL effects. This might hint that the problem

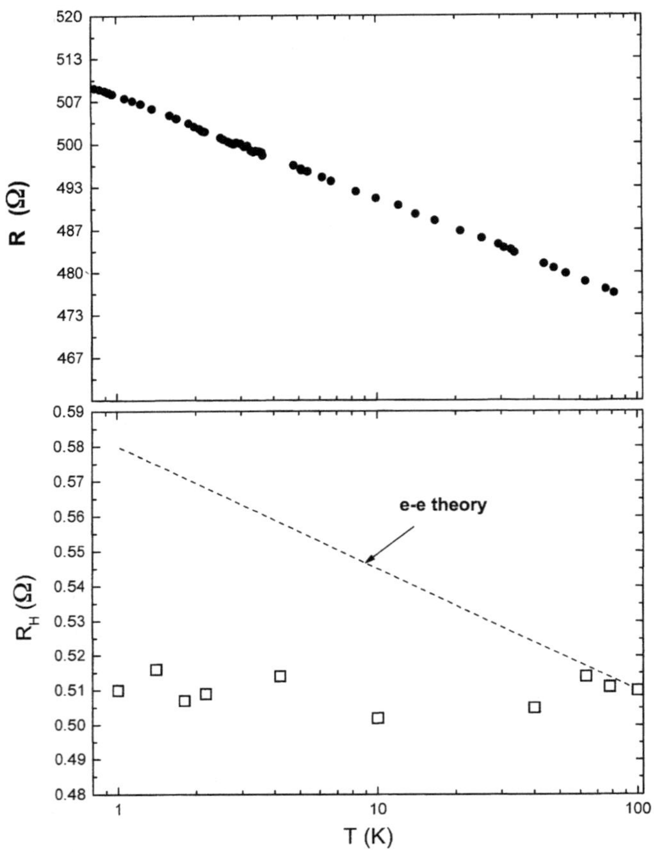

Fig. 5. The dependence of the Hall coefficient R_H (measured with H=0.05 T) on temperature for the In_2O_{3-x} film used in Fig. 1 (bottom graph). The dashed line is the dependence expected by the EEI model. The upper graph depicts the temperature dependence of the longitudinal resistance for comparison

is related to the interplay between the effects of disorder and electron correlations. It would appear that the contribution of WL and EEI effects may not be simply additive, at least as far as the Hall effect is concerned. To our knowledge, this problem is still unresolved.

So far, we discussed what happens on the metallic side and far from the MIT. The observation that R_H depends on H however seems to be more general. It has been seen in several systems near the MIT for both metallic and just-insulating samples [5]. When R_H depends on H one may be tempted to extrapolate the data to zero field in a similar vein as done with $\sigma(0)$. In contrast with the behavior of $\sigma(T)$, the reason for the change of R_H with H is not understood, namely, the function $R_H(H)$ is not known. This sheds serious doubts on the procedure of extrapolating R_H to zero field. This is unfortunate. The behavior

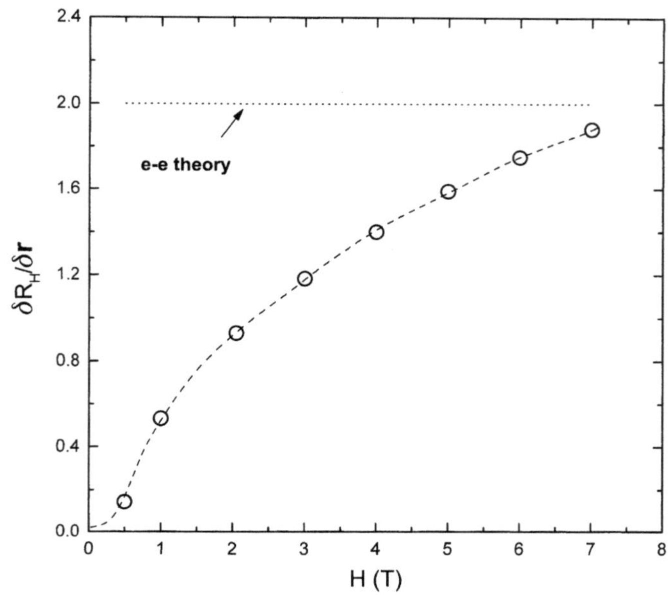

Fig. 6. The change of R_H relative to the change in resistivity ρ over a temperature range of 1.4-4.2 K as function of the magnetic field. Note that δR_H goes to zero when $H \to 0$ and $\delta R_H/\delta\rho$ approaches asymptotically the value 2 expected by the EEI picture. The sample is a 1200 Å film of In_2O_{3-x}

of the Hall coefficient near the MIT has a fundamental value. It also has a "pedagogical" value. For diffusive systems, we tend to associate R_H^{-1} with n, the density of charge carriers in the system. This may allow us to view the sample conductance G as composed of carrier-concentration n, and "mobility" so that if we know G and R_H we can say something about the carriers mobility. Such a naive deconvolution of G cannot be used for an interacting system. A more suitable description of G is given by the Einstein relation, $G \propto e^2 D \frac{dn}{d\mu}$ where $dn/d\mu$ is the thermodynamic DOS. (For an interacting system, the latter is usually quite different from the single-particle DOS discussed before). In this more general scheme a measurement of R_H reflects on $dn/d\mu$ and so we can use it to answer the following question:

At the MIT, where G goes from a finite value to zero, is it due to a vanishing DOS, or is it due to $D = 0$? The former case is what happens, e.g., in the Mott transition while $D = 0$ and $dn/d\mu > 0$ is expected for the non-interacting Anderson transition.

To determine which scenario actually occurs in a given situation one has to measure both G and some other property that reflects on $dn/d\mu$. The difficulties that we encountered with the Hall coefficient measurements motivated us to look for another technique – the field effect.

The underlying physics of the field effect (FE) is quite simple. The sample is prepared in a MOSFET-like configuration where the active channel is the system to be measured. This is typically a thin film separated by an insulating spacer from a metallic layer (gate). The sample conductance G is then measured as function of the voltage V_g between it and the gate electrode. Since this configuration is essentially that of a plate capacitor, the change in the sample carrier concentration n follows V_g namely, $\delta n \propto \delta V_g$. The FE amounts to measuring $G(V_g)$ over some range in V_g. This technique, much like the Hall effect, may be used to measure $dn/d\mu$ in thin metal films and semiconductors. Let us now see what happens to the system as it crosses over from the diffusive to the insulating regime using the FE technique.

In Fig. 7 we compare the dependence of G on the gate voltage V_g for a In_2O_{3-x} sample with two different degrees of disorder. The sheet resistance $R_\square$ at $T = 4.11$ K is taken to be a measure of the disorder ($R_\square$ plays the role of ρ in $2D$). As shown in this figure, the behavior of $G(V_g)$ is quite different for samples with $R_\square < h/e^2$ and $R_\square \gg h/e^2$. For samples with $R_\square < h/e^2$ the FE is 'normal' in the sense that $\Delta G > 0$ ($\Delta G < 0$) for $\Delta V_g > 0$ ($\Delta V_g < 0$) respectively. Namely, when extra charge is added to the system, its conductance increases, and when charge is extracted from the system the conductance decreases. In fact, from the slope of $\Delta G(V_g)$ we can estimate the density of carriers and it agrees with the value obtained from (low field) Hall effect measurements. Samples with $R_\square \gg h/e^2$ on the other hand, exhibit an anomalous $G(V_g)$. In addition to the anti-symmetric, normal component, $G(V_g)$ shows a symmetric increase around the value of V_g at which the sample was equilibrated. Before discussing the anomalous component it is important to note that the normal component of the FE is *always* there and its slope is fairly unchanged as one crosses from the diffusive to the insulating regime. This means that $\partial n/\partial\mu$ is non-zero in the insulating phase as expected for the Anderson transition. We thus achieved an answer to our question. However, nature has now confronted us with a new one – what is the reason for the anomalous component of the FE in the insulating regime? We shall now address this problem.

The important clue that is revealed in the experiments is that the magnitude of the anomalous component depends on the rate at which V_g is swept. In particular, when the gate voltage is swept sufficiently slow this component vanishes leaving only the anti-symmetric part. The cusp in $G(V_g)$ is therefore a manifestation of a *non-equilibrium* phenomenon characteristic of glasses. Secondly, no such cusp appear when $R_\square \leq h/e^2$. Recall that in two-dimensional systems $R_\square \approx h/e^2$ marks the transition from the diffusive to the strong-localization (insulating) regime. The observation that that these glassy effects are observable only in the insulating phase immediately suggests an intimate contact with its electronic nature.

We thus find a qualitative change in the behavior of diffusive versus that of insulating samples – the latter are non-ergodic, and show features that are characteristic of glasses. This feature of the insulating phase has been anticipated on theoretical grounds[6] but experimental evidence for it was obtained only

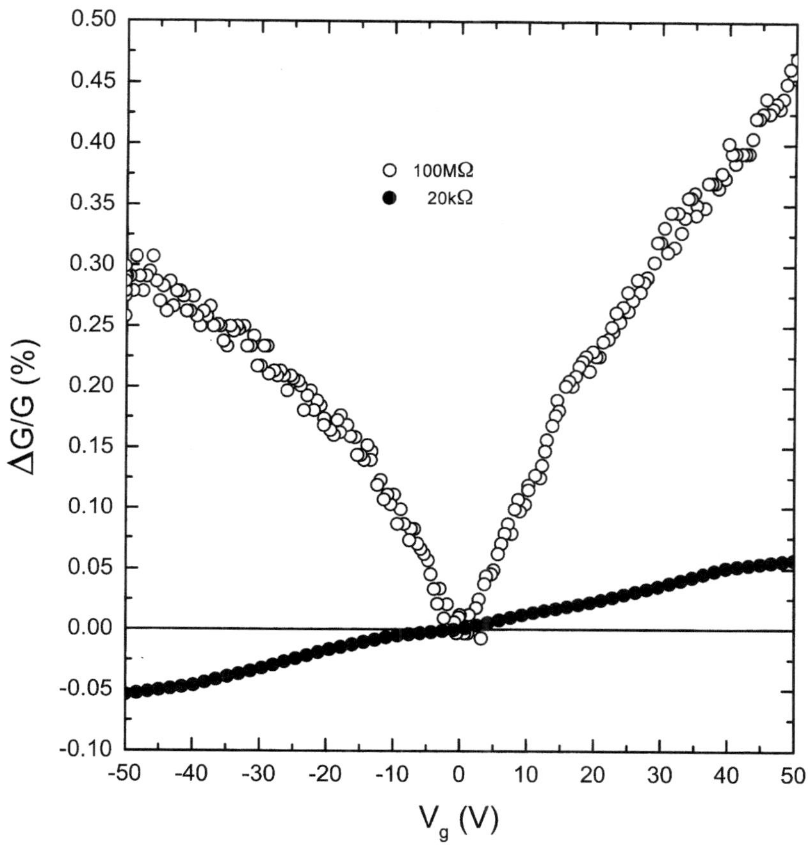

Fig. 7. The field effect of two In_2O_{3-x} samples measured at T=4.11 K. The sample with $R_\square$=20kΩ is diffusive and exhibits a normal FE. The sample with $R_\square$=100 MΩ is insulating and exhibits the anomalous cusp (see text). Both samples are 100 Å thick and they are separated from a gate electrode by a 100 μm cover glass

recently. In the following, some of the experiments that confirm the glassy nature of Anderson insulators will be shown and discussed.

One of the most fascinating attributes of glasses is the memory effect they exhibit. Such memory effects are observable in FE experiments on insulating films. As noted above, the anomalous increase of G appears around the gate voltage $V_{\rm g}^{\rm eq}$ at which the sample was allowed to equilibrate. This results in a minimum at $G(V_{\rm g}^{\rm eq})$ whether the sweep of $V_{\rm g}$ is affected by going from $V_{\rm g}^{\rm eq}$ to either side or going across $V_g^{\rm eq}$ continuously. In other words, the system remembers that it was allowed to equilibrate at $V_{\rm g}^{\rm eq}$.

This memory effect is more clearly illustrated with the help of the two-dip-experiment (TDE). The TDE involves the following procedure (c.f., Fig. 8 as an

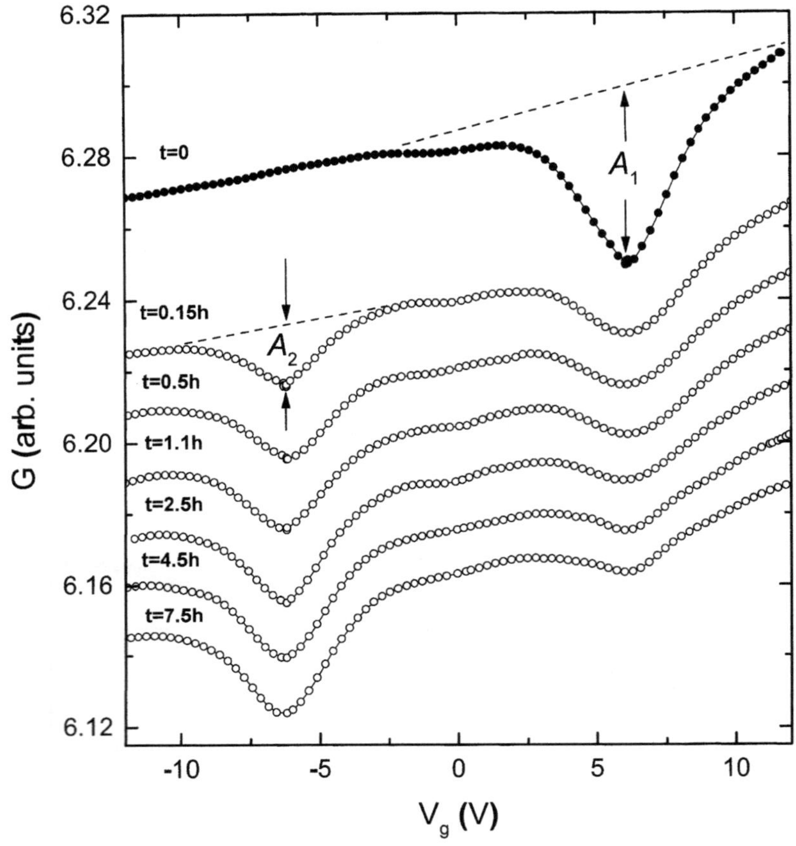

Fig. 8. An illustration of the TDE. The right dip (A_1) occurs at the cool-down value of V_g^0 and decays in time. The left dip (A_2) occurs at the value to which V_g is switched at time $t = 0$ and grows with t. The sample is amorphous indium-oxide with $R_\square$=3.8 MΩ measured at T=4.11 K. The curves are displaced along the y-axis for clarity

example). The sample is cooled to the measuring temperature with a voltage V_g^0 held at the gate, and is allowed to equilibrate for several hours. Then, a $G(V_g)$ trace is taken by sweeping V_g across a voltage range straddling V_g^0. The resulting $G(V_g)$ exhibits a local minimum centered at V_g^0. At the end of this sweep, a new gate voltage, V_g^n is applied and maintained between subsequent V_g sweeps that are taken consecutively at latter times (measured from the moment V_g^n was first applied). Each such sweep reveals two minima, one at V_g^0 (with magnitude A_1) that fades away with time and the other at V_g^n (with magnitude A_2) whose magnitude increases with time. Let us focus on the "old" minimum, which the system "remembers" for many hours (Fig. 8). We can understand such a feature is if e-e interactions are involved. Note that upon setting a particular V_g, say V_g^0 the system will eventually establish the corresponding chemical potential.

In the presence of interactions (and disorder), a dip in the tunneling density of state $N(\varepsilon = V_{\rm g}^{0})$ will be created and once V_g is moved to another value, say $V_{\rm g}^{n}$ a similar modulation will start to build up at $N(\varepsilon = V_{\rm g}^{n})$ while the one around V_g^0 will slowly fade away with time. It seems plausible to assume that the modulation in $N(\varepsilon)$ translates into an energy dependent G. This seems a natural way to explain the occurrence of the dips in $G(V_{\rm g})$ although the details of how $N(\varepsilon)$ (being a single-particle property) relates to G are by no means well understood. It is interesting to note the similarity in shape between the cusp in the FE experiments and the tunneling DOS in figure 2 as well as the curves obtained by simulations [7]. These similarities however, may be fortuitous. The cusp observed in FE experiments cannot be interpreted as a single-particle DOS because during these measurements at least some relaxation must take place. In a sense, the cusp may be referred to as a "non-equilibrium DOS" presumably reflecting the underlying effects of the electron–electron correlations.

Another aspect that can be studied with the TDE is the dependence of the characteristic relaxation time τ on various parameters. Monitoring the TDE pattern amounts to studying the dynamics of the "forming" of a cusp at a newly imposed $V_{\rm g}^{n}$ and the "healing" of an "old" cusp at V_g^0. A characteristic-time may be defined as the time at which the amplitude of the cusp at $V_{\rm g}^{n}$ just equals the amplitude of the cusp at $V_{\rm g}^{0}$. This τ may be obtained by interpolation between $G(V_g)$ scans taken at different times, it is experimentally well-defined, and it is fairly independent of the particular relaxation law. Using this method for determining τ, the following facts were observed:

- The relaxation time increases with disorder and with an applied magnetic field [8]. The effect of both factors was attributed to constraints on the basic transition rates between localized sites.
- Doping the system with spin-orbit scatterers resulted in a shorter τ[9].
- There seems to be no temperature dependence of the relaxation time over the temperature range where the TDE could be measured with reasonable accuracy (0.6–12 K).

The lack of temperature dependence of τ is intriguing. In this regard, the electron glass is quite different from "ordinary" glass that shows a dramatic increase of τ at low temperatures. This property gives yet another indication that the phenomena are not associated with dynamics of atoms. The effect of spin-orbit scattering complements the magnetic field dependence. As first described by Kamimura [10], a strong magnetic field forces the spins to be aligned thereby eliminating transitions from singly-occupied to singly occupied states (and due to detailed balance, from doubly-occupied to unoccupied states). This leads to constraint on available transitions and slows the relaxation process. The presence of spin-orbit scatterers has the opposite effect because the constraint on the spin-state are relaxed (in a similar vein that disorder relaxes constraint imposed by momentum conservation). To our knowledge, current models of the electron glass do not treat spin effects, although they are evidently relevant. It is interesting to note that effects of magnetic fields on glassy dynamics were also found in 'ordinary' glasses at ultra low temperatures [11]. In either of these cases, the magnetic field effects are difficult to reconcile with atomic motion.

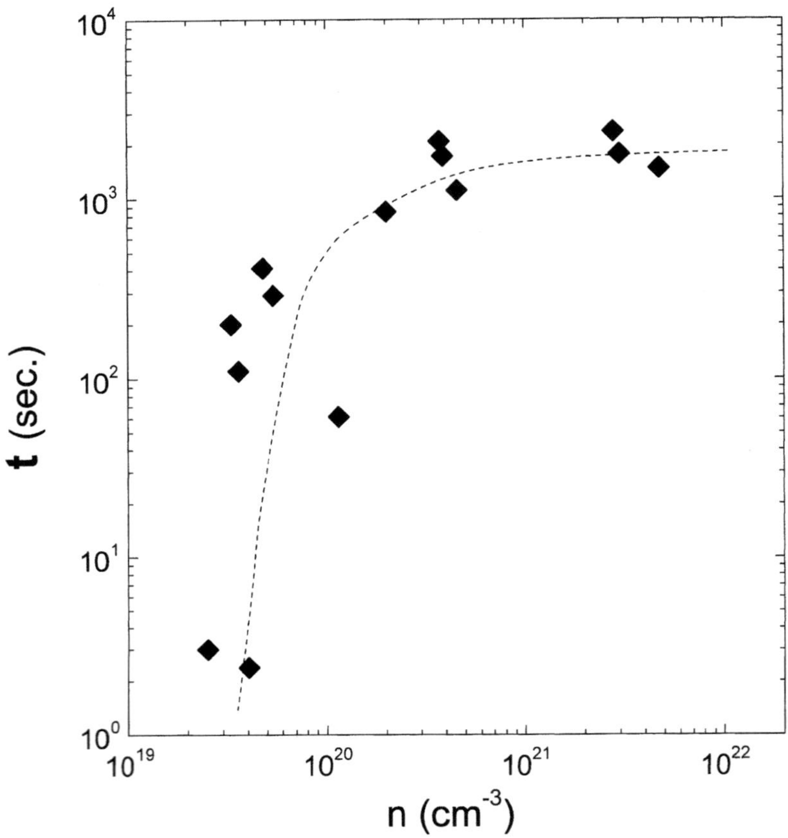

Fig. 9. The dependence of the relaxation time τ defined through the TDE for several amorphous indium-oxide samples with different composition as function of carrier concentration n. Note the sharp drop in τ for $n < 10^{20}$ cm^{-3}. The dashed line is a guide to the eye

The parameter that affects τ most dramatically is the carrier concentration n. In Fig. 9 we show the dependence of the relaxation time on n for several amorphous indium-oxide samples. These samples were measured under the same conditions (*i.e.*, temperature, and rate of V_g-sweep) and all had similar $R_\square$. They differed in terms of n, which was changed in the preparation process by adjusting the In/O ratio. Since their $R_\square$ was fine-tuned to be similar, their disorder was also different (*i.e.*, disorder in the series increases with n). The steep dependence of $\tau(n)$ in Fig. 9 however, is mostly due to n rather than to disorder. This was checked by performing TDE on several samples with fixed composition and changing the disorder by thermal annealing. The latter did result in an increase of τ with $R_\square$ (disorder) as noted above. However, this change was rather small in comparison with the orders of magnitude change of τ observed in Fig. 9

(typically, varying $R_\square$ between 1 MΩ to 1 GΩ increased τ by 20%). Therefore, the fast increase of τ with n is essentially due to the increase in the carrier concentration. As argued by Vaknin et al [12], this fact is consistent with the presence of e-e interactions. The rationale for that is as follows. If interactions play a role in slowing down the relaxation the dependence on n is natural. This is so because for a medium that lacks screening, the strength of e-e interaction is $e^2 n^{1/3}/\kappa$ (κ is the dielectric constant), and therefore interactions are more important the higher is the concentration. That the contribution of interactions becomes more important with increased density is the rule in most many-body problems, as e.g., in gases. A well-known exception to this rule is the case of the metal due to the peculiar property of screening.

The question is what are the constraint imposed by the interactions that hinder energy relaxation processes. A possible answer for this question may be found in the computer work performed by Baranovskii et al. [13]. These authors found that the "true" ground state of an interacting system could not be reached by just single-particle transitions. Rather, the system energy, after being initially reduced by fast single-particle transitions, "hangs" at a state (termed "pseudo-ground-state") having higher energy than the equilibrium state (and thus exhibiting higher conductance). Further reduction in energy must involve correlated, many-particle transitions. The latter are exponentially rare events and this leads to a sluggish, ever-slowing process. Such considerations suggest that in the limit of a dilute system relaxation would become much faster (as indeed observed in our experiments, c.f., Fig. 9). This can be understood as follows. The unscreened Coulomb interaction has no characteristic spatial scale but its magnitude decreases with the inter-particle separation r. Electrons that are separated by $r > L$ with L given by $e^2/\kappa L \approx \hbar^2/(2m^*\xi^2)$ are essentially uncorrelated. Here ξ is the localization length and m^* is the effective mass. The physical reason is that once the disorder energy $\hbar^2/(2m*\xi^2)$ exceeds the Coulomb repulsion the electrons distribution is controlled by the random potential of the ions rather than by the long-range Coulomb interaction. The condition for being in the dilute phase is thus: $n_c L^3 \approx 1$ namely $n_c^{1/3} \approx \kappa\hbar^2/(2m^* e^2 \xi^2)$. Note that this condition depends only weakly on disorder because $\kappa = \kappa(\xi)$ and near the metal insulator transition $\kappa \propto \xi^2$ [14]. Taking $\kappa \approx 10$, $m^* \approx 10^{-27}$ gm, and $\xi \approx 10$ Å as typical values one gets $n_c \approx 10^{19}$ cm^{-3} which is in good agreement with our findings (Fig. 9).

Obviously, a more realistic model needs to be developed to treat this problem. However, it is clear even from our crude picture that a long τ is favored by high-density systems such as In_2O_{3-x} and granular metals and less expected in lightly doped semiconductors. This may be the reason why non-ergodic effects of the kind described above have not been observed in impurity-band systems.

Finally, the electron glass has been shown to exhibit aging [15], which is a common feature in many glasses. This fascinating phenomenon is in essence another memory effect: the relaxation law for the excess conductance $\Delta G(t)$ reflects the time the system has spent under some modified external conditions. It turns out the electron glass exhibits the simplest form of aging involving no free parameters. Moreover, the phenomenon may be observed over convenient time scales and wide temperature range making it an attractive candidate for a study of this fundamental aspect of non-ergodic systems.

In summary, we have reviewed some of the transport phenomena observed near the MIT. On the metallic side quantum interference effects, such as orbital magneto-resistance, are observed in agreement with WL theories. These effects do no disappear as the system crosses over to the insulating side of the MIT. In fact, quantum interference effects are sometimes more prominent in the insulating regime than they are in metals [16]. These issues were discussed in the next chapter by M. Sanquer.

Effects due to electron–electron interactions are observed on the metallic side and are well described by current theories except for the behavior of the Hall coefficient, which remains enigmatic. In the insulating regime, e-e interactions lead to non-ergodic behavior typical of glasses such as slow relaxations and dependence on history. These effects are peculiar to the insulating state and disappear at the transition to the diffusive regime.

This research was supported by a grant administered by the Israeli academy for Science and Hunmanities and the German Israeli Bi-national Science Foundation.

References

1. For a review see: P.A. Lee, T.V. Ramakrishnan: Rev. Mod. Phys., **57**, 287 (1984) S. Kobayashi, F. Komori: Prog. Thoer. Phys. Suppl. **84**, 224 (1985)
2. For recent experiments on this issue see: S. Bogdanovich, M. P, Sarachik, and R. N. Bhatt: Phys. Rev. Lett. **82**, 137 (1999); S. Waffenschmidt, C. Pfeiderer, and H. v. Löhneysen: Phys. Rev. Lett. **83**, (1999)
3. I. Schwartz, S. Shaft, A. Moalem, Z. Ovadyahu: Phil. Mag. **B50**, 221-227 (1984)
4. Y. Imry, Z. Ovadyahu: Phys. Rev. Lett. **49**, 841 (1982)
5. R.C. Dynes: Surf. Sci. **113**, 510 (1982)
6. D.W. Koon, T.G. Castner: Solid State Commun. **64**, 11 (1987) and references therein
7. J.H. Davies, P.A. Lee, T.M. Rice: Phys. Rev. Lett. **49**, 758 (1982) M. Pollak: Phil. Mag. **B50**, 265 (1984) M. Pollak, M. Ortuno: Sol. Energy Mater. **8**, 81 (1982) Sh. Kogan: Phys. Rev. **B57**, 9736 (1998)
8. C.C. Yu: Phys. Rev. Lett. **82**, 4074 (1999)
9. Z. Ovadyahu, M. Pollak: Phys. Rev. Lett. **79**, 459 (1997)
10. Z. Ovadyahu: unpublished
11. A. Kurobe, H. Kamimura: J. Phys. Soc. Jpn. **51**, 1904 (1982)
12. P. Strehlow, M. Wohlfahrt, A.G.M. Jansen, R.Haueisen, G.Weiss, C.Enss, S.Hunklinger: Phys. Rev. Lett. **84**, 1938 (2000)
13. A. Vaknin, Z. Ovadyhau, M. Pollak: Phys. Rev. Lett. **81**, 669 (1998)
14. S.D. Baranovskii, A.L. Efros, B.L. Gelmont, B.I. Shklovskii: J. Phys. C Solid State Phys. **12**, 1023 (1979) E.I. Levin, V.L. Nguyen, B.I. Shklovskii, A.L. Efros: Zh. Eksp. Theor. Fiz. **92**, 1499 (1987) [Sov. Phys. JETP **68**, 1081 (1989)]
15. Y. Imry, Y. Gefen: Phil. Mag. **B50**, 203 (1984)
16. A. Vaknin, Z. Ovadyhau, M. Pollak: Phys. Rev. Lett. **84**, 3402 (2000)
17. For a review see: Z. Ovadyahu: Waves in Random Media, **9**(2), 241 (1999)

Interference Effects in Disordered Insulators

M. Sanquer

CEA-DSM-DRFMC-SPSMS, CEA-Grenoble, France

Abstract. This is an introduction to the problem of quantum interferences in disordered insulators, and their implication in the magnetoconductance at low temperature. In particular, we will show that a positive exponential magnetoconductance in the variable range hopping regime is related to a variation of the localization length with a magnetic flux applied. We also describe the crossover to the strongly insulating case, i.e. the case of insulators with moderate disorder, and comment on possible role of interaction on the magnetoconductance.

1 Introduction: Disordered Insulators, Localization Length, Interactions and Interferences

Many materials exhibit a low conductivity at low temperature, intermediate between a metallic conductivity and a pure dielectric behavior. Very often these materials are unintentionally disordered and contain a low density of carriers. A typical example is a moderately doped semiconductor, where the residual disorder comes from the random location of dopants and the density of carriers is in the range $10^{16}cm^{-3} - 10^{19}cm^{-3}$. Others examples are the low mobility inversion 2DEG, the conducting polymers, the High Tc compounds, the granular metal films, etc... At high temperature, when quantum effects are unimportant, these materials show a low Boltzmann conductivity σ_0 which corresponds to $k_{\mathrm{F}}\ell \simeq$ few units, estimated through the relation:

$$g_0(L) = \sigma_0 L^{d-2} = \frac{e^2}{h} k_{\mathrm{F}}\ell (k_{\mathrm{F}}L)^{d-2} \tag{1}$$

$g_0(L)$ is the conductance, k_{F} is the Fermi wave vector and ℓ is the elastic mean free path. $k_{\mathrm{F}}\ell \simeq 1$ means also that the distance between electrons is comparable to the distance between impurities. For d =2, $k_{\mathrm{F}}\ell \simeq 1$ means that the Boltzmann conductance is $G_{2D} = \frac{e^2}{h}$. As one decreases the temperature, the conductance of these materials generally decreases also, and for some of them, the conductance tends to zero, down to some residual conductance due to finite size effects. These materials are named insulators. They are classified in the same category as materials which present only a residual conductance at room temperature, the large gap insulators. As far as we are concerned with the low temperature situation, the difference between gaps induces a change of the typical scale for tunneling and localized orbitals extension. The localization length ξ is the extension of the electronic wave function, and also the characteristic decay length for the exponential dependence of the conductance in an insulator: $g \propto \exp^{-L/\xi}$. For a pure tunnel barrier ξ is approximately given by:

$$\xi = \frac{\hbar}{(2m^*U)^{1/2}} \tag{2}$$

U is the effective height of the barrier. If U is small, ξ is large.

Introducing disorder inside the barrier creates impurity levels and the tunnel conductance increases very much. The microscopic distinction between a highly disordered dielectrics and a highly disordered metal tends to vanish progressively as the disorder increases and $k_F\ell \simeq 1$.

The moderately disordered insulating case is covered in the celebrated book of Efros and Shklovskii [1], and well represented by the lightly doped semiconductor case. Each electron is strongly bounded to one donor at once and the overlap between electronic wave functions on different donors is very weak. In lightly doped semiconductors the localization length is the Bohr radius: $\xi \simeq a_B$. At finite temperature, in long enough samples, the transport is by variable range hopping (see later on). The unscreened attraction between one electron jumping from donors to donors ionized states and the hole left behind it results in a pseudogap in the density of states at the Fermi level. This reduces the conductance of a lightly doped semiconductor at very low temperature (Efros-Shklovskii gap)[1].

If one increases the dopant concentration (and so the disorder), a Mott transition appears. The seminal idea of Mott was to say that above the critical concentration, given in 3D by:

$$r_S = \left(\frac{3}{4\pi N_c}\right)^{1/3} \simeq 2.5a_B \tag{3}$$

others electrons screen the attractive electron-hole interaction, unbinding the electron–holes pairs.

Nevertheless, above the Mott critical concentration, carriers could be still localized by disorder. In this state, the distance between electrons $\simeq r_S \simeq k_F^{-1}$ is comparable to the mean free path ℓ and to the screening length, but smaller than the localization length. There is a finite, continuous density of states at the Fermi energy, which is located below the mobility edge (i.e. the energy level in the tail of the conduction band which separates localized and extended states).

In this regime, where the localization length ξ is large, ξ depends on the dimensionality and shape of the sample: in a quasi 1D sample (long sample of transverse cross-section S):

$$\xi = (\beta\mathcal{N} + 2 - \beta)\ell \tag{4}$$

β is a basic symmetry dependent parameter ($\beta = 1$ if there is time-reversal symmetry and spin rotation symmetry, $\beta = 2$ if there is no time reversal symmetry, and $\beta = 4$ is there is no spin rotation symmetry), and $\mathcal{N} = Sk_F^{d-1}$ is the number of electronic channels, S is the transverse cross-section) [2]. At strictly 1D (when $\mathcal{N} = 1$, that is when the cross-section is comparable to the Fermi wave length), $\xi = 2\ell$. But if $\mathcal{N}$ is large, ξ is much larger than ℓ. Equivalently at 1D (for large $\mathcal{N}$) ,

$$\xi \simeq \beta \frac{h/e^2}{2R_{\text{sqr}}} \times W \tag{5}$$

W is the width of the sample, the sheet Boltzmann resistance R_{sqr} (evaluated at high temperature) includes the thickness of the sample. At higher dimension $d \geq 2$, the dependence of ξ on β (that influences the magnetoconductance) is still an open problem. Neglecting this point, at d=2 one has:

$$\xi \simeq \ell \quad \exp\left(\frac{h/e^2}{R_{\text{sqr}}}\right) \tag{6}$$

In 1D or 2D, ξ is eventually large but always finite. At d=3 on the contrary, there is a real metal–insulator transition (MIT), at which the localization length diverges. For instance if one increases n, the concentration of carriers in a doped semiconductor:

$$\xi = a_{\text{B}}(\frac{n}{n_{\text{c}}} - 1)^{-\nu} \tag{7}$$

where n_{c} is the critical concentration ($n \leq n_{\text{c}}$) and ν a phenomenological exponent. Approaching the MIT from the insulating side, the localization length is very large as compared to ℓ.

Whatever is the dimensionality, the fact that ξ is large as compared to ℓ has consequences both on interferences and on interactions.

First, considering interferences, there is a spatial range between ℓ and ξ, where electron trajectories form loops. One expects an analog of the weak localization effect in the insulating state, which is discussed in the Sect. 4. For lengths larger than ξ and up to the sample length at zero temperature or up to the hopping length at finite temperature (see Sect. 2), the trajectories will be forward-directed: traversing a moderately disordered insulator biased by a voltage, electrons do not make loops as in the diffusive regime. They minimize the number of diffusions, describing forward-directed paths (FDP). They accumulate a phase along these paths, and the phase-averaging over a large number of paths depends in a subtle way if a magnetic flux is applied or not (see section "FDP model"below)[3]. A positive magnetoconductance is expected in disordered insulators at low temperature and low magnetic field, which results from a symmetry breaking effect, but differs from the weak localization in diffusive metals.

Secondly, considering interactions, the fact that ξ is larger than the mean distance between electrons ($\ell \simeq k_{\text{F}}^{-1}$) means a strong screening of the Coulomb interaction and of the electrostatic disordered potential. In some sense the (screened) disorder potential is smooth in this case [4]. The long range Coulomb interaction is screened between charges on different localization domains, but also inside a given localization domain. The on-site charging energy is nevertheless important, and it is tempting to treat it by analogy to a charging energy in a quantum dot. Many experiments on mesoscopic insulators show in fact Coulomb blockade oscillations near the MIT [5–8]. Any MIT in disordered systems generates spatial inhomogeneities and percolation, that has attracted great attention until recently [9]. There could be also a tendency to create electrons "lakes" because one gains energy to charge a region where there is a large local density of electrons as compared to regions with a small local density. In fact a high density electron gas is more compressive, because of exchange and correlation terms. So adding electrons (with a control gate for instance) will amplify the local fluctuations of the electron density and create the "lakes".

Interaction effects are still a very debated field near the MIT. It is outside the field of this chapter (see chapter by Zvi Ovadyahu), which is mainly concerned with interferences effects. As in the diffusive regime, interferences effects are revealed by low field magnetoconductance measurements. The next section describes the variable range hopping regime and reviews theories to explain the positive magnetoconductance seen at low temperature, including the interference mechanism amongst the forward directed paths. Then, I describe the interferences effect near the transition, when the localization length is large. We will see that the localization length itself is modified by application of a magnetic field. Finally I will discuss what kind of interaction effects can influence the magnetoconductance in insulators.

2 Variable Range Hopping

At zero temperature and small voltage, in an infinitely long insulator the conductance is strictly zero. In a thin insulator, there is conductance by direct tunneling (typically for thickness comparable to few nanometers). If dilute impurities are introduced in the gap, the dominant transport process is the resonant tunneling through these impurities [10]. At low concentration the resonant tunneling is by one impurity located just in the middle of the insulating barrier; if the concentration of impurities is increased further the resonant tunneling is by two impurities located symmetrically with respect to the middle of the barrier at a quarter of the thickness from the electrodes. Increasing even further the concentration makes the situation more complex, with resonant tunneling through series of impurity levels.

Now, at finite temperature, electrons can emit or absorb thermal excitation and change their energies. If the temperature is larger than the typical energy spacing between neighboring impurity levels Δ, electrons travel through the entire sample by hopping from neighbors to neighbors, relaxing their energy at each hop (see Fig. 1). This is the nearest neighbors (or fixed range) hopping regime. If the temperature decreases below Δ, the most favorable hop is not necessary the shortest ones, but results from the optimization between long range tunneling and thermal activation cost (Variable Range Hopping, VRH). For a long hop, there is elastic scattering by intermediate impurity levels. This problem has been solved first by Mott [11] and many aspects like the macroscopic limit [12], the dimensionnality dependence [13], the sensitivity to interactions (Efros-Schklovskii pseudogap, with exponent $1/2$ in equation 8 for the temperature dependence) [1], the competition with resonant tunneling [14], the mesoscopic

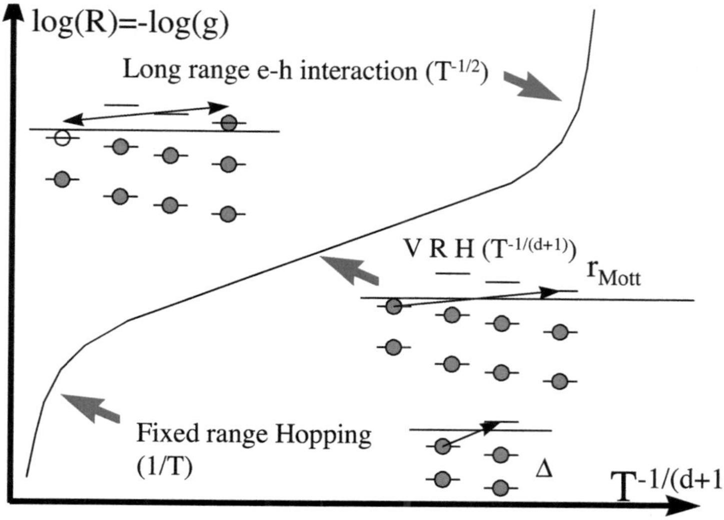

Fig. 1. Schematic temperature dependence of the resistance (or conductance) in a disordered insulator at finite temperature (d=2,3). At temperature above the mean level spacing on impurity levels Δ, carriers hop between nearest neighbors. At lower temperature, the transport is via variable range hopping, with an increasing hopping length r_{Mott} as temperature decreases (see text). At still lower temperature, the long range unscreened Coulomb interaction between the hopping electron and the hole left behind it produces a pseudo-gap at the Fermi level (Efros–Schklovskii regime)

uncoherent effects [15,16] and the interferences effects[3] have been addressed in an enormous literature. For the case of macroscopic 2D and 3D VRH without interaction, one finds:

$$g \propto \exp(-\frac{E_{\mathrm{Mott}}}{k_{\mathrm{B}}T}) = \exp(-\frac{r_{\mathrm{Mott}}}{\xi}) = \exp\left(-\left(\frac{T_0}{T}\right)^{1/1+d}\right) \tag{8}$$

with

$$k_{\mathrm{B}}T_0 = \frac{\alpha}{n(E_{\mathrm{F}})\xi^d}$$

and $\alpha \simeq 1.66$ (3D) – 14 (2D). r_{Mott} is the typical hopping length, diverging as $T^{-1/d+1}$ as temperature decreases. E_{Mott} is the typical energy mismatch for each hop, decreasing as $T^{d/d+1}$ as temperature decreases. Note that r_{Mott} replaces the total length of the sample in the expression of the tunnel conductance at zero temperature: $g \propto \mathrm{e}^{-L/\xi}$. The temperature dependence of the conductance gives T_0, i.e. the localization length ξ, once α and the density of states $n(E_{\mathrm{F}})$ are known.

3 Positive Weak Field Magnetoconductance in the Hopping Regime

In lightly doped semiconductors, a high field negative exponential magnetoconductance is generally observed and attributed to the shrinking of orbitals (so to say, an effective decrease of the Bohr radius) by the magnetic field [1]:

$$\frac{\Delta G}{G} \propto 1 - \exp(\mathcal{B}B^2) \tag{9}$$

where $\mathcal{B}$ is a temperature dependent prefactor which depends on the temperature dependence of the conductance effectively observed (VRH (equation 8) or Efros-Schklovskii). The exponential magnetoconductance comes from the high sensitivity of the conductance to the overlap between the localized wave functions. This is essentially a strong magnetic field limit, where the magnetic length $L_H = \sqrt{\hbar/eH}$ is comparable to a_{Bohr} and smaller than ℓ.

At low magnetic field and low temperature this effect is overcome by a contribution with the opposite sign observed in many materials, in the variable range hopping regime, for instance: In_2O_{3-x} [17], InP [18], GaAs:Si, [2,19,20], or GaAs 2DEG [21]. 2DEG or doped Silicon on the other hand does not show this weak field positive magnetoconductance (spin–orbit scattering (SOS) or interaction effect? We shall see later on). This positive magnetoconductance increases very much as temperature decreases, and is anisotropic for the case of $2d$ films.

Several models have been proposed to explain the weak field positive magnetoconductance.

A first one is based on magnetic moments which could exist on localized states and could be aligned by a magnetic field. This reduces the diffusion [22,23] and produces a positive magnetoconductance proportional to the square of the magnetization M: $\frac{\Delta G}{G} \propto M^2 \propto \left(\frac{B}{T}\right)^2$. But quantitative comparison with experiments gives a very large average magnetization on localized states: $\mu_{\mathrm{eff}} \simeq 10 - 100\mu_{\mathrm{B}}$.

A second model is based on the shift of the mobility edge in a 3d insulator [24,25] with magnetic field. The localization length increases (without SOS) with application of magnetic field, through the change of the mobility edge:

$$\frac{n_{\mathrm{c}}(H) - n_{\mathrm{c}}(0)}{n_{\mathrm{c}}(0)} = A\left(\frac{eHn_{\mathrm{c}}(0)^{-2/3}}{\hbar}\right)^{1/2\nu} \tag{10}$$

A is universal, does not depend on temperature but only on the symmetry class (i.e. the β parameter).[24–26]. A is negative without spin–orbit scattering (the case of spin–orbit scattering has not been treated explicitly in this model). With the help of equations (6) and (8), an exponential positive magnetoconductance is predicted in this model, which is probably not essentially different from interference models discussed hereafter when $\xi \gg \ell$.

A third model is also based on the high sensitivity of the localization length to energy. But it relies on the Zeeman splitting, which shifts the mobility edge differently for the two spin orientations. This decreases the distance between the

Fermi energy and the mobility edge for one spin orientation, giving a positive magnetoconductance (evaluated below for the exponent 1/2 in VRH due to the Efros–Schklovskii gap) [27]:

$$\frac{2\sigma(H)}{\sigma(0)} = \exp\left[\left(\frac{T_0}{T}\right)^{1/2}\left(1-\left(1-\frac{1}{2}\frac{g\mu_{\rm B}H}{E_{\rm c}-E_{\rm F}}\right)^{td/2}\right)\right] + \exp\left[\left(\frac{T_0}{T}\right)^{1/2}\left(1-\left(1+\frac{1}{2}\frac{g\mu_{\rm B}H}{E_{\rm c}-E_{\rm F}}\right)^{td/2}\right)\right] \quad (11)$$

Here t is the exponent for the localization length versus energy E: $\xi \propto (E_{\rm c}-E)^t$, $E_{\rm c}$ being the mobility edge. Because this model relies on the Zeeman energy, it predicts an isotropic magnetoconductance in films, contrarily to models based on interferences.

Two other models consider specific cases: A fourth model for the positive hopping magnetoconductance is developped by Raikh and Glazman for the case of a smooth potential [4]. Following the authors, the effect could be very large near the classical percolation threshold of the MIT.

A fifth model concerns the disorder broadened Landau levels 2DEG (in this sense it is also a large field limit), where the magnetic field increases the density of states at the Fermi level [28]. This occurs because the magnetic field decreases the overlap of the neighboring donors, and increases the spreading of the band tails. The result is a positive magnetoconductance, for small field, before the shrinking of orbitals at larger field.

A sixth model is based on quantum interferences originally developped by Nguyen Spivak and Schklovskii (NSS) [3]. A simplified version of the model, with only one intermediate scatterer has been developed by Schirmacher [29]. For one intermediate scatterer only, the magnetoconductance cannot be larger than a factor ln(2). In many experimental cases, the magnetoconductance is much larger and increases without saturation as temperature decreases. Subsequent works [30–32] on the NSS model have shown that it predicts a change of the localization length, by interferences between forward directed paths over large distance, not between loops inside the localization domain. Contrary to the second and third models, the change of the localization length is not related to the distance to the mobility edge, and its variation with the magnetic field. Because this model describes many experiments satisfactorily, let us describe it in more details. Then we will consider two last models, which are also based on a change of the localization length by an interference effect, but adapted to the regime where $\xi \gg \ell$.

The Forward Directed Paths Model

The basic hypothesis of this model is to neglect diffusion loops in the strongly localized regime. Inside a single Mott hop there is intermediate *elastic* scattering by localized levels outside the Mott energy window $E_{\rm Mott}$ around the Fermi energy. There is various scattering paths and phase accumulation along these paths. To solve the problem let us consider only the shortest paths. These paths are forward directed. Other paths are longer, and correspond to an exponentially smaller diffusion amplitude. The problem of interferences amongst forward directed paths is still very complex. For an interesting overview of this problem, as well as a detailed presentation, the reader is referred to the reference [30].

The seminal idea of the model is to start with an Anderson Hamiltonian on a lattice:

$$\mathcal{H} = \mathcal{H}_0 + \mathcal{V} = \sum_i \epsilon_i a_i^+ a_i + \sum_{i,j} V_{ij} a_i^+ a_j \tag{12}$$

with $\epsilon_i = +W$ or $\epsilon_i = -W$ with equal (or slightly different) probabilities, and $V_{ij} = V$ only for (i,j) nearest neighbors. The case of strong localization is considered: $V \ll W$. The inverse localization length is the sum of a local contribution $\xi_0^{-1} = 2\ln(\frac{\sqrt{2}W}{V})$ and a contribution coming from the interference terms, which grows as the extension of the phase coherent space, and which is calculated by the authors using the locator expansion: starting from a wave function Φ corresponding to an electron of energy E localized on an initial state, the wave function for an electron localized at the final state Ψ^+ is obtained perturbatively and iteratively as:

$$|\Psi^+\rangle = |\Phi\rangle + \frac{1}{E - \mathcal{H}_0 + \mathrm{i}\delta}\mathcal{V}|\Phi\rangle + \frac{1}{E - \mathcal{H}_0 + \mathrm{i}\delta}\mathcal{V}\frac{1}{E - \mathcal{H}_0 + \mathrm{i}\delta}\mathcal{V}|\Phi\rangle + ... \tag{13}$$

(δ tends to zero at the end of the calculation). The conductance from the initial to the final state is proportional to the term:

$$\langle\Psi^+|\Psi^+\rangle = \langle\Phi|G(E)|\Psi^+\rangle = \sum_{\Gamma}\prod_{i_\Gamma}\frac{V\mathrm{e}^{\mathrm{i}A}}{E - \epsilon_{i_\Gamma}} \tag{14}$$

where $G(E)$ is the Green function and the sum is over all the paths Γ connecting the initial and final localized states. The forward directed paths approximation permits to considerably simplify this expression:

$$\langle\Phi|G(E)|\Psi^+\rangle = \left(\frac{V}{W}\right)^t \times \sum_{\Gamma'}\prod_{i_{\Gamma'}} V\mathrm{e}^{\mathrm{i}A}\nu_{i_{\Gamma'}} = \left(\frac{V}{W}\right)^t \times J(A,t) \tag{15}$$

with $\nu_{i_{\Gamma'}} = \epsilon_i/W$ equals 1 or -1, and the sum is now only over the forward directed paths Γ' of length t. t is the shortest length between initial and final states (in units of distance between equally spaced impurities) (see Fig. 2). In reality, at finite temperature t is the Mott hopping length. A is the Gauge potential. Paths differing by only one intermediate diffusion accumulate a phase difference $\Delta A = \phi/\phi_0$ due to a magnetic flux ϕ per "plaquette" (ϕ_0 is the quantum of flux). All the information on interferences is contained in $J(A,t)$, which has been calculated by the authors of reference [30]. If $A = 0$ (no magnetic field), $J(A,t)$ is a large sum of $+1$ or -1 terms. With magnetic flux, $J(A,t)$ is a large sum of complex numbers. After averaging the distribution of $J(A,t)$ over disorder (i.e. over the realizations of $\nu_{i_{\Gamma'}}$'s), the mean magnetoconductance is obtained.

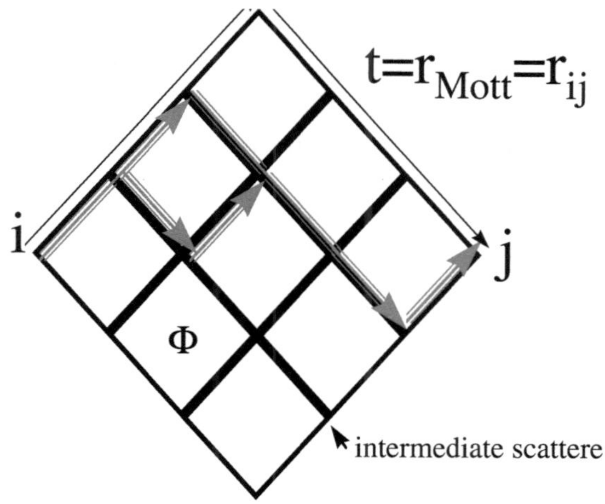

Fig. 2. The Forward Directed Paths Model: Between the initial and final states, which are separated by the Mott hopping length r_{Mott} in the variable range hopping regime, there is intermediate elastic scattering by impurity levels outside the Mott energy window E_{Mott} from the Fermi level. The model starts with an Anderson Hamiltonian on a lattice, as described in the text. Only the interferences amongst the shortest paths of length t are considered, with or without a magnetic flux Φ per plaquette. Two such paths are indicated on the schema (see [30])

Main results are summarized below: in the case of no spin–orbit scattering, there is a strong positive magnetoconductance which is indeed a (non-universal) increase of the localization length: as the Mott hopping length (i.e. t) increases at low temperature, the exponential positive magnetoconductance becomes larger. The typical magnetic field for the effect depends in a non universal way on the Mott hopping length (so on the temperature). Also the variance of the conductance distribution, which is log-normal (its logarithm is Gaussian distributed), decreases with the magnetic field. A strong spin–orbit scattering, which randomizes the spin orientation at each intermediate scattering produces an increase of the localization length in zero magnetic field, as compared to the case of no spin–orbit. But application of a magnetic field does not change further the localization length. Instead a small, non-exponential positive magnetoconductance is obtained [30].

4 Interference Effects Inside the Localization Domain. Time Reversal Symmetry Sensitivity of the Localization Length

The preceding section deals with the case of strong localization, where the localization length is less or comparable to the distance between scatterers. If this is not the case, one cannot neglect backward scattering i.e. diffusion loops even in the localized regime, and the forward directed paths approximation fails.

In 2D or 3D, as said before, the effect of a magnetic flux on the (large) localization length is an open problem. Let us consider now two models which address the interference problem in the limit $\xi \gg \ell$.

Bouchaud et al. [33] have proposed an heuristic approach of the problem which is applicable to d=1,2,3. The idea of the model is to modify the diffusion equation to introduce in a self-consistent way an exponential cutoff $\mathrm{e}^{-t/\tau}$ with $\tau = \tau_{\mathrm{e}}(\xi/\ell)^2$ for long diffusion loops of time t (τ_{e} is the elastic diffusion time).

The model permits to extend the well-known semiclassical diffusion picture from the metallic diffusive regime to the localized regime. It gives physical insight on possible effects of time reversal and spin reversal symmetry in the localized regime, by extension of the popular weak-localization theory. The following results are obtained: At d=1, ξ is doubled (without SOS). By analogy with weak localization, a magnetic flux suppresses the enhancement of the backscattering, due to time reversal conjugated diffusion loops. The wave function amplitude is halved at the center of the localization domain, and by normalization argument, is extended by a factor of two in space. At d=2:

$$\xi(B)/\ell \simeq (\tanh(b)/2)^{1/2} \times (\xi(0)/\ell)^2 \tag{16}$$

where $b = B/B^* = \frac{B\ell^2 e}{2\hbar}$. The localization length increases very much with magnetic field, with a non-universal behaviour which depends on $\frac{\xi(0)}{\ell}$. At d=3, the ratio $\frac{\xi(B)}{\xi(0)}$ could diverge at the mobility edge or be of order 1 in the strongly insulating regime. The case of spin–orbit scattering has been considered in reference [34]: strong SOS changes the sign of the magnetoconductance, by analogy with the weak localization case.

Results in 1D obtained by this semiclassical model are in agreement with those obtained by Random Matrices Theory approaches, at least for the case without spin–orbit scattering (4). Random Matrices Theory is a very powerful tool to investigate the transmission through a complex disordered metal or insulator, represented by its complex transmission matrix [2]. It is particularly adapted to the quasi 1D case. The fundamental result of (4) – the β dependence of ξ – has been obtained with this formalism, which addresses interference effects in disordered insulators in a rigorous but global way, emphasizing the role of symmetries obeyed by the transmission matrix, without semiclassical arguments. Equation 4 predicts a doubling of the localization length when the time reversal symmetry is broken (for large $\mathcal{N}$ without spin–orbit): in fact without magnetic flux $\beta = 1$, whereas when a quantum of flux passes through the localization domain, $\beta = 2$. This time reversal symmetry breaking effect is on the basis of the weak localization in the diffusive regime too. It has received two experimental illustrations in the insulating regime of GaAs:Si wires [35,36]. In a series of long GaAs wires, a geometry which permits to average out mesoscopic fluctuations, Gershenson et al. [35] observed a mean magnetoconductance in the variable range hopping regime fully compatible with equation (4) in the case of a relatively large number $\mathcal{N}$ of channels, i.e. a doubling of ξ. The typical magnetic field to produce this doubling is such that a quantum of flux passes through the localization domain $\xi \times W$, where $W \leq \xi$ is the width of the wire (perpendicularly to the field). This temperature independent field is in agreement with the predictions based on interference effects inside the localization domain and in contrast with the forward directed paths models.

In a mesoscopic GaAs wire with a gate, which permits to vary continuously $\mathcal{N}$ between 1 and 10, Poirier et al. [36] obtain a magnetoconductance again compatible with equation 4, after averaging numerically the observed mesoscopic fluctuations: for $\mathcal{N} \simeq 1$, the mean positive magnetoconductance is small. In this regime, forward directed paths approaches can equally describe the small magnetoconductance because ξ and ℓ are comparable. But for larger $\mathcal{N}$, both ξ and the positive magnetoconductance increase very much, in quantitative accordance with equation 4. The localization length becomes comparable to the phase breaking length for $\mathcal{N} \simeq 10$, and a transition to a diffusive regime is observed for larger $\mathcal{N}$ (finite size MIT). It would be very interesting to perform such an experiment in a sample with strong spin–orbit scattering.

5 Interaction and Spin Effects on the Magnetoconductance

Before concluding it is necessary to consider what could be the effect of interaction and correlations on the magnetoconductance of disordered insulators. To clarify this question let us concentrate on localized states which accommodate 0,1 or 2 electrons only, and consider the intra-site Hubbard term U. The effect of U is that energy levels which are singly occupied at energy of order U below the Fermi energy, participate to transport. In fact if a second electron is added on the same orbital (with opposite spin), the one electron level is lifted by U and aligned with the Fermi level. This model supposes that the mean spacing between single electron levels Δ is of order (or larger than) U, such that localized states could be charged (with two electrons).

Two effects appear, which give opposite signs for the magnetoconductance: First, the doubly occupied state corresponds to a wave function with a larger localization length [37]. Moreover, generally speaking, the overlap between localized wave functions could depend on the spin of the carriers. For instance, if one considers two electrons interacting via the Hubbard term on a lattice with random potential, wave functions for electrons with parallel spins are less localized than electrons with antiparallel spins [38]. This comes from a gain in exchange energy. This effect could produce a positive magnetoconductance, because the field tends to align the spins by the Zeeman effect[38]. The spin-dependent localization length effect has not received yet a conclusive experimental evidence.

If one neglects any spin dependence of the localization length, a second effect of the magnetic field is a negative isotropic magnetoconductance, associated with the Zeeman effect. It has been treated in the VRH case by Kamimura [37], and in the resonant tunneling case by Glazman and Matveev [39]. As said before, transport through doubly occupied states occurs with electrons carrying antiparallel spins. By polarizing the spins with a magnetic field, transport through these states is forbidden and the conductance decreases. The associated magnetoconductance is isotropic, negative and saturates above a magnetic field which depends essentially on $x = \mu_B H / k_B T$. In the case of resonant tunneling through one impurity, one has [39]:

$$\frac{G(B)}{G(0)} = \frac{e^{2x} \ln(1 + e^{-2x}) + e^{-2x} \ln(1 + e^{2x})}{2 \ln 2} \tag{17}$$

This behaviour has been observed in experiments on very thin disordered Silicon barriers [40].

6 Conclusions

In summary, it is clear that interferences effects are much less known in disordered insulators than in disordered metals. Roles of the magnetic flux and of the spin–orbit scattering are not yet definitely understood in these materials. Many mesoscopic aspects, which have been very well studied in diffusive metals have not been yet investigated in disordered insulators. Nevertheless these materials are widely represented in solid state physics and sometimes extremely important in applications. A considerable difficulty arising in these materials is the competition between disorder and interaction, which is far from being understood nowadays.

References

1. B.I. Shklovskii, A.L. Efros: *Electronic Properties of Doped Semiconductors*, Springer series in Solid State Sciences 45 (Springer-Verlag 1984)
2. J.L. Pichard: p397 'Random transfer matrix theory and conductance fluctuations' in *'Quantum coherence in mesoscopic systems'*, ed. B. Kramer, NATO ASI Series, Plenum Series B: Physics Vol 254 (1991) J.L. Pichard et al.: Phys. Rev. Lett. 65, 1812 (1990) J.L. Pichard, M. Sanquer: Physica A **167**, 66 (1990)
3. V.L. Nguyen, B.Z. Spivak, B.I. Shklovskii: JETP Lett. **41**, 42 (1985)
4. M.E. Raikh, L.I. Glazman: Phys. Rev. Lett. **75**, 128 (1995)
5. J.H.F. Scott-Thomas et al.: Phys. Rev. Lett. **62**, 583 (1989)
6. H. Van Houten et al.: 'Coulomb blockade oscillations in semiconducting structures' in *Single Charge Tunneling*, H. Grabert, M.H. Devoret eds. Plenum, New York (1991)
7. H. Ishikuro, T. Hiramoto: Applied Phys. Lett. **71**, 3691 (1997)
8. M. Sanquer et al.: Phys. Rev. B **61**, 7249 (2000)
9. for instance: S. Ilani et al.: Phys. Rev. Lett. **84**, 3133 (2000) and reference herein
10. A.I. Larkin, K.A. Matveev: JETP **66**, 580 (1987)
11. N.F. Mott: J. Non-Cryst. Solids **1**, 1 (1968)
12. V. Ambegaokar, B.I. Halperin, J.S. Langer: Phys. Rev. B **4**, 2612 (1971)
13. J. Kurkijarvii: Phys. Rev. B **8**, 922 (1973)
14. M.Y. Azbel, A. Harstein, D.P. DiVincenzo: Phys. Rev. Lett. **52**, 1641 (1984)
15. P.A. Lee: Phys. Rev. Lett. **53**, 2042 (1984)
16. M.E. Raikh, I.M. Ruzin: Sov. Phys. JETP **68**, 642 (1989)

17. F.P. Milliken, Z. Ovadyahu: Phys. Rev. Lett. **65**, 911 (1990)
18. J.P. Spriet, G. Biskupski, H. Dubois, A. Briggs: Phil. Mag. **B54**, L95 (1986)
19. F. Tremblay et al.: Phys. Rev. B **39**, 8059 (1989)
20. R. Rentzsch, K.J. Friedland, A.N. Ionov: Phys. Stat. Sol b **146**, 199 (1988)
21. F.W. Van Keuls et al.: Phys. Rev. B **56**, 1161 (1997)
22. Y. Toyozawa: J. Phys. Soc. Jpn. **17**, 986 (1962)
23. K. Yosida: Phys. Rev. **107**, 396 (1957)
24. B.L. Altshuler, A.G. Aronov, D.E. Khmel'nitskii: JETP Letters **36**,195 (1982)
25. D.E. Khmelnitskii, A.I. Larkin: Solid State Comm. **39**, 1069 (1981)
26. B. Shapiro: Phil. Mag **B50**, 241 (1984)
27. H. Fukuyama, K. Yosida: J. Phys. Soc. Jpn. **46**, 102 (1979)
28. M.E. Raikh et al.: Phys. Rev. B **45**, 6015 (1992)
29. W. Schirmacher: Phys. Rev. B **41**, 2461 (1990)
30. E. Medina, M. Kardar: Phys. Rev. B **46**, 9984 (1992)
31. H.L. Zhao and al.: Phys. Rev. B **44**, 10760 (1991)
32. U. Sivan, O. Entin-Wohlman, Y. Imry: Phys. Rev. Lett. **60**, 1566 (1988)
33. J.P. Bouchaud: J. Phys. I France **1**, 985 (1991)
34. J.P. Bouchaud, D. Sornette: Europhys. Lett. **17**, 721 (1992)
35. M.E. Gershenson et al.: Phys. Rev. Lett. **79**, 725 (1997)
36. W. Poirier, D. Mailly, M. Sanquer: Phys. Rev. B **59**, 10856 (1999)
37. H. Kamimura: Prog. Theoret. Physics Suppl. **72**, 206 (1982)
38. M. Eto: Phys. Rev. B **48**, 4933 (1993)
39. L.I. Glazman, K.A. Matveev: JETP Lett. **48**, 445 (1988)
40. D. Ephron, Y. Xu, M.R. Beasley: Phys. Rev. Lett. **69**, 3112 (1992)

High Resolution NMR of Biomolecules

J.H. Prestegard

Complex Carbohydrate Research Center, University of Georgia, Athens, GA, USA

Abstract. New demands for structural information on biomolecules, particularly proteins, have been created by recent advances in sequencing of genomes, including that of humans. While X-ray crystallography remains the primary source of structural information on biomolecules, nuclear magnetic resonance (NMR) methods provide a complementary approach that is applicable in solution and doesn't require crystallization. Utilization of these methods is greatly facilitated by the production of higher field magnets. Higher fields have led not only to better resolution and sensitivity, but have spawned new approaches that include transverse relaxation optimized spectroscopy and the use of residual dipolar couplings.

Over the last two decades the availability of magnets with homogeneous and stable fields of increasing strength has had a tremendous impact on application of nuclear magnetic resonance (NMR) Spectroscopy to problems in structural biology. In the area of protein structure determination by high resolution methods, the near factor of two increase in magnetic field strength has led to increases in resolution and sensitivity that validate the predicted B_0 and $B_0^{7/4}$ field dependence of these important parameters. When combined with other innovations in experimental protocols, these increases have moved the limits of applicability of NMR to biomolecular structure determination from less than 10 kDa (90 residues for proteins) to more than 40 kDa (360 residues for proteins) [1]. A move to even higher fields will help with resolution and sensitivity in much the same ways as it has in the past (21 T magnets for NMR are just becoming available), however, there are also indications that entirely new types of structural data and new methods of data acquisition will be accessed at higher fields. This is an opportune time to review the impact of high magnetic field technology on NMR spectroscopy of biomolecules and to discuss some of the prospects for the future.

There are currently nearly 2500 structures in the protein data bank (PDB) that have been determined by NMR [2]. This may seem a large number, but the pressures to increase this further are enormous. The pressure for expansion of efforts in protein structure determination comes partly from the genome sequencing activities of recent years. Sequencing of various genomes, including the human genome, has made hundreds of thousands of gene products (proteins that the sequences code) new targets for investigation [3,4]. Based on structures of these proteins, the development of new drugs, new biosynthetic processes, new materials, and new fundamental understanding of natural processes, all seem easily within grasp. For most organisms, however, only half of the sequenced genes bear a recognizable relationship to ones with previously known function, making even the choice of a target for investigation complex. One solution is the determination of structures on a massive scale; structural genomics efforts have sprung up world wide [5]. It is now well demonstrated that knowledge of structure can greatly facilitate identification of function [6].

X-ray crystallography will produce most of the needed structures (the 2500 NMR structures in the PDB represent only about 15 % of the total number of structures), but NMR has some important complementary assets; it doesn't require that a protein crystallize, as do X-ray methods, and proteins can be analyzed in an aqueous solution medium, which for most proteins, better approximates the environment in which they function. As more individual protein structures are characterized, applicability in an aqueous medium will also become more important as protein–protein interactions are investigated. This doesn't mean NMR will become the method of choice. Even with the upper limit for molecular weight above 40 kDa, the majority of proteins and protein–protein complexes will be inaccessible to NMR. This puts new ways of studying larger systems at a premium.

In order to understand the limitations of NMR based structure determination methodology it is important to review underlying principles of currently applied experiments. Virtually all of the 2500 NMR structures in the PDB have been determined using data from Nuclear Overhauser Effects (NOEs). In these experiments a resonance belonging to a spin nucleus with a high magnetogyric ratio (particular protons in the protein) is perturbed. This is usually done periodically using a pair of radio frequency (rf) pulses. The first causes magnetization associated with the resonance to rotate by 90° from the z axis (parallel to the laboratory field) to a transverse axis. During the delay between the pulses magnetization precesses in the transverse plane at frequencies characteristic of the chemical environment of the proton (its chemical shift). Only when the magnetization is perpendicular to the applied rf field of the second pulse is the magnetization fully perturbed by complete inversion. The delay is incremented so that the second of the two 90° pulses inverts magnetization at the characteristic frequency of the proton's chemical shift. A fixed delay follows the pair of rf pulses during which perturbed magnetization partially recovers to its equilibrium state. It does so by exchange of magnetization with proximal spins. The efficiency of this relaxation process depends on $1/r^6$, where r is the distance between the pair of spins. Observation of precessing magnetization at each incremented delay and subsequent Fourier transformation in two dimensions leads to a two dimensional data set in which the intensity of cross peaks connecting resonances in the two dimensions give distance restraints that can be used for structure determination.

Since a typical protein can contain a thousand or more protons, resolving all cross peaks can be a challenge. This is met partly by operation at higher fields, as frequencies of chemical shift differences scale linearly with field, while most resonance widths are independent of field. It is also met by expansion of spectra to three and even four dimensions. Sites connected by chemical bonds to protons of interest are enriched with other magnetic isotopes, such as ^{15}N at

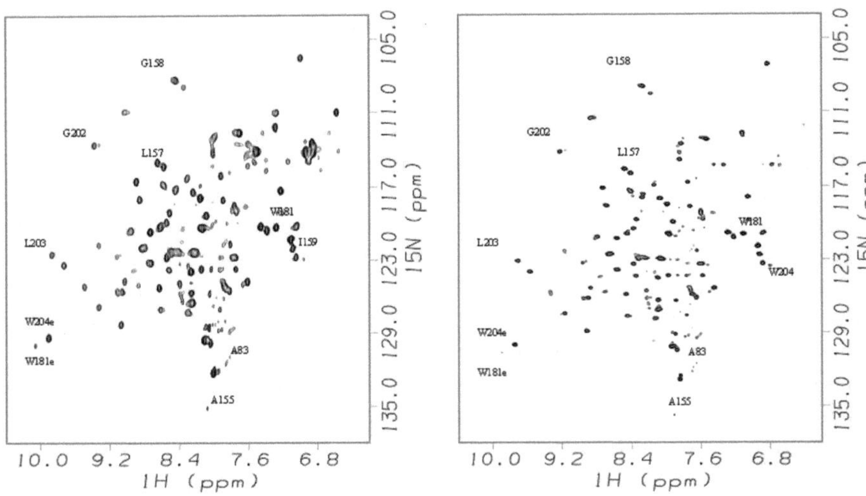

Fig. 1. HSQC (*left*) and TROSY (*right*) spectra of an $^{15}N,^{2}H$, labeled sample of mannose binding protein (53 kDa trimer). The sample is approximately 0.3 mM in aqueous buffer at 25°C and each required about 6 hr of data acquisition. Note the resolution improvement in the TROSY spectrum

backbone amide sites; this allows magnetization of protons to be modulated by the chemical shifts of these new sites before detection. The last two dimensions detected form planes that are referred to as HSQC (heteronuclear single quantum coherence) spectra. These give very simple foot prints with approximately one cross peak for each residue of the protein backbone (only proline has no ^{15}N–^{1}H pair and only arginine, glutamine, and asparagines have extra amide pairs). The NOE crosspeaks are then on columns rising from each of these HSQC cross peaks in a third dimension. An example of an HSQC spectrum for an 17.5 kDa protein that forms a 53 kDa trimer is shown in the left panel of Fig. 1. Given that 5–10 NOE cross peaks may be on each column rising from the HSQC plane, one can appreciate that a protein of this size is near the limit of applicability for structure determination by the NOE experiments described.

Recently a method referred to as transverse relaxation optimized spectroscopy (TROSY) has been introduced that could push the application of many experiments used in biomolecule structure determination beyond 100 kDa [7]. This spectroscopy depends on a field dependent relaxation interference effect that appears to optimize somewhat over 900 MHz. The origin is illustrated in Fig. 2 for an N–H amide bond. The ^{15}N spin relaxes via fluctuations in the magnetic fields that come from two sources; one is shielding of the nucleus from the externally applied magnetic field by the local electronic structure (a field dependent chemical shift anisotropy (CSA) contribution), and the other is a dipolar field from the moment of the directly bonded proton (a field independent dipolar contribution). Both fluctuate as the molecule tumbles. The effects of the CSA

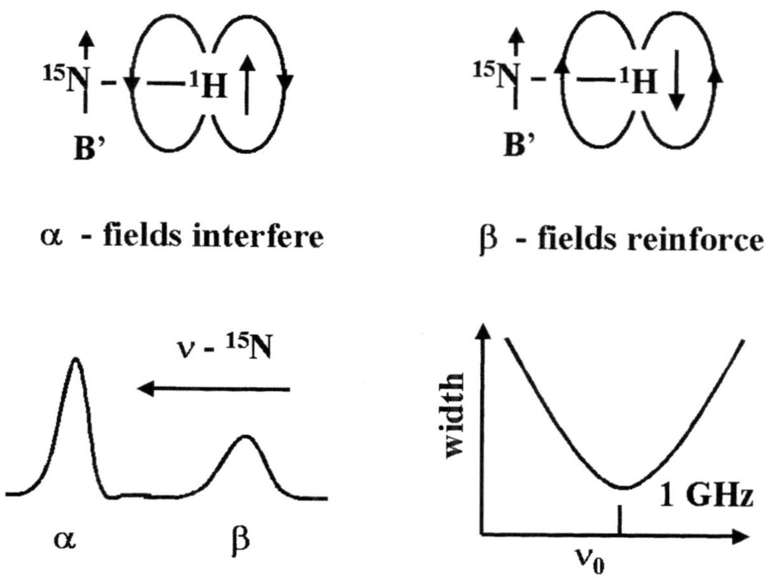

Fig. 2. Origin of the TROSY effect. The upper quadrants illustrate additivity of ^{15}N chemical shift anisotropy effects and dipole field effects at amide nitrogens with bonded protons in α and β spin states respectively. The reinforcement and interference lead to different line widths for ^{15}N resonances with attached protons in the α and β states as illustrated in the lower left quadrant. Since the chemical shift anisotropy effect is field dependent, maximum interference and minimum width for the narrower line occurs at a specific field strength as illustrated in the lower right quadrant

contribution are always the same, but the spin proton can be in two states (α and β), having opposite projections of the proton's magnetic moment on the external magnetic field. Providing that the principal axis of the CSA tensor is approximately parallel to the N–H bond, the dipolar effect will add to the CSA effect in one spin state and subtract from the CSA effect in the other spin state. Although normally HSQC spectra use decoupling strategies to remove splittings of the cross peaks in both dimensions, the two states can be observed as separate crosspeaks in an HSQC spectrum if the decoupling is removed, one cross peak for the α state and one for the β state. In TROSY, only the line where effects cancel is collected, resulting in a spectrum with much attenuated spin relaxation and, hence, narrower lines. Because the narrowing of lines adds to the normal increases in resolution obtained by operating at higher magnetic fields, significantly improved resolution is obtained. A comparison of a TROSY spectrum to an HSQC spectrum is given in Fig. 1B. There is a clear improvement in resolution of the cross peaks coming from ^{15}N–^{1}H pairs in our 53 kDa example.

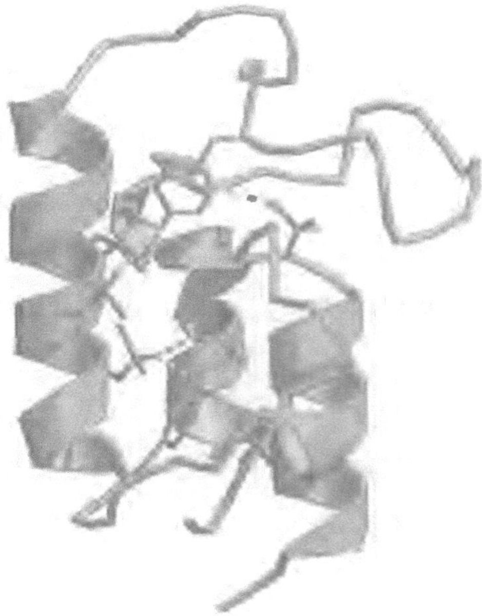

Fig. 3. NOE based NMR structures of proteins depend on short range interactions. Illustrated is the NMR structure of acyl carrier protein, a protein of 9000 Da. NOE derived constraints most important in positioning the three alpha helices are those between side chain protons in the interior of the protein (*dotted lines*)

Despite the success of NOE based structure determination strategies, there are limitations. The $1/r^6$ distance dependence is very steep. This means that an easily detected NOE for a pair of protons at 3 Å will be reduced by a factor of 64 at 6 Å and will be barely detectable above spectral noise. This is unfortunate because it is the long range NOEs (those between protons on non-sequential residues) that are most useful in defining a protein structure. This is illustrated in Fig. 3 that shows close approach of pairs of protons at the ends of residue side chains in the interior of a small protein, acyl carrier protein. The implications are that resonances for all protons, including those at the tips of side chains, must be assigned before NOEs can be interpreted. This means a lot of effort, because assignment strategies are keyed to first passing magnetization from site to site along the polypeptide backbone, before working out to the side chains. For cases where positioning of atoms in the protein backbone is of primary interest, it would be very useful to have an alternative source of structural information that did not depend on close approach of sequentially separated elements. Such is the case in structural genomics applications where a protein fold library would be the basis for building new structures by homology modeling [5]. It is also the case in many drug discovery strategies where perturbation of N–H backbone cross peaks in HSQC spectra is a primary source of screening data [8].

One source of distance independent structural information is residual dipolar coupling [9]. The origin of this coupling is actually the same magnetic dipole interaction that gives rise to NOEs and the TROSY effect, except that we want to examine the actual contribution to stationary state energy levels, and resonance frequencies corresponding to transitions between energy levels, rather than the fluctuations that lead to relaxation toward equilibrium. The equation governing dipole–dipole contributions to the splitting of a resonance (in Hz), is given as follows:

$$D_{ij}^{\rm res} = -\left(\frac{\mu_0}{4\pi}\right)\frac{\gamma_i\gamma_j h}{2\pi^2 r_{ij}^3}\frac{\langle 3\cos^2\theta - 1\rangle}{2} \tag{1}$$

Here r_{ij} is the internuclear separation, γ_i,γ_j, are the magnetogyric ratios of the interacting nuclei, and θ is the angle between the internuclear vector and the magnetic field. We regard the internuclear distance as known in this application, and our structural information comes from θ. The problem is that our timescale of observation is relatively slow and we see only the average of $(3\cos^2\theta - 1)$. In a solution where orientations are sampled uniformly (to a first approximation), the average actually goes to zero and the contributions to splittings of resonances are not usually observed.

In practice residual dipolar coupling experiments employ field orientable liquid crystal media to effect departure from isotropic sampling [10]. A medium showing reasonable compatibility with soluble proteins is a bicelle medium, which is composed of discoidal fragments of lipid bilayer membranes dispersed in an aqueous buffer. Proteins adopt low levels of order (less than one part in 1000) largely through collisions with the bicelles. An example of the type of splitting achieved is shown in Fig. 4. The spectra are pieces of HSQC spectra in ordered and isotropic states, in which the normal decoupling of protons was not done in the vertical dimension. The differences in splittings, which correspond to dipolar contributions, provide easily collected structural information.

The conversion of such contributions to splittings into a structure can take many routes [11,12]. However, analysis in terms of an order tensor along lines practiced in liquid crystal NMR proves particularly useful. In this case (1) is rewritten in terms of order parameters, S_{kl}, and direction cosines that relate an interaction vector to an arbitrarily chosen molecular frame, yielding

$$D_{ij}^{\rm res} = -\left(\frac{\mu_0}{4\pi}\right)\frac{\gamma_i\gamma_j h}{2\pi^2 r_{ij}^3}\sum_{kl} S_{kl}\cos(\alpha_k)\cos(\alpha_l). \tag{2}$$

Indices k, l run over x, y, z, giving nine order parameters that can be arranged in an order matrix. Only five of these parameters are independent since the matrix is symmetric and traceless. So, measurement of five or more splittings in the frame of a molecular fragment can lead to an order matrix determination specifically for one fragment. Diagonalization of the order matrix produces a picture of how the principal ordering frame looks from each fragment. Since fragments in a single rigid molecule must share the same ordering frame, a probable molecular structure can be assembled by rotating fragments to superimpose principal alignment frames. Some complementary data is required to remove degeneracies

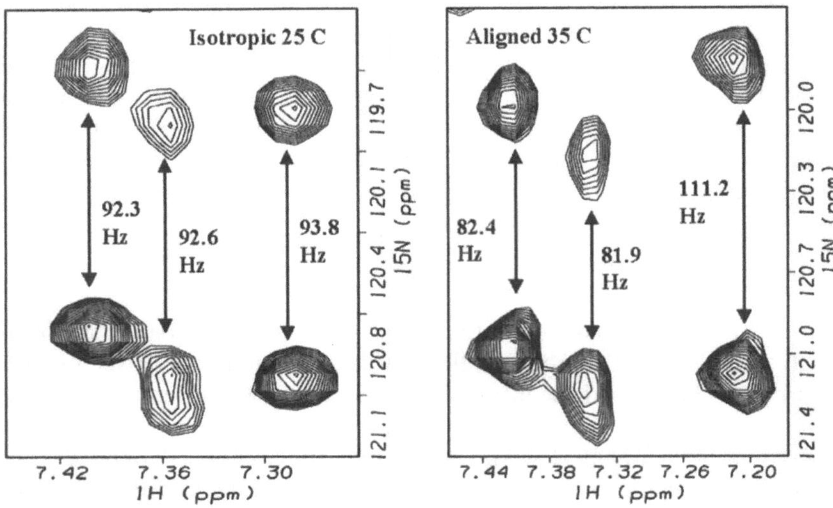

Fig. 4. Dipolar couplings in mannose binding protein. The data were acquired on a 0.3 mM sample in an aqueous dispersion of bicelles in approximately 12 hr. Temperature changes from 25°C to 35°C cause a transition from an isotropic to aligned phase. Differences in couplings in the two panels correspond to residual dipolar couplings

associated with the insensitivity of dipolar data to axis inversion and translation, however, there are now several examples of assembling protein structures largely on the basis of residual dipolar data. Some have taken fragments to be individual peptide planes of the protein [13,14], and some have taken fragments to be secondary structure elements [15]. There have also been some applications to assembly of domains in multiple domain proteins in which domains of known structure are taken to be the molecular fragments [16,17].

A special role that high field magnets may play in the above applications has not yet been fully exploited. Liquid crystal media are not always compatible with proteins targeted for study. Some proteins stick to lipid bilayers, for example, leading to complex, strongly coupled spectra and unacceptably broad lines. A simpler practice may be to rely on the inherent tendency of molecules to order in a magnetic field. All molecules have a magnetic susceptibility that leads to an induced magnetic moment when the molecule is placed in a magnetic field. These susceptibilities are seldom isotropic. So, moments of different size arise for different molecular orientations. These moments then have different energies of interaction with the external magnetic field and distributions of orientation are not uniform, but are governed by a Boltzmann distribution. Normally the departure from a uniform distribution is very small, and dipolar contributions to splittings are unobservable (we actually rely on the cooperativity of liquid crystals to scale up the departures in the applications previously described). However, for field oriented isolated molecules, the principal order parameters in (2) can be replaced by terms that depend both on the asymmetry of the susceptibility tensor ($\Delta\chi$, for example), and on magnetic field (B_0) squared (Eq. (3) for S_{zz}); this implies that splittings will become observable at sufficiently high magnetic fields.

$$S_{zz} = \Delta\chi \left[\frac{B_0^2}{15\mu_0 kT} \right] . \tag{3}$$

The first application of residual dipolar couplings to a protein was in fact done with the highest field high resolution magnet available at the time (17.5 T) and a paramagnetic protein (cyanometmyoglobin) that had a particularly large anisotropic magnetic susceptibility [18]. There have been applications to other paramagnetic proteins in the meantime [19] and even some to diamagnetic proteins and nucleic acids [20]. The dependence on field strength squared means that a move to 21.1 T magnets will make work on a wider variety of biomolecules feasible. There may also be benefits in that the actual level and direction of orientation now relates to a molecular susceptibility tensor and the electronic properties of the molecule under study [21].

In summary, the availability of high magnetic fields is having, and will continue to have, a large impact on studies of biomolecular systems. This comes about for a variety of reasons, that include well-established improvements of resolution and sensitivity with magnetic field strength, but also new field dependent phenomena, such as TROSY and residual dipolar couplings. The hope is that a move to even higher fields will uncover even more phenomena that will be valuable in satisfying new demands for structural and functional data on biomoleucles.

Acknowledgements

This work was supported by grants from the National Science Foundation, MCB9726344, and the National Institutes of Health, GM33225. I thank Drs. Junfeng Wang and Nitin U. Jain for their assistance with the manuscript.

References

1. G. Wider, K. Wuthrich: Current Opinion in Structural Biology **9**, 594 (1999)
2. H.M. Berman et al.: Nucleic Acids Research **28**, 235 (2000)
3. J.C. Venter et al.: Science **291**, 1304 (2001)
4. E.S. Lander et al.: Nature **409**, 860 (2001)
5. S.K. Burley: Nature Structural Biology, November, 932 (2000)
6. J. Skolnick, J.S. Fetrow, A. Kolinski: Nature Biotechnology **18**, 283 (2000)
7. R. Riek, K. Pervushin, K. Wuthrich: Trends in Biochemical Sciences **25**, 462 (2000)
8. P.J. Hajduk, R.P. Meadows, S.W. Fesik: Quarterly Reviews of Biophysics **32**, 211 (1999)
9. J.H. Prestegard, H.M. Al-Hashimi, J.R. Tolman: Quarterly Reviews of Biophysics **33**, 0000 (2001)
10. N. Tjandra, A. Bax: Science **278**, 1111 (1997)

11. G.M. Clore, M.R. Starich, C.A. Bewley, M.L. Cai, J. Kuszewski: J. Am. Chem. Soc. **121**, 6513 (1999)
12. J.A. Losonczi, M. Andrec, M.W.F. Fischer, J.H. Prestegard: J. Mag. Reson. **138**, 334 (1999)
13. F. Tian, H. Valafar, J.H. Prestegard: J. Am. Chem. Soc. **123**, 0000 (2001)
14. M. Zweckstetter, A. Bax: J. Am. Chem. Soc. **123**, 9490 (2001)
15. C.A. Fowler, F. Tian, H.M. Al-Hashimi, J.H. Prestegard: J. Mol. Biol. **304**, 447 (2000)
16. M.W.F. Fischer, J.A. Losonczi, J.L. Weaver, J.H. Prestegard: Biochemistry **38**, 9013 (1999)
17. C.A. Bewley, G.M. Clore: J. Am. Chem. Soc. **122**, 6009 (2000)
18. J.R. Tolman, J.M. Flanagan, M.A. Kennedy, J.H. Prestegard: Proceedings of the National Academy of Sciences of the United States of America **92**, 9279 (1995)
19. I. Bertini et al.: J. Biomolecular NMR **21**, 85 (2001)
20. N. Tjandra, S. Grzesiek, A. Bax: J. Am. Chem. Soc. **118**, 6264 (1996)
21. B.F. Volkman et al.: J. Am. Chem. Soc. **121**, 4677 (1999)

High-Resolution Solid-State NMR

Dominique Massiot

Centre de Recherches sur les Matériaux à Haute Température, CNRS,
1D, Avenue de la Recherche Scientifique, 45071 Orléans cedex 2, France

Abstract. High-resolution solid state NMR and its application to the characterization of solid state inorganic, organic, or hybrid materials, is undergoing rapid developments on both methodological and hardware points of view. Starting from a general presentation, we describe different methods that give access to the high resolution spectra; we then give examples of new methods that enable more detailed structural description through the selective reintroduction of anisotropic interactions (quadrupolar, chemical shift anisotropy, dipolar and scalar couplings). A special emphasis is given to the recent improvements obtained at high, very-high and ultra-high magnetic fields.

1 Introduction

During the last decades high-resolution solid-state nuclear magnetic resonance (NMR) spectrometry has become a widely used tool for the characterization of dielectric and diamagnetic materials. This is mainly linked to its capability of characterizing the local environment of the observed nuclei (selected by their Larmor frequency) while maintaining some resolution, even in disordered, amorphous or glassy materials. In that respect NMR is a rather unique technique providing both isotropic and anisotropic characterization of the local environment for typically light nuclei (H, D, Li, Na, Si, P, Al, O), even in powdered samples.

2 Solid-State NMR: A General Presentation

Dealing with high-resolution solid-state NMR techniques for dielectric and diamagnetic powdered materials, we shall always consider a dominant Zeeman interaction, induced by an homogeneous applied magnetic field H_0. All other interactions will be considered (and accounted for) as perturbations. The power of solid-state NMR comes from its ability, not only to provide the possibility of acquiring resolved high resolution spectra, but also to enable the description of more complex experiments combining the coherent manipulation of multi-spin systems both in real space (magic angle spinning (MAS) of the sample) and in spin space (radio-frequency pulses). Numerous nuclei posses a nuclear spin ($I > 0$) are thus potentially observable by NMR, their respective sensitivity will be a function of their gyro-magnetic ratios, their natural abundances and the strength of the perturbing interactions. The NMR active nuclei can be separated in two different classes: the $I = 1/2$ nuclei, a two level system with one transition,

Table 1. The different interactions relevant to high-resolution solid-state NMR of diamagnetic and dielectric solids

Interaction (typical magnitude)	Physical background	Information (field dependence)
Scalar or J coupling (100s of Hz)	Coupling through binding electrons	Chemical bond (field independent)
Dipolar coupling (up to several kHz)	Interaction between neighboring spins	Distances (field independent)
Chemical Shift Anisotropy (up to 10s of kHz)	Shielding of principal field by electrons	Coordination number Bond angles and distances (proportional to the field)
Quadrupolar interaction $I > 1/2$ (up to several MHz or more)	Interaction between electric field gradient and nucleus quadrupolar momentum	Geometry independent at 1st order, prop. $1/B_0$ at 2nd order

and the quadrupolar nuclei ($I > 1/2$), a much more complex $2I+1$ levels system with $2I$ single quantum transitions and possible multi-quantum (MQ) transitions. After a brief description of the dominant interactions we shall describe the tools that are used in modern high-resolution solid-state NMR experiments.

2.1 The Dominant Interactions in High-Resolution Solid-State NMR

The different interactions that have to be considered when dealing with NMR experiments on dielectric and diamagnetic solids are presented in Fig. 1 and Table 1, together with an indication of their typical magnitude and physical meaning. Depending on their respective strength, compared to the Zeeman interaction which is in the range of tens to several hundreds of MHz (and possibly GHz), the interaction will induce spectral modifications that will be accounted for by a first order development in perturbation (scalar and dipolar couplings, or chemical shift Anisotropy) or both first and second order developments (quadrupolar interaction). Each of these interactions has different physical background and thus could provide different point of view on the structure of the sample, providing that they could be reliably measured. Modern high-resolution solid-state NMR methods are able to separate these different components using spatial (mechanical) and spin manipulations. These methods first aim at obtaining the best resolution, averaging out the anisotropic parts of the interactions (Sects. 3–5). In a second step (Sect. 5), the obtained resolved spectrum can then be correlated to another spectral signature using multiple dimension experiments which selectively reintroduce detailed isotropic or anisotropic information providing a more comprehensive characterization of the sample structure or dynamics: number and nature of first neighbors, spatial proximity at different length scales, chemical bonding, and geometry of the local environment [1–3]...

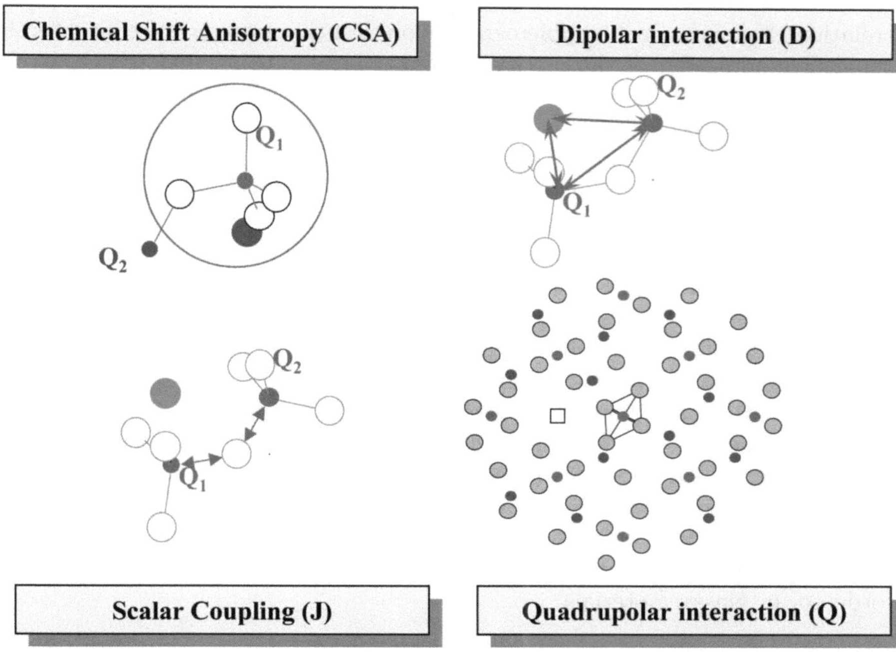

Fig. 1. Schematic representation of the four types of interactions that have to be considered in solid state NMR of diamagnetic and dielectric solids

2.2 The Tools of Modern High-Resolution Solid-State NMR

Being concerned with real materials, organic, inorganic or hybrid with different ordering states (crystalline, amorphous, glassy), we typically deal with powders as opposed to single crystals. Consequently, since the interactions mentioned in Table 1 are anisotropic, a simple NMR spectrum, acquired on a static sample is usually broad, possibly featureless and thus uninformative. This is because the spectral response of each site in a given crystallite depends upon its orientation in the principal magnetic field and the spectrum is the sum of the contributions of all the different sites for the randomly distributed crystallites. This is in contrast with the case of liquid state NMR where the Brownian motion averages out the anisotropic part of the interactions giving well resolved isotropic spectra of sharp lines. It is one of the unique features of solid-state NMR to be able to combine coherent averaging in both real space (MAS) and in multiple spin space (radio frequency pulses applied on the different coupled NMR sensitive nuclei). Consequently, there exist a very wide variety of possible experiments. One dimensional (1D) experiments only imply a single coherent evolution time, they start with the excitation of a nucleus (pulse and acquisition) but can also imply excitation of different nuclei or more complex manipulations of the observed nucleus spin system, thus providing filtered or selectively enhanced (*edited*) spectra. Multi-dimensional experiments imply several coherent evolution times, they result in correlation charts in which supplemental evolution times are acquired indirectly through a modulation of the observed signal. They can be used to selectively evidence the different interactions so as to characterize them individually while maintaining high-resolution encoding or selection. They also provide a unique opportunity of characterizing evolutions that could not be observed directly such as multiple quantum evolutions [1–3]. These methods are today rapidly growing in efficiency and can address problems ranging from local to mesoscopic scales: backbone structure of solid macromolecules, structure and polymerization of glasses, phase relations or separation.

An example of the quest for high resolution and further characterization of anisotropic interactions in solid-state NMR of materials is given in Fig. 2. We present spectra of ^{27}Al (a quadrupolar nucleus exhibiting second order quadrupolar broadening) in inorganic solid state materials ranging from crystalline to amorphous. The isotropic ^{27}Al signatures are obtained in an indirect dimension of a spectrum, which combines (correlates) magic angle spinning and multiple quantum evolution in the two dimensional MQ-MAS experiment. This experiment provides both isotropic and anisotropic characterization of the Al environments (coordination number raging from 4 to 6) in model crystalline, partly disordered, or glassy materials.

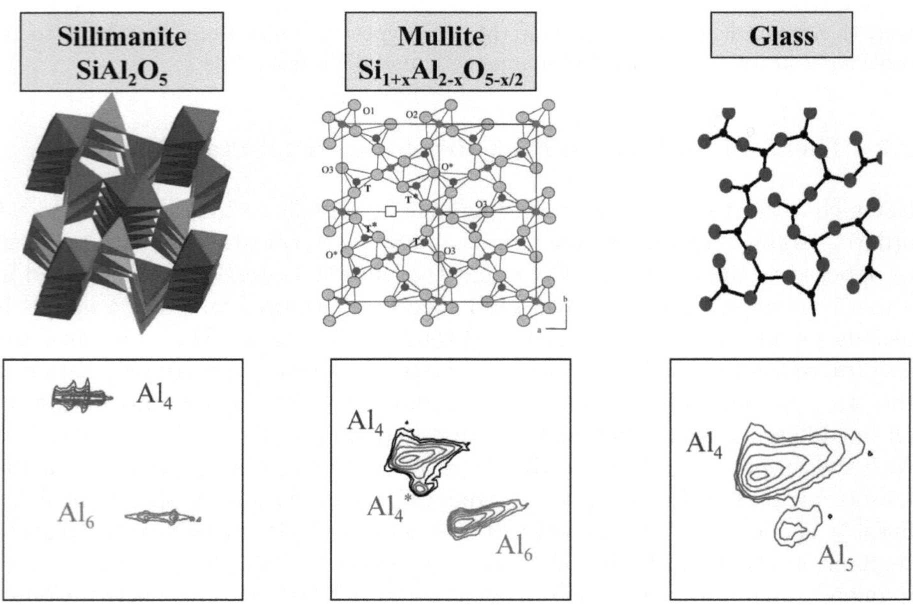

Fig. 2. High resolution ^{27}Al MQ-MAS NMR spectra of three alumino-silicates: a crystalline compound (sillimanite), a partly disordered crystalline compound (mullite) and a glass. Each sample shows different types of Al environments that can be evidenced using a MQ-MAS experiment at usual principal field (9.4 T)

3 Getting to High-Resolution, Averaging Out Anisotropic Signatures

Depending on the complexity of the spin system considered and thus on the type and multiplicity of interaction, the methods and possibilities for obtaining a high-resolution spectrum may vary from simple to more complex. We first consider the case of a single interaction and its averaging by mechanical manipulation of the sample using magic angle spinning. In a second step, MAS will be combined with proton dipolar decoupling to obtain high resolution for nuclei surrounded by protons (organic solids). The specific cases of the strongly coupled quadrupolar nuclei (second order shift and broadening) and of strongly coupled ^{1}H multispin homonuclear systems will be specifically dealt with in this section and in Sects. 4 and 5.

3.1 Magic Angle Spinning: MAS

The idea that is the main source of resolution for solid-state NMR was proposed very early (1959). It consists in rapidly spinning the sample around the C_3 symmetry axis of a cube having the H_0 field parallel to one edge. This axis makes an angle of θ_m=54.7° with the principal field and is called the magic angle. In the simple cases of a single dominant first order interaction (which depends on $3\cos^2\theta - 1$) this results in a *coherent averaging* of the broad static spectrum which is modulated into a set of sharp spinning sidebands, finally keeping only a sharp isotropic signature when going to very high spinning rates. This is illustrated with ^{31}P spectra in Fig. 3. In static conditions the spectrum is broad and unresolved. Ultimate resolution of three sharp lines is obtained at high MAS rate keeping only with the trace of the CSA tensor giving the isotropic spectrum. At intermediate spinning rate, the spectrum is modulated into spinning sidebands which intensities are characteristic of the anisotropic part of the CSA tensor. In that case, the chemical shift anisotropy can be measured from the spinning sideband intensities [4]. The case of quadrupolar nuclei ($I < 1/2$) is more complex: their signature consists in $2I$+1 allowed single quantum transitions (instead of one for an isolated $I = 1/2$ nucleus) and quadrupolar interaction can be strong enough (several MHz) to give rise to second order perturbation terms (shift and shape) which cannot be averaged out by MAS alone. These second order terms are proportional to the inverse of the principal field (or to its square when expressed in ppm) and resolution thus strongly depends on the intensity of the applied principal field. An example of this situation is given in Fig. 3 with the spectra of the crystalline Yttrium Aluminum Garnet (YAG: $Y_3Al_5O_{12}$). This cubic structure contains two types of Al sites; the Al_{VI} site undergoes small quadrupolar interaction and nearly no second order shift and broadening, it gives rise to a sharp line for the central transition and a set of sharp spinning sidebands for its symmetric outer transitions. The Al_{IV} site undergoes a ten times larger quadrupolar interaction that gives rise to the second order broadening and the shift of its line [12]. This second order effect decreases

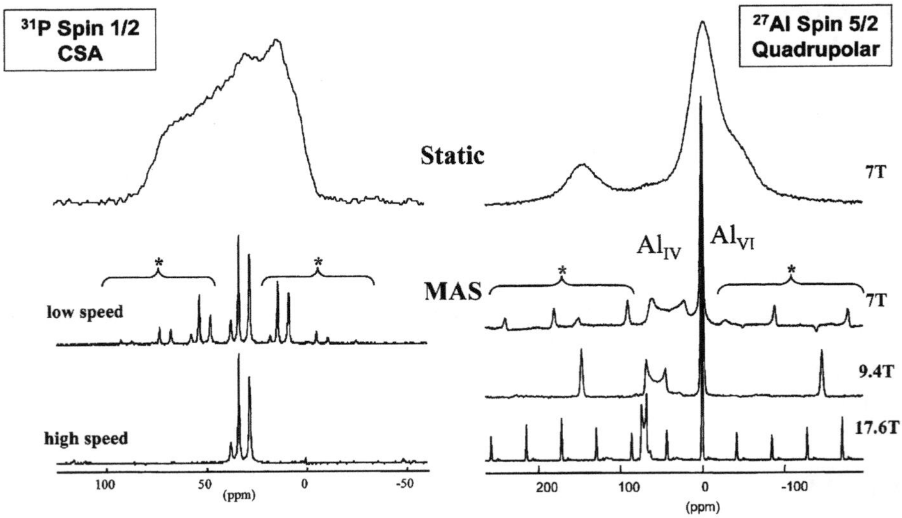

Fig. 3. Resolution improvements obtained using magic angle spinning (MAS) in the case of a single dominant interaction: CSA in the case of ^{31}P (I=1/2) and quadrupolar interaction in the case of ^{27}Al (I=5/2). The second order quadrupolar broadening of the Al_{IV} site depends upon the principal field [12]. ∗ indicates spinning sidebands

when going to higher principal fields. In this section and in Sect. 4 we shall discuss the possibilities offered to average out second order broadening in the case of quadrupolar nuclei.

3.2 Combining MAS and Dipolar Decoupling

The problem of averaging out the anisotropic signature of the different interaction is often more complex in real systems due to multiple interactions. This is especially the case of organic or protonated solids in which we would like to obtain the isotropic chemical shift signature of a nucleus (typically ^{13}C, ^{29}Si or ^{31}P) which is coupled by dipolar interaction to a many body proton bath (^{1}H proton is among the most sensitive nuclei and thus generates strong heteronuclear and homonuclear dipolar couplings). While infinitely fast MAS rate (hundreds of kHz) would provide resolution, the actual spinning rate that can be achieved mechanically are in the range of a few kHz up to 35 kHz for commercial probes and 50 kHz for the highest reported experiments. In such cases high resolution can only be obtained by combining the spatial manipulation of the sample with dipolar decoupling consisting in an irradiation of the dipolar coupled spin system during the acquisition (decoupling of ^{1}H during acquisition of X: ^{13}C, ^{29}Si or ^{31}P). This is illustrated in Fig. 4 with the example of a solid state phosphonate observed by ^{31}P NMR [13]. The static spectrum is broad and unresolved due to the combination of ^{31}P chemical shift anisotropy sites and dipolar couplings between ^{31}P and neighboring ^{1}H. By implementing dipolar decoupling of the ^{1}H

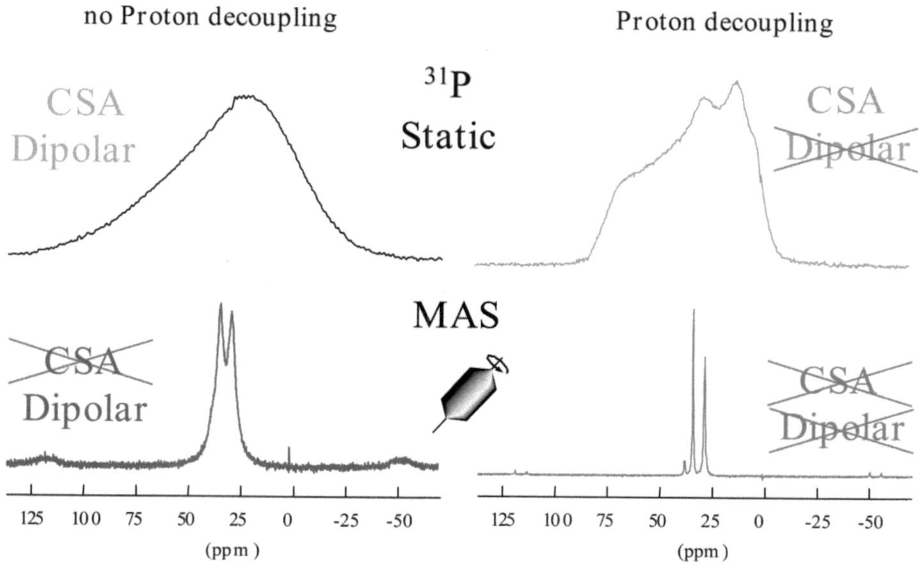

Fig. 4. Resolution improvements obtained combining magic angle spinning (MAS) and ^{1}H decoupling during acquisition. Proton decoupling (right hand panel) removes the heteronuclear dipolar interaction while chemical shift anisotropy is averaged out by MAS (typically 10 kHz). The example taken is an hybrid organic/inorganic Zn phosphonate, observed by ^{31}P NMR [13]

(during acquisition of the ^{31}P spectrum) we obtain a spectrum in which only remains the CSA of ^{31}P still giving a broad line with now well resolved discontinuities. By applying high speed MAS alone the CSA is averaged out and we obtain a partly resolved spectrum showing two overlapping lines. The finally resolved spectrum is obtained by combining dipolar decoupling and MAS. It has to be remarked that both manipulation of the sample: mechanical and in the spin space are or can be coherent and could give rise to interferences (constructive or destructive). This will be exploited in experiments aiming at reintroducing the anisotropy of the interactions that we want to average out for the moment to obtain high resolution.

4 Half-Integer Quadrupolar Nuclei

4.1 Magic Angle Spinning of Quadrupolar Nuclei

As shown in Sect. 3.1, MAS alone cannot provide ultimate high resolution by its own. This is due to the fact that second order quadrupolar terms (as well as second order cross terms when significant) require more complex trajectories or symmetries to be averaged out. In the general case, the spectral contribution for a crystallite oriented by α an β in the rotor frame making an angle θ with the principal field takes the following form (for an infinitely fast spinning rate):

$$\nu_{\langle m,m-l\rangle} = \nu_0 + \frac{\nu_Q^2(1+\eta^2/3)}{\nu_0} \sum_{l=0,2,4} C_{m,I}^l B_l^{(2)}(\alpha,\beta) P_l(\cos\theta) \quad (1)$$

For a first order interaction ℓ ranges from 0 (isotropic term) to 2 but the sum extends to 4 for second order interaction. Magic angle is the solution of $P_2(\cos\theta) = 0$ where P_2 is the second order Legendre polynomial and thus enables the averaging of first order interaction but there exist no angle satisfying $P_\ell(\cos\theta) = 0$ for both $\ell = 2$ and 4.

A mechanical implementation of a double orientation rotation (DOR) combining a first rotation at MAS angle and a second one at a $P_4(\cos\theta) = 0$ condition (70.12° or 30.56°) is difficult but has been realized [5] and opens the possibility of obtaining ultimately resolved spectra for second order broadened spectra. Fig. 5 shows an example of a DOR spectrum compared to static and MAS with the ^{27}Al spectra of $CaAl_2O_4$. The six sites, that were overlapping in the MAS spectra, are resolved in the DOR spectrum.

There exist other solutions for obtaining high resolution spectra for quadrupolar nuclei; they make use of multiple dimensional experiments and we shall describe them in the next section: DAS, MQ-MAS and STMAS.

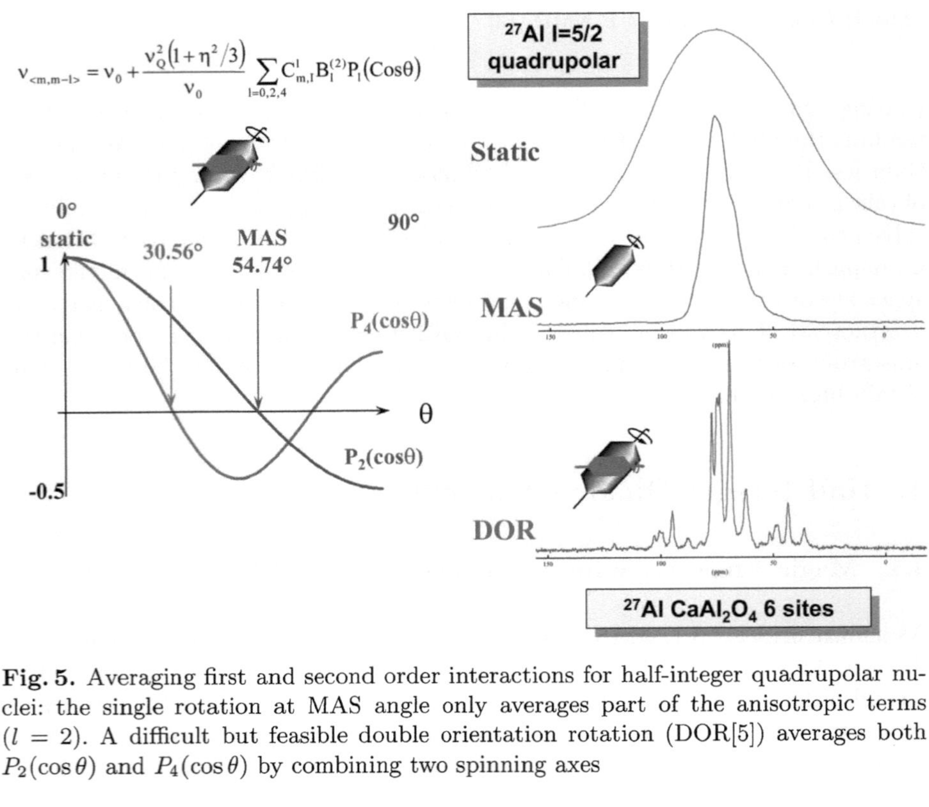

Fig. 5. Averaging first and second order interactions for half-integer quadrupolar nuclei: the single rotation at MAS angle only averages part of the anisotropic terms ($l = 2$). A difficult but feasible double orientation rotation (DOR[5]) averages both $P_2(\cos\theta)$ and $P_4(\cos\theta)$ by combining two spinning axes

Table 2. The different two-dimensional high-resolution experiments for half integer quadrupolar nuclei, reporting their respective advantages and disadvantages *(in italics)*

DAS	Central $\langle -1/2, 1/2\rangle$ to central transition with different spinning angles
1988 [6]	*Technically demanding. Time necessary to change spinning angle (~30ms)*
MQ-MAS	Symmetric multiple quantum $-m \longrightarrow m$ transition to central $\langle -1/2, 1/2\rangle$ transition
1995 [7]	*Excitation and conversion of the multiple quantum coherences is limited for high quadrupolar couplings*
STMAS	Outer 1Q satellite $\langle \pm m, \pm(m-1)\rangle$ transitions to central $\langle -1/2, 1/2\rangle$ transition
2000 [8]	*Very sensitive to experimental settings (magic angle, spinning rate). Sensitive to other second or higher order effects*

4.2 High Resolution for Quadrupolar Nuclei: DAS/MQ-MAS/STMAS

Returning to equation (1), it can be remarked that, while the angular dependence of the second order terms are identical for all transitions (single and/or multiple quantum transitions) for a given ℓ value $\mathrm{B}_l^{(2)}(\alpha, \beta)$, they undergo different scaling factors $C^l_{m,I}$. Consequently, an experiment that would correlate the spectrum of two different transitions would provide a separation of the different lines. Finally, a proper processing of the data (shearing transformation) could lead to a correlation between the usual MAS spectrum and an isotropic dimension in which all second order broadening terms ($\ell = 4$) would cancel. The position of the line in the isotropic dimension being only ruled by the interplay of $\ell = 0$ terms coming from CSA and quadrupolar interaction (second order shift). This can be described as a coherence echo generated in the multiple-level system of a single quadrupolar nucleus ($2I+1$ levels and $2I$ transitions). This idea was first described in 1988 and gave rise to several different experiments: Dynamic angle spinning or DAS in 1988 [6], multiple quantum magic angle spinning or MQ-MAS in 1995 [7], and, finally, satellite transition magic angle spinning or STMAS in 2000 [8]. Fig. 6 presents the typical pulse sequences and the example of the ^{87}Rb MQ-MAS spectrum of $RbNO_3$ [14]. The correlation between the 3Q and the 1Q (left-most) spectrum gives three ridges for the three different Rb sites of the crystalline $RbNO_3$, it can be processed to correlate the isotropic MQ-MAS to the usual 1Q MAS dimension and each individual site can be modeled to extract its quadrupolar parameters. Another example of MQ-MAS spectra was given in Fig. 2 with the ^{27}Al spectra of alumino-silicates, enabling the spectral separation of different coordination environments, even in partly disordered samples or glasses. The description of MQ-MAS by L. Frydman in 1995 has revolutionized the field of high-resolution experiment for half-integer quadrupolar nuclei, leading to numerous applications in solid state inorganic materials like ceramics, zeolites, macro or mesoporous materials.

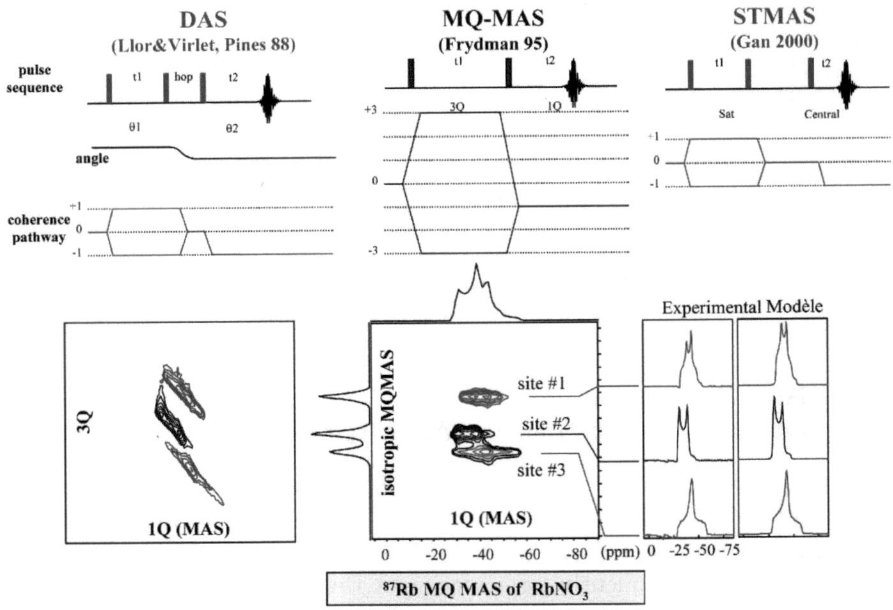

Fig. 6. Schematic representation of pulse programs and coherence pathways for the DAS, MQ-MAS and STMAS experiments. Example of the MQ-MAS spectrum of $RbNO_3$ [14]. The left-most spectrum is the correlation of 3Q and 1Q spectra; the central spectrum is the sheared spectrum in which the vertical dimension contains the resolved isotropic spectrum correlated to the MAS 1Q spectrum. Each line can be individually modeled to measure its quadrupolar tensor parameters

5 Protons in Solid State Materials

5.1 Magic Angle Spinning for Proton NMR

As stated above, the case of proton ^{1}H NMR in the solid state is not as simple as the case of other "*heteronuclei*", because of the combination of its sensitivity and its abundance in organic or hybrid materials. In most cases protons are strongly coupled between themselves by dipolar multi-spin interaction, thus constituting a so-called *proton bath*. In that case the proton spectrum becomes partly *homogeneous* and, under magic angle spinning its spectrum only sharpens progressively without breaking into well defined sharp spinning sidebands. Different approaches can be taken to enhance resolution in ^{1}H solid-state NMR experiments. Isotopic dilution by ^{2}H reduces the homonuclear dipolar coupling but implies chemical preparation of the sample. Homonuclear dipolar decoupling of the proton bath by application of tailored radio frequency pulse is another possibility which will be examined in the following section. For now, we shall just consider very fast MAS at speeds of 30 kHz and over (typical order of magnitude

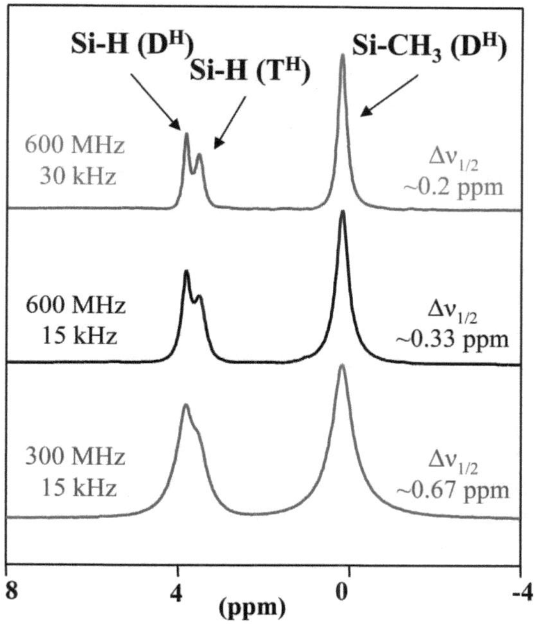

Fig. 7. High-resolution ^{1}H NMR spectrum obtained at high fields for solid state samples already showing some intrinsic mobility. The resolution is enhanced by combining higher principal fields and spinning rates [15]

of H–H homonuclear dipolar coupling). In fact the picture may not be as complex as described above if significant motion is still present in the solid sample, as in the case of polymers or gels. In such cases the combination of atomic scale mobility and MAS can already provide resolution that will be enhanced by going to higher principal fields because line separation is constant in ppm while line-width remains constant in frequency units. An example is given in Fig. 7 with the spectra of a siloxane based gel for which ^{1}H spectrum with sub-ppm line-width is obtained at high field and high spinning rate (30 kHz) [15]. In the case of less mobile materials, it is still possible to obtain significant resolution improvements with this approach. The simple observation of proton spectra under high field and high spinning rates becomes to be a very useful tool for characterization especially when combining very high fields and spinning rates [9].

5.2 Higher Resolution for Proton NMR with Homonuclear Decoupling

The strong homonuclear dipolar couplings may be averaged out (or reduced) using suitable pulse sequences. However, this requires *windowed* sequences during which the spectrum is acquired between the pulses of the decoupling sequence. An alternative way of obtaining comparable results and/or to use *windowless* sequences is to acquire the high resolution spectrum as the indirect dimension of a multidimensional experiment. Different schemes have been described for obtaining homonuclear decoupling but one of the most robust (and popular in the latest times) is the Lee Goldburg scheme which relies on irradiation of the spectrum off resonance (at the magic angle). Different implementations have been proposed and the experiment is usually carried out as a frequency-switched Lee Goldburg experiment where the offset is switched positive and negative at each duty cycle [10]. This indirect acquisition of the resolved proton spectrum can be further used in other multiple dimension experiments such as heteronuclear correlations (see Sect. 6.3).

Figure 8 illustrates the resolution improvement obtained by combining moderate MAS rate (10 kHz) and homonuclear decoupling in the case of a hybrid crystalline phosphonate [16]. The simple MAS spectrum provides no or few resolution. The two-dimensional spectrum correlates the central part of the MAS spectrum (horizontally and top) to the proton decoupled spectrum, acquired indirectly in the second dimension (vertically and left). The obtained resolution allows the separation of aliphatic, phenyls and hydroxy protons.

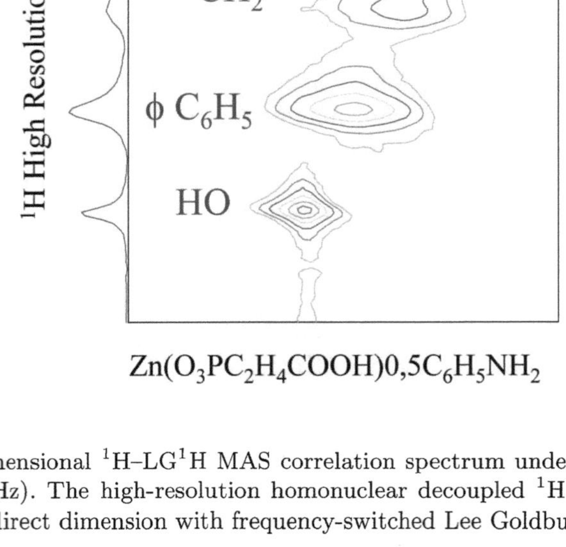

Fig. 8. Two dimensional ^{1}H–LG^{1}H MAS correlation spectrum under moderate spinning rate (10 kHz). The high-resolution homonuclear decoupled ^{1}H spectrum is obtained in the indirect dimension with frequency-switched Lee Goldburg scheme [16]

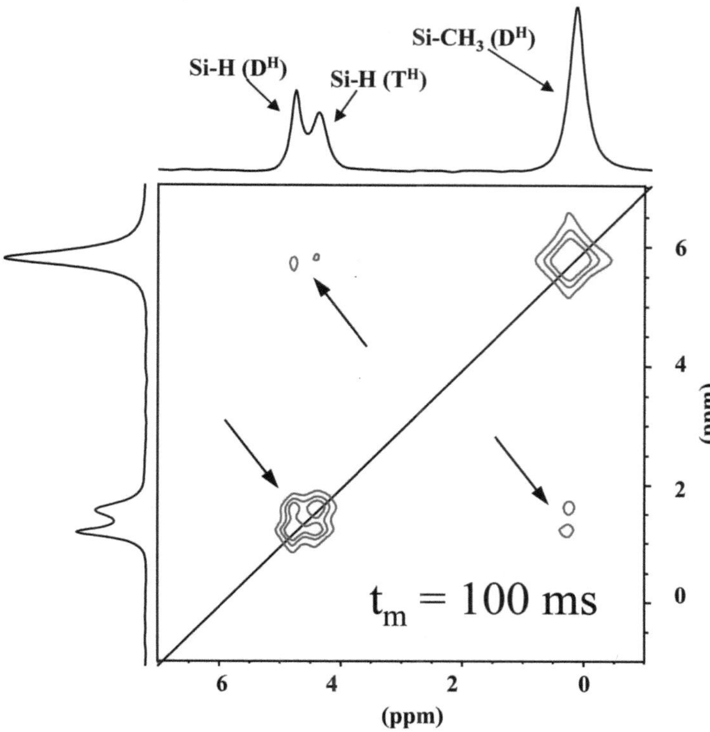

Fig. 9. Homonuclear exchange experiment; the 600 MHz 35 kHz ^{1}H MAS spectrum is self-correlated with a mixing time of 100 ms. The off-diagonal cross peaks are due to residual ^{1}H spin diffusion. Adapted from [15]

6 Reintroducing Anisotropic Interactions, Maintaining High Resolution Selection

While seeking for high resolution, we did our best to eliminate the anisotropic signatures of the different interactions. The following step is to retrieve this information while keeping a high-resolution spectral encoding. This is typically done in multiple dimensions experiments or in spectral editing pulse sequences. The simple solution is to take benefit of the imperfections of the averaging mechanisms (mechanical rotation at limited spinning rates or limited efficiency of the pulse sequences) like in the simple exchange experiment. The more complex way is to tailor excitation sequences so as to selectively reintroduce the different interaction through different mechanisms like rotary resonances, taking advantage of their different symmetry properties [3]. Among the different interactions, dipolar coupling is one of the most important as it is a direct function of the distance between the interacting spins.

6.1 Exchange by Spin Diffusion

The exchange experiment is a simple solid-state 2D experiment. It just involves storage of the magnetization along the principal field during a *mixing time* which can range from microseconds up to seconds (when relaxation times are long enough). The two dimensions are identical (homonuclear correlation) and, in absence of exchange (or for a null mixing time) all magnetization appears on the diagonal. If a process allows exchange of magnetization during the mixing time, the two dimensional chart shows off-diagonal cross peaks correlating the signals of the sites involved. Exchange can typically arise from mobility (reorientation or chemical exchange) or spin diffusion. For the example given in Fig. 7, all the different lines are correlated for a mixing time of 100 ms [15]. The cross peaks building curves will be a signature of the spatial proximity of the different resolved building units. This experiment takes benefit of the high resolution obtained through combination of high fields and high spinning rates as shown in Fig. 6 and discussed in Sect. 2.4. It can be improved by reintroducing the dipolar interaction during the mixing time so as to cancel the averaging due to magic angle spinning (radio frequency driven recoupling) and enhance the intensity of the cross peaks.

6.2 Dipolar Double Quantum–Single Quantum Correlations

A more selectively informative class of experiments can be designed to correlate the spectrum of pairs of like nuclei coupled by dipolar interaction (double quantum coherences, DQ) to the resolved MAS spectrum of individuals (single quantum, SQ). Double quantum coherences can only be observed indirectly and are generated and converted back with different types of sequences involving rotor synchronized pulses [3]. Fig. 10 gives two examples of application to ^{31}P NMR of crystalline and glassy inorganic solids. The ^{31}P spectrum DQ/SQ (double quantum / single quantum) spectrum of the crystalline compound ZrP_2O_7 enables the identification of 13 pairs of inequivalent P sites and one pair of equivalent P sites (P_2O_7 dissymmetric and symmetric units respectively) which finally results in the identification of a total of 27 inequivalent P sites [17]. From the DQ/SQ spectrum of a PbO/P_2O_5 glass it becomes possible to identify dimers, chain-end groups, and middle chain groups. A complete analysis of the 1D and 2D (DQ/SQ) spectra of a series of glasses evidences the polymerization of these inorganic glasses [18]. Providing different excitation schemes, it is also possible to efficiently generate double-quantum coherences using J scalar coupling, even in the solid state. This provides unambiguous assignment of the chemically bonded units, instead of spatially close units in the case of dipolar double quantum excitation.

6.3 Heteronuclear Correlation Through Bond or Through Space

The necessary step to obtain a heteronuclear correlation spectrum is to transfer coherence from one family of nuclei to another one. This is usually carried

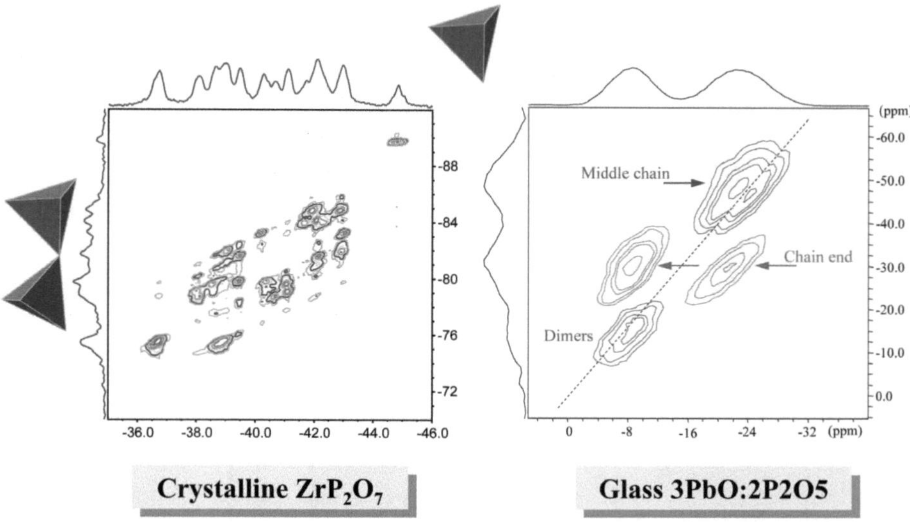

Fig. 10. Double quantum / single quantum ^{31}P spectra of a crystalline compound (ZrP_2O_7 with 27 inequivalent P sites (14 different P_2O_7 dimers) adapted from [17] and a PbO–P_2O_5 glass with identification of dimers, chain-end groups and middle chain units adapted from [18]. The vertical dimension is the double quantum dimension (pairs of coupled P), the horizontal dimension is the double quantum filtered MAS spectrum

out with the well known cross-polarization experiment which uses the through space residual dipolar interaction. As recently shown, this can also be achieved efficiently using the through bond scalar J-coupling. Both types of experiments are represented in Fig. 11 showing $^{1}H/^{13}C$ correlation spectra of tyrosine [11]. For both spectra the ^{1}H proton dimension consists in a high resolution homonuclear decoupled dimension obtained using Lee Goldburg decoupling (see Sect. 5.2). The MAS HMQC (a), closely related to the HMQC (heteronuclear multiple quantum correlation) liquid experiment, uses scalar J coupling while the HETCOR (b) experiment uses the classical dipolar cross polarization. All carbon sites appear in spectrum (b), receiving magnetization from their neighboring protons across space and from dipolar spin diffusion. In contrast, only the carbons directly bounded to protons appear in spectrum (a). Carbons 1, 4 and 9 do not appear in the J-spectrum and carbons 2, 3, 5, 6, 7 and 8 only receive intensity from their chemically bounded protons. This increased selectivity results in an increased resolution and opens the way to an unambiguous assignment of the different lines of both ^{13}C and ^{1}H spectra, even in rather complex organic molecules or macromolecules.

6.4 Isotropic–Anisotropic Correlations

When going to high resolution NMR of powders we lost the anisotropic signature of the different interactions (dipolar, chemical shift or quadrupolar inter-

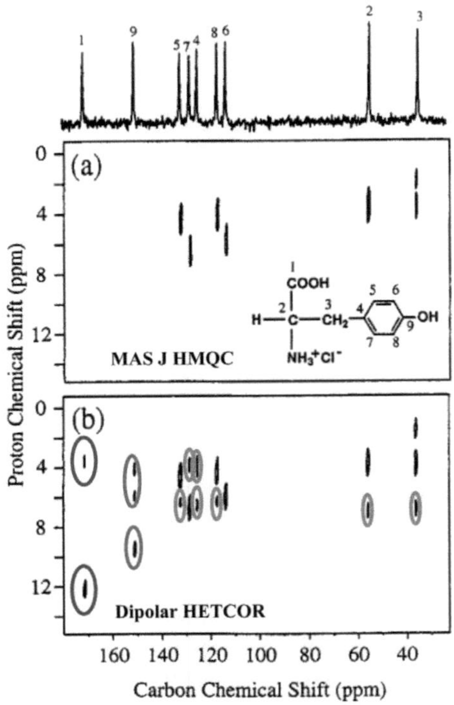

Fig. 11. Comparison of through bond and through space $^1H/^{13}C$ correlations. Spectra of tyrosine correlated through chemical bound by scalar J coupling (*a*: MAS-*J*-HMQC) and through space by dipolar coupling (*b*: HETCOR). Circled lines correspond to neighboring $^1H/^{13}C$ pairs not directly linked by a chemical bond. Adapted with permission from [11]

actions). Several experiments aim at reintroducing this anisotropic signatures while maintaining (or gaining) high resolution. Lots of different experiments have been proposed for the different interactions and lead to correlation of the high resolution spectrum with sideband patterns, scaled and/or modified image of the anisotropy, especially (but not only) in the case of $I = 1/2$ dipolar nuclei. The case of second-order broadened half-integer quadrupolar nuclei is somehow different as MAS only cannot give high resolution by itself and the already mentioned two-dimensional methods for half-integer quadrupolar nuclei (DAS, MQ-MAS and STMAS, see Sect. 4) also provide this kind of isotropic/anisotropic correlations.

7 Very High Magnetic Fields

Going through the different aspects of the quest for high resolution in solid state NMR, we already mentioned that the strength of the principal field is of major importance in two main cases: the strongly coupled bath of protons (Sect. 5) and the second order broadened half integer quadrupolar nuclei (Sect. 4). In both cases the high principal field, combined with high MAS rate allows acquisition of simple spectra showing enhanced resolution obtained with simpler pulse sequences (typically single pulse to begin with). A very convincing example of this gain in resolution is given in Fig. 12 with the ^{27}Al MAS spectra of the $9(Al_2O_3)$-$2(B_2O_3)$ compound. At low or moderate principal fields (<10 T) there exists no resolution in the MAS single pulse spectrum due to severe overlapping of the second order broadened shapes of the four different Aluminum sites. Combination of DOR, MQ-MAS or STMAS experiments can provide an increased resolution with the separation of the four different sites, at the cost of long experimental times and difficult quantitative interpretation. While currently available NMR magnets can provide fields up to 21.1 T, the National High Magnetic Field Laboratory (Tallahassee, Fl, USA) has developed resistive and hybrid magnets able to provide 25 T at ~ppm resolution and 40 T at ~5 ppm resolution for MAS spectra, respectively. The decisive advantage of very high fields is to directly provide an ultimate resolution in very simple acquisition conditions (apart from the principal field itself) thus keeping with an easy quantitative interpretation of the obtained spectrum. In the case of our test compound, at 40 T the second-order broadening which causes overlap at lower fields has vanished and most of the line width is due to principal field instability during the acquisition of the free

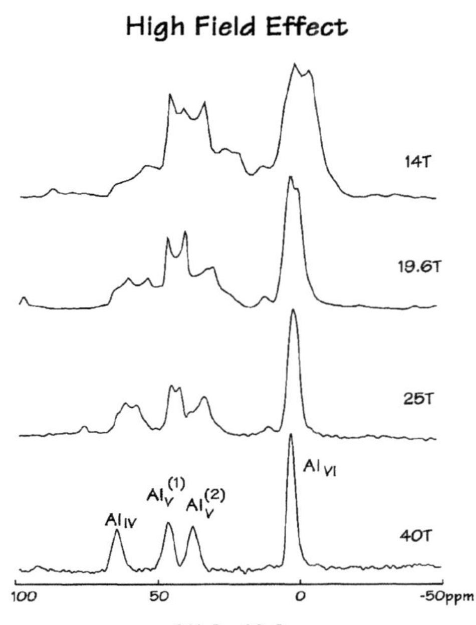

Fig. 12. ^{27}Al MAS spectra of $9(Al_2O_3)$-$2(B_2O_3)$ compound showing an example of the dramatic increase of resolution obtained at very/ultra high fields for second-order broadened quadrupolar nuclei. At 40 T the second-order broadening nearly vanishes and we obtain a spectrum of sharp lines in a few seconds. Adapted from [19]

induction decay signal. The two spectra at 25 and 40 T have been acquired with 16 scans (16 seconds acquisition time). Of course we do not expect this 40 T very unique experiment to become routine immediately, but these results can give us the flavor of what very high resolution could be for solid-state NMR, providing improvements of field stability and homogeneity. This would be a major improvement in the field of application of solid-state NMR to the characterization of solid state organic, inorganic or hybrid materials.

8 Conclusion

In this paper we focused on the recent methodological developments in the field of high-resolution solid-state NMR, as used for structural characterization of inorganic, hybrid or organic solids, in crystalline or amorphous states. The spin systems that we consider are complex; they involve chemical shift anisotropy, dipolar coupling, scalar couplings and quadrupolar interaction. The first aim is thus to obtain high resolution by removing or averaging these anisotropic interactions. In that perspective we emphasized the two difficult cases of strongly coupled protons in organic or hybrid materials and second-order broadened quadrupolar nuclei. In both cases the availability of very-high or ultra-high principal fields, combined with fast or very fast spinning rates, proves to strongly enhance resolution. The second step is to selectively retrieve the anisotropic interaction signature while keeping with high-resolution encoding in multi-dimensional experiments. This implies the design of complex experiments combining sample spinning and radio-frequency irradiation which is becoming a research field by itself and opens the possibility for a very accurate description of short-range and medium-range structures. The newly designed experiments promote hardware improvements, and rapidly migrate from pure theory to real routine application.

References

1. *Encyclopedia of Nuclear Magnetic Resonance*, ed. D.M. Grant, R.K. Harris: (Wiley, Chichester 1995)
2. K. Schmidt-Rohr, H.W. Spiess: *Multi-dimensionnal solid state NMR and polymers* (Academic Press, New York, 1994)
3. M. Levitt: *Spin Dynamics. Basics of Nuclear Magnetic Resonance* (Wiley, Chichester 2001)
4. J. Herzfeld, A.E. Berger: J. Chem. Phys. **73**, 6021 (1980)
5. Y. Wu, BQ Sun, A. Pines, A. Samoson, E. Lippmaa: J. Magn. Reson. **89**, 297 (1990)
6. A. Llor, J. Virlet: Chem. Phys. Lett. **152**, 248 (1998) P.J. Grandinetti: *'Dynamic-Angle Spinning and Applications'*. In *Encyclopedia of Nuclear Magnetic Resonance*, ed. D.M. Grant, R.K. Harris, (Wiley, Chichester 1995)
7. L. Frydman, J.S. Harwood: J. Am. Chem. Soc. **117**, 5367 (1995) A. Medek, J.S. Harwood, L. Frydman: J. Am. Chem. Soc. **117**, 12779 (1995)
8. Z. Gan: J. Am. Chem. Soc. **122**, 3242 (2000) Z.Gan: J.Chem. Phys **114**, 10845 (2001)

9. A. Samoson, T. Tuherm, Z. Gan: Solid State NMR **20**, 130 (2001)
10. E. Vinogradov, P.K. Madhu, S. Vega: Chem. Phys. Lett. **314**, 443 (1999)
11. A. Lesage, D. Sakellariou, S. Steuernagel, L. Emsley: J. Am. Chem. Soc. **120**, 13194 (1998)
12. D. Massiot, C. Bessada, J.P. Coutures, F. Taulelle: J. Magn. Reson. **90**, 231 (1990)
13. D. Massiot, S. Drumel, P. Janvier, M. Bujoli-Doeuff, B. Bujoli: Chem. Mater. **9**, 6 (1997)
14. D. Massiot, B. Touzo, D. Trumeau, J.P. Coutures, J. Virlet, P. Florian, P.J. Grandinetti: Solid State NMR **6**, 73 (1996)
15. F. Babonneau, V. Gualandris, J. Maquet, D. Massiot, M.T. Janicke, B.F. Chmelka: J. Sol-Gel Sci. and Techn. **19**, 113 (2000)
16. D. Massiot, B. Alonso, F. Fayon, F. Fredoueil, B. Bujoli: Solid State Sci. **3**, 11 (2001)
17. I.J. King, F. Fayon, D. Massiot, R.K. Harris, J.S.O. Evans: Chem. Commun. 1766 (2001)
18. F. Fayon, C. Bessada, J.P. Coutures, D. Massiot: Inorg. Chemist. **38**, 5212 (1999)
19. Z. Gan, P. Gorkov, T.A. Cross, A. Samoson, D. Massiot: J. American Chemical Society **124**, 5634 (2002)

High Frequency EPR Spectroscopy

D. Gatteschi

Department of Chemistry, University of Florence, UdR INSTM,
Luca Pardi, IFAM, CNR, Pisa, Italy

Abstract. EPR has traditionally been used in order to obtain structural information on transition metal compounds, with exciting frequencies in the range 9–35 GHz.The recent availability of high magnetic field has prompted the use of higher frequencies. In this contribution the advantages of using High-Field-High-Frequency EPR (HF EPR) experiments are reviewed. After a brief introduction aiming to recall the fundamentals of EPR spectroscopy, a short description of the experimental apparatus needed to perform HF EPR measurements is provided. The remaining sections report selected examples showing how much information can be obtained by HF EPR spectra. They range from individual ions with integer spin to molecular clusters. Particular attention is devoted to the so called Single Molecule Magnets, SMM, i.e. to molecular clusters which show slow relaxation of the magnetization at low temperature. This effect is due to Ising type magnetic anisotropy which has been efficiently monitored through HF EPR spectroscopy.

1 Introduction

EPR spectroscopy has been an excellent tool for the investigation of the structures and of the electronic structures of paramagnetic centres, both organic radicals and transition and rare earth ions [1–6]. EPR spectra can be recorded in solids, in single crystals, in polycrystalline powders, in frozen solutions and in fluid solutions and the contributions range from physics to chemistry, biology, mineralogy, geology etc. A sample is set in a static magnetic field, which can be varied, to meet the resonance conditions in the presence of an oscillating magnetic field provided by a microwave generator. The oscillating field may correspond to either a continuous wave, CW, or to a pulsed field. In general we will refer to CW experiments. Pulsed experiments will explicitly be taken into consideration in the contribution by Fuchs to this same book.

At the simplest level of approximation an isotropic paramagnetic centre characterised by $S= 1/2$ exposed to an external field H parallel to z gives rise to two split levels, $E(M_S) = M\ \mathrm{g}\mu_{\mathrm{B}}H$, where $M = \pm\ 1/2$, μ_{B} is the Bohr magneton, and g is a parameter which depends on spin and orbital contributions and therefore contains structural information. The resonance condition is expressed by:

$$h\nu = g\mu_{\mathrm{B}}H \tag{1}$$

Conventional spectrometers use the so called X-band frequency (9 GHz) in such a way that the resonance conditions for a free electron ($g = g_e$= 2.0023) are met at ca. 0.3 T. The reason for this is given by the easy availability of the microwave sources and the easily achieved fields even with electromagnets. However this set up is far from being ideal in many cases. First of all the spectral resolution can be improved using higher frequencies and higher fields. Let us suppose to have two systems characterised by g_1 and g_2, respectively. The difference in the resonance fields, $H_1 - H_2$, will be higher for higher used frequency ν, therefore, assuming that no instrumental broadening occurs, increasing the frequency will increase the spectral resolution. This is particularly important for organic radicals for which the orbital contributions are usually small and the g values are close to the free electron value.

A typical example, showing the dramatic increase in resolution obtained on increasing frequency, are the spectra of the radical $P700^{\bullet+}$ reported [7] in Fig. 1. The radical is the primary electron donor, present in Photosystem I, the reaction centre responsible of ferredoxin reduction [8]. The shape of the spectra will be discussed below, in Sect. 2. They correspond to the first derivative of a normal absorption spectrum. It is apparent that at high frequency three features are present, corresponding to three different g values, while at low frequency only one g value is observed. In fact the high field spectra provided three g values, namely g_1= 2.00317(7), g_2= 2.00260(7), g_3= 2.00226(7). The corresponding resonant fields are H_1=0.34735, H_2=0.34745, H_3=0.34750, at 9.74 GHz, and H_1= 15.5821, H_2=15.5865, H_3=15.5892, at 436.94 GHz.

In order to introduce another important advantage of HF EPR spectroscopy it is necessary to consider the EPR spectra of systems containing more than one unpaired electron, S >1/2. In general the $2S$+1 M levels belonging to the S multiplet in symmetry lower than spherical, will have different energies, even

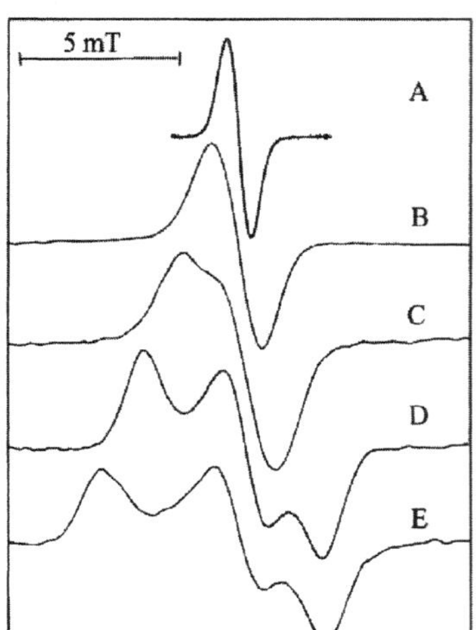

Fig. 1. EPR spectrum of the light induced $P700^+$ signal at varying microwave frequencies: (A) 9.74 GHz; (B) 108.45 GHz; (C) 216.91 GHz; (D) 3235.32 GHz; (E) 436.94 GHz. After [7]

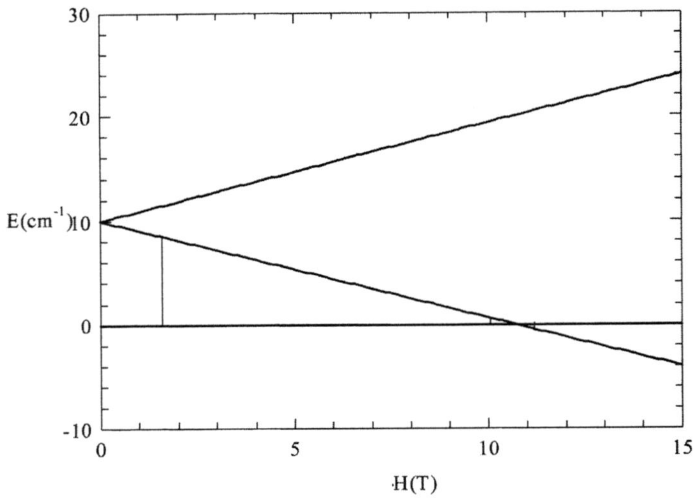

Fig. 2. Energy levels for a triplet with axial zero field splitting (D= 10 cm^{-1}) in the presence of a varying magnetic field parallel to z. The marked transitions correspond to EPR experiments at 9.5 and 250 GHz, respectively

in the absence of an applied magnetic field (zero field splitting, ZFS), as shown in Fig. 2 for a triplet. If ZFS is much larger than the microwave quantum $h\nu$, the resonance requires extremely high field to be observed. Many paramagnetic transition metal ions show large ZFS's and have been considered as EPR silent, because no signal can be detected at X-band frequency. This is particularly true for systems containing an even number of unpaired electrons, because in systems with an odd number of unpaired electrons Kramers theorem demands that the minimum degeneration of the levels is two. In other words there will always be pairs of degenerate levels in zero field and transitions between the two levels obtained by an external magnetic field will in principle always be possible at relatively small fields. However going to higher frequencies may in principle overcome the inconvenience [9,17]. Increasing the quantum $h\nu$ decreases the required field for inducing transitions, thus making the spectra observable. The case pictured in Fig. 2 corresponds to a frequency of 9.5 and 250 GHz, respectively, with a zero field splitting of 10 cm^{-1}.

In the last years high fields have become much easier to achieve taking advantage of superconducting magnets therefore several groups have built new high frequency high field EPR spectrometers in order to extend the possibility of investigation of matter using this magnetic resonance technique. High is a relative term, therefore it must be defined what exactly is meant for high frequency EPR. Generally it is agreed that the realm of HF-EPR begins at 95 GHz, and we will stay with this definition. We wish to show here briefly which have been the major results so far achieved and indicate which are the perspectives of HF

EPR. We will focus essentially on metal ions, while organic radicals will be more extensively treated in the contribution of Fuchs to this same book.

The organisation of the paper will be the following: the principles of EPR spectroscopy will be briefly reviewed in Sect. 2, the instrumental requirements for HF EPR spectroscopy will be given in Sect. 3, while Sect. 4 will be devoted to examples of HF EPR spectra, subdivided in radicals, single ions and clusters. Finally Sect. 5 will discuss some of the perspectives of HF EPR spectroscopy.

2 EPR in a Nutshell

At the simplest level of approximation we may imagine to be interested at the EPR spectra of a system with an orbitally non-degenerate ground state, characterized by a spin S. The appearance of the spectra depends on three main type of interactions, namely that of the electron spin with the external magnetic field (Zeeman interaction), that of the electrons between themselves, which can be dipolar in nature or mediated by spin orbit coupling (fine structure), and that of the electron spin with the moment of magnetic nuclei (hyperfine or superhyperfine structure). A convenient way for describing the spectra is that of using the spin hamiltonian formalism in which all the orbital variables are neglected and only the spin coordinates are taken into consideration:

$$\mathcal{H} = \mu_B \mathbf{B}.\mathbf{g}.\mathbf{S} + \mathbf{S}.\mathbf{D}.\mathbf{S}. + \Sigma_k \mathbf{S}.\mathbf{A_k}.\mathbf{I_k} \tag{2}$$

where $\mathbf{B}$ is the external magnetic field, $\mathbf{g}$, $\mathbf{D}$, and $\mathbf{A}_k$ are tensors describing the Zeeman, fine and hyperfine interactions, respectively. (2) is the lowest possible level of approximation: higher order terms must be included in several cases but we will neglect them for the time being for the sake of simplicity.

The $\mathbf{g}$ tensor can be conveniently expressed as [18,19]:

$$\mathbf{g} = g_e \mathbf{I} + \lambda \mathbf{\Lambda} \tag{3}$$

where $\mathbf{I}$ is the identity matrix, $\lambda = \pm\zeta/2\mathrm{S}$, ζ is the spin orbit coupling constant. The + sign applies for configurations less that half full and the − sign for configurations more than half full. $\Lambda_{\alpha\beta}$ is given by:

$$\Lambda_{\alpha\beta} = \sum_n \frac{\langle g|L_\alpha|n\rangle \langle n|L_\beta|g\rangle}{E_g - E_n} \tag{4}$$

where $|\mathrm{g}\rangle$ is the ground state, the sum is over all the $|\mathrm{n}\rangle$ excited states, and E_g, E_n are the corresponding energies. Since spin orbit coupling is small for organic radicals, and the energy difference between ground and excited state is large the correction to the free electron value is usually small, as anticipated in Sect. 1. Matters are different for transition metal ions, where spin orbit coupling becomes important, and the separation between the ground state and the excited states is relatively small. Therefore relatively large deviations of the g values from the free electron value must be expected. Further, given the low symmetry

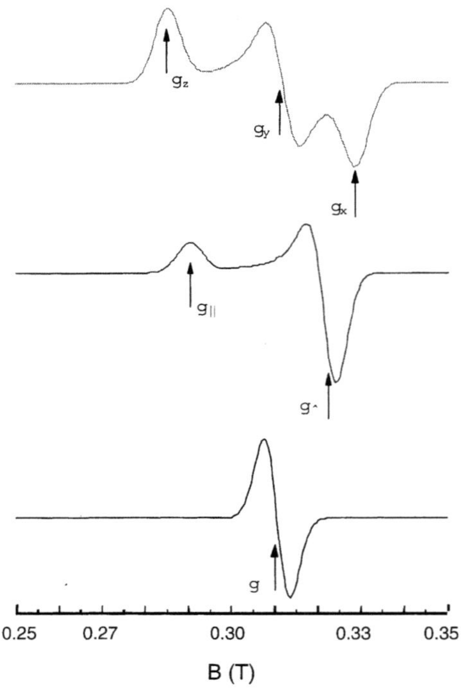

Fig. 3. Polycrystalline powder EPR spectra for a system with S= 1/2. Lowest: isotropic **g**, middle: axial **g**, highest: rhombic **g**

environment in which the metal ions may be found the **g** tensors may become largely anisotropic. Spin orbit coupling effects are even larger for rare earth ions. Further in these systems the valence f electrons are relatively well shielded from the external environment in such a way that large non-quenched orbital contributions are present also in relatively low symmetry compounds.

The spectra can be recorded on single crystals, on polycrystalline powders, and in frozen solutions. An important feature of EPR spectroscopy is that it is possible to measure spectra in isotropic environments, like powders and solutions, and nevertheless to obtain information on the anisotropy of the **g** tensors. In Fig. 3 are shown some typical polycrystalline powder EPR spectra of a system with $S = 1/2$. The lowest spectrum corresponds to an isotropic case, the intermediate one to an axial distortion, while the highest one corresponds to a completely rhombic situation.

The fine structure tensor has two contributions. The dipolar interaction between the electrons is typically of a few gauss. That mediated by spin orbit coupling is given by:

$$\mathbf{D} = \lambda^2 \mathbf{\Lambda} \tag{5}$$

where $\mathbf{\Lambda}$ is defined in (3,4). The dipolar term is dominant for organic radicals, while the second is dominant for transition metal ions. The fine structure tensor **D** is in general slightly rearranged to be traceless. In this scheme in the coor-

Table 1. Resonance fields of the free electron at several frequencies.

Resonance frequency (GHz)	Resonance field (T)
9	0.3234
35	1.2578
95	3.1441
200	7.1876
300	10.7814
500	17.9690

dinate frame in which **D** is diagonal it is possible to express the fine structure using only two parameters D and E defined by the hamiltonian:

$$H = D[S_z^2 - S(S+1)/3] + E(S_x^2 - S_y^2) \tag{6}$$

where without loss of generality it may be assumed that $-1/3 \leq E/D \leq +1/3$. E is zero for axial symmetry, while D is zero in cubic symmetry. D can be of a few cm^{-1} for many transition metal ions.

The appearance of the spectra of systems with more than one unpaired electron, characterized by a spin S, depends on the relative importance of the Zeeman and ZFS terms. They are particularly simple in the high field approximation, when the effect of the zero field splitting can be calculated as a perturbation on the Zeeman energies. For a given orientation of the external magnetic field the line is split into $2S$ components. In axial symmetry the resonance fields for the $M \to M+1$ transitions are given by:

$$H(M \to M+1) = (g/g_e)[H_0 + D'(M/2)(3cos^2\theta - 1)] \tag{7}$$

where H_0 is the resonance field for the free electron at the given frequency, $D' = D/(g_e\mu_B)$, and θ is the angle of the external magnetic field with the unique axis. Some H_0 fields for selected frequencies are given in Table 1.

3 Instrumentation

It has been said that an universal design of an High Field EPR spectrometer does not exist. It is actually true that, especially for frequencies above 150 GHz and toward the THz, the different kind of application dictates the instrumental requirements. However one can ideally draw a general scheme of an EPR spectrometer which covers all the frequencies of use, say from 1 GHz to 1 THz. The drawing in Fig. 4 is representative of such a scheme for a Continuous Wave-EPR (CW-EPR) spectrometer and includes a source, a transmission line, a sweepable magnet, a detection scheme, a structure containing the sample, that may or may not include a resonator, and a combination of devices which control the thermodynamic parameters of the experiment, that is: temperature and pressure and other experimental devices such as additional sources for multiple irradiation etc. Each different part of the spectrometer will be summarily described and the

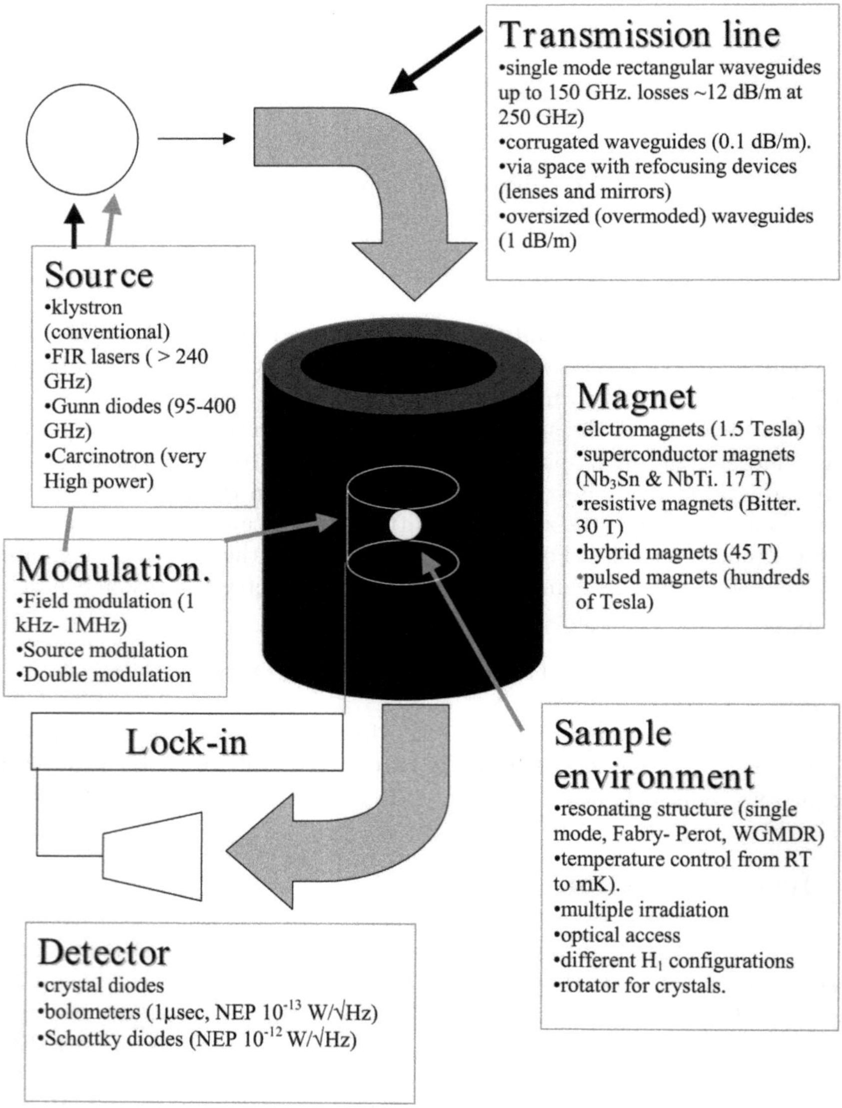

Fig. 4. Block scheme of a general EPR spectrometer with description of the devices adapted for the different microwave bands from conventional to sub-millimeter EPR

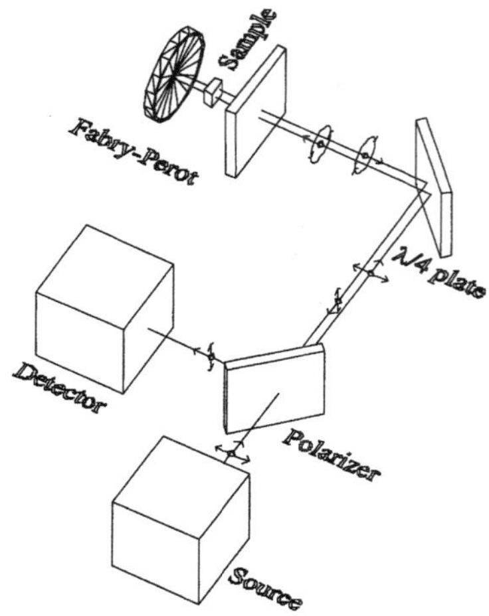

Fig. 5. Schematic representation of a millimeter wave quasi-optical bridge (see text)

solutions appropriate for the different frequency regimes will be illustrated with particular emphasis for the high frequencies. A very well known practical and successful realization of the scheme reported in Fig. 4 is the conventional EPR spectrometer working at X or Q band (for instance Varian or Bruker) depicted in Fig. 5. Conventional EPR fully exploits the microwave technology with the use of hollow rectangular metallic waveguides and resonant cavity crystal diodes are generally used as detectors in a homodyne detection scheme. For a general review of the conventional experimental EPR techniques the reader is addressed to the book of Poole [20].

3.1 Sources

Sources can be of different nature. Historically tubes such as klystron at X and Q band, which were easily available due to their common use in the military industry as mw sources for RADAR technology were used. Today solid state sources such as InP and GaAs Gunn effect diodes are more convenient. They generate radiations in the range from 1 to 100 GHz at relatively high power (10–50 mW). Non linear devices such as Schottky diodes are available which can multiply the fundamental frequency to the sixth harmonic thus covering the frequency range up to 600 GHz ca. Gunn diodes have also good performances as far as phase and frequency noise are concerned and they can be easily phase and/or frequency locked. At higher frequencies toward the THz regime FIR lasers become more convenient in term of emitted power. The latter can be successfully optimized for millimeter wave range, but their principal drawback is their poor amplitude stability that requires the implementation of complex stabilization circuitry. The laser are moreover practically non tunable devices while solid state source can in general be tuned in a frequency interval ranging from few hundred of MHz to several GHz. On the other hand FIR laser show a very good spectral purity. Other sources are backward wave oscillator and gyrotrons. The low power versions of the latter are devices that can deliver several Watt of power, and are thus convenient for very high power applications. Comparative studies on the use of sources of different kind in HF-EPR applications are to our knowledge still missing.

3.2 Detectors

Two of the figures of merit for detectors that are generally considered are: the noise equivalent power (NEP) and the response time. The NEP is the minimum power hitting the detector that gives rise to a signal- voltage equal to the noise-voltage from the detector. It is measured in $W/\sqrt{\ }$ Hz. The response time of a detector measures the ability to respond to short radiation pulses or, in other terms, its modulation frequency limits and is obviously measured in seconds. For cw applications a very convenient choice of detector is represented by liquid helium cooled fast InSb hot-electron bolometers. These have very good figures of merit concerning their sensitivity and response time. NEP is of the order of 10^{-13} $W/\sqrt{\ }$ Hz while they show a response time of the order of 1 μs, which limits the modulation frequency to 1 MHz. Schottky diodes are used as detectors in different detection schemes described in literature. They are generally faster (response time of the order of nanoseconds) but have lower sensitivity, with NEP of the order of 10^{-12} W/γ Hz.

3.3 Magnets

In the range of frequencies of interest for HF-EPR (say from 95 to 1 THz) two kinds of vertical bore magnets are generally used: superconductor magnets (up to 17 Tesla) and resistive magnets (up to 30 Tesla). The parameters characterizing a magnet are: maximum reachable field, field stability, homogeneity, bore diameter, maximum sweep rate and stray field. The homogeneity of superconductor magnets for EPR is generally of some ppm. Bore diameter range from few cm to 88 mm. Homogeneities of the same order of magnitude have been reached for resistive magnets optimized for high resolution applications. Since resistive (Bitter) magnets require the use of a massive stream of cooling water, mechanical vibrations are often critical. The development of high field magnets for EPR followed that for NMR. The main difference between the NMR and EPR magnets is in the fact that the latter need to be sweepable at constant homogeneity. These are somehow contradictory constraints, since an easily sweepable magnet requires a low inductance, while for highly homogeneous one a high inductance is requested. A compromise is generally made between the imposed technological constraint. For static fields higher than 35 Tesla hybrid magnets are available in few specialized facilities. These are a combination of a resistive coil with a coaxial superconducting coil. Even higher fields can be reached for very short times (fractions of seconds) by means of long (60–80 T) and destructive pulse magnets (800–900 Tesla). Superconducting coils can be operated either sweeping the main coil or, when existent, by using a smaller coil (sweep coil) keeping the main coil persistent. The latter configuration has the advantage to reduce the liquid helium boil off and to avoid the problems linked to the hysteresis of the main coil. The use of split coil and horizontal bore magnets have been described in literature.

3.4 Submillimeter Bridges

The instrumental development of HF-EPR spectroscopy is mainly devoted to the implementation of high frequency bridges including the different function present in the typical microwave bridge depicted in Fig. 5. At frequencies higher than 200 GHz conventional microwave techniques are not convenient. For example the theoretical insertion loss of a single-mode rectangular wave-guide is of the order of 3 dB/m and 12 dB/m at 100 and 250 GHz respectively. Since the microwave technology is no more convenient at higher frequencies, a totally different approach has been introduced in the millimeter and sub-millimeter wavelength range. The simplest way to perform EPR at very high frequencies and field is to transmit the radiation using overmoded metallic wave-guides. These are generally brass or copper pipes with an internal diameter of 8–12 mm. Their insertion loss is of the order of 1 dB/m up to 440 GHz. Using this transmission technique the single pass or traveling wave probe- head can be realized. These offer several advantages and few drawbacks. They are intrinsically very broad band, simple and cheap to realize and manipulate. On the other hand they generally do not use resonator and thus their sensitivity is quite poor. EPR spectrometer have been realized using this technique when the multifrequency requirements are more important than sensitivity considerations. When the sensitivity becomes a critical factor a more sophisticated approach is required. The quasi-optical (q-o) techniques [21] have been first introduced and fully exploited in recent years in HF-EPR spectroscopy. Geometrical optics corresponds to a ray description of radiation which ignores its wave-like nature. This description is generally inappropriate when the wavelength is not small compared with that of the optical devices used (lenses, mirrors, polarizers etc. at 300 GHz = 1 mm). Quasi-optic is both a formalism and a technique which is appropriate to describe and treat the propagation of electromagnetic radiation when the microwaves techniques and geometrical optics are inadequate. The radiation waves are described by a Gaussian beam, which is a modified plane wave whose amplitude decreases not-monotonically as one moves radially from the optical axis [22]. Several elegant applications of the q-o methods have been presented in literature [23–26]. A simple example is reported in Fig. 3 [27]. The block scheme of the q-o bridge shows the path of the radiation from the source to the Fabry Perot Resonator and to the detector. The bridge works in reflection mode and is based on polarization coding of the EPR signal for optimal detection. The radiation delivered by the source is horizontally polarised and passes unchanged through the polarizer. The quarter wave plate converts it into circularly polarized radiation which is delivered to the F-P resonator. The latter is coupled to the transmission line through a electroformed mesh that works also as a mirror. The radiation coming out of the resonator is circularly polarised in the opposite sense compared to the incoming radiation, therefore the quarter wave plate converts it into vertically polarized radiation which is totally reflected by the polarizer and directed into the detector. The main advantage of such setup is the fact that the polarization coded EPR signal is detected on a small background. Different solution have been proposed using q-o techniques and the case illustrated above has to be considered only a simple example between many other. The different solution that have been proposed differ in the detection scheme, in the transmission line, and in the kind of used resonating structure. As mentioned above Fabry Perot resonators can be conveniently used. These are devices made up by a couple of coaxial mirrors which work as interferometers. The multliple reflections within the two mirrors give rise to multiple bean interference and the transmitted and reflected power shows a seires of maxima and minima. For normal incidence of the radiation the maxima of the power correspond to the condition:

$$\mathbf{d} = \frac{\mathbf{m}\lambda}{2} \tag{8}$$

where d is the spacing between the two mirrors l is the wavelength and m is an integer. The maxima/minima correspond to resonance condition in which the radiation is concentrated in the cavity. Figure of merit for F-P resonators are their free spectral range $\Delta\nu$ defined as the frequency difference between two successive maxima and their finesse:

$$\mathbf{F} = \frac{\Delta\nu}{\Delta\nu_{1/2}} \tag{9}$$

where $\Delta\nu_{1/2}$ is the width of the peak at half maximum, and is equivalent to the Q factor in conventional microwave cavities. Single mode cavities are currently used up to 150 GHz ca and have been recently implemented in HFEPR applications at higher frequencies (360 GHz). Other probes make use of q-o techniques without a resonator. They have been commercialised by Kyospin and make use of a simple reflection arrangement with a bucket containing the sample. The use of dielectric resonators in the whispering gallery modes has been recently proposed [28–31]. The latter are intrinsically broad band devices with very promising Q factors. The use of these resonators in very high frequency is limited up to now to the large insertion losses of the tapered dielectric waveguides necessary to excite the resonant modes of these evanescent field devices. Very active in the development and/or the application of q-o techniques beyond the cited laboratory at Cornell University are the Laboratories in St Andrews [25,32] Berlin [26] Tallahassee [33,34] and Grenoble.

4 EPR Spectra of Systems with Integer Spin

One of the main advantages of HFEPR is that of allowing to measure the spectra of systems with integer spin, which are often impossible to be observed in

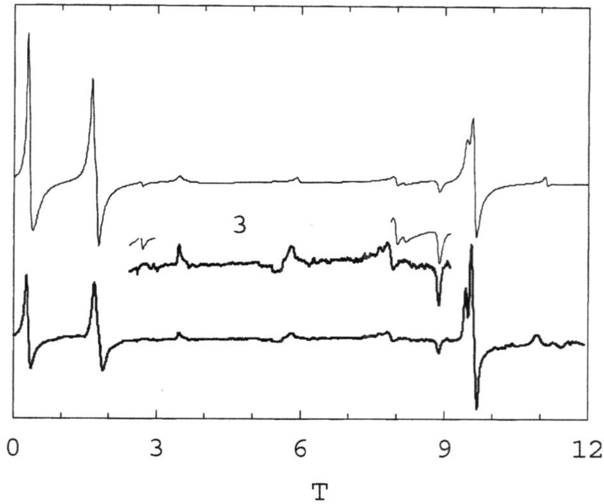

Fig. 6. Polycrystalline powder EPR spectra of $Mn(dbm)_3$ at 349.3 GHz and 15 K. The free electron resonates at ca 12.46 T. After Ref. [35]

conventional EPR spectrometers, because the large zero field splitting requires the use of extremely high fields to see any spectrum at all. A typical example of this kind of behavior is manganese(III). This is a d^4 ion, which in octahedral symmetry has a ground ^5E_g state, which is unstable against Jahn-Teller distortions [19]. In general vibronic coupling determines a tetragonally elongated coordination which removes the orbital degeneracy. The low symmetry components of the ligand field are responsible for a large zero field splitting of the $S = 2$ ground state. The $\Delta M = \pm 1$ transitions can occur only at very high field if a 9 GHz frequency is used like in a X-band spectrometer, but at 200 GHz plenty of transitions are observed. For these reasons the spectra reported in the literature of manganese(III) compounds are extremely few up to the end of the "90"s when the advent of HF EPR spectrometers dramatically changed the situation.

One of the first molecular compounds of manganese(III) to be reported [35] is $Mn(dbm)_3$, sketched below:

The manganese(III) ions are bound to six oxygens of the three dbm ligands, with four relatively short and two relatively long bond distances. In fact the usual tetragonal elongation associated to the Jahn-Teller distortion is observed. The EPR spectra recorded at 200 GHz are shown in Fig. 6. In order to fit the spectra a computer simulation was performed, which yielded the following spin hamiltonian parameters: D=-4.35 cm^{-1} and E=0.26 cm^{-1} with $g_x=g_y=1.99$ and $g_z=1.97$. They were well reproduced by ligand field calculations.

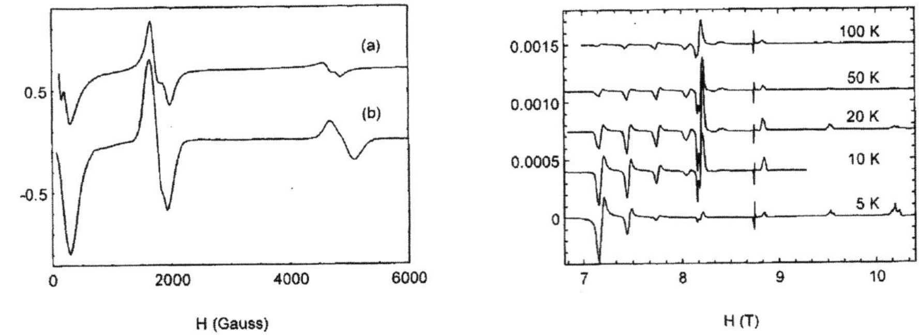

Fig. 7. Polycrystalline powder EPR spectra of Cu_6 at (a) 9 GHz, (b) 245 GHz. After Ref. [36]

5 Molecular Clusters

Molecular clusters have been much studied in the last few years both for their interest in biological systems and for single molecule magnets.

A particularly elegant example showing the advantages of HF compared to conventional EPR spectroscopy is given by a cyclic cluster comprising six copper(II) ions, Cu_6, d^9, S= 1/2 as schematized below [36].

The metal ions are not directly bound, but they are connected by pairs of oxygen bridges. The Cu-O-Cu angles are close to 90° and the superexchange coupling between the copper(II) ions is ferromagnetic in such a way that the ground state has $S = 3$. The polycrystalline powder EPR spectra of Cu_6 recorded at X band are shown in Fig. 7a. It is apparent that a transition close to zero field is present, indicating that a zero field splitting separation between M levels is close to the energy quantum of 0.3 cm^{-1} of the spectrometer. A simulation program provided the best fit spin hamiltonian parameters, $g_{\parallel}= 2.055$, $g_{\perp}= 2.221$, D=0.3 cm^{-1}. The corresponding spectra recorded at 245 GHz are shown in Fig. 7b. A fine structure is clearly observed, with sets of lines approximately equidistant corresponding to crystallites with the unique axis or the molecular plane parallel to the external field, respectively. The separation between the parallel lines is twice that of the perpendicular lines, in agreement with the high field limit of the EPR spectra (7), and it provides immediately the $2D/g\mu_B$ parameter, which agrees very well with that obtained by the laborious fitting of the X-band spectra.

Another important information which is contained in the HF-EPR spectra is the sign of the zero field splitting parameter which can be obtained by the temperature dependence of the intensity of the signals. It is apparent that the relative intensities of the transitions dramatically change on decreasing temperature as shown in Fig. 7b. In fact the external transitions (either low- or high field) increase their relative intensities with respect to the other transitions. This is due to the fact that at 245 GHz the Zeeman energy corresponds to ca. 12 K.

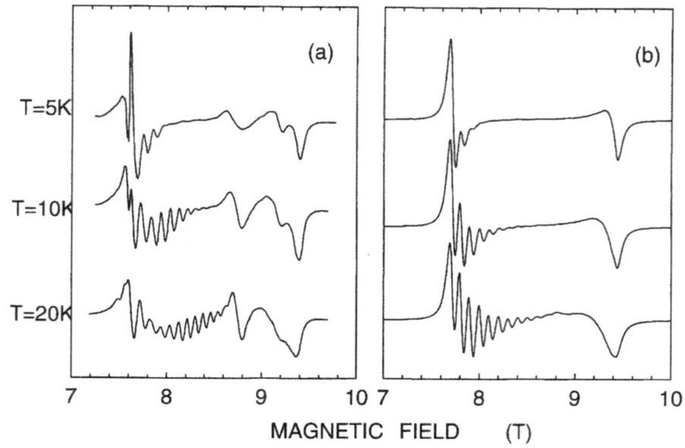

Fig. 8. Polycrystalline powder EPR spectra of Mn_{10} at 200 GHz. The spectra were measured on loose powders. Orientation effects were taken into account through a gaussian distribution of crystallites. (a) experimental; (b) calculated

Going to low temperature, the only transition corresponding to the lowest lying energy state in the presence of the external magnetic field will be observed, due to Boltzmann depopulation effects. In particular it is observed that the low field transition corresponds to a perpendicular transition, while the high field transition corresponds to a parallel transition. This means that the anisotropy is of the easy plane type, or the zero field splitting parameter D is positive. In the case of an easy axis type magnetic anisotropy (negative D) the pattern of transition intensities would be reversed, with the parallel transition observed at low field and the perpendicular one at high field.

Among the many advantages of HF EPR it is also important to mention some problems. An important one is associated to the use of high fields, which generate high magnetic anisotropies which give rise to torques which tend to orient the polycrystalline powders in the spectrometers. This is clearly shown in the spectra reported in Fig. 8 [37]. They correspond to a cluster comprising ten manganese ions, Mn_{10}, which was reported to have a ground state with $S = 12$ by magnetization measurements. The HF-EPR spectra confirmed that the ground state corresponds to $S = 12$. However the spectra are completely different if recorded on pellets or on loose powders. The latter show clear orientation effects, which may also give rise to incorrect assignment of the spectra if not recognized. On the other hand it must be stressed that it may be an advantage to be able to orient the powders, thus obtaining pseudo single crystal spectra. The orientation of the crystallites is best avoided by embedding the powders in wax.

A molecular cluster which has recently been suggested to be a model for one of the species present in Photosystem II has been characterized also by HF EPR spectroscopy [38] [39]. Photosystem II is the center responsible of the catalytic water oxidation performed in the photosynthetic process. One of the active species is a cluster comprising four manganese ions, which in the catalytic process are oxidized in steps, one electron at a time [40,41]. One of these intermediate forms has been found to give rise to a signal at $g = 4.2$ in a spectrum recorded at X-band frequency. The shape of the signal can be compatible both with an isotropic signal or with the perpendicular feature of an axial spectrum. Now, since the cluster comprises manganese(III) and manganese(IV) ions it is beyond doubt that the low resonance field cannot be associated with a large orbital contribution. Rather, signals in the same spectral region are usually observed in systems with $S > 1/2$, an odd number of unpaired electrons and a large zero field splitting. Let us take into consideration for instance a system with $S = 3/2$. An axial ligand field will split the S manifold into two Kramers doublets, $\pm 1/2$ and $\pm 3/2$, separated by $2D$. If D is larger than the microwave quantum only the transitions between the split components of the $\pm 1/2$ will be observed. With the external field parallel to the z axis the energies of the two levels will be $\pm\ g\mu_B H/2$, while for the field perpendicular to the axis the energies will be $g\mu_B H$. For the parallel orientation the resonance field is given by $H = h\nu/(g\mu_B)$, while for the perpendicular orientation $H = h\nu/(2g\mu_B)$. If $g = 2$ the resonance for the latter will be observed at ca. 4. A similar situation is met with $S = 5/2$, which in the case of completely rhombic splitting, $E/D = 1/3$ gives rise to an isotropic signal at $g = 4.3$. The transition involves the split components of the middle Kramers doublet. Magnetization measurements on the natural system provided evidence for the latter hypothesis, suggesting the zero field splitting parameters $D = 1.7\ (-1.1)\ \mathrm{cm}^{-1}$, $E/D = 0.25$ [42].

Another approach for discussing the structural features of the active cluster is that of investigating model compounds. Armstrong et al. reported a tetranuclear cluster comprising three manganese(IV) and one manganese(III), Mn_4, whose structure is sketched in Fig. 9 [43]. The temperature dependence of the magnetic susceptibility showed that the ground state is $S = 5/2$, as suggested for the natural cluster. The X-band EPR spectra showed a signal at $g = 4.2$, again in agreement with the presence of a ground $S = 5/2$ state with large magnetic anisotropy. HF EPR spectra recorded at 195 GHz showed [44] a larger number of transitions, showing that the zero field splitting is smaller than the exciting frequency. Indeed the spectra were fit with $D = -1.11\ \mathrm{cm}^{-1}$, $E/D = 0.24$.

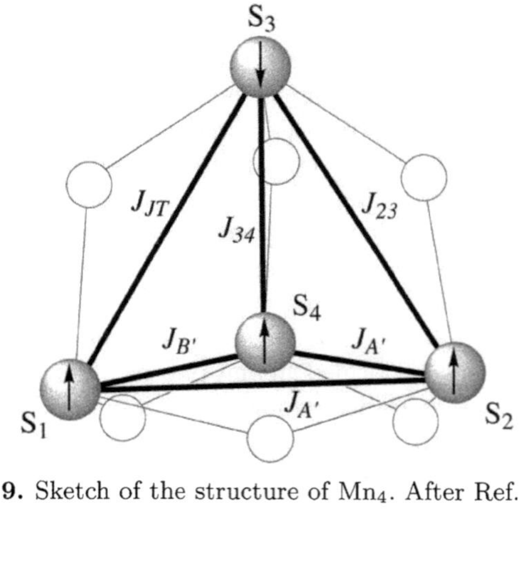

Fig. 9. Sketch of the structure of Mn_4. After Ref. [38]

6 Single Molecule Magnets

One of the fields where the use of HF-EPR spectroscopy has provided dramatic advantages for the understanding of the magnetic properties is that of the so-called single molecule magnets, SMM [45,46]. These are molecules which contain a large, but finite, number of interacting magnetic centers, with a large ground spin state, and a large easy axis type magnetic anisotropy. The first consequence is that the relaxation of the magnetization becomes extremely long at low temperature, the molecules behaving like superparamagnets [47]. Below the blocking temperature they give rise to magnetic hysteresis, like bulk magnets, whose origin is in the slow relaxation. Rather unusually this classic behavior is associated also with large quantum effects which have their origin on the small size of the molecules [48,49]. The SMM's are therefore been almost ideal systems for testing the theories of quantum behavior in mesoscopic matter. Beyond quantum tunnelling effects, other exotic phenomena like the Berry phase in magnets have been unequivocally observed in these SMM's [50].

The first SMM is a cluster comprising twelve manganese ions, with the structure sketched in Fig. 10 [51]. The eight external ions are manganese(III), each with four unpaired electrons, and $S = 2$, while the four internal ones are manganese(IV), each with three unpaired electrons. The ground state has $S = 10$, and is characterized by a large zero field splitting, with a negative D, i.e. by an easy axis type magnetic anisotropy. The information on the nature of the ground state zero field splitting came from HF EPR spectra. The highest frequency spectra reported so far are shown in Fig. 11 [52]. They correspond to an exciting frequency of 525 GHz, and even at 30 K they show clear polarization effects due to the fact that the Zeeman energy is ca. 25 K. The features at low field correspond to transitions of crystallites with the static magnetic field parallel to the tetragonal axis of the cluster, while the high field features correspond

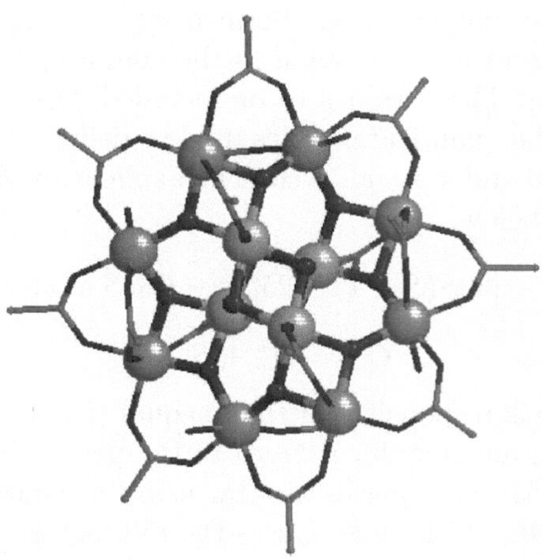

Fig. 10. Sketch of the structure of Mn_{12}. After Ref. [51]

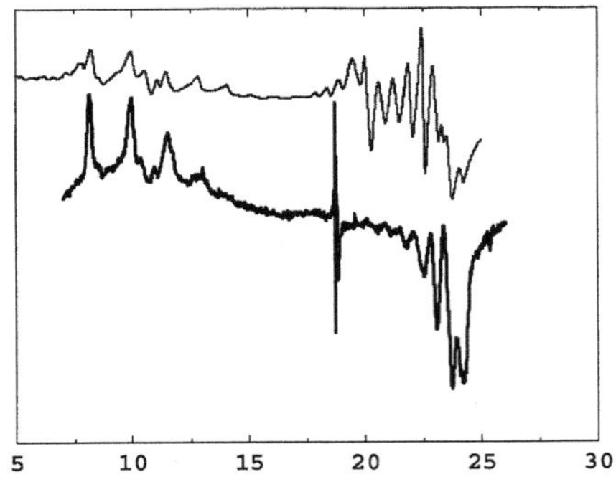

Fig. 11. Polycrystalline powder EPR spectra of Mn_{12} recorded at 525 GHz. After Ref. [52]

to transitions of crystallites with the static field perpendicular to the unique axis. This is a clear indication of a negative D parameter, as shown in Sect. 2. A closer look to the spectra clearly indicates that the separation between the neighboring lines, either parallel or perpendicular, is not constant. This is an indication that the second order magnetic anisotropy terms, D and E, are not sufficient to describe the properties of Mn_{12}, and higher order perturbation terms must be added. Considering only even order terms the first terms to be added correspond to fourth order terms. The best suited way of including higher order terms to describe the zero field splitting of S multiplets is to use the so-called Stevens operator equivalents [53]:

$$\Sigma_{n,k} B_n^k O_n^k \tag{10}$$

where the B_n^k are parameters to be obtained from the analysis of experimental data and O_n^k are operators of power n in the spin angular momentum. k is restricted to $0 \leq k \leq$ n. The k values to be included depend on the symmetry of the system and on the quantization axes used to define the basis set. For tetragonal symmetry $k = 0$ and 4 are allowed. The explicit forms of the corresponding operators are given below:

$$O_4^0 = 35S_z^4 - [30S(S+1) - 25]S_z^2 - 6S(S+1) + 3S^2(S+1)^2 \tag{11}$$

$$O_4^4 = (S_+^4 + S_-^4)/2 \tag{12}$$

The O_4^0 operator is responsible of the unequal spacing of the parallel transitions, while O_4^4 gives an irregular pattern to the perpendicular transitions. The analysis of the spectra was performed by using a computer program, implemented by Weihe [54], which diagonalizes the full matrix of 21 states associated to the S= 10 ground state. The results are shown in Table 2.

Beyond HF-EPR spectroscopy information on the zero field splitting of the Mn_{12} cluster have been obtained also by inelastic neutron scattering [56] and

Table 2.

D	−0.46(2)	−0.457(2)	−0.47	−0.46
B_4^0	$-2.2(2)\times10^{-5}$	$-2.33(4)\times10^{-5}$	-1.5×10^{-5}	-2.19×10^{-5}
B_4^4	$\pm4(1)\times10^{-5}$	$\pm3.0(5)\times10^{-5}$	-8.7×10^{-5}	
Ref	[55]	[56]	[57]	[58]

by zero field EPR spectroscopy [58]. The latter is a quasi-optic sub-millimeter backward-wave- oscillator technique, BWO. The BWO technique is operated in the frequency-scanning mode. The scanning is carried out at different points by varying the BWO supply voltage by steps. The frequency resolution of the apparatus $\Delta\nu/\nu$ is $10^{-4}-10^{-5}$. Three transitions were observed in the frequency range 0-35 cm^{-1}, corresponding to the transitions $\pm10 \rightarrow \pm9, \pm9 \rightarrow \pm8$, and $\pm8 \rightarrow \pm7$. Also in this case the irregular spacing of the levels confirmed the need of fourth order terms.

The temperature dependence of the relaxation time of the magnetisation has been found to follow a thermally activated behaviour, with a barrier Δ= 60 K [47]. A simple theoretical model developped by Villain et al. [59], which assumes a generalised Orbach mechanism for the relaxation of the magnetisation, using only second order zero field splitting terms suggests that the barrier for the reorientation of the magnetisation in the case of negative D is given by:

$$\Delta = DS^2 \tag{13}$$

Comparing with the values of Table 2 it is apparent that the height of the barrier calculated using the experimental parameters is higher than 60 K. A possible reason of the deviation may be due to the detail of the mechanism of relaxation. The simple theoretical model assumes that in order to relax the system must pass from the lowest $M = -10$ to $M = -9$, then to $M = -8$ up to $M = 0$. However it is also possible that for M levels relatively high in energy the system can relax by jumping directly from, say, -2 to $+2$, without reaching the top of the energy levels. This cross-cut can be considered as a quantum tunnelling effect, and indeed evidence for this has been found in the stepped hysteresis of the magnetization observed at low temperature [48,49].

The quantum tunneling effects are determined by the admixture of $+M$ and $-M$ states which is brought about by transverse magnetic fields, as can be generated by zero field splitting terms. Assuming tetragonal symmetry only the B_4^4 parameters give an admixture of states. In first order perturbation states differing in M by ±4 can be admixed, and several theoretical models have been tested using the experimental values obtained through HF EPR spectroscopy.

Mn_{12} is not unique: beyond the long list of similar compounds obtained by changing the carboxylate groups another compound has been widely investigated in the frame of mesoscopic magnetic properties. This is a cationic cluster comprising eight iron(III) ions, $S = 5/2$, whose structure is shown in Fig. 12 [60,61]. It has a ground $S = 10$ state, like Mn_{12}, but the symmetry is lower than axial. Again HF-EPR has been extremely useful to provide information on the zero

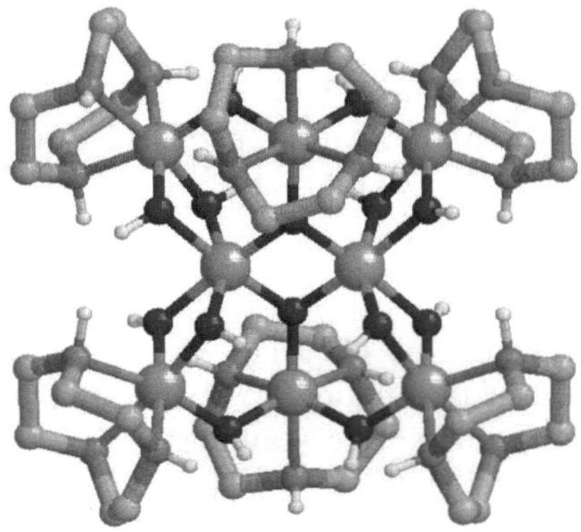

Fig. 12. Sketch of the structure of Fe_8. After Ref. [65]

field splitting of the ground state. At the simplest level of approximation they have been interpreted using the hamiltonian (6) [62], with $D = -0.191$ cm^{-1}, $E/D = 0.168$. Single crystal spectra showed also the direction of the principal axes [63]. The spectra corresponding to the easy axis, reveal that the separation between neighboring lines is not regular, again demanding the introduction of fourth order terms. These were also independently obtained from INS experiments [64]. The line widths depend on the M value of the level involved in the transition, approximately as M^2. This has been tentatively attributed to strains which give a distribution of D values in the crystal.

The smaller D and the larger E/D ratio of Fe_8 compared to Mn_{12} suggest that the barrier is lower for the latter, and that the relaxation mechanism is more efficient. Both these points have been borne out by experiment, thus giving confirmation to the models of relaxation. A further confirmation on the role of transverse field for the relaxation of the magnetization is provided by the EPR spectra of derivative of Fe_8 in which some of the bromide counter anions have been substituted by perchlorate anions, Fe_8PCl. The polycrystalline powder EPR spectra of the latter immediately reveal (Fig. 13) that the transverse field is larger, because the splitting of the perpendicular transitions at high field is higher in the perchlorate than in the bromide derivative. This prediction is confirmed by the observed relaxation times of the magnetization.

7 Conclusions and Perspectives

High field EPR spectroscopy has only recently started to become wide spread, but the number of possible applications has rapidly increased. In this short contribution we have only touched some examples, which we hope were significant, but certainly it must be expected that many more are possible. For instance we have not touched low dimensional materials for which in perspective HF EPR can provide useful information. Improvements must be expected on the

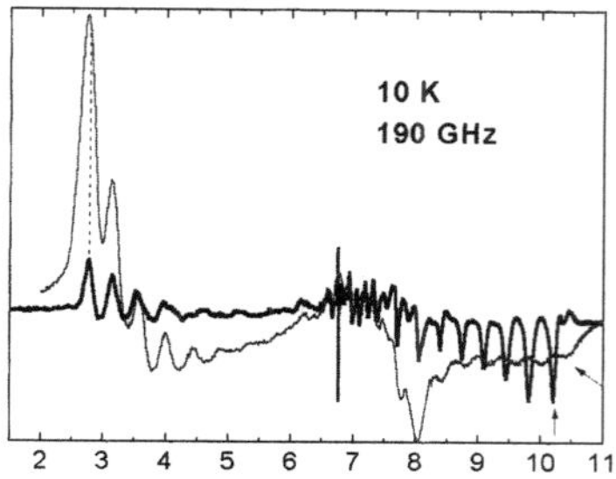

Fig. 13. Polycrystalline powder EPR spectra at 195 GHz of Fe_8 and Fe_8PCl

instrumental side, where the sensitivity so far is far from optimal. Also pulsed techniques can provide a wealth of information, if properly exploited, and double resonance techniques, like ENDOR can enjoy large advantages associated with the higher spectral resolution.

Acknowledgements

The financial support of the EU Network SENTINEL is gratefully acknowledged.

References

1. A. Abragam, B. Bleaney: *'Electron Paramagnetic Resonance of Tansition Ions'* (Dover, New York, 1986)
2. A. Bencini, D. Gatteschi: *'EPR of Exchange Coupled Systems'* (Springer-Verlag, Berlin, 1990)
3. J.R. Pilbrow: *'Transition Ion Electron Paramagnetic Resonance'* (Clarendon Press, Oxford, 1990)
4. F.E. Mabbs, D. Collison: *'Electron Paramagnetic Resonance of d Transition Metal Compounds'* (Elsevier, Amsterdam, 1992)
5. J.E. Wertz, J.R. Bolton: *'Electron Spin Resonance: Elementary Theory and Practical Applications'* (McGraw Hill, New York, 1972)
6. A. Bencini, D. Gatteschi: *'Inorganic Electronic Structure and Spectroscopy'*, eds. E.I. Solomon, A.B.P. Lever (Wiley, New York, 1999), Vol. I, p. 93
7. P.J. Bratt et al.: J Phys. Chem. B **101**, 9686 (1997)
8. M.C.W. Evans, J.H.A. Nugent: *'The Photosynthetic Reaction Center'*, eds. J. Deisenhofer, J.R. Norris (Academic Press, San Diego, 1993), Vol. 1, p. 391
9. A.L. Barra et al.: Angewandte Chemie, International Edition in English **36**, 2329 (1997)
10. D.P. Goldberg et al.: J. Am. Chem. Soc. **119**, 8722 (1997)
11. P.J. Van Dam, A.A.K. Klaassen, E.J.Reijerse, W.R.Hagen: J. Magn. Reson. **130**, 140 (1998)
12. J.Telser, L.A. Pardi, J.Krzystek, L.C.Brunel: Inorg. Chem. **37**, 5769 (1998)

13. P.L.W. Tregenna-Piggott et al.: Inorg. Chem. **38**, 5928 (1999)
14. L.A. Pardi et al.: Inorg. Chem. **39**, 159 (2000)
15. J. Bendix, H.B. Gray, G. Golubkov, Z. Gross: J. Chem. Soc, Chem. Commun. 1957 (2000)
16. J. Mrozinski et al.: J. Mol. Struct. **559**, 107 (2001)
17. J. Limburg et al.: Inorg. Chem. **40**, 1698 (2001)
18. B.R. McGarvey: Transition Metal Chemistry **3**, 89 (1966)
19. J.S. Griffith: *'The Theory of Transition Metal Ions'* (Cambridge University Press, Cambridge 1961),
20. C.P. Poole: *'Electron Spin Resonance, A Comprehensive Treatise on Experimental Techniques'*, (Dover, Mineola, New York 1996)
21. J.C.G. Lesurf: *'Millimetre- Wave Optics, Devices, and Systems'*, (Dover, Bristol 1996)
22. K.A. Earle, D.E. Budil, J.H. Freed: Adv. Magn. Reson. and Opt. Reson. **19**, 253 (1996)
23. K.A. Earle, J.H. Freed: Appl. Magn. Reson. **16**, 247 (1999)
24. K.A. Earle, D.S. Tipikin, J.H. Freed: Rev. Sci. Instrum. **67**, 2502 (1996)
25. G.M. Smith, J.C.G. Lesurf, R.H. Mitchell, P.C. Riedi: Rev. Sci. Instrum. **69**, 3924 (1998)
26. M.R. Fuchs, T.F. Prisner, K. Mobius: Rev. Sci. Instrum. **70**, 3681 (1999)
27. J.H. Freed: Annual Rev. Phys. Chem. **51**, 655 (2000)
28. G. Annino et al.: Appl. Magn. Reson. **19**, 495 (2000)
29. G. Annino et al.: J. Magn. Reson. **143**, 88 (2000)
30. G. Annino et al.: Rev. Sci. Instrum. **70**, 1787 (1999)
31. G. Annino, M. Cassettari, I. Longo, M. Martinelli: Chem. Phys. Lett. **281**, 306 (1997)
32. G.M. Smith, E.J. Milton: International Journal of Remote Sensing **20**, 2653 (1999)
33. A.K. Hassan et al.: Appl. Magn. Reson. **16**, 299 (1999)
34. A.K. Hassan et al.: J. Magn. Reson. **142**, 300 (2000)
35. A.L. Barra et al.: Angewandte Chemie, International Edition in English **36**, 2329 (1997)
36. E. Rentschler et al.: Inorg. Chem. **35**, 4427 (1996)
37. D.P. Goldberg et al.: J. Am. Chem. Soc. **115**, 5789 (1995)
38. C.E. Dube et al.: J. Am. Chem. Soc. **121**, 3537 (1999)
39. W.H. Armstrong et al.: Abstracts of Papers of the Am. Chem. Soc. **220**, 414-INOR (2000)
40. R.D. Britt: *'Oxygenic Photosynthesis: The Light Reactions'*, eds. D.R. Ort, C.F. Yocum (Kluwer Academic Publishers, Dordrecht 1996), p. 137
41. V.K. Yachandra, K. Sauer, M.P. Klein: Chem. Rev. **96**, 2927 (1996)
42. O. Horner et al.: J. Am. Chem. Soc. **120**, 7924 (1998)
43. C.E. Dube et al.: J. Am. Chem. Soc. **121**, 3537 (1999)
44. A.L. Barra: private communication
45. G. Christou, D. Gatteschi, D.N. Hendrickson, R. Sessoli: Mrs Bulletin **25**, 66 (2000)
46. A. Caneschi et al.: J. Magn. Magn. Mater. **200**, 182 (1999)
47. R. Sessoli, D. Gatteschi, A. Caneschi, M.A. Novak: Nature (London) **365**, 141 (1993)
48. L. Thomas et al.: Nature (London) **383**, 145 (1996)
49. J.R. Friedman, M.P. Sarachik, J. Tejada, R. Ziolo: Phys. Rev. Lett. **76**, 3830 (1996)
50. W. Wernsdorfer, R. Sessoli: Science **284**, 133 (1999)
51. T. Lis: Acta Crystallog. B **36**, 2042 (1980)

52. A.L. Barra, D. Gatteschi, R. Sessoli: Phys. Rev. B-Condens. Matter **56**, 8192 (1997)
53. A. Abragam, B. Bleaney, *'Electron Paramagnetic Resonance of Tansition Ions'*, (Dover, New York, 1986)
54. C.J.H. Jacobsen, E. Pedersen, J. Villadsen, H. Weihe: Inorg. Chem. **32**, 1216 (1993)
55. A.L. Barra, D. Gatteschi, R. Sessoli: Phys. Rev. B-Condens. Matter **56**, 8192 (1997)
56. I. Mirebeau et al.: Phys. Rev. Lett. **83**, 628 (1999)
57. S. Hill et al.: Phys. Rev. Lett. **80**, 2453 (1998)
58. A.A. Mukhin et al.: Europhys. Lett. **44**, 778 (1998)
59. J. Villain, F. Hartman-Boutron, R. Sessoli, A. Rettori: Europhys. Lett. **27**, 159 (1994)
60. K. Wieghardt, K. Pohl, I. Jibril, G. Huttner: Angewandte Chemie, International Edition in English **23**, 77 (1984)
61. C. Delfs et al.: Inorg. Chem. **32**, 3099 (1993)
62. A.L. Barra et al.: Europhys. Lett. **35**, 133 (1996)
63. A.L. Barra, D. Gatteschi, R. Sessoli: Chemistry-a European Journal **6**, 1608 (2000)
64. R. Caciuffo et al.: Phys. Rev. Lett. **81**, 4744 (1998)
65. K. Wieghardt, K. Pohl, I. Jibril, G. Huttner: Angewandte Chemie, International Edition in English **23**, 77 (1984)

Pulsed-High Field/High-Frequency EPR Spectroscopy

Michael Fuhs and Klaus Möbius

Institut für Experimentalphysik, Freie Universität Berlin,
Arnimallee 14, 14195 Berlin, Germany

Abstract. Pulsed high-field/high-frequency electron paramagnetic resonance (EPR) spectroscopy is used to disentangle many kinds of different effects often obscured in continuous wave (cw) EPR spectra at lower magnetic fields/microwave frequencies. While the high magnetic field increases the resolution of G tensors and of nuclear Larmor frequencies, the high frequencies allow for higher time resolution for molecular dynamics as well as for transient paramagnetic intermediates studied with time-resolved EPR. Pulsed EPR methods are used for example for relaxation-time studies, and pulsed Electron Nuclear DOuble Resonance (ENDOR) is used to resolve unresolved hyperfine structure hidden in inhomogeneous linewidths. In the present article we introduce the basic concepts and selected applications to structure and mobility studies on electron transfer systems, reaction centers of photosynthesis as well as biomimetic models. The article concludes with an introduction to stochastic EPR which makes use of another concept for investigating resonance systems in order to increase the excitation bandwidth of pulsed EPR. The limited excitation bandwidth of pulses at high frequency is one of the main limitations which, so far, made Fourier transform methods hardly feasible.

1 Introduction

Electron Paramagnetic Resonance (EPR) spectroscopy has been developed for applications in various fields of physics, chemistry and biology for about 40 years. Until 10 years ago most EPR spectrometers worked at magnetic fields of 0.34 T and corresponding microwave frequencies of 9.5 GHz, measuring the absorption of continuously irradiated microwaves (continuous wave, cw-EPR). Nowadays, the two major trends are the development of advanced pulsed techniques and the use of high magnetic fields (HF-EPR) from 3.4 T/95 GHz up to 12.8 T/360 GHz or even higher. In particular, pulsed EPR at high magnetic fields offers various advantages in investigating for example protein structure and dynamics. However, at high fields the concomitant increase in microwave frequencies imposes strong technical limitations on the possibility of using pulsed techniques. Nevertheless, in the last years the number of pulsed spectrometers at 3.4 T/95 GHz and 5 T/140 GHz has been rapidly increasing and pulsed spectrometers at 7 T/180 GHz, 10 T/280 GHz and 12.8 T/360 GHz are currently being developed. Generally, there is no hard criterion for defining which magnetic field fulfills the condition for being "high field". It depends very much on sample and application. High-field EPR is used for increase in spectral resolution, increase in sensitivity for small samples, information on molecular motion at shorter time scales, and the accessibility of large zero-field fine-structure splittings.

EPR is usually applied to systems containing only one or two electron spins. Their resonance lines are often inhomogeneously broadened. The questions and also the reasons for using pulsed instead of cw-EPR are different compared to Nuclear Magnetic Resonance (NMR), although the theoretical treatment is similar. Pulsed high-field/high-frequency EPR is mainly used to measure relaxation times, which gives insight in the dynamics of the spin carrying site and its surroundings. Another goal is to measure hyperfine couplings of nuclei using Electron Nuclear DOuble Resonance (ENDOR) or Electron Spin Echo Envelope Modulation (ESEEM), which gives insight in the structure of the spin environment. Depending on experimental conditions, pulsed methods may provide a sensitivity advantage, sometimes due to underlying theoretical principles, sometimes due to technical reasons. For NMR pulsed techniques are standard by now and are applied to systems containing many nuclear spins which are somehow coupled but nevertheless distinguishable by their chemical shift or gyromagnetic moment. Sophisticated multidimensional pulsed methods often allow, for example, the measurement of selected couplings and distances. Even though in solid-state NMR the spectra of disordered samples might be inhomogeneously broadened by anisotropic interactions, Magic Angle Spinning (MAS) and special pulse sequences allow spectral resolution to a high degree. In NMR it has been demonstrated how advanced pulsed techniques, compared to cw-techniques, can enormously increase information content and sensitivity. In contrast, the thousandfold higher EPR frequencies lead to technical complications which restrict the feasibility of these sophisticated pulse methods.

2 Interactions Measured by EPR

EPR spectra are sensitive to many different static and dynamic magnetic interactions. Furthermore, most interactions depend on the orientation of the paramagnetic centers with respect to the magnetic field and are, therefore, described by anisotropic interaction tensors. Only for systems which are fast tumbling in solution and thereby exhibit only averaged interactions, or for single crystals, one observes single narrow resonance lines. However, most of the samples are measured in disordered states (frozen solutions or powders) and, therefore, exhibit inhomogeneous line broadening due to overlap of different orientations and anisotropic interactions. Additional inhomogeneous line broadening occurs due to unresolved electron spin - nuclear spin interactions (hyperfine structure). Modulation of magnetic interactions which may be caused by spatial fluctuations, thereby modulating anisotropic magnetic interactions, leads to relaxation effects. The transversal T_2 relaxation is related to the intrinsic width of a resonance line, the homogeneous line broadening. Often it is a hard task to disentangle all the different effects visible in EPR spectra and to relate them with information on structure and dynamics.

However, these effects scale differently with magnetic field strength which leads to the strategy of multifrequency EPR and Zeeman magnetoselection. With magnetoselection one selects specific orientations of molecules by resolving their different resonance frequencies due to anisotropic Zeeman interaction even in disordered samples. Furthermore, with the aid of pulsed EPR, it is possible to disentangle homogeneous and inhomogeneous broadening and thereby to gain information on dynamics from the relaxation behavior. Double resonance methods such as ENDOR allow to measure hyperfine interactions even when they are unresolved in inhomogeneously broadened EPR spectra.

One has to distinguish single spin ($S = 1/2$), high spin ($S \geq 1$) and many electron systems (two or more weakly coupled $S = 1/2$ spins). For $S = 1/2$ spin systems the most important interactions are the electron Zeeman and the hyperfine interactions. For the other two cases one has to consider also exchange and dipolar coupling. In the following we give a short introduction to the magnetic interactions of $S = 1/2$ spin systems and refer to standard EPR text books, such as [1,2], for further reading.

2.1 Electron Zeeman and Hyperfine Interactions

The Hamiltonian $\hat{\mathcal{H}}$ for the EPR relevant splittings of a $S = 1/2$ system in a static magnetic vector field $\boldsymbol{B}_0$ is given by

$$\hat{\mathcal{H}} = \mu_B \boldsymbol{B}_0 \,\bar{\bar{\mathrm{G}}}\, \hat{\boldsymbol{S}} + \sum_k \hat{\boldsymbol{S}} \,(\bar{\bar{\mathrm{A}}}{}'_k + a_{\mathrm{iso}k})\, \hat{\boldsymbol{I}}_k \,, \tag{1}$$

where μ_B is the Bohr magneton, $\bar{\bar{\mathrm{G}}}$ is the anisotropic G tensor, $\hat{\boldsymbol{S}}$ the electron spin operator, $\hat{\boldsymbol{I}}_k$ the spin operator of nuclear spin k, $\bar{\bar{\mathrm{A}}}{}'_k$ its anisotropic hyperfine interaction tensor and $a_{\mathrm{iso}k}$ is its isotropic Fermi contact interaction. The G tensor contains the g value for the free electron, g_e, and contributions from the spin-orbit coupling of not completely quenched angular momentum. Because the quenching of angular momentum is related to the electric field the spin carrying molecule is exposed to in its environment, the G tensor is anisotropic and contains information on geometric and electronic structure in an integrated form. Therefore, for quantitative interpretations of measured G tensors it is often necessary to use complex quantum chemical calculations. Contrary to this, the hyperfine splittings can be directly linked to electronic structure. The isotropic coupling $a_{\mathrm{iso}k}$ is given by $(8\pi/3)\, g\mu_B\, g_n\mu_n\, |\Psi(r_k)|^2$ and, therefore, proportional to the probability of finding the electron with wave function Ψ at the position r_k of nucleus k. $\bar{\bar{\mathrm{A}}}{}'$ contains the dipolar interaction of electron and nuclear spins. By measuring the angular dependence of the interaction energy related to $\bar{\bar{\mathrm{A}}}{}'$, which is given by

$$\frac{\mu_0}{4\pi}\mu_e\mu_n \,\frac{3\cos^2\alpha - 1}{r^3} \tag{2}$$

with μ_e, μ_n the magnetic moments of electron and nuclear spins, one can, for instance, determine distance and orientation of hydrogen bonds of quinone anion

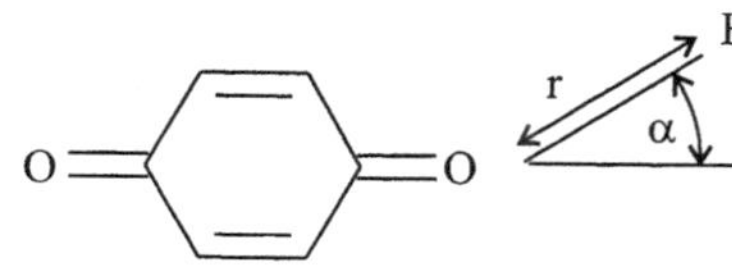

Fig. 1. Hydrogen bond to the oxygen of a quinone anion radical

radicals in various environments (Fig. 1, see Sect. 5.2). This technique can also be successfully applied to proteins with paramagnetic centers such as Cu atoms, for which complementary NMR studies are blind up to 5 Å around the paramagnetic center.

Diagonalizing (1) by first order perturbation theory and neglecting the nuclear Zeeman term one obtains the following orientation dependent g and A'_k values

$$g(\phi,\theta) = \sqrt{\frac{\boldsymbol{B}_0}{B_0}\,\bar{\bar{\mathrm{G}}}\bar{\bar{\mathrm{G}}}\,\frac{\boldsymbol{B}_0}{B_0}}\,, \qquad A'_k(\phi,\theta) = \sqrt{\frac{\boldsymbol{B}_0}{B_0}\,\bar{\bar{\mathrm{A}}}'_k\bar{\bar{\mathrm{A}}}'_k\,\frac{\boldsymbol{B}_0}{B_0}}\,. \tag{3}$$

For technical reasons in EPR one keeps the frequency of the irradiating microwaves fixed and varies instead the B_0 value. The resonance fields are given by

$$B_{\mathrm{res}} = \frac{h\nu}{g(\phi,\theta)\mu_B} - \frac{1}{g(\phi,\theta)\mu_B}\sum_k \left(A'_k(\phi,\theta) + a_{\mathrm{iso}k}\right) m_{I_k}\,. \tag{4}$$

Here m_{I_k} is the magnetic quantum number of nuclear spin k. The five EPR hyperfine transitions of a spin system with $S = 1/2$ and four equivalent nuclei with $I = 1/2$ are shown in Figs. 2a and b. While in EPR one measures the electron spin transitions, in ENDOR one monitors the nuclear spin transitions, excited by an additionally applied radiofrequency (rf) field, via its influence on the EPR amplitude. While the number of EPR resonance lines doubles for each additional non-equivalent nucleus, the number of ENDOR lines increases only additively by two. Also for equivalent nuclei, in ENDOR the number of resonance lines is significantly reduced (Fig. 2b). This leads to the resolution advantage of ENDOR for hyperfine structure which in EPR produces overlapping resonance lines.

2.2 High-Field EPR

As already discussed above, the spectra of disordered samples consist of the overlapping resonance lines of different molecular orientations. Only when the shift in resonance for different molecular orientations is larger than the linewidth of the individual lines, it is resolved in the spectra. When the linewidth contribution due to unresolved hyperfine structure, ΔB_{hfi}, is dominant resolution is enhanced at higher magnetic fields/microwave frequencies. This is because the hyperfine splitting is independent of B_0 while the g resolution scales with B_0. This is shown for a quinone anion radical by spectral simulations at different magnetic fields/microwave frequencies, see Fig. 3a. At 3.4 T/95 GHz the principal values of the G tensor are well resolved and, by performing additional ENDOR, ESEEM or relaxation time measurement at these resonance fields, one selects only those molecules with their respective principal axis of the G tensor oriented along $\boldsymbol{B}_0$. The 'high field' condition to be fulfilled for good magnetoselection is therefore

$$\frac{\Delta g}{g_{\rm iso}} B_0 > \Delta B_{\rm hfi} \, . \tag{5}$$

At higher fields, resolution is also enhanced for nuclear Larmor frequencies. Therefore, high-field ENDOR is often applied to study systems which contain different sorts of nuclei. At lower fields, the ENDOR spectra, centered at the nuclear Larmor frequency of the respective nucleus, are overlapping while at higher fields they are well separated.

3 Pulsed EPR

There are several ways to investigate a spin system by EPR. With continuous wave (cw) EPR one measures the response of continuously irradiated microwaves while sweeping the magnetic field. This is equivalent to the sweep of the microwave frequency at fixed B_0. On resonance, the microwave energy is partially absorbed. Another possibility is to excite the spin system with a single pulse and to monitor the response. One observes a rotating magnetization and its decay due to dephasing of the individual spins (T_2 relaxation) and due to relaxation to thermal equilibrium (T_1 relaxation). This is called Free Induction Decay (FID). The Fourier transform of the FID signal is equivalent to the cw-EPR spectrum.

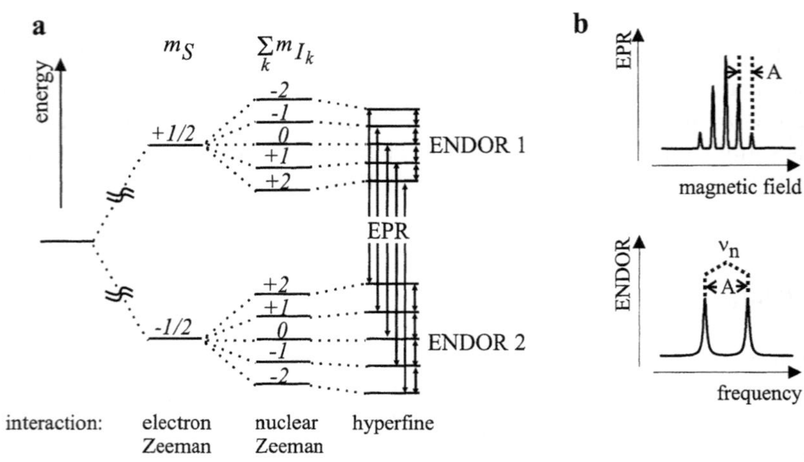

Fig. 2. EPR and ENDOR resonance transitions in a system of one electron with $S = 1/2$ and four equivalent nuclei with $I = 1/2$. $\Delta m_S = \pm 1$ selects the EPR transitions, $\Delta m_I = \pm 1$ selects the ENDOR transitions. (**a**) energy levels. (**b**) EPR and ENDOR spectra

Using more than one pulse, because of the non-linearity of spin systems, electron spin echo (ESE) phenomena appear. Pulsed Fourier transform methods offer the advantage of higher sensitivity. Moreover, sophisticated methods using several pulses help to disentangle the various interaction parameters hidden in cw-EPR spectra. This leads to the field of multidimensional magnetic resonance. Particularly interesting is the combination of high magnetic fields/high microwave frequencies with its high G and high time resolution and pulsed methods. However, the high frequency also leads to limitations of microwave technology and restriction to very basic (field swept) pulsed methods. Some of these field swept methods we introduce in the following. A third method for investigating a spin system by subjecting it to a broadband noise excitation, we will introduce in Sect. 6. For further reading we refer to standard text books [1] for the basic principles and to [3,4] for a thorough treatment of more sophisticated methods.

3.1 Motion of Spins in Magnetic Fields

Generally, the motion of spins in magnetic fields is described with the Liouville equation for the density matrix. The observed magnetization $\langle \boldsymbol{M} \rangle$ is given by the expectation value of $-g\mu_B\hat{\boldsymbol{S}}$. Restricting oneself to non-coupled spin systems, however, the Bloch equations are a good approximation for many purposes:

$$\frac{dM_x}{dt} = \frac{g\mu_B}{\hbar}(\boldsymbol{B}\times\boldsymbol{M})_x - \frac{M_x}{T_2} \tag{6a}$$

$$\frac{dM_y}{dt} = \frac{g\mu_B}{\hbar}(\boldsymbol{B}\times\boldsymbol{M})_y - \frac{M_y}{T_2} \tag{6b}$$

$$\frac{dM_z}{dt} = \frac{g\mu_B}{\hbar}(\boldsymbol{B}\times\boldsymbol{M})_z + \frac{M_0 - M_z}{T_1} \tag{6c}$$

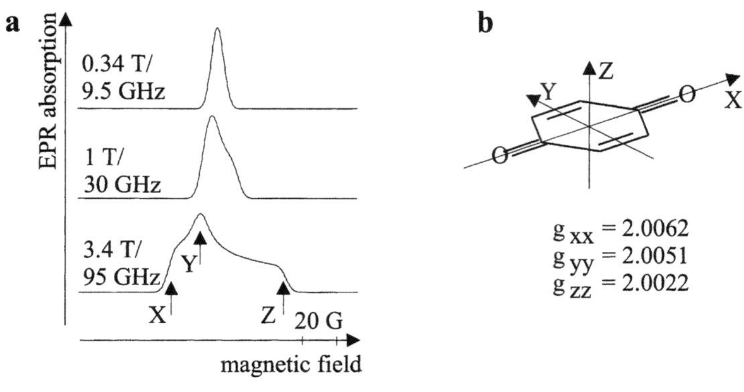

Fig. 3. EPR on disordered samples of quinone anion radicals. (**a**) spectral simulations at different magnetic fields/microwave frequencies. The arrows at the bottom spectrum indicate the orientations which dominantly contribute to the EPR absorption at the respective resonance positions. (**b**) structure of a quinone anion radical with the principal axes of the G tensor and its principal values

The torque of the magnetization in the magnetic field acts on the spin angular momentum which is related to the magnetization by the gyromagnetic factor $g\mu_B$. The z component of $\boldsymbol{B}$ is the static magnetic field B_0, the x and y components contain the magnetic component of the circular polarized microwave field with frequency ω_1, i.e. $\boldsymbol{B} = (B_1 \cos\omega_1 t, B_1 \sin\omega_1 t, B_0)$. The Bloch equations may be used to illustrate the different physical meanings of the two relaxation times: The magnetization of the z component decays with the spin-lattice relaxation time T_1 to its thermal equilibrium value, $(0, 0, M_0)$. This process is connected with energy transfer from or to the lattice. The decay of the transversal components is governed by $T_2 = (1/T_2' + 1/2T_1)^{-1}$. The processes responsible for T_2' lead to a dephasing of individual spin packets and do not transfer any energy from the spin states to the lattice.

Without microwave, the magnetization precesses about $\boldsymbol{B_0}$ with the Larmor frequency $\omega_0 = g\mu_B B_0$. A microwave field has maximum effect on $\boldsymbol{M}$ when ω_1 is equal to the Larmor frequency ω_0. In order to treat the effect of the microwave field, it is convenient to work in a coordinate system rotating with the microwave frequency ω_1 about $\boldsymbol{B_0}$. In this reference frame $\boldsymbol{B_1}$ is stationary along the y_{rot} axis. The magnetization rotates about an effective magnetic field

$$\boldsymbol{B}_{\mathrm{eff}} = \boldsymbol{y}_{\mathrm{rot}} B_1 + \boldsymbol{z}\,\frac{\omega_0 - \omega}{\omega_0}\,B_0 \tag{7}$$

and, therefore, on resonance ($\omega_1 = \omega_0$) it rotates about the y_{rot} axis. A microwave pulse with duration τ_p flips the magnetization of resonant spins with flip angle

$$\alpha(\tau_p) = g\mu_B B_1\,\tau_p\,. \tag{8}$$

3.2 FIDs, Spin Echoes and Field Swept Methods

Following a $\pi/2$ pulse, the magnetization is rotating in the $x_{\mathrm{rot}}y_{\mathrm{rot}}$ plane with $\omega_0 - \omega_1$. In NMR it thereby induces a current in the receiver rf coils, in EPR the coil is replaced by a microwave resonator. However, due to static inhomogeneities of $\boldsymbol{B_0}$ and stochastic fluctuations of local magnetic fields, individual spin packets have different Larmor frequencies. They dephase, thereby reducing the magnetization.

So far, we have assumed that the microwave pulses have the same effect on all the spins in the sample. This is the prerequisite for performing Fourier transform EPR where a single excitation shot has to contain all spectral information. However, not all spins are on resonance and whether our assumption is true or not depends on the excitation bandwidth of the pulses. Figures 4a and c depict the x_{rot}, y_{rot} and z component of the magnetization of an EPR spin ensemble off resonance with $\omega_0 - \omega_1 = 10$ MHz, following 10 ns and 80 ns pulses, respectively. The pulses have different B_1, and according to (8) the flip angle is $\pi/2$. While the 10 ns pulse flips the magnetization almost totally to the $x_{\mathrm{rot}}y_{\mathrm{rot}}$ plane, the 80 ns pulse flips only a small fraction of the magnetization. Figures 4b and d show the bandwidth of the two pulses. Their FWHM bandwidths are 100 and 12.25 MHz (3.6 and 0.5 mT), respectively. It is evident that the spins 10 MHz off

a b c d

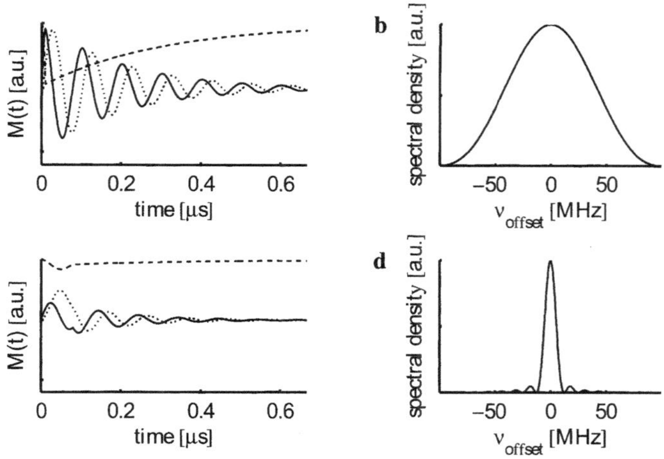

Fig. 4. Magnetization (left) after $\pi/2$ pulses in the frame rotating with ω_1 and pulse excitation bandwidth (spectral density, right). The dashed line is M_z, the solid line $M_{x_{\mathrm{rot}}}$, the dotted line $M_{y_{\mathrm{rot}}}$. The spins are 10 MHz off resonance, $T_1 = T_2' = 0.3\,\mu$s. (**a, b**) $\tau = 10$ ns, $B_1 = 0.89$ mT. (**c, d**) $\tau = 80$ ns, $B_1 = 0.11$ mT. The distortions in **c** during the first 80 ns are caused by the long microwave pulse

resonance are not fully excited by the 80 ns pulse. This is generally the problem in high-field/high-frequency EPR. The $\pi/2$ pulse lengths e.g. at 95 GHz are generally longer than 40 ns, often about 80 ns and even longer[1]. EPR spectra of typical organic radicals with resolved G tensors are often broader than 5 or 10 mT (140 or 280 MHz). Therefore, Fourier transform EPR is not possible for these samples. Nevertheless, pulsed methods allow to burn holes in the resonance lines and, by sweeping B_0, one can scan step by step the response to the pulses for all spins in the sample. Some of the field swept pulsed methods are shown in Fig. 5.

A second pulse with flip angle π, following the first $\pi/2$ pulse at time τ, refocuses the magnetization to the electron spin echo (ESE) at time 2τ (Fig. 5a). In field swept methods the echo intensity is integrated for each magnetic field step. The integrated echo intensity over the magnetic field is basically equivalent to the absorption cw-EPR spectrum. Compared to cw-EPR spectra, ESE detected EPR spectra are insensitive to phase distortions and saturation effects. However, only the spins subject to the same static perturbations before and after the π pulse refocus, while stochastic perturbations lead to an echo decay with relaxation time T_2. Therefore, using different τ it is possible to probe T_2 (Fig. 5b). The lineshapes of ESE detected spectra depends on τ when T_2 varies over the

[1] Freed and coworkers at Cornell have started to use high power extended interaction oscillators (EIO) at 95 GHz which allow them to use much shorter pulse lengths. However, the costs of EIOs at such high frequencies are very high (private communication). Before EIOs have been also used at 140 GHz [5]

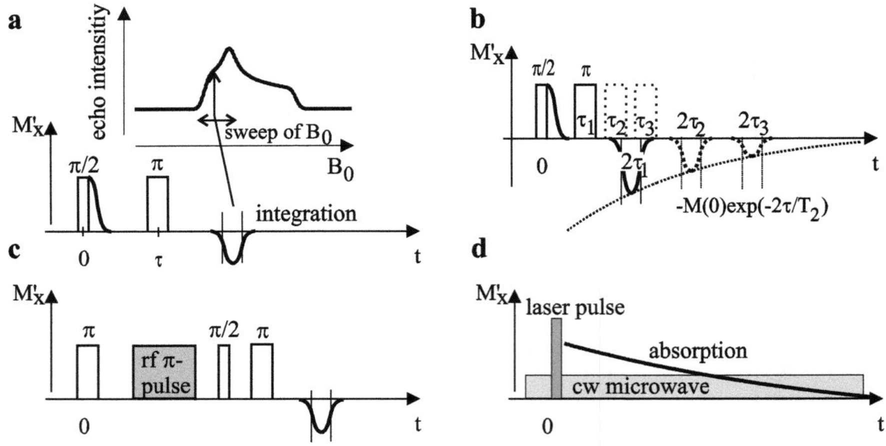

Fig. 5. Frequently used methods for pulsed high-field EPR. The measured magnetization is shown by bold line. The relevant part is integrated as indicated by the two vertical lines. (**a**) Electron-Spin-Echo (ESE) detected EPR. The integrated echo intensity is monitored over a magnetic field sweep. (**b**) Measurement of two-pulse echo decay. The echo intensity decays with relaxation time T_2 and might be modulated due to hyperfine structure (ESEEM). (**c**) Davies ENDOR. (**d**) Time-resolved EPR (TREPR)

spectrum (see Sect. 5.1). Additionally, the echo decay might be modulated due to hyperfine structure. This leads to Electron Spin Echo Envelope Modulation (ESEEM). ENDOR can be measured in cw as well as in pulsed mode. Because pulsed ENDOR has some technical advantages, it is mainly used in high-field EPR. Figure 5c shows the Davies ENDOR pulse sequence. The first π pulse inverts the electron spin magnetization. This inversion is partially destroyed when the second rf pulse is resonant with a nuclear spin transition. The following spin echo sequence probes the resulting magnetization, which is monitored during the sweep of the frequency of the rf pulse. Time-resolved EPR (TREPR) shown in Figure 5d is not a pulsed method, but it needs, nevertheless, very high time resolution less than 50 ns. This is normally also provided by pulsed EPR spectrometers. In this experiment one probes the absorption of cw-microwave by transient paramagnetic intermediates generated after photo excitation. If relaxation times are slow enough, the transient states can be probed also by a spin echo sequence.

4 Instrumentation

The basic setup of an EPR spectrometer is straight forward. One needs a magnet, a transmitter, which provides the microwave field, pulsed and cw, a microwave resonator and a receiver, which has to have the required time resolution. The optimum setup depends on the desired application and purpose. In recent years, a lot of different high-field/high-frequency spectrometers have been developed, pulsed as well as cw. The spectrometers vary in frequency, in time resolution, in sensitivity and in microwave output power, which determines the $\pi/2$ pulse length. A good overview on pulsed high-field EPR instrumentation and sensitivity consideration is given in [6,7].

The sensitivity depends mainly on the resonator design and construction and on the noise figure of the detector. The most sensitive setups use cylindrical cavities as resonators and heterodyne detection. They use microwave mixers which convert the high-frequency signal to an intermediate frequency between 0.5 and 10 GHz. Sensitivity is limited by the noise figure of the microwave mixer (typically 9 dB at 95 GHz), the amplitude and phase noise of the microwave source which is transmitted to the detector via the local and signal arms of the microwave bridge. Furthermore, sensitivity scales linearly with the quality factor Q of the resonator (typically $Q = 2000$ at 95 GHz for cylindrical cavities), and quadratically with the filling and conversion factors. For pulsed EPR, the conversion factor c becomes more important than the filling factor, because it gives the ratio of incident microwave power to the magnetic field strength of the microwave at the center of the resonator, $c = B_1/\sqrt{QP_{\mathrm{mw}}}$. Therefore, the higher the conversion factor, the shorter the pulse lengths τ_p (see (8)) for the available microwave power. This is crucial for samples with short T_2 relaxation times. In an ESE experiment (Fig. 5a) the delay time τ between the two pulses cannot be too short because otherwise the echo at 2τ overlaps with the second pulse or is buried in the spectrometer dead time after the second pulse (typically $10-50$ ns at 95 GHz). On the other hand, for high sensitivity of pulsed EPR 2τ should be shorter than T_2. For this reason it is very important to provide high microwave power output. At 95 GHz, typical output powers are between 4–200 mW. The shortest $\pi/2$ pulse lengths are about 40 ns. As mentioned above, the excitation bandwidth of these pulses is not sufficient for most samples and, consequently, it is often not possible to perform Fourier transform high-field EPR.

On the one hand, when going to higher fields/frequencies the filling and conversion factors increase leading to higher sensitivity for small samples such as single crystals. However, there are many technical problems related with high microwave frequencies, for instance output power and noise figures of the microwave components, and the theoretical sensitivity increase can hardly be realized. Furthermore, the losses in standard rectangular waveguides become enormous and one has to use quasioptical setups instead. A cylindrical cavity at 95 GHz has a diameter of only 3 mm. The sample diameter is about 0.2 mm for aqueous and 0.8 mm for samples with low dielectric absorption. These values scale with wavelength, and at very high frequencies one has to use Fabry-Perot resonators instead of single-mode cavities, or even no resonators at all. In such cases the conversion factor drastically reduces and pulsed experiments become much more difficult to perform. The mechanical complications, related to the short wavelength, induce further noise by vibrations, particularly of the microwave coupling. This is a very sensitive component because it matches resonator impedance to the waveguide impedance. It has to be variable because resonator impedance may change with sample and temperature.

Pulsed ENDOR setups need additional rf coils in order to apply the rf π pulse. The Larmor frequency of protons in 3.4 T/95 GHz spectrometers is 140 MHz. The main problems are penetration depth through the resonator walls which, therefore, have to be slotted, matching of the rf circuit in order to allow high power pulses over a bandwidth of at least 20 MHz, mechanical and heating problems. The latter two points make pulsed ENDOR at high fields more feasible than cw-ENDOR.

5 Applications to Photosynthetic Reaction Centers and Model Systems

Photosynthetic reaction centers in cell membranes consist of a protein backbone and embedded cofactors. During the primary steps of photosynthesis after photo excitation of chlorophyll electron donors (P), an electron is transfered over a chain of cofactors which leads to charge separated states and, thereby, to trans-membrane potentials. Thus, light energy is converted with high quantum yield, first to electric and afterwards to chemical energy. EPR has been used since long in the investigation of structure and dynamics of photosynthetic reaction centers of bacteria and green plants. It was, for example, applied to the doublet radical of the primary electron donor $P^{+\bullet}$ of the bacterial reaction center of *Rhodobacter sphaeroides*, to doublet radicals of the two quinone acceptors, $Q_A^{-\bullet}$, $Q_B^{-\bullet}$, and to the intermediate radical pair state during electron transfer, $P^{+\bullet}Q_A^{-\bullet}$. Of special interest is the geometric structure, i.e. the orientation and distance of the cofactors with respect to each other and to the membrane, their dynamics in their respective binding pocket, and its relation to electron transfer characteristics. For an exhaustive overview of EPR work related to photosynthetic reaction centers we refer to [8]. In the following we want to illustrate the application of relaxation time and ENDOR measurements to determine structure and dynamics of $Q_A^{-\bullet}$ in the binding pocket of bacterial reaction centers, as well as the application of time-resolved EPR to structural studies of model systems for photosynthetic electron transfer.

5.1 Molecular Motion Studied by Relaxation-Time Measurements

The 2D spectrum in Fig. 6a shows the echo decay, measured by the pulse sequence in Fig. 5b, with different pulse separation times τ and magnetic field values [9]. Because the echo decay has different time constants at different magnetic field positions, 1D spectra at different pulse spacing times τ have different line shapes (Fig. 6b, compare solid and dotted lines). Because at 3.4 T the high-field condition (5) is fulfilled for $Q^{-\bullet}$, the different echo decays can be assigned to specific molecular orientations. In the figure the resonance positions for the orientations along the principal axes of the G tensor are indicated. Magnetoselection works well for the g_{xx} and g_{zz} orientation while intermediate orientations overlap at the g_{yy} field position. The echo decays are mono exponential, with the transversal relaxation time T_2 as time constant (see Sect. 3). The T_2 values extracted from the 2D spectrum are shown in Fig. 6c.

The transversal relaxation (T_2 processes) is caused by stochastic modulation of resonance frequencies. Two quantities characterize the relaxation: (i) the time scale of the stochastic process, and (ii) the depth of the frequency modulation. Under certain conditions, a stochastic process is described with a correlation function $G(\tau)$, which is the probability of finding the system at time $t = \tau$ in the same state as at time $t = 0$. For Markov processes, $G(\tau)$ decays exponentially with the correlation time constant τ_r which is a measure for the time scale of the modulation. The mathematical treatment of relaxation processes for the case of fast motion, i.e. $\tau_r^{-1} \gg$ modulation depth in frequency units, was described by Redfield [10] and covered in most EPR books (see e.g. [1]). For slower motions the treatment becomes more difficult. The treatment of anisotropic small-angle fluctuations and its relation to ESE-detected EPR spectra was discussed in [11–14].

It is assumed that the Q_A is subject to small-angle fluctuations in its binding pocket. The depth of the modulation of the resonance frequency depends, therefore, on the amplitude of the fluctuations and on the functional relationship of fluctuation angle and resonance frequency.

In the case of $Q_A^{-\bullet}$, the modulation in resonance frequency is caused by the dependence of the effective g value on molecular orientation. The relaxation is described in the fast-motion regime [9] and T_2 is related to the rms average fluctuation amplitude, $\langle \delta g^2 \rangle$, by

$$1/T_2 = \mu_B^2 \tau_c B_0^2 \langle \delta g^2 \rangle . \tag{9}$$

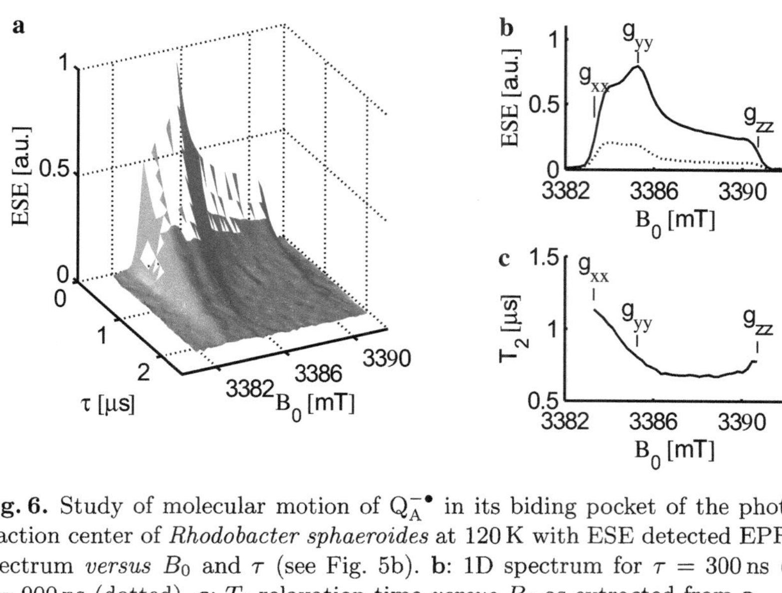

Fig. 6. Study of molecular motion of $Q_A^{-\bullet}$ in its biding pocket of the photosynthetic reaction center of *Rhodobacter sphaeroides* at 120 K with ESE detected EPR [9]. **a**: 2D spectrum *versus* B_0 and τ (see Fig. 5b). **b**: 1D spectrum for $\tau = 300$ ns (solid) and $\tau = 900$ ns (dotted). **c**: T_2 relaxation time *versus* B_0 as extracted from **a**

Because of magnetoselection the dependence of T_2 on resonance position reveals fluctuation amplitudes for different molecular orientations. Because of the quadratic dependence of $1/T_2$ on B_0, the resolution for motional anisotropies is increased at higher fields. One can simulate spectra such as in Fig. 6c and thereby relate them with motional anisotropy of the Q_A acceptor and with characteristics of the binding pocket in the protein. It was shown that, because at 120 K T_2 is largest at the g_{xx} field position, Q_A is fluctuating around the x axis [9]. Of special interest are motional anisotropies due to H-bonding [9], the comparison of different models for anisotropic motion and the comparison of the motional anisotropies of the two different quinone acceptors, $Q_A^{-\bullet}$ and $Q_B^{-\bullet}$ [15]. Furthermore, studying the temperature dependence reveals changes of motional modes and changes in H-bond lengths [15].

5.2 Hydrogen Bonding Studied by ENDOR

The three ENDOR spectra of $Q_A^{-\bullet}$ shown in Fig. 7 have been measured using the Davies ENDOR pulse scheme from Fig. 5c. Three different EPR resonance positions have been selected corresponding to the molecular orientations of the principal axes of the G tensor [16]. By means of this magnetoselection, it is possible to determine the isotropic values and the principal values of hyperfine tensors and their orientation with respect to the G tensor. By comparing the results with ENDOR on quinone radicals in different environments [16] and measurements at different magnetic fields/frequencies, it was possible to assign the ENDOR lines indicated by arrows in Fig. 7 to a strong H-bonded proton. Using (2), it was possible to suggest the partners involved in the H-bond as well as their distance.

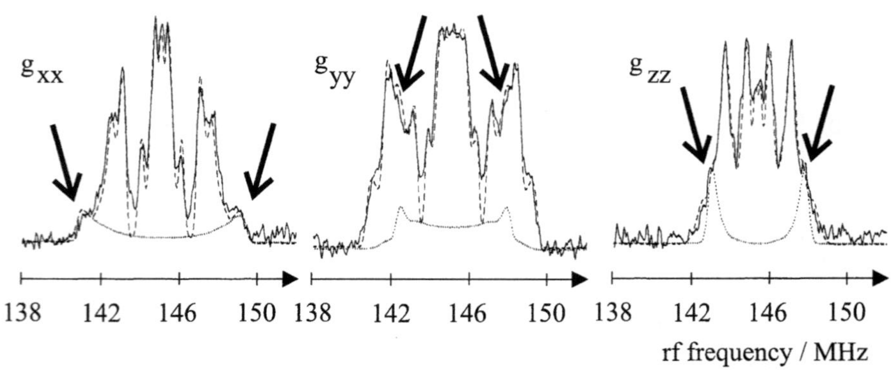

Fig. 7. ENDOR spectra of quinone anion radical $Q_A^{-\bullet}$ in the binding pocket of the photosynthetic reaction center of *Rhodobacter sphaeroides*. The picture is adapted from [16]. The ENDOR spectra are measured at the three different EPR resonance fields indicated by the arrows in Fig. 3 (bottom). From spectra simulations (dotted lines) the hyperfine couplings (indicated by arrows) could be deduced. They are related to hydrogen bonds of the quinone to amino acid residues of its binding pocket

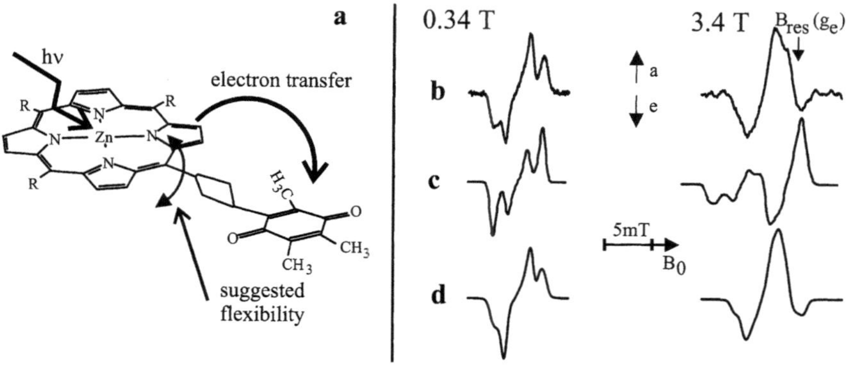

Fig. 8. TREPR on radical pairs $P^{+\bullet}Q^{-\bullet}$ of a model system for photosynthetic electron transfer [17]. (**a**) structure of the porphyrin (P)- cyclohexylene bridge- quinone (Q) system. The molecular flexibility indicated was suggested from the TREPR results. (**b**) TREPR spectra in frozen solution (ethanol/toluene) 2 µs after the laser pulse. (**c**) simulations as weakly coupled correlated radical pair (small exchange interaction $J_mathrmex$). (**d**) simulation as strongly coupled radical pair with orientation dependent T_1 relaxation

5.3 Electron Transfer and Spin Dynamics Studied by TREPR

Time-resolved EPR (TREPR) is used to observe states immediately after their creation in chemical reactions or after photo excitation (see [18,19]). Generally, these states are not in thermal equilibrium but spin polarized, i.e. the population of the states does not reflect Boltzmann distribution. The EPR intensity depends on both transition probability and population difference. Therefore, resonance intensities on spin polarized systems may be drastically enhanced and appear in absorption as well as in emission.

TREPR was successfully applied to study paramagnetic intermediates of the photosynthetic photo cycle, such as the $P^{+\bullet}Q_A^{-\bullet}$ state. In contrast to freely diffusing radicals in solution, the two radicals of this radical pair have fixed distance and orientation. The dipolar and exchange coupling are small compared to the differences in local magnetic fields (G tensors), and the radicals are, therefore, weakly coupled. One observes the resonance lines of the individual radicals split by the dipolar coupling. In this case, the spectra can be well described using a static correlated coupled radical pair (CCRP) model [20]. The higher the magnetic field and microwave frequency, the more the spectrum is determined by the anisotropic G tensors and the higher is the accuracy with which distance and orientation of the radical-pair partners can be determined.

In Fig. 8 TREPR experiments on the radical pairs of biomimetic model systems for photosynthetic electron transfer are shown [17]. The system consists of a porphyrin electron donor which, after photo excitation, transfers one electron to the quinone acceptor covalently linked to the porphyrin by a cyclohexylene bridge. After the porphyrin is photo-excited to its singlet state, inter-system-crossing populates the triplet state from which electron transfer takes place, i.e.

$$\mathrm{PQ} \rightarrow \mathrm{P}^{S1}\mathrm{Q} \rightarrow \mathrm{P}^{T}\mathrm{Q} \rightarrow \mathrm{P}^{+\bullet}\mathrm{Q}^{-\bullet} \rightarrow \mathrm{PQ}\,. \tag{10}$$

By varying donors and acceptors and investigating longer chain systems (for the porphyrin - quinone type systems up to two porphyrins and two quinones) one can study spin dynamics and the conditions for highly efficient electron transfer [21–23]. The spectra have been obtained by measuring the absorption of cw-microwave after the laser pulse (see Fig. 5d). It is also possible to detect the transient magnetization using ESE detected EPR, sometimes even with higher sensitivity [24].

The spectra of the dyad system in Fig. 8b at 0.34 T could be qualitatively simulated using the CCRP approach (Fig. 8c). However, the simulation of the TREPR spectrum at 3.4 T shows large differences to the experimental spectrum. From this it could be shown that, in contrast to the radical-pair in the natural photosystem, the radical-pair partners in the model system are strongly coupled to a triplet state. Moreover, the spectral characteristics are due to spin lattice (T_1) relaxation effects combined with fast electron recombination [17]. The relaxation effects are associated to anisotropic small-angle fluctuations and intra molecular flexibility. This model led to the simulations shown in Fig. 8d. It could be concluded that the life time of the charge separated state, i.e. the quantity which has to be optimized for artificial photosynthesis with high quantum yield, is limited by these small-angle fluctuations and concomitant relaxation effects.

6 Fourier-Transform EPR Using Broad-Band Stochastic Excitation

One of the main problems of high-field EPR is the lack of sufficient microwave power. Therefore, the excitation bandwidth of the pulses is often too small for conventional Fourier-transform EPR (see Sect. 3.2). We now present an alternative method for investigating linear and non-linear responses of spin systems. The bandwidth of microwave irradiation is increased by using broad-band noise sources. Noise can be generated by fast stochastic or pseudostochastic phase switching of the microwave irradiation. The stochastic resonance method was introduced to NMR in 1970 by Ernst [25] and Kaiser [26] who have shown that the sensitivity is the same as that of pulsed magnetic resonance, while the required power is reduced by the ratio of pulse length and spin lattice relaxation time. The spectral information is obtained by cross-correlating the excitation with the response of the spin system. This is theoretically equivalent to dispersive infrared spectroscopy in Michelson type interferometers.

Figure 9a shows the first 200 ns of a pseudostochastic phase switching sequence used for stochastic high-field/high-frequency EPR [27]. The FWHM spectral bandwidth of this sequence is 250 MHz, similar to the bandwidth of a 4 ns pulse. While by using a single pulse the whole excitation energy has to be concentrated at once, by using the pseudostochastic excitation the excitation energy

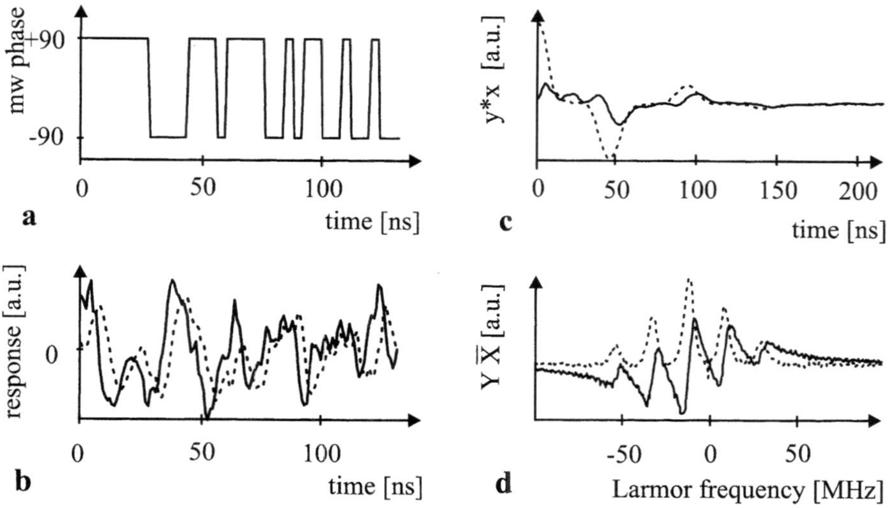

Fig. 9. Fourier-transform EPR at 3.4 T/95 GHz on the nitronyle-nitroxide radical [27] using broad-band stochastic excitation. (**a**): First 200 ns of a 1 μs long pseudostochastic excitation sequence $x(t)$. The phase of the microwave is modulated according to this sequence. The FWHM bandwidth is 250 MHz (**b**): Response $y(t)$ of the spin system (solid: in phase, dashed: out-of-phase component). (**c**): cross-correlation of x and y, $x*y$. The first 5 ns have been reconstructed using linear prediction. (**d**): Fourier transform of $x * y$. The dashed line is equivalent to the absorption spectrum of the two coupled nitrogens, the solid line is equivalent to the dispersion spectrum

is spread over time. Thereby the required peak power is reduced as discussed by Ernst and Kaiser (approximately, the time average of the power must be the same). As an example of stochastic EPR, the response of a nitronyle-nitroxide molecule to the applied pseudostochastic sequence is shown in Fig. 9b. It was digitized, and the cross-correlation with the excitation, shown in Fig. 9c, was performed by computer. The cross -correlation is equivalent to the FID, and its Fourier transformation, shown in Fig. 9d, is equivalent to the EPR spectrum. Using quadrature detection, the spectra are obtained in absorption as well as in dispersion (dotted and solid lines).

Although a highly esthetic and intellectually stimulating approach, the experimental realization of stochastic EPR is difficult. In contrast to standard pulsed EPR, many problems are created by the simultaneous excitation and detection. Therefore, it is necessary to decouple excitation and detection modes using bimodal resonators. Furthermore, for samples with relaxation times shorter than 100 μs, the currently available microwave power of 5 dBm is still below optimum. However, so far it could be shown, that although technically difficult due to the high time resolution required, the experimental bandwidth corresponds to that of the excitation and that the sensitivity is equal to the theoretically expected sensitivity. Based on the obtained results it could be estimated under which conditions the particularly interesting two dimensional (2D) high-field

Fourier-transform EPR experiments are feasible [27]. These types of measurements, known as COSY type experiments in NMR, could for example reveal spin couplings in two-spin systems such as radical pairs. The stochastic approach to multidimensional NMR was discussed and realized by Blümich and Ziessow (see [28,29]).

7 Conclusion

Pulsed high-field EPR is used now in many areas of spectroscopy in biology, chemistry and physics. One rapidly growing field of application is the study of structure and dynamics of proteins and (bio)organic molecules. A small selection of it has been presented in this article. Even when EPR methods may be not so generally applicable and have to be designed for the respective project, the obtained information is extremely valuable and complementary to X-ray crystallography and NMR. Different from X-ray crystallography, with EPR one can probe the molecular complexes in their functional states and environments, and studies of their dynamic behavior are also possible. Furthermore, the environment of paramagnetic centers is probed in a more direct way and it is even possible to investigate cofactors of electron transfer systems "in action". Contrary to NMR, in EPR cw and pulsed methods will continue to coexist. In concluding it is fair to state that time-resolved and pulsed multifrequency EPR – in particular at high magnetic fields – has matured over recent years to present protein X-ray crystallographers and protein NMR spectroscopists with a powerful new ally for determining structure and dynamics of large biosystems.

Acknowledgements

It is a pleasure to acknowledge the cooperation of Thomas Prisner and Martin Rohrer (now at University Frankfurt) as well as of Harry Kurreck, Alexander Schnegg and Martin Fuchs from the FU Berlin during the projects touched in this article. Financial support by the Deutsche Forschungsgemeinschaft and Volkswagenstiftung is gratefully acknowledged.

References

1. N.M. Atherton: *'Principles of Electron Spin Resonance'*, (Ellis Horwood, Chichester 1993)
2. A. Carrington, A.D. McLachlan: *'Introduction to Magnetic Resonance'*, (Harper & Row, New York, Evanston, London 1969)
3. R.R. Ernst, G. Bodenhausen, A. Wokaun: *'Principles of Nuclear Magnetic Resonance in One and Two Dimensions'*, (Clarendon Press, Oxford 1987)
4. A. Schweiger, G. Jeschke: *'Principles of Pulse Electron Paramagnetic Resonance'*, (Oxford University Press, Oxford 2001)
5. T.F. Prisner, S. Un, R.G. Griffin: Isr. J. Chem. **32**, 357 (1992)
6. T.F. Prisner: Adv. Magn. Opt. Reson. **20**, 245 (1997)

7. *'Special Issue on High-Field EPR'*, Appl. Magn. Reson. **16**, 106 (1999)
8. A.J. Hoff, J. Deisenhofer: Phys. Rep. **287**, 1 (1997)
9. M. Rohrer, P. Gast, K. Möbius, T. Prisner: Chem. Phys. Lett. **259**, 523 (1996)
10. A.G. Redfield: *'The Theory of Relaxation Processes'*, (Academic Press, London, 1965) Vol. 1 of Adv. in Magn. Res. p.1
11. G.L. Millhauser, J.H. Freed: J. Chem. Phys. **81**, 37 (1984)
12. S.A. Dzuba, Y.D. Tsvetkov, A.G. Maryasov: Chem. Phys. Lett. **188**, 217 (1992)
13. S.A. Dzuba: Spectrochimica Acta A **56**, 227 (2000)
14. S. Grimaldi, F. MacMillan, T. Ostermann, B. Ludwig, H. Michel, T. Prisner: Biochemistry **40**, 1037 (2001)
15. A. Schnegg, M. Fuhs, M. Rohrer, W. Lubitz, T. Prisner, K. Möbius: *'Molecular Dynamics of Q_A^- and Q_B^- in Photosynthetic Bacterial Reaction Centers Studied by Pulsed W-band High-field EPR'*, in preparation
16. M. Rohrer, F. MacMillan, T.F. Prisner, A.T. Gardiner, K. Möbius, W. Lubitz: J. Phys. Chem. B **102**, 4648 (1998)
17. M. Fuhs, G. Elger, A. Osintsev, A. Popov, H. Kurreck, K. Möbius: Molec. Phys. **98**, 1025 (2000)
18. K. McLaughlan: *'Time-Resolved EPR'*. In Advanced EPR, Applications in Biology and Biochemistry, ed. A.J. Hoff (Elsevier, Amsterdam, 1989) p.345
19. D. Stehlik, K. Möbius: Annu. Rev. Phys. Chem. **48**, 745 (1997)
20. P.J. Hore: *'Analysis of Polarized Electron Paramagnetic Resonance Spectra'*. In Advanced EPR, Applications in Biology and Biochemistry, ed. A.J. Hoff (Elsevier, Amsterdam, 1989) p.405
21. H. Kurreck, M. Huber: Angew. Chem. Int. Ed. Engl. **34**, 949 (1995)
22. A. Berg, Z. Shuali, M. Asano-Someda, H. Levanon, M. Fuhs, K. Möbius, R. Wang, C. Brown, J.L. Sessler: J. Am. Chem. Soc. **121**, 7433 (1999)
23. P. Piotrowiak: Chem. Soc. Rev. **28**, 143 (1999)
24. T.F. Prisner, A. van der Est, R. Bittl, W. Lubitz, D. Stehlik, K. Möbius: Chem. Phys. **194**, 361 (1995)
25. R.R. Ernst: J. Magn. Reson. **3**, 10 (1970)
26. R. Kaiser: J. Magn. Reson. **3**, 28 (1970)
27. M. Fuhs, T. Prisner, K. Möbius: J. Magn. Reson. **149**, 67 (2001)
28. B. Blümich, D. Ziessow: J. Chem. Phys. **78**, 1059 (1983)
29. B. Blümich: Prog. NMR Spectrosc. **19**, 331 (1987)

MIX
Papier aus verantwortungsvollen Quellen
Paper from responsible sources
FSC® C105338

If you have any concerns about our products,
you can contact us on
ProductSafety@springernature.com

In case Publisher is established outside the EU,
the EU authorized representative is:
Springer Nature Customer Service Center GmbH
Europaplatz 3, 69115 Heidelberg, Germany

Printed by Libri Plureos GmbH
in Hamburg, Germany